# Photonic Aspects of Modern Radar

For a complete listing of the *Artech House Optoelectronics Library*,
turn to the back of this book

# Photonic Aspects of Modern Radar

Edited by

Henry Zmuda
Edward N. Toughlian

Artech House
Boston • London

**Library of Congress Cataloging-in-Publication Data**
Photonic Aspects of Modern Radar/Henry Zmuda, Edward N. Toughlian, editors
Includes bibliographical references and index.
ISBN 0-89006-670-1
1. Radar. 2. Photonics. 3. Microwave devices. 4. Signal processing—
digital techniques. I. Zmuda, Henry, 1955– .
TK6580.P56 1994 94-2362
621.3848–dc20 CIP

**A catalogue record for this book is available from the British Library**


**685 Canton Street**
**Norwood, MA 02062**



International Standard Book Number: 0-89006-670-1
Library of Congress Catalog Card Number: TK6580.P56 1994

10 9 8 7 6 5 4 3 2 1

# Contents


Preface xiii

**Section I Introduction**

Chapter 1 Introduction 1

**Section II Introduction to Radar Systems**

Chapter 2 Radar Fundamentals for Optical Engineers 3
- 2.1 Introduction 3
- 2.2 The Relationship Between Optics and Radar 4
- 2.3 Types of Radar Systems 5
- 2.4 Radar Functions 7
- 2.5 Radar Block Diagram and Subsystem Functions 9
- 2.6 Radar Range Equation 13
- 2.7 The Detection of Signals in Noise 24
- 2.8 Radar Target Models 28
- 2.9 Signal-to-Noise Enhancement: Matched Filtering and the Integration of Radar Returns 30
- 2.10 Detection of Radar Targets in Clutter for Ground-Based Radar 36
- 2.11 Detection of Radar Targets in Clutter for Airborne or Space-Based Radar 45
- 2.12 Conclusion 47
- References 48

Chapter 3 System Considerations 51
- 3.1 Anatomy of a Radar System 51
- 3.2 Types of Radar Systems 53
- 3.3 Factors Influencing the Design of a Radar System 56
- 3.4 Preliminary Design Process: an Example 57
- References 64

**Section III Fundamentals of Photonics**

Chapter 4 Foundations of Optical Engineering 67
4.1 Introduction 67
4.2 Two-Dimensional Fourier Transform 67
4.3 Fundamentals of the Diffraction Theory 71
4.4 Transfer Function of a Thin Lens 74
4.5 Gaussian Beam Optics 77
4.6 Holography and Matched Filtering 77
4.7 Passive Components 80
4.8 Optical Heterodyne Process 84
References 86

Chapter 5 High-Speed Semiconductor Laser Sources and Detectors for Microwave Fiber-Optic Applications 87
5.1 Lasers 87
5.1.1 Introduction 87
5.1.2 Quantum Well and Strained-Layer Quantum Well Lasers 88
5.1.3 Narrowband Modulation Beyond Bandwidth Cutoff by Mode-Coupling Enhancement 95
5.1.4 Intermodulation 98
5.1.5 Intensity Noise 100
5.1.6 Modulation Efficiency and Signal Throughput 107
5.1.7 Microwave Phase Noise 113
5.1.8 SNR of Fiber Links Consisting of Fabry-Perot and Distributed Feedback Lasers 118
5.1.9 Conclusions 122
5.2 Detectors 124
5.2.1 Introduction 124
5.2.2 High-Speed Waveguide PIN Photodiodes 131
5.2.3 Down-Converting Photodetector by Photoconductive Mixing 147
5.2.4 Conclusions 157
Acknowledgments 158
References 158

Chapter 6 Spatial Light Modulators for Optically Controlled Phased Array Radar Signal Processors 163
6.1 Typical OCPAR Signal-Processing Architectures 164
6.2 SLMs: Structures, Principles, and Limitations 166
6.3 Characterization of SLMs 169
6.4 Specific SLM Architectures 172
6.4.1 Elector-Optic SLMs 172
6.4.2 Photorefractive Materials And Devices 175
6.4.3 Liquid Crystal SLMs 179

6.4.4 Optoelectronic Bistable Optical Devices 185
6.4.5 Deformable Mirror Devices 186
6.4.6 Membrane Mirror Light Modulators 187
6.4.7 Magneto-Optic SLMs 191
6.5 Future Directions 192
References 192

Chapter 7 Acousto-Optic Bragg Cell Devices 201
7.1 Introduction 201
7.2 Acousto-Optic Bragg Interaction 202
7.2.1 Acousto-Optic Effect 202
7.2.2 Interaction Geometries 205
7.3 Acousto-Optic Bragg Cells 212
7.3.1 Device Types 213
7.4 Acousto-Optic Bragg Cell Signal Processing Characteristics 218
7.4.1 Electrical-to-Optical Spatial Transfer Function 223
7.4.2 Electrical-to-Optical Intensity Transfer Function 224
7.4.3 Dynamic Range 225
7.5 Summary 226
References 226

Chapter 8 Optical Waveguide 229
8.1 Introduction 229
8.2 Optical Waveguides for Radar System Applications 229
8.3 Types of Optical Waveguide 233
8.3.1 Fiber-Optic Waveguide 233
8.3.2 Integrated Optical Waveguide 238
8.4 Optical Coupling Techniques 243
8.4.1 Butt Coupling 244
8.4.2 Tapered Couplers 245
8.4.3 Prism Couplers 246
8.4.4 Diffraction Gratings 246
8.5 Waveguide Materials 247
8.5.1 Glass Waveguides 247
8.5.2 Lithium Niobate Waveguides ($LiNbO_3$) 247
8.5.3 Semiconductor Waveguides 248
8.6 Conclusions 250
References 251

Chapter 9 Optical Integrated Circuits 255
9.1 Introduction 255
9.2 Aspects of Integration 255
9.2.1 Guided and Unguided Lightwaves 256
9.2.2 Applications of Integration 256

9.2.3 Electro-optic Materials 257
9.2.4 Semiconductor Techniques 258
9.3 Optoelectronic Integration 258
9.3.1 Photonic integrated circuits 259
9.3.2 Optoelectronic Integrated Circuits 261
9.3.3 Hybrid Optoelectronic Integration 261
9.3.4 Grafting and Related Techniques 262
9.3.5 Monolithic Optoelectronic Integration 263
9.3.6 Fiber-to-Circuit Coupling 263
9.3.7 Optical-Microwave Circuits 264
9.4 Phased-Array Antenna OEICs 266
9.4.1 Optical Command Circuits 266
9.4.2 Optical Phase Shift Circuits 266
9.4.3 Optical Time-Delay Circuits 267
9.4.5 Millimeter-Wave OEICs 270
9.5 Conclusion 275
References 275

**Section IV Optical Beamforming for Phased-Array Systems**

Chapter 10 Control Requirements for Dynamic Beam Steering Systems 277
10.1 Introduction 277
10.2 Array Pattern Characteristics and System Parameters 278
10.2.1 Patterns of Planar Arrays 278
10.2.2 Array Directivity and Sidelobe Levels 283
10.2.3 The Effects of Uncorrelated Errors in Phase and Amplitude 284
10.3 Array Feeds and Control 288
10.3.1 Analog versus Digital Beam Forming 288
10.3.2 Analog Beam Forming Feeds 289
10.4 Conclusion 296
References 296

Chapter 11 Spatial Optical Processing 297
11.1 Introduction 297
11.1.1 Background—Beam forming 299
11.1.2 Background—Photonics 304
11.2 Hardware Realizations 311
11.2.1 Differential Phase Systems 311
11.2.2 Time-Delay Systems 316
11.3 New or Enhanced Capabilities 318
11.4 Summary 320
References 321

Chapter 12 RF Fiber-Optic Links 323
12.1 Introduction 323

12.2 Fiber-Optic Link Design Criteria 324
12.3 RF-IF Mixing Architectures 326
12.4 RF/Optical Modulation Techniques 330
12.5 Optical/RF Detection Techniques 336
12.6 Fiber-Optic Link Performance Optimization 338
12.7 Fiber-Optic Link Performance Models 342
References 349

Chapter 13 Switched Fiber-Optic Delay Architectures 351
13.1 Introduction 351
13.2 Phased-Array Antenna Time-Delay Steering 352
13.3 Characterization Parameters For Fiber-Optic TTD Architectures 355
13.4 Noncompressive Fiber-Optic TTD Architectures 356
13.4.1 Optical Switch Parallel Fiber-Optic Delay Line 356
13.4.2 Laser Diode Switch Parallel Fiber-Optic Delay Line 358
13.5 Delay-Compressive Fiber-Optic TTD Architectures 359
13.5.1 Binary Fibeer-Optic Delay Line 360
13.5.2 Square-Root Fiber-Optic Delay Line 365
13.6 Delay-and Element-Compressive Fiber-Optic TTD Architectures 367
13.6.1 Wavelength-Multiplexed TTD Architectures for 1-D PAAs 367
13.6.2 Fiber-Optic Wavelength-Multiplexed TTD Architecture for 2-D PAAs 374
Acknowledgments 379
References 379

**Section V Signal Processing Systems**

Chapter 14 Optical Processing of Synthetic Aperture Radar Data 381
14.1 Introduction 381
14.2 Synthetic Antenna Principle 382
14.3 Radar Signal 385
14.4 Formation of the Holographic Image 389
14.5 Tilted Plane Processor 392
14.6 Digital Computer Competition 396
14.7 The Legacy 398
References 400

Chapter 15 Correlation Systems 403
15.1 Purpose and Mathematical Description of Optical Correlation 403
15.2 Role of Optical Imaging and Optical Diffraction in Achieving Optical Correlation 404
15.3 Fundamental Types of Optical Correlation Architecture and the Interpretation of Their Output Spatial and Time Coordinates 405
15.3.1 Spatial correlators (Space-Integrating) 405
15.3.2 Time-Domain Correlators (Time-Integrating) 412

15.3.3 Comparison of Fixed- and Moving-Window Configurations 414
15.4 Use of Time-Integrating Devices to Produce Spatial Correlation Output Functions 418
15.4.1 Process of Detection and Time Integration 418
15.4.2 Correlator Architectures Using the Other Devices, Such as Liquid Crystal Light Values 419
15.4.3 Correlator Architectures Using Photorefractive Devices 420
15.5 Adaptive Systems Using Optical Correlators 421
15.5.1 Characterization of Adaptive Systems 421
15.5.2 An Example of a Least Mean Squares Algorithm Adaptive System Using Cascaded Optical Correlators 422
15.5.3 Other Adaptive Architectures 426
References 426

Chapter 16 Acousto-Optic Processing for Receiver Technologies 429
16.1 Introduction 429
16.2 Conventional Receiver Technologies 429
16.2.1 Crystal Video Receiver 430
16.2.2 Superheterodyne Receiver 431
16.2.3 Instantaneous Frequency Measurement Receiver 432
16.2.4 Compressive Receiver 433
16.2.5 Channelized Receiver 434
16.2.6 Limitation of Receiver Techniques 435
16.3 Principle of Acousto-Optic Spectrum Analysis 435
16.3.1 Acousto-Optic Diffraction 436
16.3.2 Acousto-Optic Channelized Receiver 438
16.3.3 Acousto-Optic Compressive Receiver 442
16.4 Key Component Performance 444
16.4.1 Laser Source 444
16.4.2 Bragg Cells 444
16.4.3 Photodetector Array 447
16.5 Power Acousto-Optic Receiver 448
16.5.1 Integrating Acousto-Optic Channelized Receiver 448
16.5.2 Parallel Output Acousto-Optic Channelized Receiver 449
16.6 Coherent Acousto-Optic Receiver 449
16.6.1 Acousto-Optic IFM 454
16.7 Summary 456
References 456

Chapter 17 Fiber-Optic Delay Line Signal Processing Architectures 459
17.1 Overview, Background, and Motivation 459
17.1.1 Signal Processing 465
17.1.2 Optical Processing Versus Other Techniques 466

17.1.3 Properties of Fiber Optics 466
17.1.4 Temperture Effects 469
17.1.5 Fiber Structures 470
17.1.6 Sources and Detectors 470
17.1.7 Analog Versus Digital Signal Processing 471
17.2 Incoherent Versus Coherent Regime 473
17.2.1 Coherence Effects 473
17.2.2 Coherence Functions 473
17.2.3 Linearity 474
17.2.4 Incoherent Versus Coherent Processing 476
17.3 Fiber Filter Analysis 477
17.3.1 Z-Transform 477
17.3.2 Stable and Minimum Phase Filters 478
17.3.3 Block Diagram Representation and Mason's Rule 479
17.3.4 Elementary Signal Processing Functions 480
17.3.5 Example: Fiber-Optic Recirculating Delay Line Filter 481
17.4 Optical Amplifiers In Fiber-Optic Signal Processing 487
17.4.1 Motivations for Optical Amplification 487
17.4.2 Optical Amplifier Characteristics 488
17.4.3 Effect of EDFA Length on Filter Design 492
17.4.4 Effect of Gain on Stability 492
17.5 Fiber-Optic Filtering Architectures 493
17.5.1 General Lattice Structures 494
17.5.2 Fiber-Optic Implementation of Lattice Structures 495
17.5.3 Cascaded Fiber Lattice Structures 498
17.5.4 Amplified Lattice Filters 500
17.5.5 Theoretical Formulations 500
17.6 Bipolar, Complex, and Positive Optical Systems 508
17.6.1 Bipolar and Complex Systems 509
17.6.2 Positive Systems 510
17.7 Conclusions 515
Acknowledgments 518
References 518

**Section VI Future Opportunities and Directions**

Chapter 18 Summary, Future Opportunities, and Directions 523
References 526

List of Acronyms 529

About the Authors 533

Index 539

# *Preface*

In what amounts to an extremely short period of time, the use of optics has had a significant impact on modern technology. From fiber-optic-based telecommunications to high-power lasers for the SDI defense systems, researchers are discovering the potential of optics to provide capabilities not achievable with other technologies. Yet another area where the newly emerging field of photonics is making significant contributions is that of high-speed parallel signal processing. Perhaps the greatest beneficiary of this capability is the radar community, where, by the very nature of the radar problem, the need to perform sophisticated signal processing functions at high speed is essential.

The rate at which component technology is maturing suggests that the next generation of radar systems will contain an optical processor(s) at the subsystem level. Such an accomplishment requires an understanding of the specific and often unique requirements of radar systems. It also requires that radar specialists, from hardware designers to mathematicians responsible for developing high-speed algorithms, appreciate the capabilities as well as the limitations inherent in photonic systems. In a similar way, optical engineers must be aware of radar system needs if developmental efforts are to meet these needs. Although a wealth of information exists on both sides, it is often in an unfamiliar language, located in a wide variety of (often difficult to obtain) journals, and lacks the perspective needed to apply the knowledge to the problem at hand.

This book is the first to fulfill the specific needs of both radar and photonics engineers. Its goal is to provide an essential reference for researchers, practicing engineers, and technical managers alike. It brings together in a systematic manner all the necessary information necessary for either radar or optical engineers to come "up to speed" in an expeditious manner. In the same manner, people already working in the field will find the information contained in this volume useful for both broadening their perspectives and updating their knowledge.

The book assembles a unique blend of expertise from industry, academia, and government to provide the broadest perspective of this subject. To set the tone for the entire book, as well as provide optical engineers with a background in radar, the two chapters of Section II feature the broad subject of radar. Chapter 2 is an overview of the fundamental principles of radar operation written especially for optical engineers, and Chapter 3 considers various types of radar systems and important issues associated with radar design from the system level.

Section III provides a detailed study of the concepts and components used in the design of photonic systems. This section opens with a chapter on the theory of optical system design, or what is more commonly referred to as *Fourier optics,* and the basic ideas of diffraction theory. Examples are included that help make the theory relevant to radar systems. The chapters that follow discuss important optical components, such as lasers, detectors, electro-optic devices, spatial light modulators, acousto-optic devices, and other components commonly found in microwave optical systems. This section closes with a chapter on the well-established field of optical waveguide theory and a chapter on the currently evolving field of optical integrated circuits. The latter chapter considers the monolithic integration of optical and microwave components. Unique to the exposition given in these chapters is that all optical components are presented with the radar application in mind. Advantages that these components provide as well as the limitations they impose on RF/microwave systems are made explicit wherever possible.

To date, the aspect of radar perhaps most influenced by photonics is that of optical control of phased array antennas. Consequently the Section IV is devoted entirely to this area, starting with a chapter that provides the essential theory of phased arrays. Optical processors for antenna arrays fall into two types; spatial processors and optical-fiber-based systems. Spatial processors exploit the analogy between Fourier transform optics and far-field antenna patterns. These include phase, frequency, and time delay steering approaches. Fiber-based systems perform the necessary array signal processing by the switching of optical fiber of varying length. Characteristics of some of the more promising architectures from a practical viewpoint are presented. Optical fiber has been found to be an effective means of transmitting RF/microwave energy for all types of systems, including antenna arrays. A chapter on fiber links is thus included in this section, though its applicability extends far beyond the scope of this section and even this book.

Section V of the book examines how photonics can be used to replace some of the front-end RF/microwave components in a radar and perform some of the high-speed signal processing usually performed by a digital computer. This section opens with a chapter on one early and still outstanding success of optical signal processing in the field of synthetic aperture radar. A mathematical function frequently employed in radar processing is that of correlation. For this reason, a chapter is devoted to the fundamentals of optical systems for space- and time-integrating correlators and convolvers, including various configurations for adaptive systems and practical issues for implementation. This in turn leads to a chapter on the use of acousto-optic technology for receiver applications. The last chapter in this section studies the use of tapped optical fiber as the delay mechanism in a transversal filter architecture. Such transversal filters have long been known to be a powerful signal processing approach in radar as well as other communication systems. Optical implementations of both recursive and nonrecursive systems for null steering, notch, and bandpass filtering are examined. The use of optical amplifiers to enhance system performance is included.

The book closes with a chapter surveying the accomplishments of the field while trying to predict its future in this rapidly changing "dual-use" world. The academic researcher, industrial developer, and government project manager can use this chapter as the springboard for future planning documents and proposals.

The editors owe a debt of gratitude to many of their colleagues at Stevens and Rome Lab for their insights and varied contributions throughout the preparation of this book. We especially wish to thank Professor Dave Sumberg for his assistance. What would have been the formidable task of assembling the final manuscript of this book was made easy due in large part to the unique talents of Doug Norton. These editors owe him special thanks for responding above and beyond the call of duty as a contributor. Last but by no means least, a very warm thank you is heartily extended to Reinhard Erdmann for so many things. We're especially proud to call him a friend as well as colleague.

Henry Zmuda<br>
Ed Toughlian<br>
*September 1993*

# *Chapter 1*

# *Introduction*

***David K. Barton***
*Anro Engineering*

Radar has, since its inception, used optical phenomena as an interface between the electrical signals at the output of the receiver and the user of the radar. The primary output device for most of the past 50 years has been the cathode-ray tube display, which presented to operators a picture of that portion of the world observed by a particular radar, or a graphical display of signal amplitude as a function of one or another radar coordinate. Further, in the specialized field of synthetic aperture radar (SAR), analog optical computing performed the Fourier transform signal processing necessary to convert the phasor signals over many pulse repetition intervals to high-resolution maps of the observed surface. Until the 1970s, such optical methods were the only means of processing the mass of data involved in the SAR system. The more general application of photonics to a variety of radar problems, however, has only become possible during this last decade as new devices and techniques have been developed to perform the functions previously implemented in electronic analog or digital circuits. These new devices and techniques constitute the field of photonics, a term so new that it is not even defined in the 1992 *IEEE Standard Dictionary of Electrical and Electronic Terms.*

As this book is being written, radar is undergoing a significant change in status as a sensing instrument for both military and civilian applications. With the end of the cold war, shrinking military budgets (at least in the major world powers) have forced a reevaluation of many programs based on radar for surveillance, tracking, and guidance. New systems must accomplish more with fewer resources: power, equipment mass and volume, operating and maintenance personnel, and cost. To make radar a practical technique for large-scale civil applications such as automobile collision avoidance, cost and size must be drastically reduced. To reserve its major role in marine and air navigation, in competition with the Global Positioning System, there must be significant improvements in performance along with cost reduction. The future radar engineer must be prepared to exploit every applicable technology in meeting these challenges. The optical engineer must understand the needs of the radar system for new capabilities, or for replacement of existing electronic circuits with more effective photonics. The editors and authors of this book have herein provided both these engineering groups, and those who direct and use their efforts, with the information necessary to integrate the fields of optics and electronics and to exploit this new technology to meet future demands.

# *Chapter 2*
# *Radar Fundamentals for Optical Engineers*

***Barry Fell***
*GORCA Systems*

## 2.1 INTRODUCTION

Radar and optics are related through Maxwell's equations. Maxwell came to the realization that light was an electromagnetic wave on the basis of its velocity [1]. We know today that the difference between optics and radar is the wavelength or frequency of the electromagnetic radiation. Although the physical basis of both sciences is identical, the use of these two different parts of the electromagnetic spectrum is determined by the technology that exists for these widely separate frequency bands. It is interesting to note that one of the first experiments based on the Maxwell equations was the production of meter-long (VHF-UHF) radiation by Henrich Hertz [2]. With the discovery of the lasing action and the ability to produce "high-power" coherent optical radiation, the field of laser radar was created. Since then the relationship between optics and microwave/millimeter-wavelength radar has expanded to include using optical subsystems for ultra-high-speed signal processing [3] and antenna control and beam forming [4].

This chapter covers the fundamentals of radar systems engineering for engineers and scientists engaged in optical research and applications. Its goal is to provide an understanding of the basic principles of microwave radar design and to discuss new and innovative approaches to adaptive processing for rejection of clutter (unwanted returns) and radio frequency (RF) interference. A similar paper was written twelve years ago [5] for a Society of Photo-optical Instrumentation Engineers (SPIE) symposium sponsored by the U.S. Army Advanced Ballistic Missile Defense Agency (ABMDA), currently the U.S. Army Strategic Defense Command (SDC) in Huntsville, Alabama. The theme of that symposium was the use of optical technology for application to missile defense radar systems. That mission still exists today in the form of theater missile defense (TMD).

Therefore, the subject matter of this book is very timely and important for fostering the application of high-performance optical systems to enhance existing capabilities for microwave radar, and in some cases make new capabilities possible. As indicated above, recent advances in optical radar (ladar or lidar) [6,7] and millimeter-wave radar [8] provide additional system applications that require the advanced optical technologies discussed in this book.

This chapter presents an overview of radar and radar systems and discusses the basics of radar design, including a description of the radar range equation, radar target characteristics, the radar environment, detection theory for targets in the presence of thermal noise, and methods for detecting targets in clutter. The important subject of parameter estimation by radar is described, and is followed by a discussion of advanced topics in radar applications.

## 2.2 THE RELATIONSHIP BETWEEN OPTICS AND RADAR

Today the relationship between optics and radar goes beyond being different frequency domains described by Maxwell's equations. This is clearly demonstrated by the publication of this volume.

There are two primary areas where radar and optics technology cross. The first is the application of optical technology to radar components, and the second is the use of optical systems in a radarlike mode (i.e., the introduction of lidar systems).

The development of optical technology has opened the door to cost-effective implementation of electronically steered phased-array radar antenna systems [9]. In addition, the ultrawide bandwidth afforded by optical hardware has led to optical processors that can provide very large signal processing throughput [10]. Therefore, the primary areas of optical subsystems in radar are implementations of phased-array systems and high-speed signal processing systems.

The importance of optical system implementation of phased-array radar is to perform the array control functions which provides the phase information and amplitude weights for each phased-array element or subarray [11]. In addition, signal information normally transmitted by an RF feed can now be transferred along lightweight optical fibers. This not only reduces the weight and depth of the array aperture, but also increases system reliability and permits construction of conformal arrays, which can take on the shape of the surface on which they are mounted. Another advantage of optical fibers is their inherent immunity to external sources of electromagnetic interference and signal intercept [12].

Optical systems have traditionally provided the initial implementation for very-high-speed signal processors [13]. Even with the advent of high-speed digital processors, current optical processors still provide two to three orders of magnitude greater processing speed for the same weight, size, and power consumption [14]. One of the special requirements for high-speed optical processing systems is the need for precise mechanical stability due to the short wavelength of light. This requirement to maintain precise mechanical tolerance throughout the optical processor system translates into good mechanical design and precise temperature control. These requirements can sometimes make an optical processor too large or too delicate to be used in airborne or spaceborne systems, as well as in ground-based systems that may be subject to severe mechanical vibration and shock. In addition, optical processors have suffered from restrictions on dynamic range. However, recent advances in optical processors have made progress in overcoming this problem.

Lidar (light detection and ranging) systems were first introduced in the late 1960s and early 1970s. Their initial use [15] was applied to:

- Surveying,
- Tactical laser rangefinders,
- Tactical target designators,
- Beam rider missile guidance systems,
- Precision instrumentation radar,
- Command guidance radar,
- Platform attitude determination,
- Geodetic satellite and lunar ranging systems,
- Doppler laser radar, and
- Laser atmospheric probes.

More recently [16], lasers have been used as illuminators for smart bombs and precision munitions delivery and as replacement for X-band and $K_u$-band police radars. Chapter 18 describes more recent advances and uses of laser radar.

The traditional frequency bands allocated for microwave radar are VHF from 137 to 144 MHz and from 216 to 225 MHz, UHF from 420 to 450 MHz and from 890 to 940 MHz, L band from 1,215 to 1,400 MHz, S band from 2,300 to 2,550 MHz and from 2,700 to 3,700 MHz, C band from 5,255 to 5,925 MHz, X band from 8.5 to 10.7 GHz, $K_u$ band from 13.4 to 14.4 GHz and from 15.7 to 17.7 GHz, K band from 23 to 24.25 GHz, and the millimeter-wave bands at 35 GHz and 95 GHz.

## 2.3 TYPES OF RADAR SYSTEMS

The basic functions of a radar are the detection of targets and the measurement of target parameters. Figure 2.1 illustrates the measurements made directly by a typical radar system. As shown, these measurements include range to the target, the direction of the target with respect to the pointing angle or broadside of the radar antenna, and the rate at which the line-of-sight radar range to the target is changing. In addition to these direct measurements, the radar can also derive values for target velocity and angle rates. Recent advances in radar signal processing also permit radars to produce images of ground scenes (i.e., synthetic aperture radar (SAR) [17]) and images of airborne and surface ship targets (i.e., inverse synthetic aperture radar (ISAR) [18]).

Radars have been installed in ground-based, airborne, shipboard, and spaceborne environments. Radars have been operated in both arctic and tropical regions, as well as

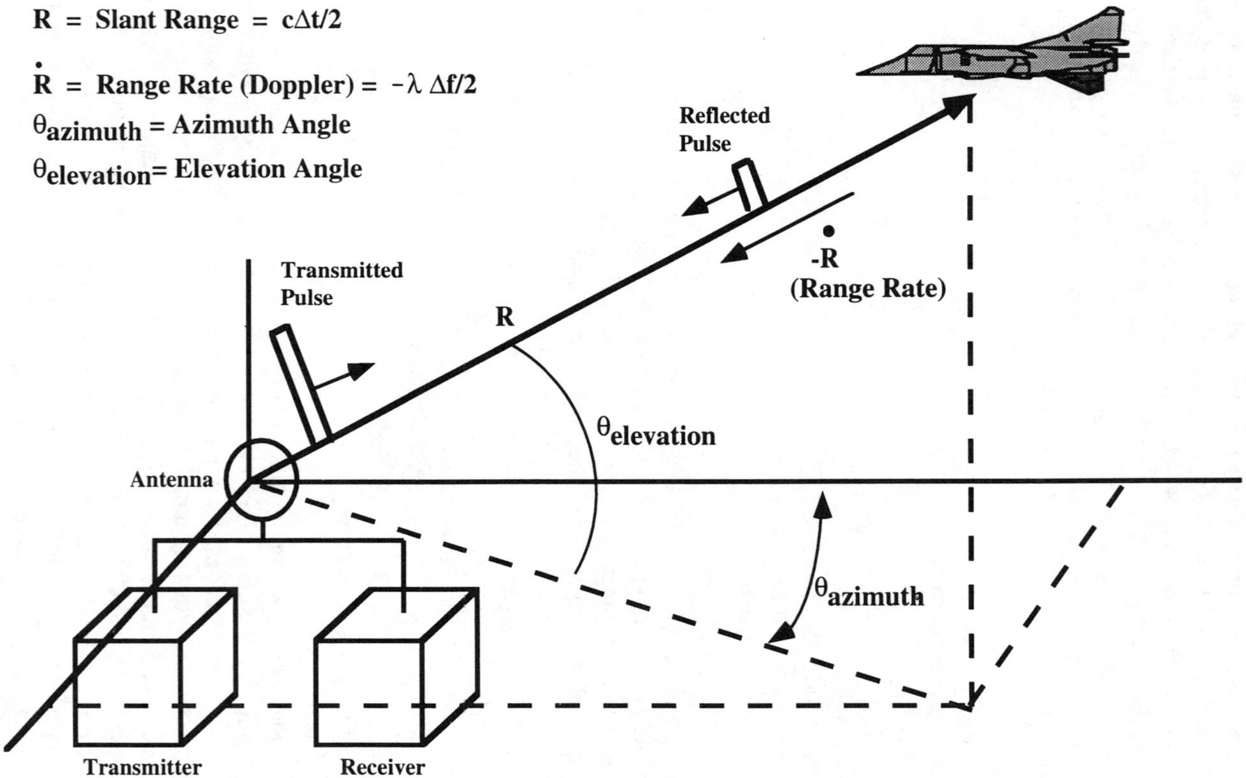


**Figure 2.1** Parameters measured by a typical radar system.

in moderate and temperate zones of the earth. Radars have been taken to the moon [19], and as this article was being written, a radar was circling the planet of Venus and mapping its surface.

Ground-based radars are used for air traffic control at airports and en route tracking of commercial and private aircraft. Ground-based radar is also used for monitoring the weather and tracking the progress of severe weather such as hurricanes and thunderstorms. Ground-based radars have been used extensively within the defense department to provide: long-, medium-, and short-range detection of airborne targets; detection of personnel at high-security installations; precise tracking information on airborne targets for antiaircraft gun and missile systems; high-resolution imaging systems; and precision tracking of earth-orbiting satellites. The earliest measurements of the moon's surface were done with terrestrial radar [20]. These measurements were used to characterize the lunar surface prior to manned lunar voyages. Recent earth-based studies have used radar to determine features of all the planets, some of their moons, the sun, and even some of the asteroids [21]. These radar measurements have also been used to further verify the general theory of relativity.

Shipboard radars perform functions ranging from the detection of other surface ships and the detection and tracking of airborne targets, to fully automated systems, such as the U.S. Navy AEGIS defense system, which performs airborne target detection, target and missile tracking, and missile guidance.

Airborne radars are used on commercial airliners for detection of severe weather and for monitoring the airspace in the forward sector of the aircraft for other commercial or private airplanes. Military aircraft carry tactical radars for detection and tracking of airborne targets, high-resolution mapping, and imaging of the ground using SAR. The Joint Stars (JSTARS) airborne radar system detects moving ground targets. The U.S. Navy E-2C and U.S. Air Force E-3 Airborne Early Warning and Aircraft Control System (AWACS) perform airborne surveillance, detection, and tracking of aircraft, as well as detection and tracking of surface shipping.

Spaceborne radars have been used for rendezvous of spacecraft, descent to the lunar surface, earth sensing, and high-resolution mapping of the moon and Venus. Currently, the Jet Propulsion Laboratory (JPL) at the California Institute of Technology is using an SAR orbiting Venus to determine the surface features of the planet.

## 2.4 RADAR FUNCTIONS

The primary functions of radar are detecting a target and estimating target parameters. Other specialized functions of radar systems have included classification and identification of ground targets, ships, and airborne targets.

Target detection is achieved by performing surveillance of a region of space and sampling each range, angle, and Doppler resolution cell to determine the presence of a target. The type of detection processing depends on if the surveillance region is in free space or if clutter or interference is present. Detection performance in free space is determined by the ratio of energy received at the radar ($E$) to the noise spectral density ($N_0$) measured at the radar receiver, or $E/N_0$. This ratio is equal to the radar signal-to-noise power ratio $S/N$, where $S$ is the signal power returned to the radar from the target and $N$ is the root-mean-square (rms) noise power associated with the radar system.

Figure 2.1 illustrates the parameters measured by a radar system. These parameters include: range ($r$); two angles with respect to the radar antenna boresight such as azimuth ($\varphi$) and elevation ($\theta$); and range rate ($\mathrm{d}r/\mathrm{d}t$), which is related to the target Doppler shift by the relation $\Delta f_d = -2(\mathrm{d}r/\mathrm{d}\,t)/\lambda$, where $\Delta f_d$ is the shift in frequency of the returned carrier with respect to the transmitted carrier frequency, and $\lambda$ is the wavelength of the transmitted signal. The accurate estimation of these parameters is performed by differentiating these basic parameters with respect to time, angle, and frequency, respectively, and performing an interpolation procedure to further refine the measurement.

Table 2.1 lists the basic parameters measured by a radar and indicates the resolution and accuracy associated with each parameter. Parameter accuracy is measured in terms of the standard deviation of the measured quantity. As indicated in Table 2.1, the accuracy of the measurement is determined by the radar resolution associated with that parameter divided by the square root of twice the ratio of $E$ to $N_0$. The factor $k_m$ appearing in the relationship is related to the type of temporal, spatial, or Doppler weighting used in the estimation process.

Resolution is the property of being able to separate two targets' range, angle, or velocity (Doppler).* The range resolution of a radar is determined by the pulsewidth or reciprocal bandwidth of the transmitted pulse. Angle resolution is determined by the respective aperture dimension in units of the transmission wavelength. Doppler resolution is determined by the reciprocal of the total coherent dwell time of the radar on the target.

**Table 2.1**
Radar Parameters, Resolution, and Accuracy

| *Parameter* | *Resolution* | *Accuracy* |
|---|---|---|
| Range ($R$) | $\Delta R = c/B$ | $\sigma_{\Delta R} = \Delta R/[k_{\Delta R}(2S/N)^{1/2}]$ |
| Angle ($\Delta\theta$) | $\Delta\theta = \lambda/L$ | $\sigma_{\Delta\theta} = \Delta\theta/[k_{\Delta\theta}(2S/N)^{1/2}]$ |
| Range Rate ($\dot{R}$) (Doppler) | $\Delta\dot{R} = -\lambda/2T_D$ | $\sigma_{\Delta\dot{R}} = \Delta\dot{R}/[k_{\Delta\dot{R}}(2S/N)^{1/2}]$ |

Note: $c$ = speed of electromagnetic propagation; $B$ = signal instantaneous bandwidth = $1/\tau$ (for matched receiver); $\tau$ = pulse width; $\lambda$ = wavelength; $L$ = characteristic length of antenna; $T_D$ = dwell time; $S$ = target signal level at radar receiver; $N$ = RMS noise at radar receiver; $\sigma$ = standard deviation of the measured quantity $x$; $\sigma_{\Delta R}$ = range accuracy; $\sigma_{\Delta\varphi}$ = angle accuracy; and $\sigma_{\Delta\dot{R}}$ = range-rate accuracy.

*Following common usage, the Doppler frequency shift of the RF carrier due to the target velocity will be referred to simply as "doppler" for the remainder of this chapter.

## 2.5 RADAR BLOCK DIAGRAM AND SUBSYSTEM FUNCTIONS

Figure 2.2 is a generic radar block diagram showing the important functions of all radar systems. The timing and control subsystem is central to the radar operation. It controls the radar operation on a pulse-to-pulse basis. Its parameters can be set by a radar operator or by a computer that selects the specific radar pulse-to-pulse parameters based on the target information the radar derives from its environment. The waveform transmitted by the radar is developed in the waveform generator. This subsystem also generates all the low-voltage timing and frequency signals used in the radar, such as a master clock or master oscillator (MO) frequency, local oscillator (LO) frequency, and additional LOs if the receiver is multistage. The waveform is characterized by its pulsewidth or duration ($\tau$) and its bandwidth or frequency content ($B$). The product of these two factors is called the *time-bandwidth product* of the waveform ($B\tau$).

The signal from the waveform generator is sent to the transmitter, which in turn boosts the power of the basic waveform. The transmitter is characterized by its peak power ($P_{\text{peak}}$), its average power ($P_{\text{avg}}$), and its duty factor ($D$). The relationship between these parameters is

$$D = P_{\text{avg}}/P_{\text{peak}} \tag{2.1}$$

Transmitters can be oscillators that have their output signal gated by the waveform generator or power amplifiers that directly amplify the power of the input waveform signal. The amplified transmit signal is directed to the radar antenna by way of a transmit/receive (T/R) switch. The primary purpose of this switch is to protect the sensitive radar receiver from the high-power waveform produced by the transmitter.

The high power output port of the T/R switch is connected to the radar antenna. The antenna is a transducer that matches the high-power waveform that is present in an RF waveguide or transmission line to free space. The antenna launches the high-power RF energy into space and directs this energy to the desired azimuth and elevation bearing. The antenna is characterized by its directive gain ($G_D$) or effective area ($A_{\text{eff}}$). The relationship between these two parameters is

$$G_D = 4\pi A_{\text{eff}}/\lambda^2 \tag{2.2}$$

The relationship between the physical area ($A$) of an antenna and its effective area is determined by the antenna radiation efficiency ($\eta$). The relationship is $A_{\text{eff}} = \eta A$. To understand (2.2), we must consider another important characteristic of the antenna, the antenna principal beamwidths ($\theta_{\text{az}}$ and $\theta_{\text{el}}$). From diffraction theory [22], it is well known that

$$\theta_{\text{az}} = \lambda/a \tag{2.3}$$

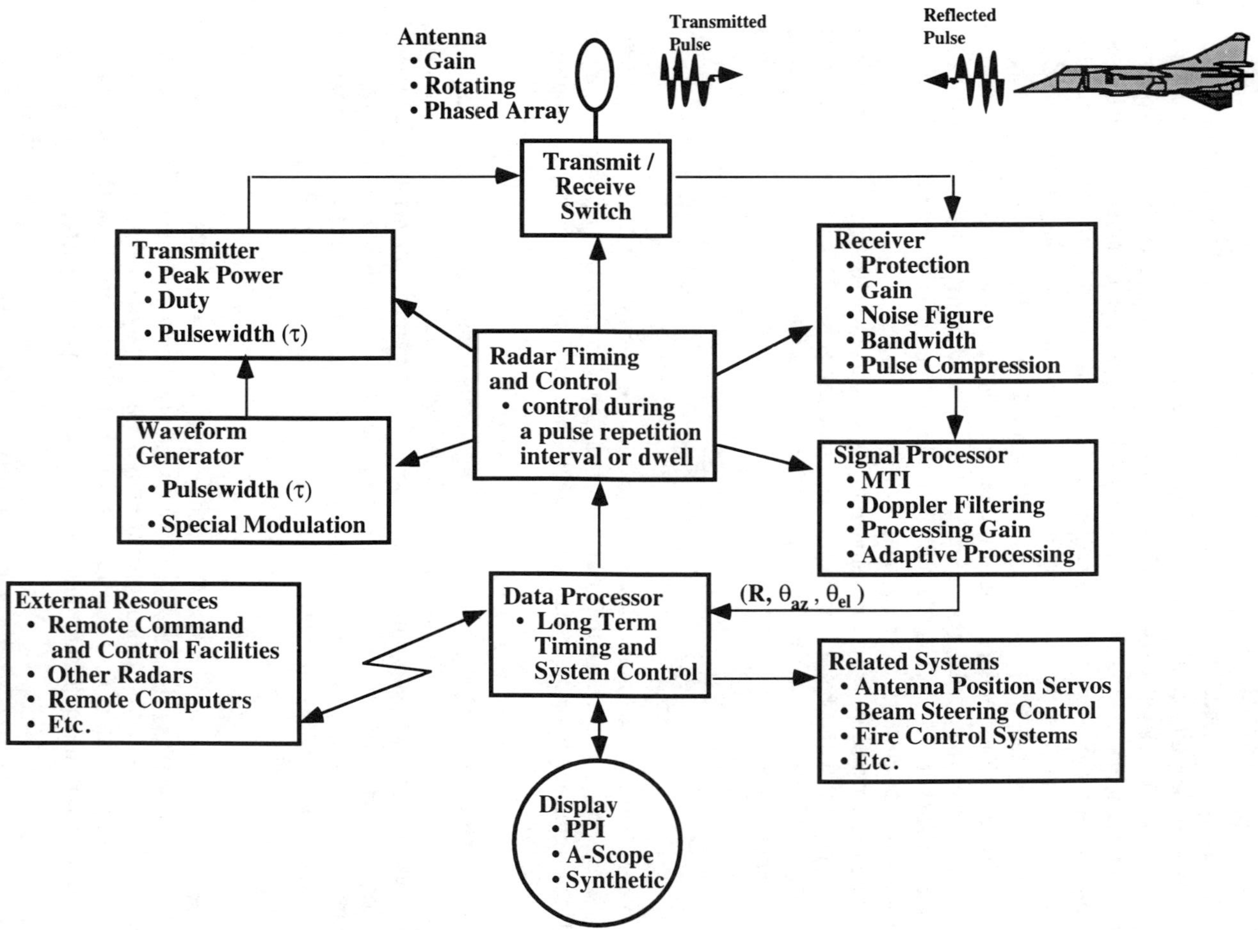


**Figure 2.2** Generic radar system functional diagram.

and

$$\theta_{el} = \lambda / b \tag{2.4}$$

where $a$ is the horizontal antenna dimension and $b$ is the vertical antenna dimension. The directive gain of the antenna represents the amount the antenna concentrates the beam. $G_D$ is determined from the relationship

$$G_D = 4\pi/(\theta_{az}\theta_{el}) \tag{2.5}$$

The antenna gain $G$ is called the *power gain* of the antenna. It includes the losses associated with the antenna and is related to the directive gain by the radiation efficiency factor $\eta$ and the reflection efficiency factor $\rho$ through the expression

$$G = \eta\rho G_D \tag{2.6}$$

Combining (2.5) and (2.6),

$$G = \eta\rho 4\pi/(\theta_{az}\theta_{el}) \tag{2.7}$$

and, from equations (2.3) and (2.4),

$$G = \eta\rho 4\pi ab/\lambda^2 \tag{2.8}$$

and, since $\eta\rho ab = A_{eff}$, we arrive at (2.2).

The radar RF energy launched by the antenna propagates through the atmosphere at the speed of light $c$ until it is reflected by a target. The target reflects the incident radiation in all directions. The relative amount of radiation scattered in a particular direction is described by the target's radar cross section $\sigma(\theta,\phi)$. Formally, the radar cross section $\sigma$ is the ratio of power reflected in the direction $\theta,\phi$ to the power density incident on the target from the direction $\theta,\phi$. The target cross section also describes the polarization characteristics of the target through the $2 \times 2$ polarization matrix

$$\sigma(\theta,\phi) = \begin{bmatrix} \sigma_{hh} & \sigma_{hv} \\ \sigma_{vh} & \sigma_{vv} \end{bmatrix} \tag{2.9}$$

where $\sigma_{hh}$ corresponds to the target cross section associated with transmission of horizontal polarization and reception of horizontal polarization. Similarly, $\sigma_{hv}$ corresponds to transmission of horizontal polarization and reception of vertical polarization. The other quantities in (2.9) are self-explanatory. The on-diagonal terms, $\sigma_{hh}$ and $\sigma_{vv}$, are called the *copolarization terms*, since the same polarization is transmitted and received. The off-diagonal terms, $\sigma_{hv}$ and $\sigma_{vh}$, are called the *cross-polarization terms*, since the received polarization is orthogonal to the transmitted polarization.

Radiation reflected back to the radar from the target is captured by the radar antenna and directed to the T/R switch. The T/R switch directs the received RF energy to the radar receiver.

The first stage of the receiver is usually a highly sensitive low-noise amplifier that minimizes the noise present in the system. This noise competes with detection of the signal returned from the target. The receiver is characterized by its noise figure (NF), which is related to the additional noise produced at the output of the receiver. More specifically, the internal receiver noise that appears at the receiver output is equal to (NF − 1)k$T_0 B G_{rec}$, where k is the Boltzmann constant, 1.38 ± $10^{-23}$, which relates kelvin temperature to thermal energy, $T_0$ is the receiver ambient kelvin temperature, and $G_{rec}$ is the receiver gain (= $S_{out}/S_{in}$). The total noise out of the receiver is given by

$$N = \mathrm{k}T_0 B \mathrm{NF} G_{rec} \tag{2.10}$$

The receiver conditions the received signal by performing functions such as pulse compression and matched filtering at the single-pulse level. This is performed by a series of converters that mix the RF signal with the LO signal to translate the target signal from RF to one or more intermediate frequencies (IF) and finally to baseband, where digital signal processing can be performed. The baseband processor consists of a product detector [23] that converts the final IF stage signal to a pair of in-phase (I) and quadrature (Q) signals that are 90 deg out of phase. The I and Q signals are each converted from analog to digital signals by a pair of analog-to-digital (A/D) converters.

The signal processor, described in Sections 2.9 and 2.10, receives one or more pulses from one or more channels of the receive subsystem and performs the detection and the initial parameter estimation process on the received energy. The signal processor performs moving target indication, pulse Doppler analysis, or adaptive processes on the returned energy, looking for the presence of a target. The signal processor may also complete the match filtering process for the radar receiver by performing processing over a series of pulses. This processing may consist of coherent or noncoherent integration of a series of pulses to enhance the detectability of the target. The detection reports and any estimation of target parameters measured by the radar are transmitted to a data processor and/or are displayed to an operator.

The data processor is used in many modern radars to improve the estimate of target parameters and develop tracks on the targets. More sophisticated uses of the data processor include establishing the target mix in the surveillance volume of the radar and controlling the waveforms the radar transmits based on the target type. Phased-array or electronically steered agile beam antenna radars normally use a sophisticated radar processor to determine what functions the radar should perform, when these functions should be performed, and what waveform and power levels should be used.

## 2.6 RADAR RANGE EQUATION

One of the most important tools for the radar systems engineer is the radar range equation (RRE). The RRE is a phenomenological equation that uses the radar parameters described above to determine power levels at the radar from targets, clutter, jammers, beacons, and so on. The RRE also describes the background noise power from the environment and the radar receiver subsystem. The information provided by the RRE permits the determination of radar performance in terms of target-detection capability and the parameter estimation accuracy of the system under operating conditions in free space, clutter, jamming, and so on.

The radar range equation in free space can be written as

$$S/N = P_{\text{peak}} G_{\text{xmt}} \sigma(\theta,\varphi) A_{\text{eff}} / \{(4\pi)^2 R^4 L \text{k} T_0 B \text{NF}\} \quad (2.11)$$

where $P_{\text{peak}}$ denotes the transmitter peak power; $G_{\text{xmt}}$ denotes the transmit antenna power gain; $\sigma(\theta,\varphi)$ denotes the monostatic radar cross section; $A_{\text{eff}}$ denotes the effective radiation capture area of the receive antenna; $R$ denotes the distance from the radar to the target; $L$ denotes the losses associated with the radar microwave hardware, antenna scanning procedure, signal propagation, target characteristics, and signal processing characteristics. The remaining parameters relating to noise, k, $T_0$, $B$, and NF, are described above. Equation (2.11) can be understood by writing it in terms of two equations:

$$S = P_{\text{peak}} G_{\text{xmt}} \sigma(\theta,\varphi) A_{\text{eff}}\, G_{\text{rec}} / [(4\pi)^2 R^4 L] \quad (2.12)$$

which describes the signal power returned to the radar from the target, and

$$N = \text{k} T_0 B \text{NF} G_{\text{rec}} \quad (2.13)$$

which describes the noise power within the radar system that is competing with the target signal. The parameter $G_{\text{rec}}$ appearing in both these equations is the receiver gain. Thus, these equations represent the radar signal and noise output referenced to the output of the radar receiver. Equation (2.1) follows from dividing (2.2) by (2.3).

Figure 2.3 illustrates the derivation of (2.12). Equation (2.14) represents the power density at a range $R$ from the radar, which is given by

$$P_{\text{peak}} / (4\pi) R^2 \quad (2.14)$$

which represents the power delivered from an isotropic antenna. This is multiplied by the transmit antenna gain to yield

$$P_{\text{peak}} G_{\text{xmt}} / (4\pi) R^2 \quad (2.15)$$

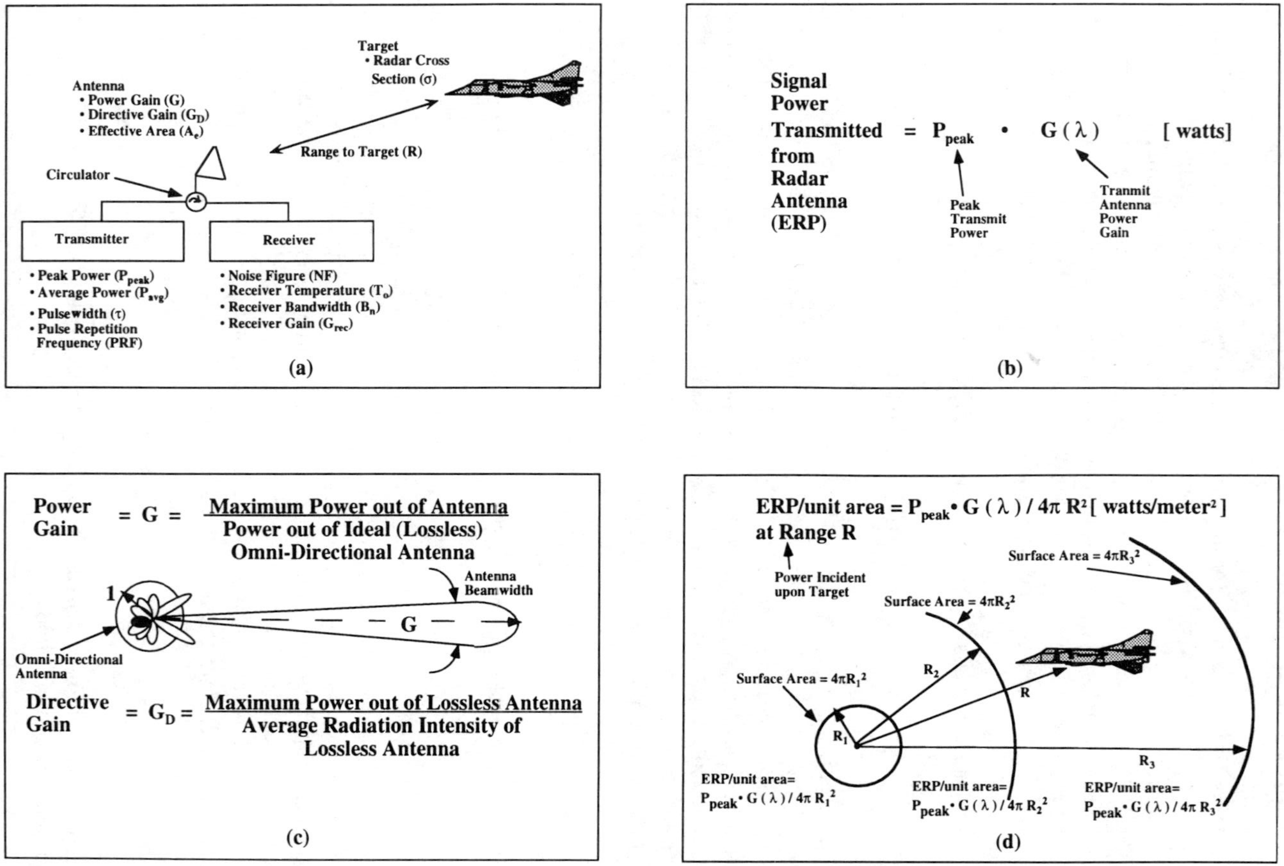


**Figure 2.3** Physical foundation of the radar range equation: (a) radar-target parameters; (b) radar effective-radiated power (ERP); (c) antenna power and directive gain; (d) power incident on radar target.

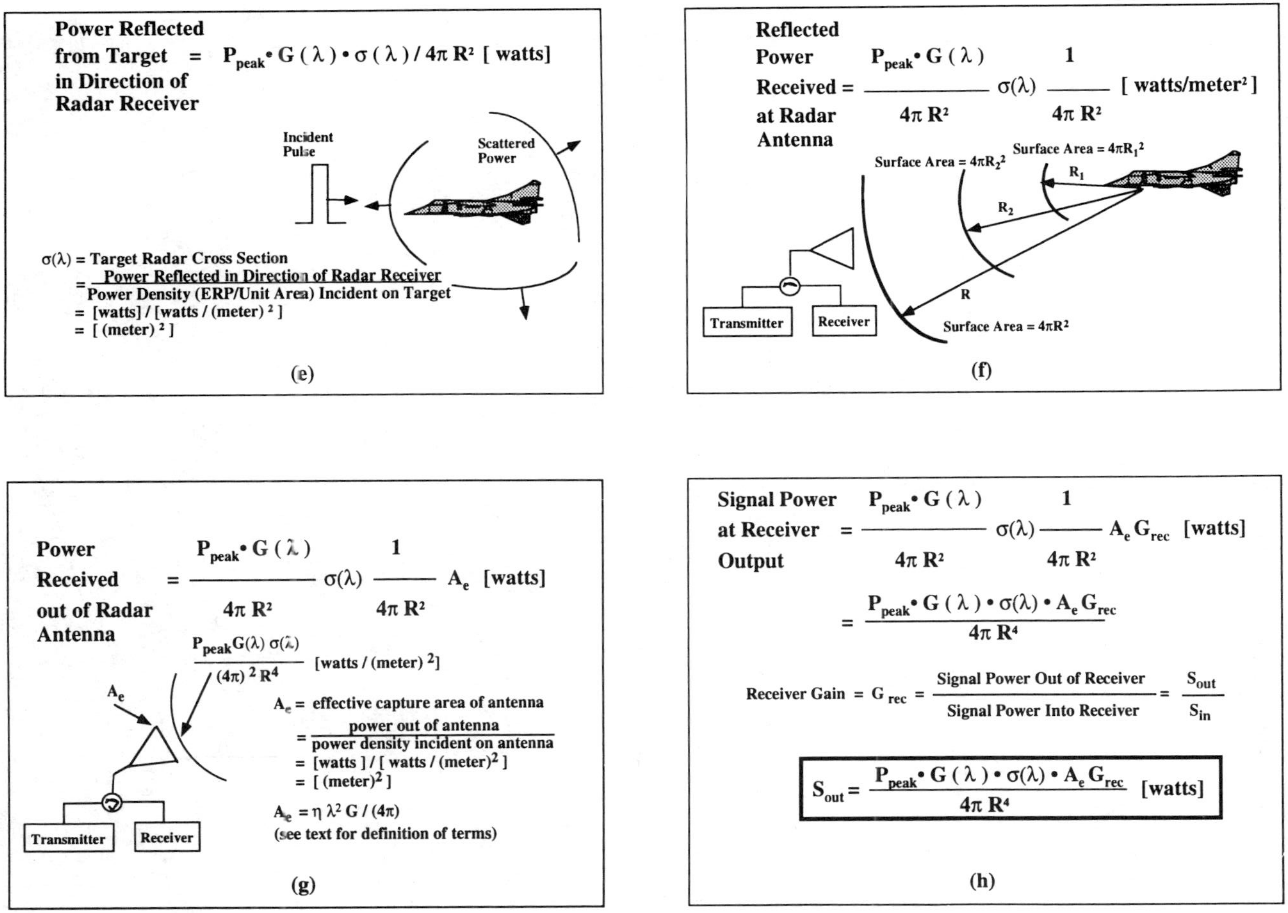


**Figure 2.3 (cont.)** Physical foundation of the radar range equation: (e) monostatic radar cross section of target, $\sigma(\lambda)$: (f) reflected power incident on radar antenna; (g) power received out of radar antenna; (h) signal power at output of radar receiver.

which represents the power density incident on the radar target at range $R$ from the radar. The power reflected from the target in the direction of the radar is given by

$$P_{\text{peak}} G_{\text{xmt}} \sigma(\theta,\varphi)/(4\pi)R^2 \tag{2.16}$$

where $\sigma(\theta,\varphi)$ is the monostatic radar cross section of the target. The power density (watts/square meter) delivered to the input of the radar antenna is given by the expression

$$P_{\text{peak}} G_{\text{xmt}} \sigma(\theta,\varphi)/(4\pi)^2 R^4 \tag{2.17}$$

The power delivered to the output of the antenna is given by

$$P_{\text{peak}} G_{\text{xmt}} \sigma(\theta,\varphi) A_{\text{eff}}/(4\pi)^2 R^4 \tag{2.18}$$

and the power present at the output of the receiver is given by

$$P_{\text{peak}} G_{\text{xmt}} \sigma(\theta,\varphi) A_{\text{eff}}\, G_{\text{rec}}/(4\pi)^2 R^4 \tag{2.19}$$

The clutter power returned to a radar can be calculated in a similar fashion.

Clutter is unwanted returns to a radar system. As shown in Figure 2.4, clutter power can be returned from the ground, the sea, or a rain storm. Ground and sea clutter is called *two-dimensional* (2-D) clutter, since the backscattered energy is determined by the antenna beamwidth and pulsewidth, as shown in Figure 2.5, or by the antenna azimuth and elevation beamwidths for high grazing angles, as illustrated in Figure 2.6. Clutter power due to rain or chaff (i.e., dipoles intentionally dropped in the surveillance region of a radar to mask a target) is called *three-dimensional* (3-D) clutter, because the clutter power is determined by the azimuth and elevation beamwidths and the radar pulsewidth. See Figure 2.7.

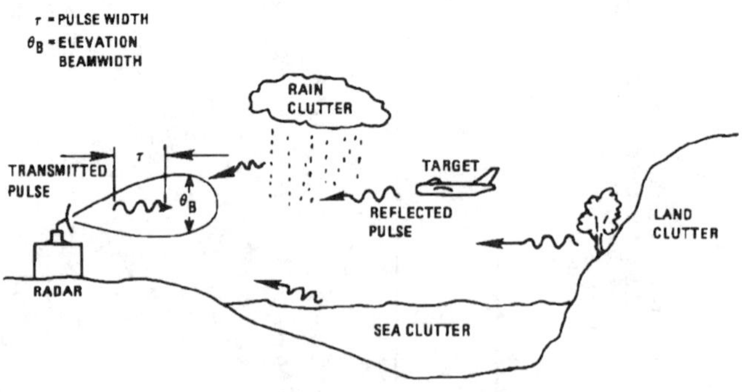


**Figure 2.4** Sources of radar clutter.

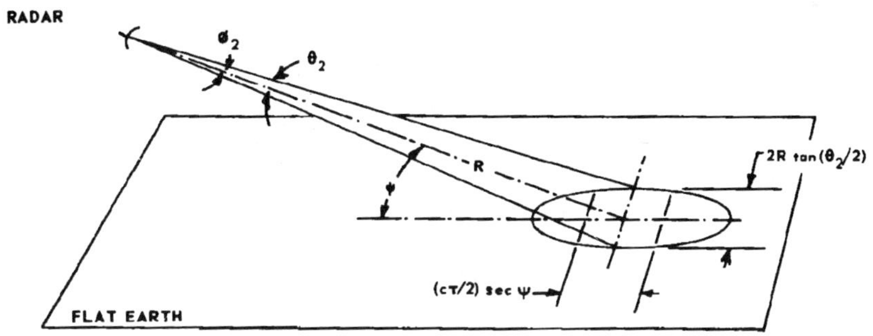


**Figure 2.5** Factors needed to calculate 2-D clutter return (low grazing angle): tan $\Psi$ < [2R sin($\phi_2$/2)]/(c$\tau$/2), pulse-length-limited case.

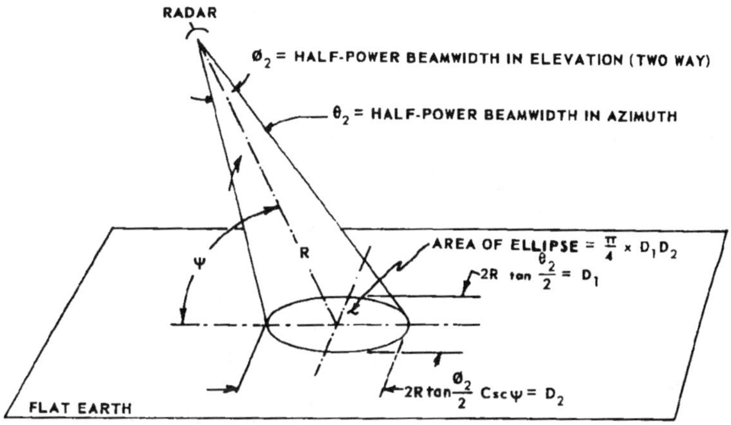


**Figure 2.6** Factors needed to calculate 2-D clutter return (high grazing angle): tan $\Psi$ < [2R tan($\phi_2$/2)]/(c$\tau$/2), beamwidth-limited case.

The equations for determining the 2-D clutter power returned to the radar are

$$C_{2\text{-D}}\ \text{LGA} = P_{\text{peak}}G_{\text{xmt}}\{R\theta_{\text{az}}(c\tau/2)\sigma_{02}\}A_{\text{eff}}\,G_{\text{rec}}/[(4\pi)^2R^4L] \tag{2.20}$$

$$= P_{\text{peak}}G_{\text{xmt}}\{\theta_{\text{az}}(c\tau/2)\sigma_{02}\}A_{\text{eff}}\,G_{\text{rec}}/[(4\pi)^2R^3L] \tag{2.21}$$

and

$$C_{2\text{-D}}\ \text{HGA} = P_{\text{peak}}G_{\text{xmt}}\{R\theta_{\text{az}}R\theta_{\text{el}}\sigma_{02}\}A_{\text{eff}}\,G_{\text{rec}}/[(4\pi)^2R^4L] \tag{2.22}$$

$$= P_{\text{peak}}G_{\text{xmt}}\{\theta_{\text{az}}\theta_{\text{el}}\sigma_{02}\}A_{\text{eff}}\,G_{\text{rec}}/[(4\pi)^2R^2L] \tag{2.23}$$

Note that LGA stands for low grazing angle and HGA stands for high grazing angle. These equations are derived in the same manner as (2.12) for the signal power returned

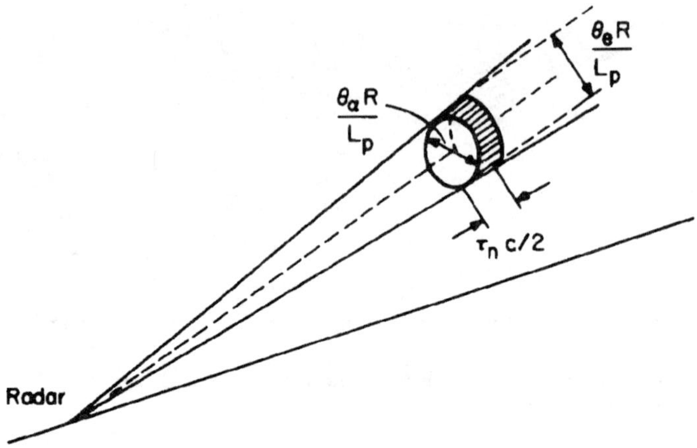


**Figure 2.7** Factors required to calculate 3-D clutter return.

from the target. The only difference is to replace the target cross section $\sigma(\theta,\varphi)$ with the expressions

$$\sigma_{2\text{-D}} \text{ clutter LGA} = R\theta_{az}(c\tau/2)\sigma_{02} \tag{2.24}$$

and

$$\sigma_{2\text{-D}} \text{ clutter HGA} = R\theta_{az}R\theta_{el}\sigma_{02} \tag{2.25}$$

where $\sigma_0$ represents the clutter reflectivity in units of $m^2/m^2$ for the land or sea clutter patch being illuminated by the radar. The value associated with $\sigma_{02}$ depends on the grazing angle, polarization of the radar signal, type of clutter being illuminated, and so on. This will be discussed more fully in Section 2.7.

The equation for determining 3-D clutter power returned to the radar is given by

$$C_{3\text{-D}} = P_{peak}G_{xmt}\{R\theta_{az}R\theta_{el}(c\tau/2)\sigma_{03}\}A_{eff}\,G_{rec}/[(4\pi)^2R^4L] \tag{2.26}$$
$$= P_{peak}G_{xmt}\{\theta_{az}\theta_{el}(c\tau/2)\sigma_{03}\}A_{eff}\,G_{rec}/[(4\pi)^2R^2L] \tag{2.27}$$

The factor $\sigma_{03}$ is the 3-D clutter reflectivity given in units of $m^2/m^3$ as described in Section 2.7.

In (2.20) through (2.27), the clutter reflectivity is multiplied by the radar 2-D or 3-D resolution cell. This resolution cell is determined by (1) the radar two-way 3-dB antenna azimuth or elevation beamwidth ($\theta_{az}$ and $\theta_{el}$) and the radar pulsewidth ($\tau$). As shown in Figure 2.5, at a low grazing angle the 2-D illuminated clutter patch has a physical area of $R\theta_{az}(c\tau/2)$, where $R$ is the radar range to the clutter patch and $c$ is the speed of light.

At a high grazing angle, the 2-D illuminated clutter patch is given by $R\theta_{az}R\theta_{el}$, as shown in Figure 2.6. Lastly, the 3-D clutter patch illuminated by the radar has a physical volume of $R\theta_{az}R\theta_{el}(c\tau/2)$, as shown in Figure 2.7.

In a similar way, the power delivered by an interference source to the radar can also be calculated. Consider the scenario shown in Figure 2.8, where a noise jammer (i.e., a device tuned to the radar frequency which produces white noise in the radar bandpass) is at a range $R$ from the radar. The noise energy present at the radar due to the jammer $J$ is given by the expression

$$J = P_{avg-J}\, G_J(B_R/B_J)A_e\, A_{sll}/[4\pi R^2 L_J] \tag{2.28}$$

where:

$P_{avg-J}$ = average power delivered by the jammer.
$G_J$ = gain of the jamming antenna.
$(B_R/B_J)$ = bandwidth ration of the radar to the jammer. This ratio has the value 0 if the jammer is not radiating within the bandpass of the radar receiver, and equals 1 if $B_J < B_R$ and the jammer radiates within the jammer bandpass. Otherwise, the ratio is equal to the ratio $B_R/B_J$.
$A_e$ = effective area of the radar receive antenna.
$A_{sll}$ = mean sidelobe level of the radar antenna measured with respect to the antenna main-beam gain, as shown in Figure 2.9.
$R_J$ = range between the radar and the jammer.
$L_J$ = RF losses of the jammer and the radar.

The radar range equation can be expressed in many different forms, depending on its application. Two very useful forms are the surveillance equation [24] and the tracking equation [25].

The surveillance equation is used to determine the radar power-aperture product needed to achieve a specified surveillance requirement. The surveillance radar equation determines the S/N at the radar in terms of the surveillance radar problem. The equation is derived from (2.11) as follows. As shown in Figure 2.10,

$$P_{peak}\tau = P_{avg}\text{PRI} \tag{2.29}$$

where $P_{peak}$ denotes the peak power of the radar; $\tau$ denotes the radar pulse width; $P_{avg}$ denotes the average power of the radar; PRI denotes the pulse repetition interval (i.e., the time between successive transmitted pulses). Also consider the relation between antenna gain and the antenna azimuth and elevation beamwidths given by (2.5). Let us define the 2-D solid angle subtended by the antenna beam as

$$\Omega_{beam} = \theta_{az}\theta_{el} \tag{2.30}$$

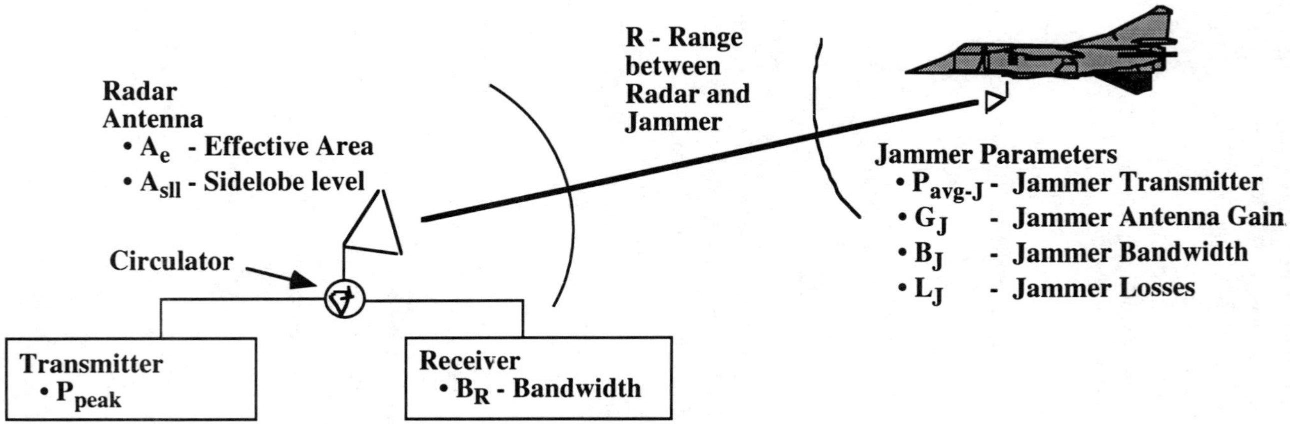


**Figure 2.8** Factors required for determining jamming power delivered to the radar.

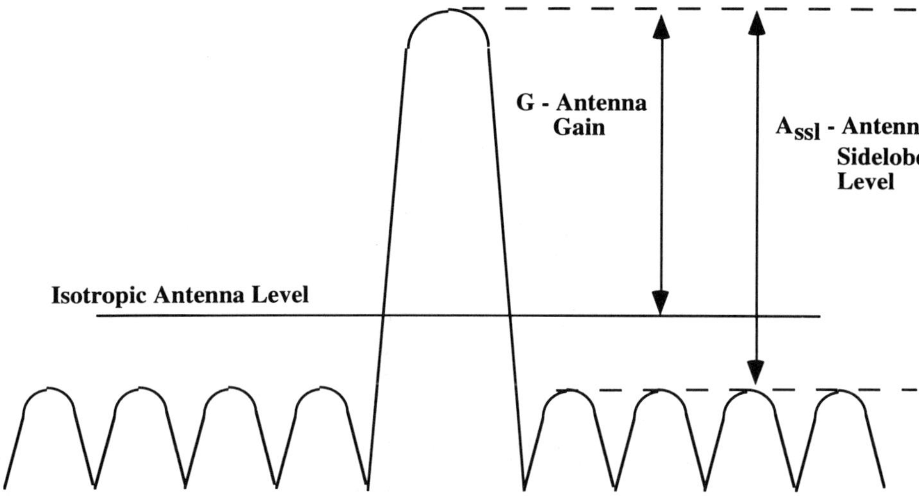


**Figure 2.9** Definition of antenna sidelobe level $A_{sll}$ and average power of a radar are related by the equation.

Let us also define the surveillance volume to be scanned by a surveillance radar by the parameter $\Omega_{\text{volume}}$. If $t_{\text{dwell}}$ and $t_{\text{search}}$ are the total dwell time and the time allowed for the radar to search the solid angle $\Omega_{\text{volume}}$, then the following relations hold:

$$t_{\text{search}}/t_{\text{dwell}} = \Omega_{\text{volume}}/\Omega_{\text{beam}} \tag{2.31}$$

and

$$t_{\text{dwell}} = n_{\text{pulses}}\text{PRI} \tag{2.32}$$

where $n_{\text{pulses}}$ is the number of pulses (or PRIs) in a single search dwell. Substituting (2.29) and (2.30) into (2.11) results in the relationship

$$S/N = [P_{\text{avg}}\text{PRI}\sigma(\theta,\varphi)A_{\text{eff}}]/(4\pi\Omega_{\text{beam}}R^4Lk T_0 B\tau \text{NF}) \tag{2.33}$$

and using relations (2.31), (2.32), and the relationship $B\tau = 1$ for a matched receiver system, (2.33) becomes

$$S/N = [P_{\text{avg}}t_{\text{search}}\sigma(\theta,\varphi)A_{\text{eff}}]/(4\pi\Omega_{\text{search}}n_{\text{pulses}}R^4Lk T_0 \text{NF}) \tag{2.34}$$

Solving this for the product $P_{\text{avg}}A_{\text{eff}}$, the average transmit power times the effective receive aperture, results in the surveillance form of the radar range equation:

$$P_{\text{avg}}A_{\text{eff}} = [(S/N)4\pi\Omega_{\text{search}}n_{\text{pulses}}R^4Lk T_0 \text{NF}]/[t_{\text{search}}\sigma(\theta,\varphi)] \tag{2.35}$$

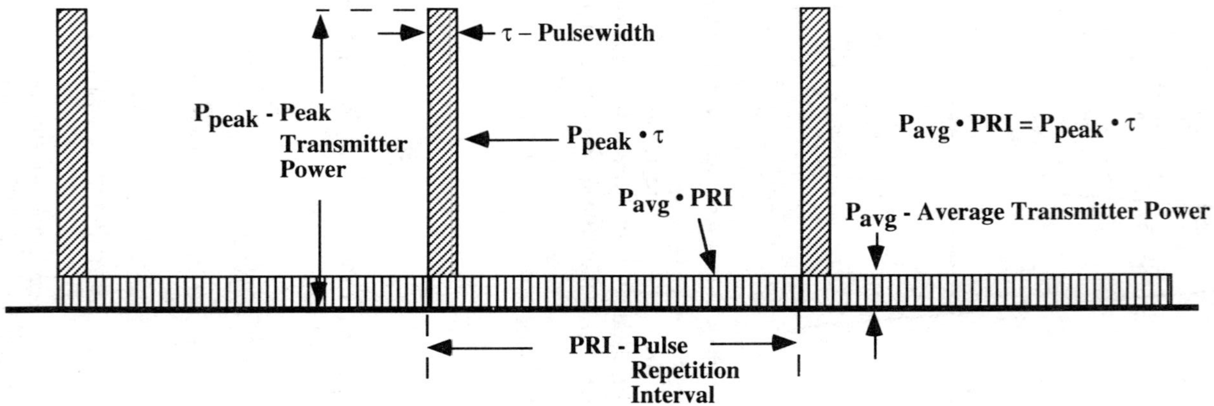


**Figure 2.10** Relationship between radar peak and average power.

The parameters on the right-hand side of (2.35) specify: (1) the surveillance design requirements for the radar ($S/N$, which is the required signal-to-noise ratio for detection, $\Omega_{search}$, which is the volume to be searched, $t_{search}$, which is the time available to search the volume, and $R$, which is the required detection range on a target with radar cross section $\sigma(\theta,\varphi)$); and (2) a limited number of radar design parameters ($n_{pulse}$, which is the number of pulses per dwell predicated on the detection and MTI requirements, and NF, which is the radar receiver noise figure). These parameters can be set by the radar designer to values consistent with current technology, economic constraints, and operational constraints of the radar. The significance of (2.35) is that once the surveillance performance of a radar is specified, the primary radar parameters that determine the radar design are the radar average power and receive antenna aperture. It is somewhat misleading, but, at first glance, (2.35) seems to indicate that surveillance performance is independent of frequency. This is not the case, however, since both the radar losses $L$ and the radar target cross section $\sigma(\theta,\varphi)$ vary with frequency. Normally, both $L$ and $\sigma(\theta,\varphi)$ increase with increasing frequency, with $L$ increasing faster than $\sigma(\theta,\varphi)$. Thus, most long-range surveillance radars are at low frequency in the UHF or $L$ band. The product $P_{avg}A_{eff}$ is called the *power-aperture product of the surveillance radar system.*

A similar equation can be developed to describe the tracking capability of a radar. Recall, from Table 2.1, that the relationship for the azimuthal and elevation angle accuracy of a radar is given by the equations

$$\sigma_{az} = \theta_{az}/\mathrm{k}_{az}(S/N)^{1/2} \tag{2.36}$$

$$\sigma_{el} = \theta_{el}/\mathrm{k}_{el}(S/N)^{1/2} \tag{2.37}$$

A measure of the overall tracking accuracy of a radar, given that angle accuracy is much poorer than range accuracy, is given by the parameter $\sigma^2_{angle} = (\sigma_{az})(\sigma_{el})$.

Use of (2.2), (2.5), and (2.11) results in

$$\sigma^2_{angle} = \frac{4\pi R^4 L\mathrm{k}TB\mathrm{NF}\lambda^4}{\mathrm{k}_{az}\mathrm{k}_{el}\sigma P_{peak}A^3_{eff}} \tag{2.38}$$

Note that in (2.38), $\sigma_{angle}$ is the standard deviation of the angle measurement and $\sigma$ is the radar cross section of the target.

This relationship indicates that angle accuracy increases as the radar peak power is increased, the antenna aperture is made larger, and the transmission wavelength is smaller.

Tracking radars are generally high-frequency radars (i.e., X band and above). The primary difficulty of high-frequency radar is increased loss due to extended propagation through the atmosphere and increased attenuation in the presence of rain or high moisture content in the atmosphere.

Hybrid radars that perform both surveillance and tracking (i.e., the U.S. Navy AEGIS, AN/SPY-1, at S band and the U.S. Army Patriot Radar at C band) are operated at midfrequency in order to provide the best compromise between surveillance and tracking performance from a single sensor.

## 2.7 THE DETECTION OF SIGNALS IN NOISE

The detection of a target signal in noise is described mathematically in terms of probability because noise is a statistical phenomenon. Noise, as discussed in Section 2.6, is an uncontrollable fluctuation in the voltage at the output of the radar receiver. Its primary sources are cosmic radiation, terrestrial atmospheric phenomena, lossy elements in the radar RF hardware, and the receiver itself.

Figure 2.11 illustrates representative pulse repetition intervals for a search radar and a tracking radar. The search radar looks for targets at all ranges, whereas the tracking radar redetects the target under track within a narrow range gate centered at the expected range of the target. In both cases, the target must be extracted from the noise background. Figure 2.12 illustrates this process. Noise is a background fluctuation with amplitude described by the Rayleigh distribution discussed below. As shown in Figure 2.12, the presence of a target is determined by a voltage threshold. This threshold can be set manually or can be automatically determined by an averaging circuit in the radar signal processor referred to as a *constant false alarm rate* (CFAR) [26]. Once again, referring to Figure 2.12, a signal exceeding the threshold can either be a true target or a large noise fluctuation. Declaring a target when the threshold crossing is due to a target signal is called a *valid detection*. Declaring a target when the threshold crossing is due to noise is called a *false alarm*. It is also possible that a target is present but does not cross the threshold. This condition is called a *missed target detection*. These events are quantified through the use of probability density functions that represent noise (Rayleigh distribution) alone and the combination of target signal and noise (Rice distribution).

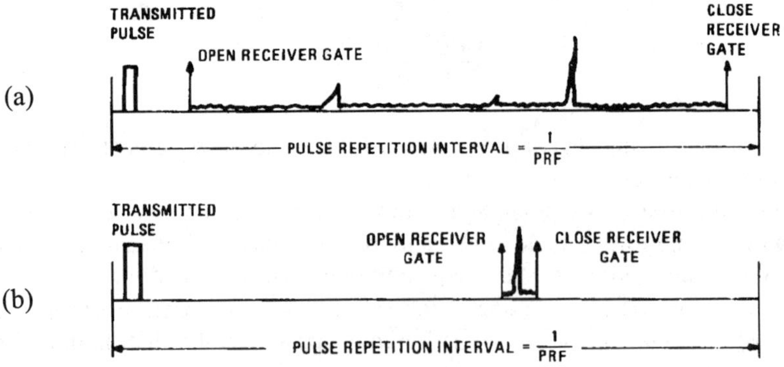


**Figure 2.11** Detection of targets in noise for (a) a search radar; (b) a tracking radar.

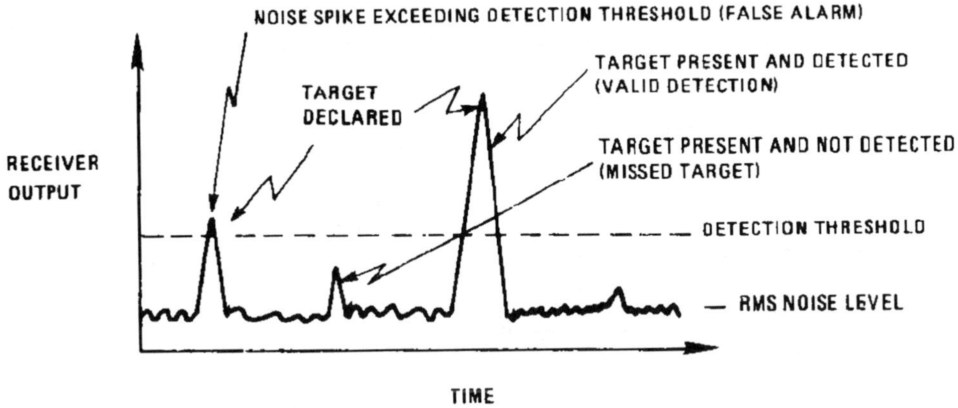


**Figure 2.12** Detection process.

Receiver noise is described by Planck's black body radiation law, which relates noise power to ambient temperature through the expression

$$u_{avg}(\nu,T) = \frac{h\nu}{e^{h\nu/kT} - 1} \tag{2.39}$$

where $u_{avg}$ is the noise energy or noise power per cycle of radiation of frequency $\nu$ at an ambient temperature $T$, $h$ is Planck's constant of action ($= 6.62 \times 10^{-34}$ J · s), and k is the Boltzmann constant relating temperature to energy ($= 1.38 \times 10^{-23}$ J/K).

For radar, $h\nu \ll kT$ and (2.39) reduces to

$$u_{avg} \approx kT \tag{2.40}$$

which represents the average noise energy (noise power per cycle) at the radar receiver output when the receiver is at an ambient temperature $T$. Table 2.2 shows representative values of $\nu$ and $T$ for microwave and laser radar. As is shown, Planck's law must be used to represent noise for infrared and laser radar detectors. The approximation described by (2.40) is valid for microwave radars from VHF up to 100 GHz (millimeter-wave radar).

As described in Section 2.6, the root-mean-square (rms) noise power at the output of a radar receiver at ambient temperature $T_0$ is

$$N = kT_0B\text{NF}G_{rec} \tag{2.41}$$

where $B$ is the bandwidth of the receiver, NF is the receiver noise figure, and $G_{rec}$ is the receiver gain. The probability density curve that represents thermal noise (i.e., a Gaussian process) at the output of a narrowband receiver (i.e., signal bandwidth is small compared

**Table 2.2**
Evaluation of Planck Distribution for Microwave and Optical Frequencies

| *Radar Band* | *Frequency (f)* | *hf/kT* | *ρ(f)* |
|---|---|---|---|
| VHF | 100 Mhz | $1.66 \times 10^{-5}$ | kT |
| L-band | 1 Ghz | $1.66 \times 10^{-4}$ | kT |
| X-band | 10 Ghz | $1.66 \times 10^{-3}$ | kT |
| mm-waves | 100 Ghz | $1.66 \times 10^{-2}$ | kT |
| Visible light (lasar radar) | $10^6$ Hz | 16.6 | hf/(exp(hf/kT) − 1)* |

Note: h = Planck's constant = $6.63 \times 10^{-34}$ joule-sec; k = Boltzmann's constant = $1.38 \times 10^{-23}$ joule/°K; T = room temperature = 290° K = 62.6° F:
*Photon noise.

to the RF center frequency) is shown in Figure 2.13. The curve is a Rayleigh distribution. The Rayleigh probability density is given by

$$p_n(v) = (v/N)\exp(-v^2/2N) \tag{2.42}$$

where $p_n(v)$ is the probability the noise voltage is between $v$ and $v + dv$, and the rms noise power is given by $N$.

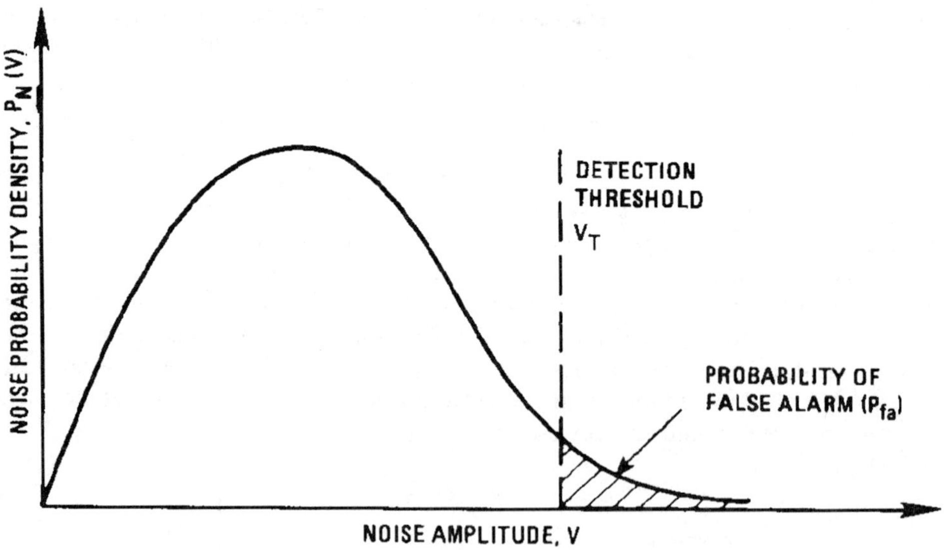


**Figure 2.13** Rayleigh distribution of noise only showing the relation between the detection threshold ($V_T$) and the probability of false alarm ($P_{fa}$).

The area under the curve between $v$ and $v + dv$ represents the probability that a noise sample will have a voltage level between $v$ and $v + dv$. As shown in the figure, once a detection threshold ($V_T$) has been set, the probability of false alarm is given by

$$P_{fa} = \int_{V_T}^{\infty} [(v/N^{1/2})\exp(-v^2/2N^{1/2})]dv \tag{2.43}$$

$$= \exp(-v_T^2/2N^{1/2}) \tag{2.44}$$

Equation (2.44) may be solved for $v_T$ in terms of $P_{fa}$ to yield

$$v_T = \sqrt{(2N^{1/2}\ln P_{fa})} \tag{2.45}$$

This permits a detection threshold to be established when a false alarm probability is established on the basis of system requirements.

Figure 2.14 shows the change in the Rayleigh distribution when a signal (single-tone sine wave) is present along with the noise. The resulting distribution is called a *Ricean distribution*, after [27]. The Ricean distribution is given by

$$p_{s+n}(v;A,N) = (v/N)\{\exp[-(v^2 + A^2)]\}I_0(Av/N) \tag{2.46}$$

where $p_{s+n}(v)$ is the probability that the voltage of a sine wave of amplitude $A$ plus white noise will be between $v$ and $v + dv$. The function $I_0$ is a modified Bessel function of the

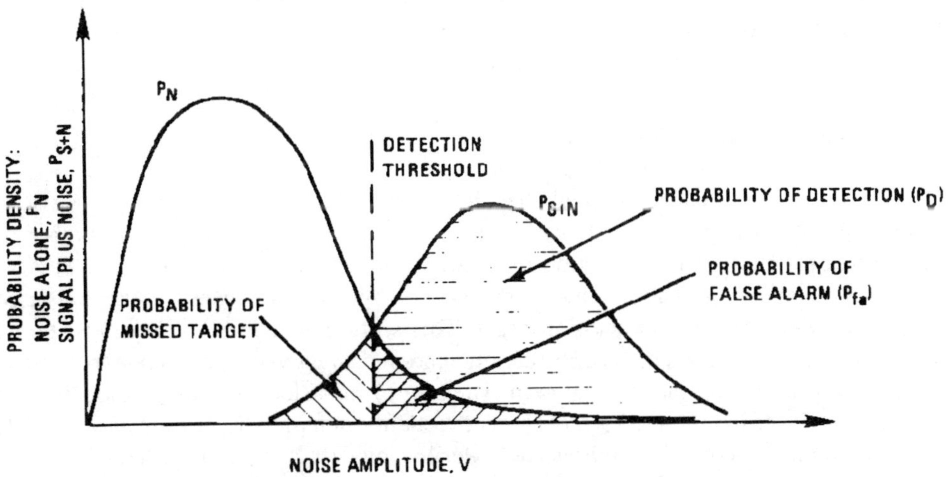


**Figure 2.14** Rice distribution of signal plus noise showing the relation between the detection threshold ($v_T$), the probability of target detection ($P_d$), and the probability of missing a target ($P_m = 1 - P_d$).

first kind of order 0 [28]. The signal-to-noise ratio (SNR), associated with the sine wave of amplitude $A$ in the presence of noise with rms noise power $N$, is given by [29]:

$$\text{SNR} = A^2/(2N) \tag{2.47}$$

As shown by the right-hand curve in Figure 2.14, the peak of the Ricean distribution moves to the right with increasing SNR. The probability of detection of a target with SNR = $A^2/(2N)$ is determined by the area under the curve to the right of $v_T$, given by $P_{s+n}$. The probability of detection $P_d$ is determined by

$$P_d(v,\text{SNR}) = \int_{v_T}^{\infty} p_{s+n}(v;A,N)\text{d}v \tag{2.48}$$

Note that the probability of not detecting a target that is present, $p_m$ (i.e., the probability of a missed target), is given by the expression

$$p_m = 1 - P_d(v,\text{SNR}) \tag{2.49}$$

The results described mathematically by (2.43) and (2.48) are summarized by the curves in Figure 2.15, which plots the probability of detection versus the SNR returned by the target as a function of probability of false alarm (i.e., the value adopted for the detection threshold $v_T$). As shown in Figure 2.15, for a given detection threshold (probability of false alarm), the greater the SNR, the greater the detection probability. Also, the smaller the probability of false alarm (i.e., the higher the detection threshold setting), the smaller the probability of detection is for a given SNR. The most comprehensive collection of detection curves versus SNR is provided by [30].

## 2.8 RADAR TARGET MODELS

Up to this point, we have assumed that detection is based on reception of a single transmitted pulse by the radar. However, it may be desirable to integrate a number of radar pulses to enhance the detection of a target. Section 2.9 describes the method of calculating detection probability based on the coherent or noncoherent integration of multiple pulses received from a radar target. This section describes various radar target models that have been used to describe the fluctuation characteristics of targets from pulse to pulse and from radar scan to radar scan. The four most widely used target models were developed by Swerling in the 1950s [31]. An excellent discussion describing these models and developing a theory that unifies the Swerling models and more generalized radar target models is given by Weinstock [32]. Table 2.3 summarizes the characteristics of the four basic Swerling target models. Swerling target models I and II describe the target reflection characteristics by an exponential distribution:

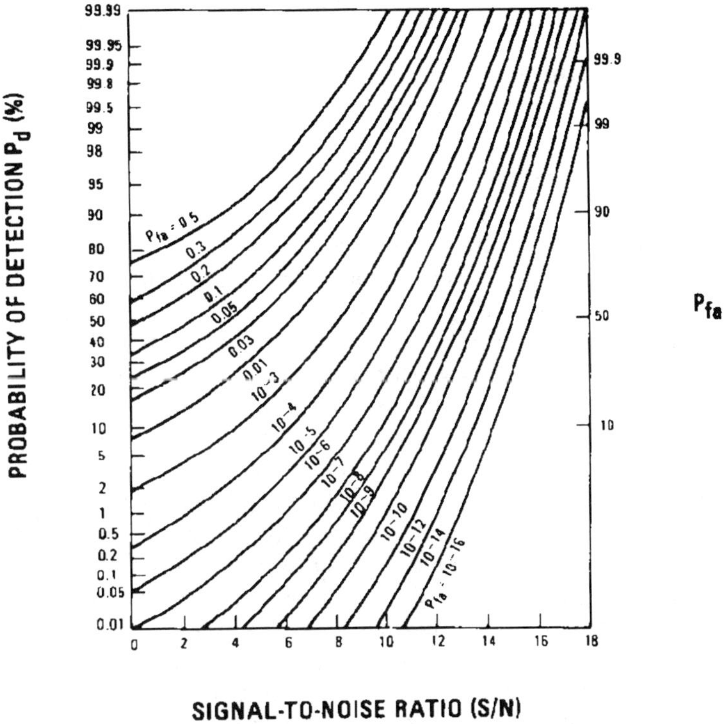


**Figure 2.15** Curves of probability of detection ($P_d$) versus SNR as a function of false alarm probability ($P_{fa}$).

**Table 2.3**
Radar Target Swerling Fluctuation Models

| *Swerling Model* | *Characteristics/Physical Significance* | *Examples* |
|---|---|---|
| I | Pulse-to-pulse correlated<br>Many small reflectors | Jet aircraft |
| II | Pulse-to-pulse uncorrelated<br>Many small reflectors | Propeller-driven aircraft or pulse-to-pulse frequency agility |
| III | Pulse-to-pulse correlated<br>One dominant reflector, many small reflectors | Missile |
| IV | Pulse-to-pulse uncorrelated<br>One dominant reflector, many small reflectors | Missile illuminated by radar with pulse-to-pulse frequency agility |

$$w(s,s_0) = [\exp(-s/s_0)]/s_0 \qquad s > 0 \tag{2.50}$$

Swerling target models III and IV describe the target reflection characteristics by a Rayleigh distribution:

$$w(s,s_0) = [\exp(-2s/s_0)] \cdot 4s/s_0^2 \qquad s > 0 \tag{2.51}$$

where $s_0$ is the mean cross section exhibited by the target.

As indicated in Table 2.3, Swerling target types I and III are considered to be correlated from pulse-to-pulse and independent from scan-to-scan. This results in the use of the distributions described by (2.50) and (2.51) in determining the probability of a target exhibiting a value of cross section equal to $s$ at each new radar scan. The cross section remains constant from pulse to pulse as the radar beam sweeps across the target during a single scan. On the other hand, Swerling target types II and IV fluctuate from pulse to pulse, and hence the probability the target cross section is $s$ is given by (2.50) and (2.51), respectively, from pulse to pulse as the radar beam sweeps across the targets.

Table 2.3 also shows the physical meaning of these four target models. Note that by changing frequency from pulse to pulse with a radar, a Swerling I or III target can be made to acquire the characteristics of a Swerling II or IV target. The change in transmitted frequency must be greater than the instantaneous bandwidth of the target.

A target that does not exhibit fluctuation characteristics is called a *Marcum* or *steady radar target*. As the instantaneous bandwidth of a radar increases, the radar resolution becomes finer and finer until it begins to illuminate only small regions of the target. When this happens, the mean target cross section $\sigma$ decreases and behaves more like a steady or constant target. The reason for this is that a narrow-bandwidth radar encompasses the entire target with a single pulse. The resulting radar return is composed of various parts of the target coherently combining to produce the return pulse. This return pulse will have a different amplitude, depending on the orientation of the target with respect to the radar, and hence the Swerling target characteristics. As the bandwidth of the radar becomes wider and the range resolution becomes smaller, only a portion of the target is illuminated within a single range cell, and the target return becomes more and more constant. This is especially true with laser radar, where ultrashort pulses are possible.

Many microwave radars use wideband waveforms to perform target identification. In these instances, it is important to use the proper target cross section to correctly analyze radar range performance.

## 2.9 SIGNAL-TO-NOISE ENHANCEMENT: MATCHED FILTERING AND THE INTEGRATION OF RADAR RETURNS

The requirement described by the relation $B\tau = 1$ is called the *matched filter relationship*. A radar receiver is a matched filter if the receiver transfer function is chosen to maximize the ratio of the peak signal power to rms noise power [33]. The derivation of the matched filter relationship between the transmitted pulse and the receiver transfer function was originally developed by North [34] and appears in many books [35,36]. The resulting transfer function is given by the equations

$$H(f) = G_{rec}S^*(f)\exp(-j2\pi f t_0) \tag{2.52}$$

in the frequency domain and

$$h(t) = G_{rec}s(t_0 - t) \tag{2.53}$$

in the time domain, where $H(f)$ is the Fourier transform of the receiver impulse response $h(t)$, $G_{rec}$ is the receiver gain, $S(f)$ is the Fourier transform of the signal $s(t)$, and $t_0$ is the time the maximum signal is present at the receiver. The form taken by (2.52), where $H(f)$ is related to the conjugate of $S(f)$, is the reason a matched filter is sometimes called a *conjugate filter*.

Physically, (2.53) indicates that the impulse response of the matched filter is shaped to be the same as the transmitted waveform, but reversed in time as shown in Figure 2.16. Figure 2.16 also shows the output of a matched filter, $s_{out}(t)$, which is given by

$$s_{out}(t) = \int s_{in}(\tau)h(\tau - t)\mathrm{d}t \tag{2.54}$$

$$= G_{rec}\int s_{in}(\tau)s(t_0 + \tau - t)\mathrm{d}t \tag{2.55}$$

However, (2.55) describes the behavior of a correlation receiver in the time domain. Therefore, the matched filter is equivalent to a correlation receiver that stores a replica

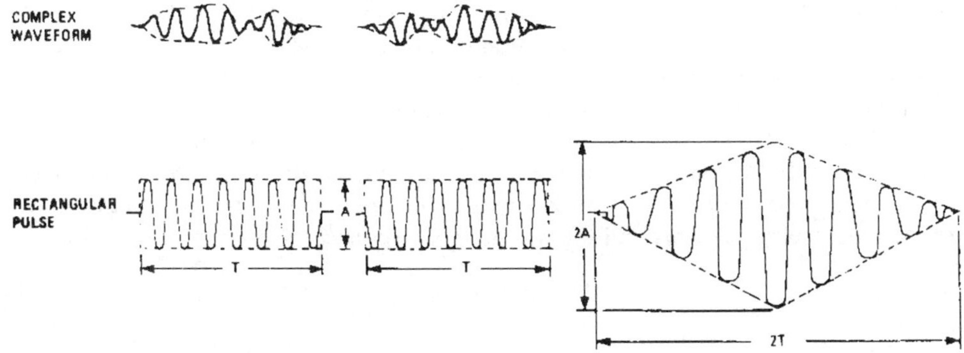


**Figure 2.16** Time domain representation of a matched filter.

of the transmitted signal and performs a convolution of the received signal with a replica of the transmitted signal.

The other interesting aspect of the matched filter shown in Figure 2.16 is that the matched filter does not preserve the shape of the transmitted waveform. It can be shown [37] that the maximum ratio of the peak signal to rms noise power $R_f$ is given by

$$R_f = 2E/N_0 \tag{2.56}$$

where $E$ is the total energy of the waveform and $N_0$ is the noise power per unit bandwidth. Equation (2.58) can also be written as

$$R_f = 2S/N \tag{2.57}$$

where $S$ is the target signal power and $N$ is the rms noise power. This follows from

$$S/N = (S/B)/(N/B) = S\tau/N_0 = E/N_0 \tag{2.58}$$

where

$$S\tau = E \tag{2.59}$$

and

$$N_0 = N/B \tag{2.60}$$

The detection probability of a target can also be increased by integrating a series of pulses returned by a target. Pulse integration is of two types: *coherent* and *noncoherent*. Coherent integration is based on preserving both the phase and amplitude of the returned signal from each pulse. Since this is performed prior to detection, coherent integration is sometimes called *predetection integration*. Noncoherent detection, on the other hand, is performed after detection and essentially adds only a fraction of each pulse amplitude on average, since the phase information for each pulse is lost after the detection process. Noncoherent integration is also called *postdetection integration*.

Figure 2.17 illustrates a pulse train transmitted by a radar. The interval between pulses is the PRI. The rate at which pulses are transmitted is the pulse repetition frequency of the radar (radar PRF). Since the rate at which pulses are transmitted is the reciprocal of the time interval between pulse transmissions, we have the relation

$$(\text{PRI})(\text{PRF}) = 1 \tag{2.61}$$

The PRI of a pulse train determines the unambiguous range $R_u$ of the radar through the relationship

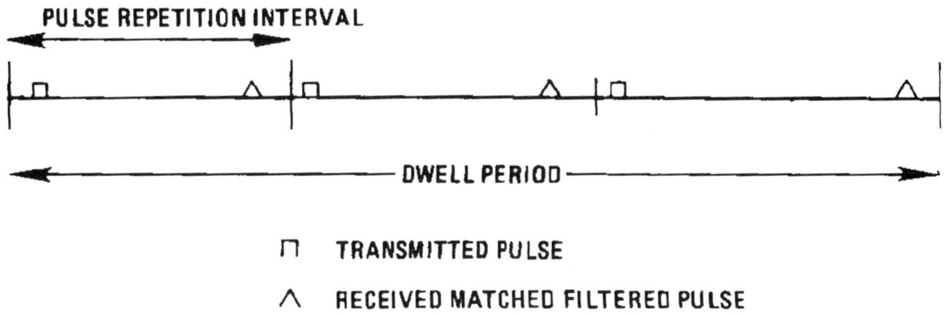


**Figure 2.17** Representative radar pulse train.

$$R_u = \text{cPRI}/2 \tag{2.62}$$

where c is the speed of light. The factor of 2 takes into account that the range measurement is based on the time it takes the pulse to travel a distance $R_u$ out to a target and then, once reflected, travel the same distance $R_u$ from the target back to the radar. Any target at a distance greater than $R_u$ will be detected in a PRI later than the one from which the pulse was transmitted; hence the name *maximum unambiguous range*.

The unambiguous Doppler of the pulse train, as will be discussed in more detail in Section 2.10, is determined by the PRF of the pulse train. Any Doppler impressed on the return pulse due to the motion of the target is only known to within a carrier phase shift of $2\pi$ radians. This results in Dopplers above the PRF folding into Dopplers between zero and the PRF. Hence, the PRF sets the limit on determining an unambiguous Doppler just as the PRI sets the limit on determining unambiguous range. However, because of the relationship between PRI and PRF, expressed by (2.61), once a system establishes an unambiguous range by setting the PRI, the unambiguous Doppler is also established as 1/PRI.

As discussed above, the integration of a series of pulse returns from a target can be accomplished either coherently or noncoherently. Figure 2.18 illustrates the coherent integration of radar pulses. To perform coherent integration of pulse returns, the radar must maintain coherence from pulse to pulse upon transmit. As shown, the amplitude and phase of each pulse is preserved upon receive. If $n$ pulses are transmitted, let the voltage amplitude and phase of the $i$th pulse be given by $A_i$ and $\phi_i$, respectively. As shown in Figure 2.18, since the phase of each pulse is known, the phase of each pulse can be used to line up the pulses from end to end. This results in an integrated signal strength of

$$A_{\text{ci}} = \sum_{i=1}^{n} A_i = nA \tag{2.63}$$

where we have assumed that the amplitude of each pulse is about the same magnitude $A$. Thus, the integrated signal power $S_{\text{ci}}$ is determined by the expression

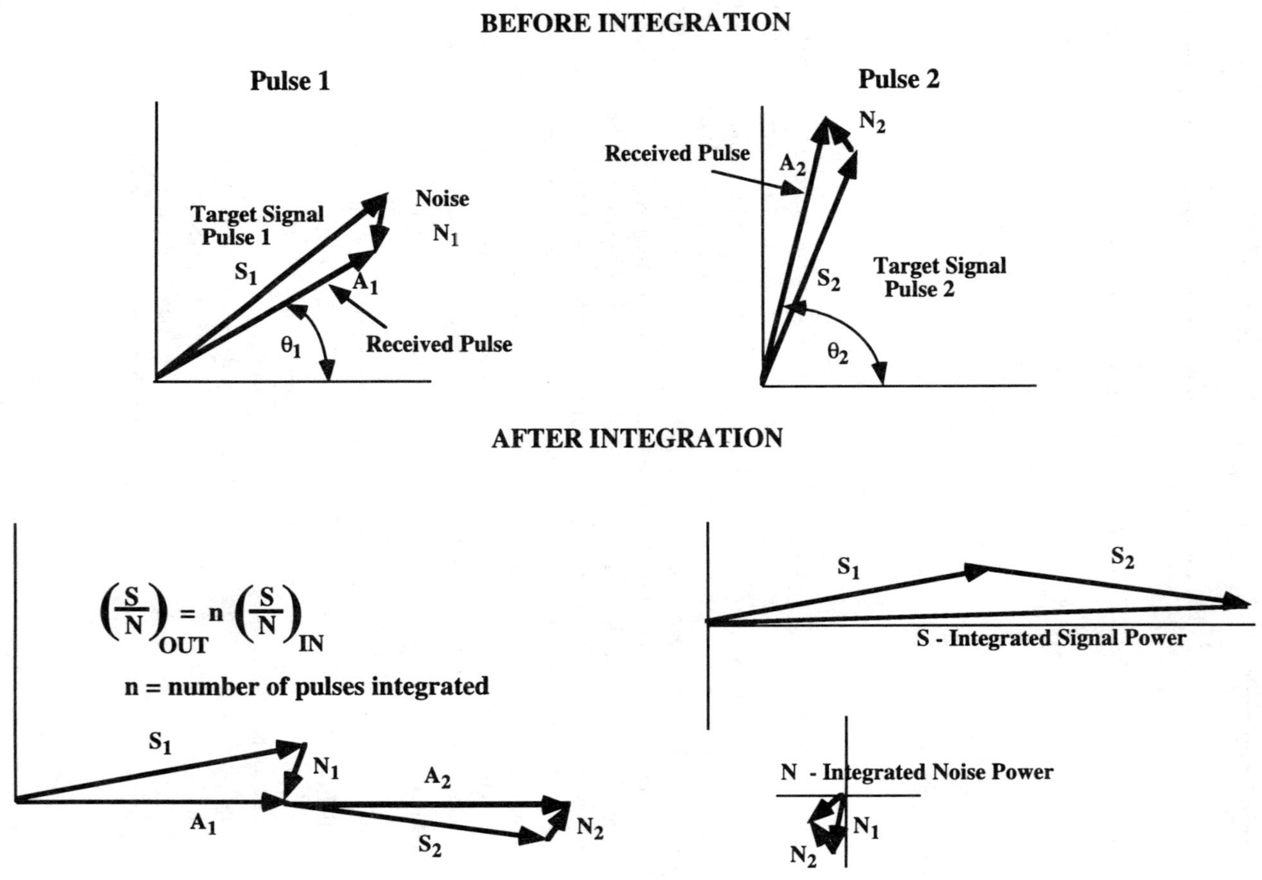


**Figure 2.18** Coherent integration process.

$$S_{ci} = (A_{ci})^2 = (nA)^2 = n^2A^2 = n^2S \quad (2.64)$$

where $S$ is the signal power returned from a single pulse. On the other hand, if the noise power from a single pulse is given by $N$, the total noise power from $n$ pulses, $N_n$, is given by the expression

$$N_n = nN \quad (2.65)$$

This result follows from noise being a random process. Hence, each noise sample is independent, and the noise power, rather than the noise voltages, adds linearly. Hence, the resulting SNR after coherent integration, $\text{SNR}_{ci}$, is given by

$$\text{SNR}_{ci} = S_{ci}/N_{ci} = n^2S\,(nN) = n(S/N) \quad (2.66)$$

Therefore, the coherent integration of $n$ pulses results in an increase in SNR of $n$.

Noncoherent integration of a pulse train is illustrated in Figure 2.19. Since noncoherent integration is performed after detection, the phase information associated with each pulse is lost to the radar. Hence, only the projection of the pulse amplitudes on a single direction in signal space, the x axis for instance, is available for integration. Figure 2.20 illustrates the result of postdetection integration and compares noncoherent integration with a perfect integrator given by the solid line (i.e., a coherent integrator) and an integrator that achieves a gain of $\sqrt{(n)}$, given by the dashed line. The $\sqrt{(n)}$, behavior is representative of the performance of a human operator viewing a cathode ray display with a persistence which is on the order of the total integration time.

The above discussion of pulse integration assumed a target cross section model which is constant over the integration period (i.e., a Marcum or steady target). Swerling developed the theory of postdetection integration with respect to the four radar target fluctuation models described in Section 2.8. An excellent discussion of these results is given in Chapters 10 and 11 of [33]. Figure 2.21 indicates the results of this theory, in which the probability of detection versus single-pulse SNR is plotted for the four Swerling target models and the Marcum steady-target model. As shown, the Swerling target models 1 and 3, which are constant from pulse to pulse and independent from scan to scan, require less SNR per pulse to achieve low probabilities of detection; whereas the Swerling target models 2 and 4, which are random from pulse-to-pulse and scan-to-scan, require less SNR to achieve high probability of detection. The physical reason for this is that when a Swerling 1 or 3 target exhibits a small radar cross section, that cross section tends to stay small during the integration period. However, the cross section of a Swerling 2 or 4 target can fluctuate widely during the integration period, and if a large value of cross section is returned on only a single pulse, it may be sufficient for target detection. As discussed above, the most comprehensive collection of $P_d$ versus SNR curves for a wide range of $P_{fa}$ is found in [30].

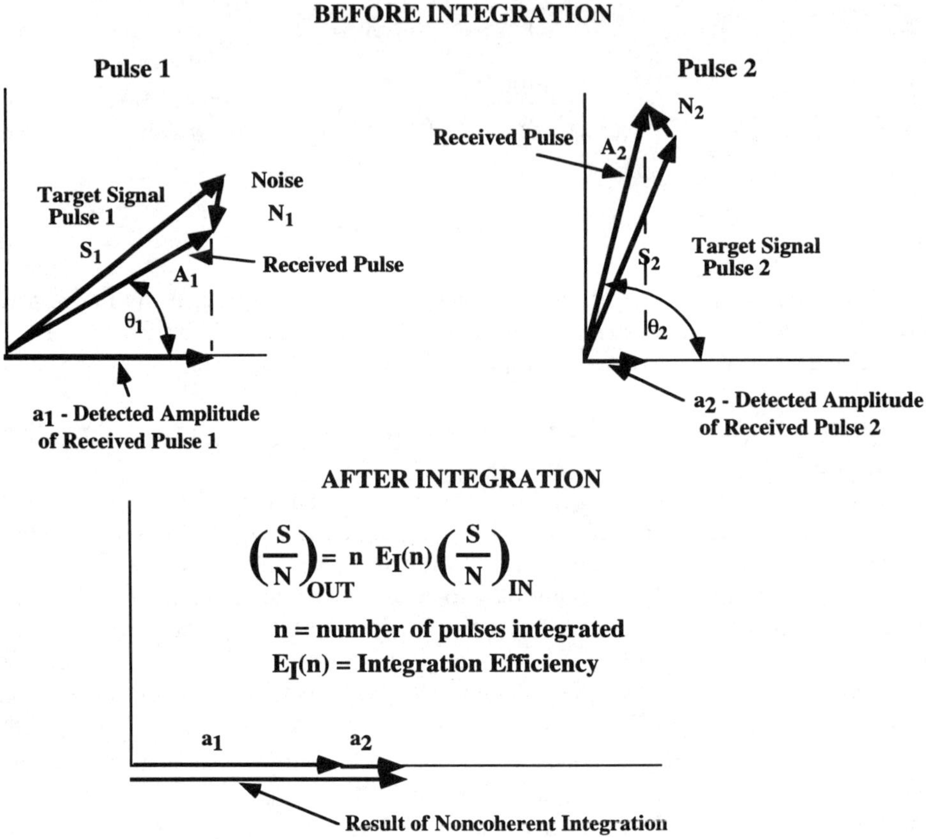


**Figure 2.19** Noncoherent integration process.

## 2.10 DETECTION OF RADAR TARGETS IN CLUTTER FOR GROUND-BASED RADAR

Figure 2.22 illustrates the origin of the Doppler effect. As a target moves towards a radar, the frequency of the carrier in the returned pulse is higher than the frequency of the transmitted carrier because the target motion advances the return of subsequent portions of the transmitted pulse. This is contrasted with a stationary target which returns all portions of the transmitted pulse with the same delay. Although the Doppler effect is characterized as a shift in frequency, in reality it is due to a continuous phase shift in time. This is the true meaning of the Doppler shift, as is seen in the following derivation.

If R(t) is the distance between the radar and target as a function of time, then

$$R(t) = R_0 + \dot{R}(t - t_0) \tag{2.67}$$

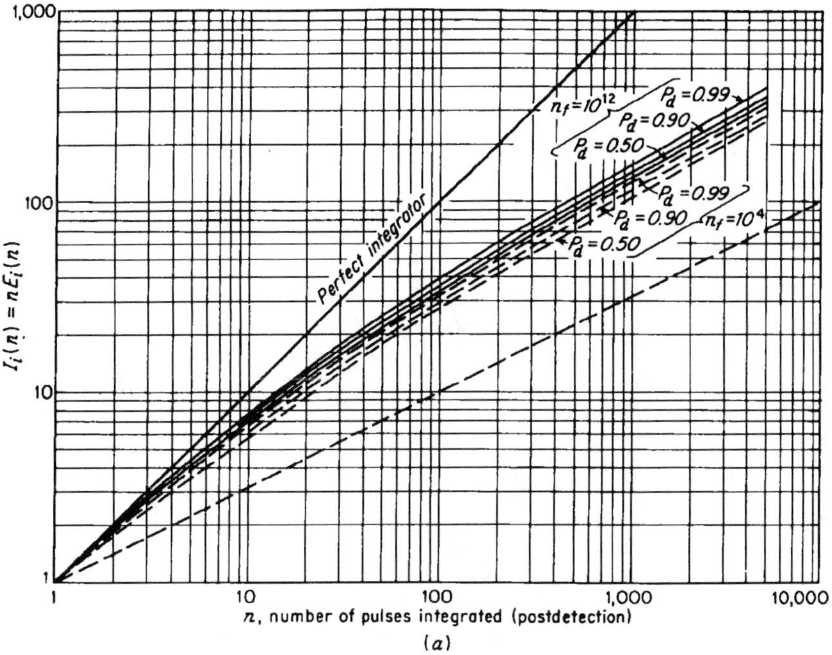


**Figure 2.20** Postdetection, noncoherent integration performance (after Marcum [38]).

where $R_0$ is the target range at $t_0$, and $\dot{R}$ is the projection of the velocity of the target in the direction of the radar range vector (i.e., the target range rate). The phase of the received target signal, $\varphi(t)$, at the radar as a function of time is given by

$$\varphi(t) = (2\pi/\lambda) \cdot 2R(t) \tag{2.68}$$

The factor of 2 takes into account the propagation of the pulse to the target and back to the radar. The factor of $(1/\lambda)$ measures the two-way distance between the target and radar in units of the transmission wavelength. The factor of $2\pi$ converts distance in units of wavelength into phase.

The waveform frequency as a function of time $f(t)$ is given by

$$f(t) = f_0 - (1/2\pi)(\mathrm{d}\varphi(t)/\mathrm{d}t) \tag{2.69}$$

The minus sign is necessary because if the phase increases in time, then the carrier frequency decreases, whereas if the phase decreases in time, then the carrier frequency increases. Therefore,

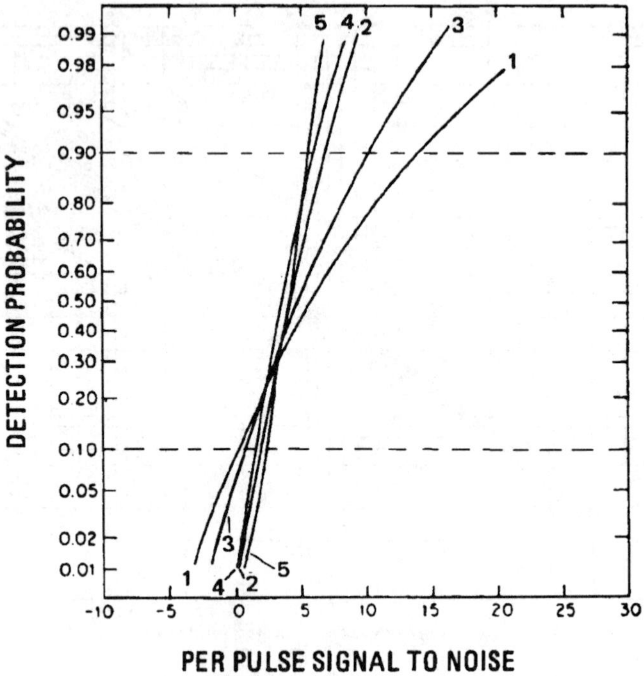


**Figure 2.21** Comparison of detection probability for fluctuating and steady targets. False alarm probability = $10^{-7}$; number of pulses integrated = 10; curve 1 = Swerling target model I; curve 2 = Swerling target model II; curve 3 = Swerling target model III; curve 4 = Swerling target model IV; curve 5 = Marcum steady-target model.

$$f(t) = f_0 - (1/2\pi)(\mathrm{d}[(2\pi/\lambda) \cdot 2R(t)]/\mathrm{d}t) \tag{2.70}$$

$$= f_0 - (2/\lambda)(\mathrm{d}[R_0 + \dot{R}(t - t_0)]/\mathrm{d}t) \tag{2.71}$$

$$= f_0 - (2\dot{R}/\lambda) \tag{2.72}$$

Hence, the Doppler shift,

$$\Delta f = f(t) - f_0 \tag{2.73}$$

is given by the well-known relationship

$$\Delta f = -(2\dot{R}/\lambda) \tag{2.74}$$

This relationship indicates that if the target is moving towards the radar ($\dot{R}$ is negative), then there is a positive shift in carrier frequency, and if the target is moving away from the radar ($\dot{R}$ is positive), then there is a negative frequency shift.

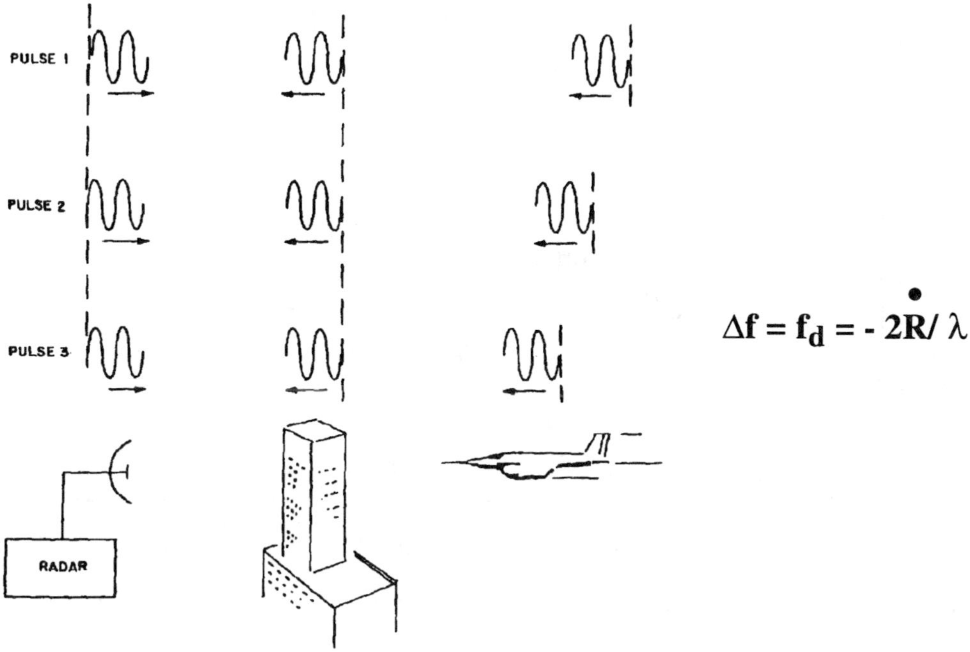


**Figure 2.22** The origin of the Doppler effect.

Equation (2.74) can be solved for the range rate in terms of the frequency shift to yield

$$\dot{R} = -\lambda\Delta f/2 \tag{2.75}$$

This is useful in translating a Doppler shift from frequency to a range rate.

It is important to understand the Doppler effect in terms of a phase shift over time. Equation (2.69) can be solved for $\Delta\varphi$ to yield

$$\Delta\varphi = 2\pi\Delta f \cdot \Delta t \tag{2.76}$$

For instance, if $\Delta f = \text{PRF}$, then the phase shift from pulse to pulse is given by

$$\Delta\varphi = 2\pi\text{PRF} \cdot \text{PRI} = 2\pi \tag{2.77}$$

This result follows from the fact that the time interval between successive pulses is by definition the PRI and the relation between the PRI and PRF given by (2.61).

Figure 2.23 illustrates the behavior of clutter in the Doppler domain as seen by a ground-based radar. As is shown, land clutter is centered about zero Doppler, since the

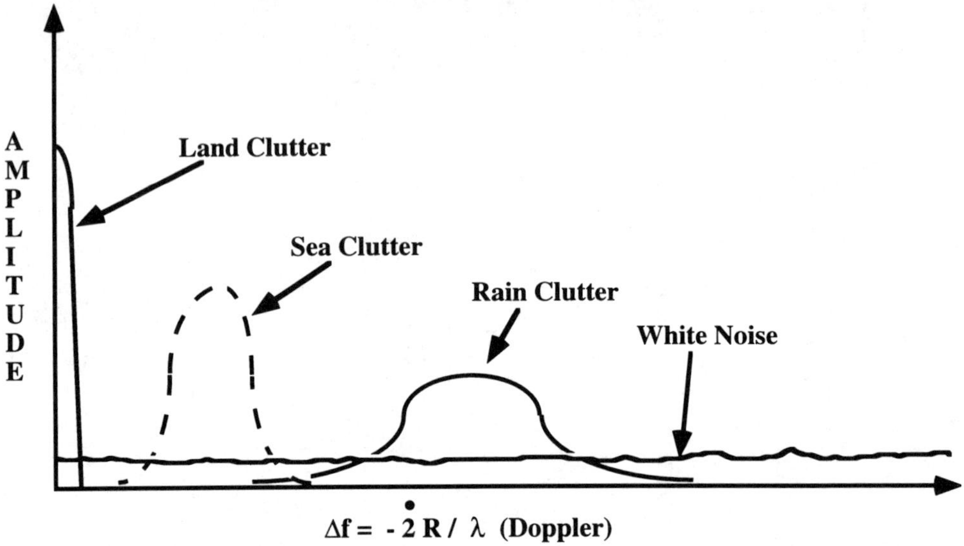


**Figure 2.23** Representative clutter response in the Doppler domain.

ground is stationary with respect to the radar. The spread that accompanies the ground clutter is due to either motion within the clutter itself or instabilities within the radar. The ground clutter motion can be due to tall grass or tree leaves blowing in the breeze. Radar instability can be due, for instance, to transmitter jitter. Sea clutter has broader width because of its inherent motion. The mean motion of sea clutter is dependent on the orientation of the radar and the mean motion of the sea surface. Rain or chaff has a spread dependent on wind shear, beam broadening, atmospheric turbulence, and rainfall velocity distribution within the beam of the radar [39]. The rainfall will have a mean velocity centered at a Doppler determined by the orientation of the radar to the rainstorm and the mean velocity of the wind in the vicinity of the rainstorm.

A ground-based radar uses the Doppler effect to separate targets of interest from clutter. Techniques that perform the function of target/clutter separation on the basis of velocity or Doppler discrimination are called *moving target indicator* (MTI) systems. MTI systems make use of the phase change from pulse to pulse to determine target or clutter Doppler. Therefore, the pulse trains transmitted by the radar must be coherent. This provides a phase reference that can be used to develop a filter in the Doppler domain of the radar.

Figure 2.24 illustrates the Doppler response of a coherent pulse train that has pulsewidth $\tau$ and an interpulse period PRI = 1/PRF. As is shown, the Doppler response of this pulse train is a (sin $u$)/$u$ envelope behavior modulating a series of spectral lines spaced by the PRF of the radar. The peak of the response is at the radar center frequency, and the nulls of the (sin $u$)/$u$ response are spaced at the center frequency plus and minus

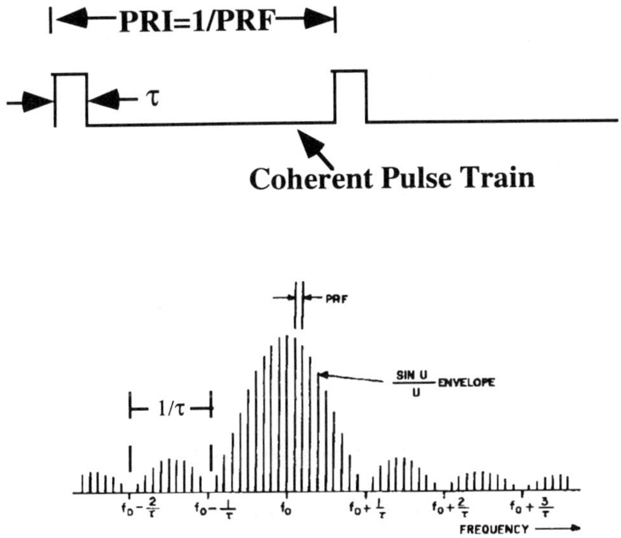


**Figure 2.24** Doppler response of a periodic pulse train.

integral multiples of the reciprocal of the pulsewidth. Just as the PRI determines the unambiguous range associated with the radar in the time domain, the radar PRF which is the reciprocal of the PRI determines the unambiguous Doppler associated with a radar system. Any radar return will produce multiple lines in the Doppler domain as shown, except the lines will be shifted by $\Delta f$, the Doppler shift. Hence, focusing on any one Doppler interval, such as the one bounded by $f_0$ and $f_0$ + PRF, provides maximum information on moving targets and clutter.

The two basic techniques used to provide MTI in ground-based radars are classic MTI based on multiple-pulse cancellation and Doppler filtering. The basic idea of multiple-pulse cancellation is shown in Figure 2.25. A two-pulse canceler is shown. Each pulse is subtracted from the previous pulse. The response of the canceler in the Doppler domain is shown. One way to interpret the behavior of the canceler in terms of the phase progression of successive pulses is shown in Figure 2.26. As is shown, if there is no phase change from pulse to pulse (i.e., a stationary target) when the pulses are subtracted, there is zero response in the filter. When the target has a range rate equal to PRF/2, the filter has maximum response because the phase change from pulse to pulse is equal to $\pi$. However, if the target Doppler is equal to the radar PRF as shown above, the phase change from pulse to pulse is $2\pi$ and once again the MTI filter has zero response. As a matter of fact, any target that has a Doppler which is a multiple of the radar PRF will produce a zero response at the filter. The range rates at which the filter produces a zero response are called *blind speeds.* A Doppler equal to the PRF is called the *first blind speed.* A Doppler equal to n·PRF is called the *nth blind speed.* Methods to operate MTI filters which provide

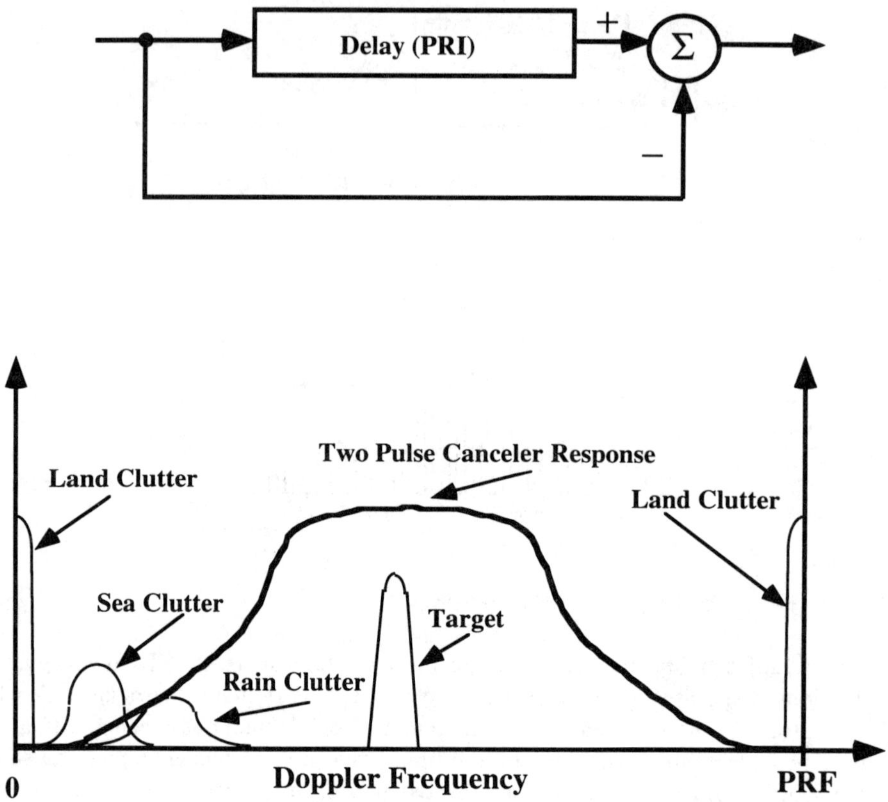


**Figure 2.25** Two-pulse MTI canceler: filter circuit diagram (top); two-pulse canceler filter response (bottom).

detection of targets moving at blind speeds include periodically changing the MTI PRF or using multiple PRFs within a single MTI dwell. These latter waveforms are called *staggered PRF waveforms* [40,41]. Cancelers that form MTI responses using *n* pulses have deeper notches at the edges. The depth of the notch is determined by the total length of the radar dwell [39–41]. Note that if the clutter falls into the nonzero response portion of the MTI filter, it will interfere directly with target detection. This is especially true of rain or chaff-type clutter.

The development of the pulse Doppler radar processor provided much more flexibility in radar operation. It overcame the problem of handling rain clutter with the MTI canceler and provided coherent integration gain on the target. The development of the fast Fourier transform algorithm and high-speed digital processors has led to the cost-effective implementation of pulse Doppler processing in many modern radars. Figure 2.27 illustrates the Doppler response of a pulse Doppler radar. The width of the individual

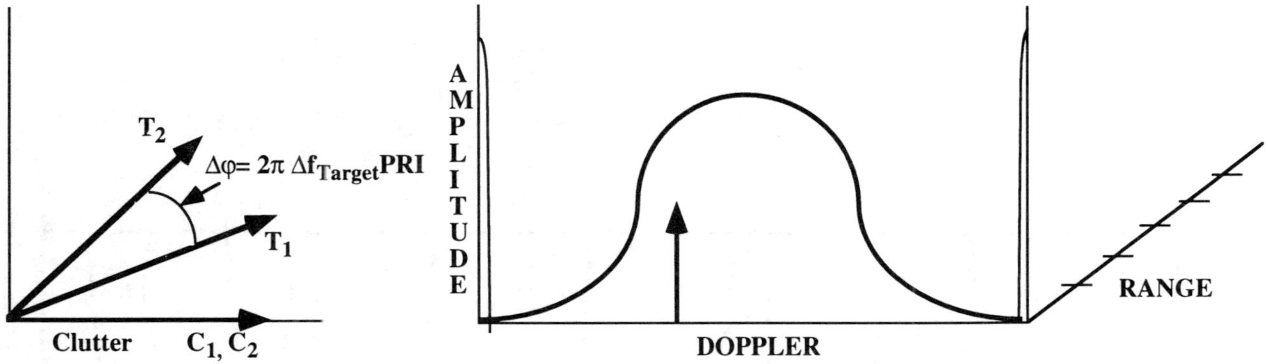


2 Pulse MTI Residue $E_r = E_2\,(r,PRI) - E_1(r,0)$

2 Pulse MTI Transfer Function $H(f) = 1 - e^{-j2\pi f_d PRI}$

2 Pulse MTI Filter Response $|H(f)|^2 = 4\sin^2\,(\pi f_d\,PRI)$

**Figure 2.26** Operation of the two-pulse canceler.

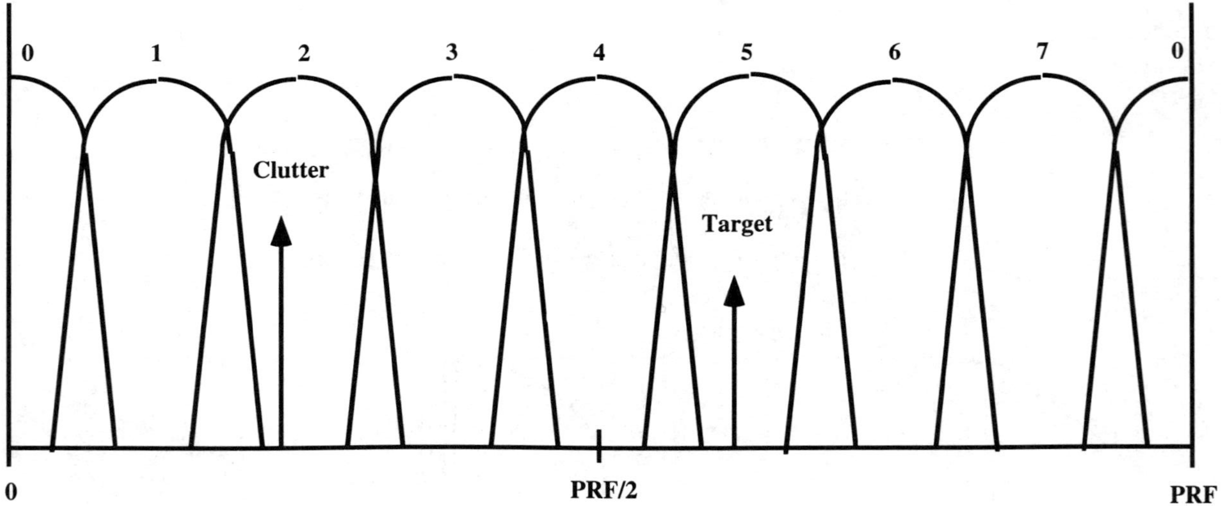


**0th - Doppler Filter Response** $X(0) = \sum A_n$

**1st - Doppler Filter Response** $X(1) = \sum A_n\, e^{-j2\pi\, n\, PRI\, (PRF/8)}$

**mth - Doppler Filter Response** $X(1) = \sum A_n\, e^{-j2\pi\, n\, PRI\, (m \cdot PRF/8)}$

**Figure 2.27** Doppler filter bank produced by a pulse Doppler radar.

Doppler filters is determined by the overall dwell time associated with the pulse Doppler waveform. Pulse Doppler operation requires coherence of the transmitted waveform.

Figure 2.28 illustrates the method of forming each filter. The filter response is developed by multiplying the sequence of returned signals by a different set of phase progressions. A phase progression that does not change for each pulse results in formation of the zero Doppler filter. A phase progression from pulse to pulse that advances the phase by $\pi$ from pulse to pulse results in formation of the central Doppler filter. Note that the formation of each Doppler filter results in a coherent integration gain equal to the number of pulses in the pulse train for that particular filter. Each filter has a $(\sin u)/u$ response with −13-dB first sidelobes. If weighting coefficients are applied to each pulse return in addition to the proper phase sequence, the sidelobes of each Doppler filter can be reduced. Weighting, however, is accompanied by a broadening of the filters and a reduction in the coherent integration gain achieved on the target.

There is still a blind speed problem at multiples of the PRF. This is overcome by using multiple PRFs while performing surveillance.

Note that the Doppler filters can be used to isolate ground, sea, rain, or chaff clutter within a single filter or small number of filters to permit target detection over a wide range of Doppler frequencies even though the clutter may have a nonzero mean velocity. One of the excellent applications of pulse Doppler processing for a ground-based radar system is the moving target detection (MTD) radar system developed for the Federal Aviation Association (FAA) airport surveillance radars by the MIT Lincoln Laboratory in the late 1970s [42–44].

## 2.11 DETECTION OF RADAR TARGETS IN CLUTTER FOR AIRBORNE OR SPACE-BASED RADAR

Airborne radar introduces a new dimension to MTI or Doppler processing. If the radar performs surveillance in the presence of ground clutter, the motion of the aircraft induces a Doppler response at the airborne sensor that, as shown in Figure 2.29, has a Doppler response that is look-direction dependent. The design of an airborne radar system that performs in a clutter environment requires consideration of low-, medium-, and high-PRF waveforms. A low-PRF waveform is one whose range is unambiguous but whose Doppler is ambiguous, as described in Section 2.10. A medium-PRF waveform is one that is ambiguous in both range and Doppler. Multiple PRFs are required to resolve the range and, if desired, the Doppler ambiguity. A high-PRF waveform is one that is unambiguous in Doppler but ambiguous in range. Excellent discussions of airborne radar design can be found in [45–47]. Airborne radar is used to: detect hostile weather patterns and permit the aircraft to navigate around them; provide limited air surveillance and tracking capability to fighter aircraft; perform wide-area surveillance capability from an airborne platform to detect low-altitude aircraft in a timely fashion; perform terrain-following and terrain-avoidance functions; and develop high-resolution ground maps using SAR for remote sensing or ground target detection.

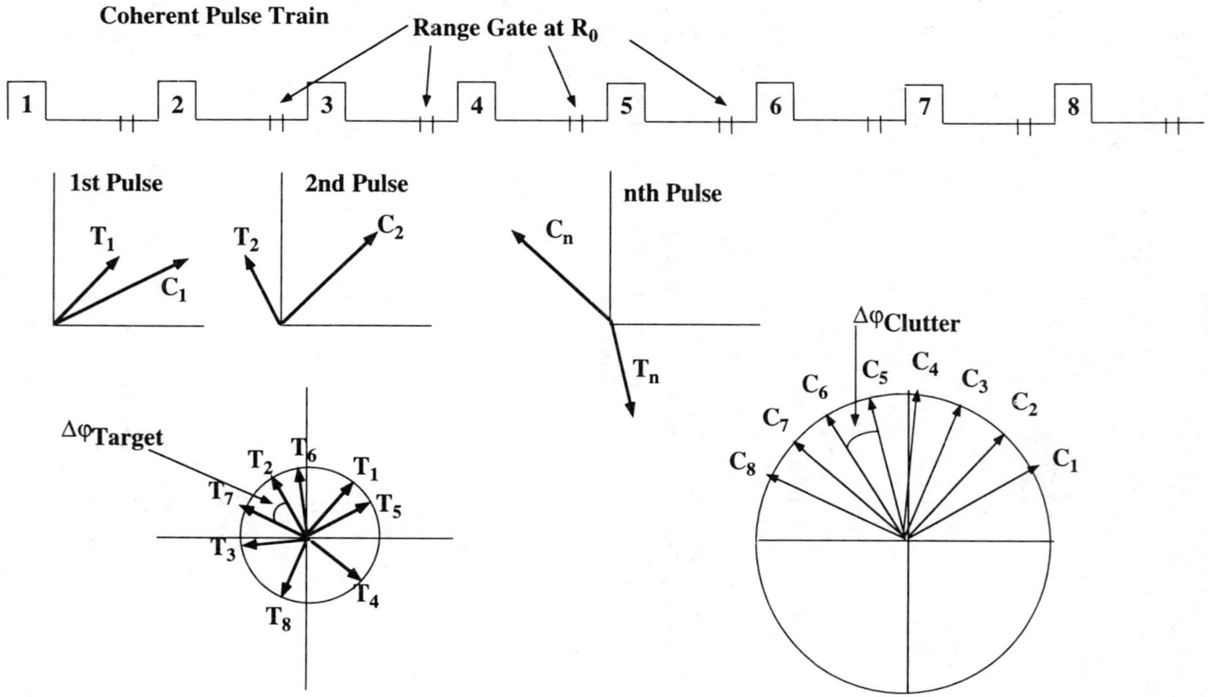


**Figure 2.28** Formation of pulse Doppler filters.

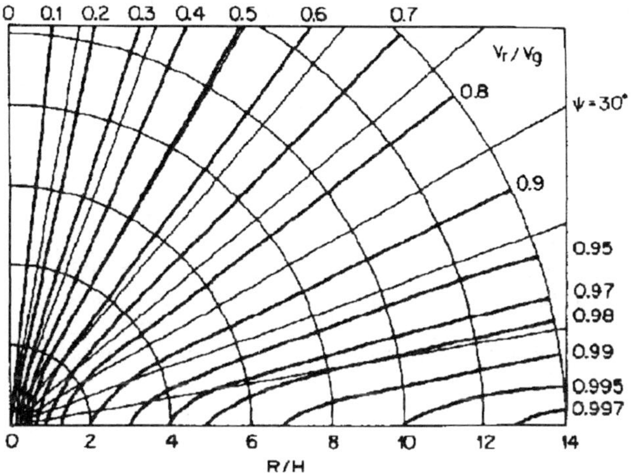


**Figure 2.29** Doppler response in an airborne is look-direction dependent.

Space-based radar has long been of interest to the military community for providing wide-area surveillance. The Soviet military developed and flew a space-based radar during the 1970s and 1980s called RORSAT, which was used to detect and track the movement of surface ships. A number of commercial space-based radars have been developed for remote sensing operation. These include SEASAT and the shuttle imaging radars (SIR), SIR A, SIR B, and SIR C.

The technical design considerations of a space-based radar are similar to those for airborne radar. The primary differences are: the speed of a space-based radar is fixed by its orbit; the system must be very reliable since it is difficult, if not impossible, to service the radar; and weight, including that of the power system, is critical to keeping the cost of the system reasonable. The best source of information on space-based radar is [48].

## 2.12 CONCLUSION

Radar systems have been operated since the early 1930s. They function to detect targets based on reflection and reradiation of electromagnetic energy from targets. In addition, radars measure target echo characteristics to determine target parameters that include range, range rate (Doppler), angle, velocity, and signature. Advanced radar techniques that include SAR permit modest-size antennas to develop ultra-high-resolution ground maps for remote sensing and the detection and classification of military targets. Much has been accomplished in the first 60 years of radar development. What occurs during the next 60 years will definitely rely on advances in optical components and processing. This book is a pioneering volume on these topics. It is difficult to cover such a comprehensive and complex topic as radar in such a limited space. For those who wish to pursue a deeper understanding of the topics discussed in this chapter, please consult the References.

## REFERENCES


[1] Maxwell, J. C., *Treatise on Electricity and Magnetism,* Vols. 1 and 2, New York: Dover, 1954.
[2] Hertz, H., *Annalen der Phyzik,* Vol. 31, 1887, p. 421.
[3] See Section V, in this volume.
[4] See Section IV, in this volume.
[5] Fell, B., ''Principles of Radar for Optical Engineers,'' *Effective Utilization of Optics in Radar Systems,* SPIE Proc., Vol. 128, September 1977, pp. 2–19.
[6] Jelalian, A. V., *Laser Radar,* Norwood, MA: Artech House, 1992.
[7] Johnson, C. M., ''Laser Radars,'' Ch. 37 in *Radar Handbook,* M. I Skolnik, ed.
[8] Bhartia, P., and I. J. Bahl, *Millimeter Wave Engineering and Applications,* New York: John Wiley-Interscience, 1984.
[9] Zmuda, H., and E. Toughlian, ''Adaptive Microwave Signal Processing: A Photonic Solution,'' *Microwave J.,* Feb. 1992.
[10] Turpin, T., and J. L. LaFuse, ''IMSYN™ Photonic Image Synthesis Applied to Synthetic Aperture Radar, Microscopy, and Ultrasound Imaging,'' *SPIE Optical Conf. Proc.*, San Diego, July 1992.
[11] See Ch. 11, in this volume.
[12] Kao, C. K., *Optical Fiber Systems: Technology, Design, and Applications,* New York: McGraw-Hill, 1982.
[13] Cutrona, L. J., ''Synthetic Aperture Radar,'' Ch. 23 in *Radar Handbook,* 1st edition, M. I. Skolnik, ed., New York: McGraw-Hill, 1970.
[14] See Ch. 15 in this volume.
[15] Cooke, C. R., ''Laser Radar Systems, Some Examples,'' *Effective Utilization of Optics in Radar Systems,* SPIE Proc., Vol. 128, Sept. 1977, pp. 103–107.
[16] Jelalian, A. V., *Laser Radar,* Norwood, MA: Artech House, 1992.
[17] See Ch. 14 in this volume.
[18] Mensa, D. L., *High Resolution Radar Cross-Section Imaging,* Norwood, MA: Artech House, 1991.
[19] Fenner, R. G., and R. F. Broderick, ''Spaceborne-radar Applications,'' Ch. 34 in *Radar Handbook,* 1st edition, M. I. Skolnik, ed., New York: McGraw-Hill, 1970.
[20] Evans and Pettengill, ''The Scattering Behavior of the Moon at Wavelengths of 3.6, 68, and 784 Centimeters,'' *J. Geophysical Research,* Vol. 68, 1964, pp. 463–447.
[21] Weinberg, S., *Gravitation and Cosmology,* New York: John Wiley, 1972, pp. 201–207.
[22] Born, M., and E. Wolf, *Principles of Optics,* 3rd edition, Oxford: Pergamon Press, 1965, pp. 392–401.
[23] Skolnik, M., *Introduction to Radar Systems,* 2nd edition, New York: McGraw-Hill, 1980, Fig. 4.21.
[24] Barton, D., *Modern Radar Systems Analysis,* Norwood, MA: Artech House, 1988.
[25] Brookner, E., *Radar Technology,* Norwood, MA: Artech House, 1977.
[26] Skolnik, M., *Introduction to Radar Systems,* 2nd edition, New York: McGraw-Hill, 1980, pp. 392–395.
[27] Rice, S. O., ''Mathematical Analysis of Random Noise,'' in *Selected Papers on Noise and Stochastic Processes,* Nelson Wax, ed., New York: Dover, 1954.
[28] Watson, G. N., *A Treatise on the Theory of Bessel Functions,* 2nd edition, New York: Macmillan, 1948.
[29] Skolnik, M., *Introduction to Radar Systems,* 2nd edition, New York: McGraw-Hill, 1980, pp. 27.
[30] Meyer, D. P., and H. A. Mayer, *Radar Target Detection,* Orlando: Academic Press, 1973.
[31] Swerling, P., ''Probability of Detection for Fluctuating Targets,'' *IRE Transactions,* Vol. IT-6, April 1960, pp. 269–308.
[32] Weinstock, W., ''Radar Cross Section Target Models,'' in *Modern Radar,* R. Berkowitz, ed., New York: John Wiley, 1965.
[33] DiFranco J. V., and W. L. Rubin, *Radar Detection,* Norwood, MA: Artech House, 1980.
[34] Turin, C. L., ''An Introduction to Matched Filters,'' *IRE Trans. on Information Science,* Vol. IT-6, No. 3, June 1960, pp. 311–329. (Based on the original concept by D. O. North, ''Analysis of Factors Which Determine Signal/Noise Discrimination in Radar,'' *RCA Laboratories, Princeton, NJ, Report,* PTR-6C, June 1943).

[35] Cook, C. E., and M. Bernfeld, *Radar Signals, An Introduction to Theory and Application,* New York: Academic Press, 1967.
[36] Whalen, A. D., *Detection of Signals in Noise,* New York: Academic Press, 1971.
[37] DiFranco and Rubin, Ch. 5 in *Radar Detection,* Norwood, MA: Artech House, 1980.
[38] Marcum, J. I., "A Statistical Theory of Target Detection by Pulsed Radar," *Mathematical Appendix--IRE Trans.,* Vol. IT-6, April 1960, pp. 145–267.
[39] Nathanson, F. E., *Radar Design Principles,* New York: McGraw-Hill, 1969.
[40] Skolnik, M., *Introduction to Radar Systems,* 2nd edition, New York: McGraw-Hill, 1980, pp. 114–117.
[41] Shrader, W. W., "MTI Radar," Ch. 17 in *Radar Handbook,* 1st edition, M. I. Skolnik, ed., New York: McGraw-Hill, 1970.
[42] Meuehe, C. E., "Digital Signal Processor for Air Traffic Control Radars," *IEEE NEREM 74 Record, Part 4: Radar Systems and Components,* Oct. 1974, pp. 73–82.
[43] Muehe, C. E., "Advances in Radar Signal Processing, *IEEE Electro '76,* Boston, May 1976.
[44] Drury, W. H., *Improved MTI Radar Signal Processor,* Report No. FAA-RD-74-185, MIT Lincoln Laboratory, April 1975.
[45] Stauderer, F. M., "Airborne MTI," Ch. 18 in *Radar Handbook,* 1st edition, M. I. Skolnik, ed., New York: McGraw-Hill, 1970.
[46] Stimson, G. W., *Introduction to Airborne Radar,* Hughes Aircraft Company, El Segundo, 1983.
[47] Long, M., ed., *Airborne Early Warning System Concepts,* Norwood, MA: Artech House, 1992.
[48] Contafio, L. J., *Space-Based Radar Handbook,* Norwood, MA: Artech House, 1989.

# Chapter 3
# System Considerations

**Paul C. Hamilton**
*Anro Engineering*

## 3.1 ANATOMY OF A RADAR SYSTEM

In this chapter, we will examine several practical aspects of radar system design. We deal with three issues in this section: the primary design drivers of a radar system's cost and complexity, the fundamental building blocks of a radar system, and the operation of a typical modern radar. Section 3.2 describes the various types of radars in operational use today and several ways in which radar can be categorized, including the important distinction between coherent and noncoherent classes of radar. Section 3.3 discusses eight key factors that influence the design of a radar. Section 3.4 builds on the material from previous sections and from Chapter 2 and is devoted to a brief review of the preliminary design process, using a two-dimensional, ground-based search radar as an example.

Radar systems may be elaborate and expensive, as in the case of the Boeing E3A AWACS, or simple and relatively cheap, as in the case of a hand-held police radar designed to measure automobile speed. The cost and complexity of a radar arc, in simplest terms, proportional to the product of average transmitter power and receiving antenna aperture. This *power-aperture product* is a function of the target radar cross-sectional area, the spatial and temporal resolution requirements of the radar, the accuracy needed in the measurement of target parameters, and the data rate required to support this information on *n* targets on each radar scan. Radar cost and complexity are also greatly influenced by the environment, both natural and manmade, in which the radar must operate, and by the reliability required of the radar system over time.

Regardless of the level of sophistication incorporated in a particular design, a modern radar system is composed of three fundamental types of elements: hardware, software, and external interfaces (including the operator). Each radar subsystem may contain one or more of these types. Figure 3.1 shows a simplified block diagram of a typical modern phased-array radar. As configured, the system might represent a 3-D multifunctional radar operating under computer control. Division of the block diagram into major functional subsystems mirrors the traditional organization of technical disciplines within the radar engineering field, obscuring somewhat the seamless digital interfaces and the central importance of the control computer and software programs that implement antenna beam steering, waveform generation, system timing, signal processing strobe generation, and display and processor control. The sequence of operations is, however, basically identical to that of any conventional pulsed radar, with the control computer serving the function of master scheduler for all radar actions. (A *radar action* is defined as the transmission and reception of one or more pulses from a given beam position.)

When a transmission is scheduled, the modulator is digitally commanded to supply a high-voltage pulse to the RF amplifier, which simultaneously receives the synthesized carrier frequency (the RF drive signal) from the exciter. The resulting high-power RF pulse is routed through the microwave transmission line (waveguide) to the duplexer, then to the antenna, where the pulse is radiated into space. The duplexer serves as a microwave switch, preventing the high transmit power level from entering the receiver during transmit, while allowing reception of the return pulse. The beam steering processor

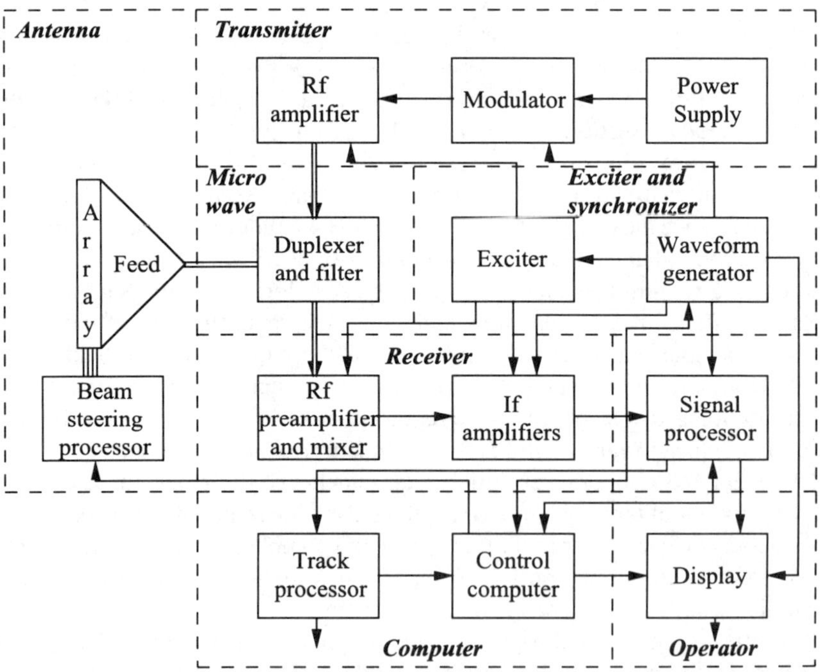


**Figure 3.1** Phased-array radar block diagram (from [1, Figure 1.1.2]).

controls the antenna phase shifter settings which establish the direction (pointing) of the transmit and receive beams. On receive, the echo pulse is routed from the antenna to the RF preamplifier and mixer, where it is translated to an intermediate frequency. The resultant signal is subsequently amplified and filtered prior to processing through the digital signal processor. The digital signal processor contains the envelope detector, which reconstructs the pulse waveform and makes it available to the display and for further processing. If the radar uses a pulsed doppler (PD) waveform, the fast Fourier transform (FFT) within the digital signal processor extracts target radial velocity data as well.

The control computer and associated software provide an inherent flexibility in radar operation by optimizing the use of radar resources according to the sensed environment (e.g., by reconfiguring receive path and gain settings to prevent saturation in response to jamming). With the advent of computer-controlled radar operations, the importance of sound system engineering, always a key issue, becomes even more critical, but from the operational point of view, no single part of the total radar system is more important than another. A faulty line of computer code or a poorly designed circuit breaker in the prime power generator in an otherwise well-designed, high-performance radar system can result in failure at a critical moment with potentially catastrophic results.

## 3.2 TYPES OF RADAR SYSTEMS

Radar systems have been categorized in several different and often confusing ways. Any taxonomy of radar tends to be somewhat arbitrary, but we can approach some logical order if we recognize two fundamental descriptors: *platform* and *function*. By platform we mean airborne, surface-based (land- or ship-based), or space-based. Radars within these categories may be further described by *primary function*: search or surveillance, tracking and guidance, navigation, surface mapping, and instrumentation. Specific radar applications within these functional categories form a third tier, listed in Table 3.1.

Radars have been, through use, further classified by specific features of the radar design, including:

- Operating frequency band: HF, VHF, UHF, L, S, C, X, $K_u$, $K_a$,V, W; V- and W-band radars are often referred to as *millimeter-wave radar*;
- Propagation mode: over-the-horizon (OTH);
- Waveform or signal processing characteristic: coherent, noncoherent, pulse, continuous wave (CW), FM/CW, PD, MTI, and pulse compression;
- Pulse repetition frequency: low (LPRF), medium (MPRF), high (HPRF); PRF designations usually apply to PD radar, where LPRF is unambiguous in range, HPRF is unambiguous in range rate (doppler), and MPRF is ambiguous in both range and doppler;
- RF bandwidth: narrowband, wideband, spread-spectrum, impulse; spread-spectrum and impulse radars are also referred to as *low-probability-of-intercept* (LPI) radar;
- Relative location of transmitter and receiver: monostatic and bistatic;

**Table 3.1**
Radar Applications

| *Function* | *Applications* |
|---|---|
| Air surveillance | Long-range early warning (including airborne early warning) |
| | Ground-controlled intercept |
| | Acquisition, for weapon system |
| | Height-finding and 3D radar |
| | Airport and air route surveillance |
| Space and missile surveillance | Ballistic missile warning |
| | Missile acquisition |
| | Satellite surveillance |
| Surface search and battlefield surveillance | Sea search and navigation |
| | Harbor and waterway control |
| | Ground mapping |
| | Intrusion detection |
| | Mortar and artillery location |
| | Airport taxiway control |
| Weather radar | Observation and prediction of precipitation and winds |
| | Weather avoidance (aircraft) |
| | Clear-air turbulence detection |
| | Cloud visibility indicators |
| Tracking and guidance | Antiaircraft fire control |
| | Surface fire control |
| | Missile guidance |
| | Range instrumentation |
| | Satellite instrumentation |
| | Precision approach and landing |
| Astronomy and geodesy | Planetary observation |
| | Earth survey |
| | Ionospheric sounding |

Source: [1, Table 1.1].

- Number of spatial dimensions resolved: 2-D and 3-D;
- Scan type: horizon, sector, raster, and conical;
- Beam shape: pencil, fan, $csc^2$, and inverted $csc^2$;
- Embedded technology: solid-state, digitally controlled, and laser;
- Antenna type: mechanically rotating, nodding, passive phased-array, active array, space-fed, corporate-fed, stacked beam, and monopulse; and
- End user: military and commercial.

This list is far from complete, but is meant to convey the several descriptions of radar in common use both within and outside of the radar community. Obviously, for any particular radar system, several categories may apply simultaneously.

An important distinction among radar types is based on the radar's ability to detect moving targets against a background of clutter, defined as echoes from objects other than the target, usually stationary and within the radar field of view. Conventional pulse radars that employ a waveform in which the RF phase relationship from pulse to pulse is uncontrolled or random are termed *noncoherent* radars. The echoes from a moving target can sometimes be distinguished from those of a stationary object by virtue of the relatively small frequency shift imparted to its spectrum. Noncoherent radars are generally not capable of detecting (or tracking) this Doppler shift, and, therefore, the clutter return will dominate. *Doppler radars*, on the other hand, represent the class of radars specifically designed to select targets and reject clutter based upon the Doppler shift in the return echo, where the Doppler shift is defined approximately as

$$f_d = \frac{2V_r}{\lambda} \text{ (Hz)} \tag{3.1}$$

Here, $V_r$ is the radial component of target velocity relative to the radar and $\lambda$ is the radar wavelength. In most cases, the Doppler shift amounts to a few kilohertz or tens of kilohertz. Because the spectra of typical radar pulses are much wider than this, the Doppler shift cannot normally be isolated and measured. Doppler radars solve this problem either by extending the pulsewidth (and thereby narrowing the bandwidth), as in the case of CW radar, or by transmitting a coherent train of narrow pulses during observation time to, as in the case of PD radar. In *coherent* pulse transmission, the phase of the RF carrier remains constant from pulse to pulse over the observation interval. Phase coherency between the transmitted signal and the receiver processing operations in the PD radar is ensured by the use of a common coherent and highly stable source.

CW radars have been widely and successfully used for acquisition of low-altitude targets and in single-target tracking radar, where the use of FM waveforms permits a relatively coarse estimate of target range. Being fully coherent, CW radars exhibit a high degree of clutter cancellation, but these radars require separate transmit and receive antennas to achieve the very high levels of isolation required to detect the weak target echoes in the presence of simultaneous powerful transmitter signals.

MTI radars are pulse radars that make use of the *delay-line canceler*, a circuit that delays a copy of the input pulse train by one interpulse period and then subtracts the copy from the original. Clutter returns, whose phase does not change from pulse to pulse, are canceled, and moving-target returns, whose phase changes continually at the Doppler rate, are not canceled and can therefore be isolated and detected.

PD radars are coherent Doppler radars like CW radars, but PD radars take advantage of the pulse waveform to allow transmission and reception from the same antenna. Reception takes place in the interval between the transmitted pulses. While PD radar may be considered a subset of the broader MTI class, technological improvements in oscillator stability and digital signal processing techniques now allow PD radar performance in strong clutter to exceed that of any other type.

Figure 3.2 is a representative block diagram of an airborne high-PRF PD radar system. For simplicity, only one receiver channel is shown, but multiple range channels

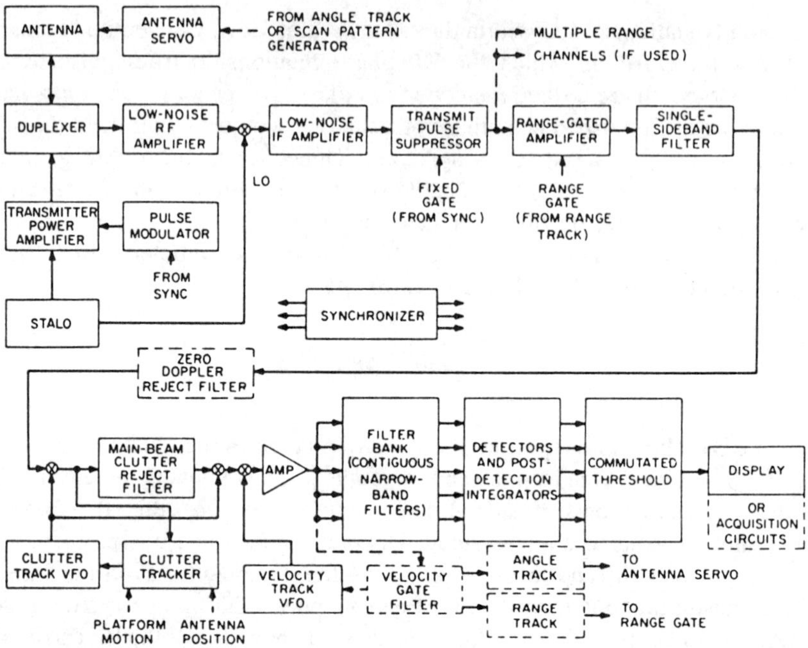


**Figure 3.2** Typical pulse doppler radar configuration (from [2, Figure 7.4.3]).

are usually incorporated to reduce the eclipsing loss inherent in range-ambiguous systems and to permit unambiguous range determination through the use of multiple PRFs during operation in a track-while-scan mode. The block diagram highlights a major and expensive component of a PD radar. If there were no clutter, the moving-target signal could be extracted from the background noise through the use of simple filters. With a clutter background, however, the many narrowband Doppler filters required can be one of the most expensive parts of the radar.

## 3.3 FACTORS INFLUENCING THE DESIGN OF A RADAR SYSTEM

Eight basic factors drive the design of any radar system.

1. *Function*: What is the role of the radar? What data must it provide to accomplish its primary function? If the radar is to operate in conjunction with other radars or communications systems, what interfaces must be provided?
2. *Performance*: Over what spatial extent must the radar provide the required data, and with what resolution, accuracy, and data rate? What are the characteristics of the targets that the radar is intended to detect and track? What is the expected distribution of targets in space, time, and velocity? What average false-alarm time is acceptable? In what environment must the radar operate, natural or manmade?

3. *Physical Constraints*: What are the maximum weight, weight distribution, and density limitations? How much prime power is available? What are the limitations in cooling capacity? What are the transportability requirements and how do these affect the mechanical and electrical design? What provisions in the mechanical and electrical design should be made for incorporating design modifications or improvements over the life span of the radar?
4. *Availability*: How long must the radar operate without failure? What is the expected operational duty cycle (on/off) of the radar? What is the mean time to repair a fault? What level of repair should or must be accomplished in the field versus at a central depot?
5. *Operability*: Is the radar intended for attended or unattended operation? What skills are required of operation and maintenance personnel? How will the required training be implemented?
6. *Test and Evaluation*: How will the performance of the radar and its satisfactory operation under the expected operational environment be verified prior to committing to production? What analytical, simulated, and physical tests are required to achieve a high level of confidence in the feasibility and practicality of the design?
7. *Life Cycle Cost*: How much funding is available for research, development, test, and evaluation? How many units will be manufactured? What is the per unit cost goal? How much funding is available for production, spares, training, operation, and maintenance over the intended lifetime of the radar?
8. *Development Time*: When is the first operational unit required? When must the design be frozen to ensure that the production schedule can be met? What radar technology is available that is consistent with the production schedule?

Although by no means complete, this list touches on virtually all areas for which more detailed specifications must be prepared if the radar development is to result in a usable product. Inherent in the successful completion of an operational radar design is the concept of *compromise*. A good radar design is one that achieves a reasonable balance among several, often conflicting, requirements and constraints, while taking advantage of available technology without incurring excessive technical or cost risk.

## 3.4 PRELIMINARY DESIGN PROCESS: AN EXAMPLE

A number of computer programs suitable for personal computer use are available to aid in the preliminary design of a new radar concept or in the evaluation of existing radar systems [3–5]. The more comprehensive of these programs include sophisticated modeling of atmospheric effects, such as clear-air and weather (rain and snow) attenuation of the radar signal, rain and chaff backscatter, refraction, multipath, and ducting effects, as well as noise jamming and ground clutter. In particular, the ability of these programs to accurately model the radar antenna pattern, its interaction with the natural and manmade environment and the target, results in estimates of radar performance more complete, realistic, and orders of magnitude faster than would be possible through manual means. For the purpose of illustration, however, we will make use of a much-simplified methodology to convey some appreciation of the many factors involved in the synthesis, or preliminary design, of a radar concept.

The example chosen is a surface-based, 2-D search radar, operating at S band (3,000 MHz) and using a mechanically scanned fan beam to provide all-azimuth early warning and detection of aircraft targets at altitudes from the radar horizon up to approximately 10 km, out to a slant range of approximately 100 km. The preliminary design process, illustrated in Figure 3.3, begins with the establishment of first-order performance requirements, which, for our example, are stated below.

*Search Coverage.* For a 2-D radar, the elevation coverage angle is equal to the elevation beamwidth. To detect targets up to a 10-km altitude at ranges near 100 km, the beamwidth must be at least 5 deg. Since the beam must scan through 360 deg, the resulting solid angle of search, in steradians (sr), is

$$\Psi_s = 2\pi\,(\sin\,\theta_{\text{el max}} - \sin\,\theta_{\text{el min}}) = 2\pi \sin 5° = 0.55 \text{ sr} \tag{3.2}$$

*Azimuth Resolution and Other Antenna Parameters.* For a large-volume search radar, an azimuth resolution of 1 deg is reasonable and consistent with modest antenna dimensions at S band. The width of the antenna aperture can be calculated from the relationship

$$w = K_q\ \lambda/\theta_{\text{az}} \tag{3.3}$$

where $K_q$ is the beamwidth constant, assumed equal to 1.22 here. The solution to (3.3) yields the nominal antenna dimensions $w \cdot d = 7\text{m} \cdot 1.4\text{m}$ for this radar, where the radar wavelength $\lambda = 0.1\text{m}$ and the beamwidths $q_{\text{az}}$ and $q_{\text{el}}$ are expressed in radians. Nominal antenna gain $G_0$, for an antenna with tapered illumination,* is related to antenna beamwidth by the relationship

$$G = 11.3/\theta_{\text{az}}\theta_{\text{el}}(\text{rad}) = 11.3/(0.0175 \cdot 0.0873) = 7397 = 38.7 \text{ dB} \tag{3.4}$$

Antenna gain in turn is related to the *effective antenna aperture* used to find the required power-aperture product by the expression

$$A_r = G\lambda^2/4\pi \tag{3.5}$$

*A uniform illumination function would provide maximum antenna gain, but the antenna sidelobes produced are often unacceptably high due to errors produced in generating this function. By tapering the illumination (i.e., applying a weighting function as a function of the vertical and horizontal distance from the center of the antenna), lower sidelobes result, but at a cost to main beam gain.

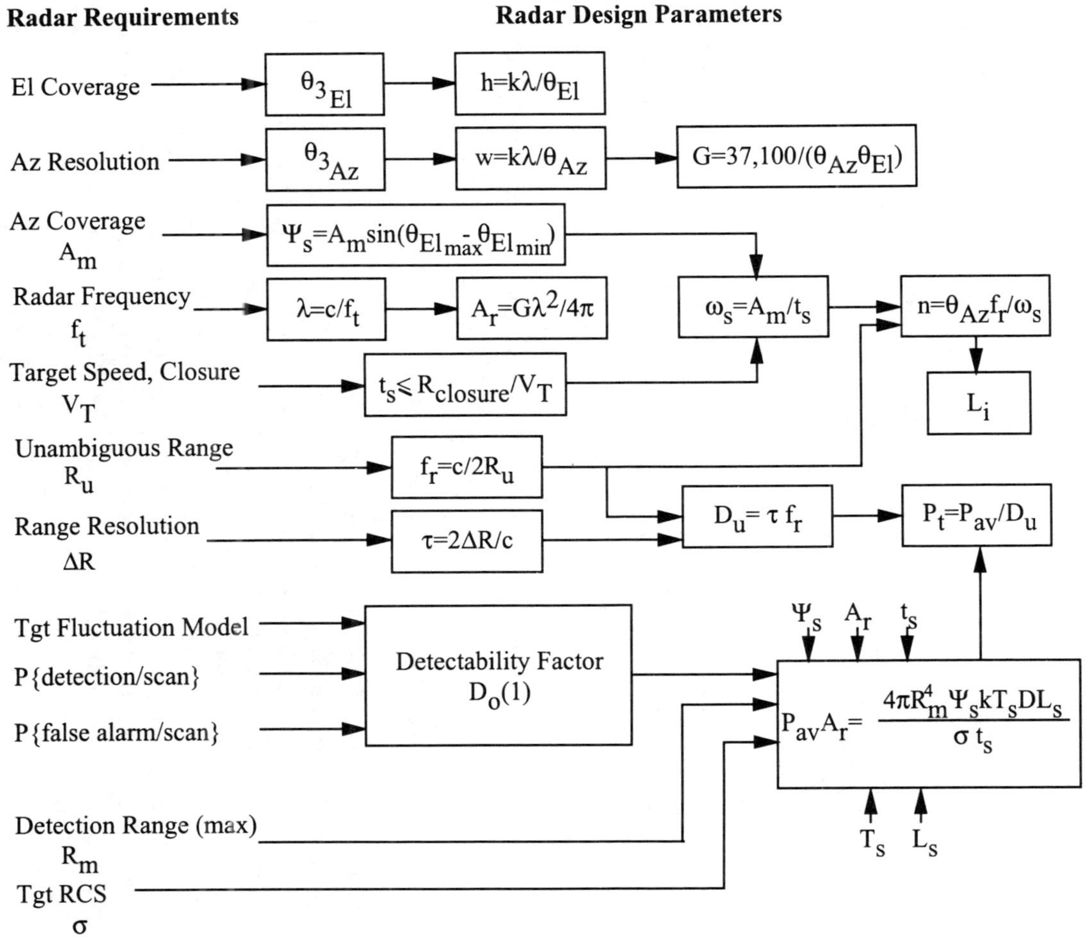


**Figure 3.3** Derivation of key radar characteristics.

which, when solved for the parameters of this S-band example, lead to a value of about 6 $m^2$. Note that this effective aperture area is different than the physical aperture area (9.8 $m^2$). The difference is due to the weighting applied to the antenna illumination function to reduce the sidelobe level of the antenna. The ratio of the actual (weighted) gain over the maximum gain is the *aperture efficiency*, which in this example is 61%.

An important requirement yet to be established is the antenna scan rate. In this example, the requirements specify the target as having a maximum speed of 330 m/s. In search radars it is customary to specify the maximum allowable range that a radially inbound target can travel between scans, relative to the radar site location. For this example, this range closure will be specified to be no more than 2.5% of the maximum detection range. If we specify that the maximum detection range versus a target with a radar cross section of 1 $m^2$ will be 85 km, then the antenna scan time, the interval between sequential scans over the 330 m/s target, must be less than or equal to 6 sec. The required antenna azimuth scan rate ($\omega_s$ in Fig. 3.3) is simply the azimuth sector coverage divided by the scan time, or 60 deg/s.

*Unambiguous Range and Range Resolution.* For a pulsed radar, the PRI, the reciprocal of the PRF, must be greater than the time it takes for the radar pulse to travel to a target and back when the target is at the maximum required detection range. If this interval is shorter than that time, then it is not possible, without the use of multiple PRFs, to determine target range unambiguously. If we arbitrarily establish that the radar must be capable of determining range to targets at 1.5 times the nominal detection range, then this value for the maximum unambiguous range, $R_u = 127$ km, can be used to set the maximum allowable PRF from

$$R_u = c/2\text{PRF} = c/2f_r \tag{3.6}$$

Solving this equation yields a requirement for the radar PRF to be less than or equal to 1,176 pulses/s. As Figure 3.3 shows, the PRF is used in conjunction with the azimuth beamwidth and the antenna scan rate to determine the number of pulses $n$ during single-scan observation time. The value of $n$ in turn allows us to calculate the amount of loss due to noncoherent integration of $n$ pulses (Table 3.3).

The desired *range resolution* ($\Delta$R) determines the radar pulsewidth $\tau$, as shown in Figure 3.3. Pulsewidth times PRF ($f_r$ in Fig. 3.3) defines the radar *duty factor* $D_u$, which is necessary to set the radar peak power required. The ability to discriminate two targets separated in range by 150m is a reasonable requirement for a search radar, leading to a pulsewidth of 1 $\mu$s.

*Target Model.* Another portion of the statement of requirements is a target model which specifies (1) the average radar cross section $\sigma$, which is the ratio of backscattered power to the power density of the incident radar signal, (2) the target fluctuation model, which describes the statistics of the time-varying target signal amplitude fluctuation, and (3) the target motion model, which describes the target linear and rotational motion with respect to the radar. We have assumed for this example that the target has an average radar cross section $\sigma$ of $1\text{m}^2$ and that the amplitude fluctuation is described by Rayleigh statistics (i.e., an exponential distribution). Peter Swerling described such targets as case 1 when the amplitude fluctuation was slow relative to a single pulse. (See, for example, [1, p. 81].) Target velocity has already been specified to be 330 m/s. Target rotational motion is assumed to be of secondary importance for this example.

*Free-Space Radar Detection Range.* Finally, we require initially that the *free-space* radar detection range, which is the range at which the target would be detected over a line-of-sight path in a vacuum, be 85 km with probability of detection $P_d = 0.9$ per scan, and with an associated false-alarm probability $P_{fa} = 10^{-6}$. Subsequently, the actual losses due to the propagation path can be used to adjust the radar parameters as calculated previously.

The derivation of the key parameters of a pulsed radar solution to these requirements is shown in a flow-down method in Figure 3.3, with the resulting nominal values summarized in Table 3.2.

The range equation is a relationship used to calculate achievable detection range for a given set of radar parameters, or parameters required to achieve a given detection range. The search form of the radar range equation states that the average power required for a given range when other major parameters are know is [1]

$$P_{avg} = \frac{4\pi R_D^4 \Psi_S \mathrm{k} T_S D_0(1) L_S}{A_r t_s \sigma} \tag{3.7}$$

**Table 3.2**
Key Radar Parameters

| *Parameter* | *Symbol* | *Value* |
|---|---|---|
| Frequency of operation (center) | $f_T$ | 3 GHz |
| Free-space detection range | $R_D$ | 85 km |
| Solid angle of search | $Y_S$ | 0.55 sr |
| Effective aperture area | $A_r$ | $5.9\text{m}^2$ |
| Antenna beamwidth, azimuth | $\theta_3$ az | 1 deg |
| Antenna beamwidth, elevation | $\theta_3$ el | 5 deg |
| Antenna gain | $G_T$ | 38.7 dB |
| Antenna dimensions | $w \cdot h$ | 7m × 1.4m |
| Scan rate | $\omega_S$ | 60 deg/s |
| Pulse width | $t$ | 1 ms |
| Observation time | $t_o$ | 0.0166 sec |
| PRF | $f_r$ | 1,176 Hz |
| Number of pulses per $t_o$ | $n$ | 20 |
| Azimuth angle resolution | $\Delta A_z$ | 1 deg |
| Range resolution | ΔR | 150 m |
| Transmission duty factor | $D_U$ | 0.0012 |
| Probability of detection | $P_d$ | 0.9/scan |
| Probability of false alarm | $P_{fa}$ | $1 \times 10^{-6}$/scan |

where $R_D$, $\Psi_s$, $A_r$, and $\sigma$ are previously defined. Additionally, $t_s$ is the search frame time, k is the Boltzmann constant equal to $1.38 \times 10^{-23}$ W · s/K, $T_S$ is the radar system input noise temperature, $D_0(1)$ is the *detectability factor* or the SNR required to detect a nonfluctuating target (Swerling case 0) with the prescribed $P_d$ and $P_{fa}$, and $L_s$ represents the combined radar system losses applicable to the search.

In the absence of a detailed design, we assume $T_S$ = 1,000K, not untypical of a radar that must operate in the presence of jamming. The product k$T_S$ represents the receiver noise power spectral density (in watts per hertz) with which the target signal must compete. The remaining unknown, the combined search loss factor $L_s$, is partly a function of the radar waveform and signal processing as discussed in Section 3.2, and, therefore, so is the solution for required average power. For a noncoherent radar, $n$ pulses received during the observation time can be integrated after they have been detected, but a loss factor $L_i$ must be included to account for the difference in SNR required over that needed if all $n$ pulses were coherently integrated. To account for the fact that the actual target has a fluctuating radar cross section (RCS) (Swerling case I), a fluctuation loss $L_f$ must be applied as well, unless pulse-to-pulse frequency diversity is used to decorrelate the target fluctuation. Table 3.3 gives a summary of typical search losses for a single-frequency, noncoherent radar.

A coherent pulse radar would not suffer the 3-dB integration loss applicable to a noncoherent radar (and it would be better equipped to detect the target in rain, chaff, or ground clutter), meaning that for this case the average power required of a coherent radar having the same characteristics would be one-half that of the noncoherent radar.

There is yet a third option to consider: a noncoherent pulse radar that incorporates frequency agility. If the radar has the ability to change frequency from pulse to pulse by an amount sufficient to decorrelate the target returns, the target fluctuation loss $L_f$ can be reduced by the factor $n_e$, the number of independent pulses per time on target. From Table 3.2, we see that there are 20 pulses per time on target (or during the observation time $t_o$). Assume that the radar transmitter shifts frequency such that only two pulses are transmitted

**Table 3.3**
Search Losses for Noncoherent Search Radar

| *Type* | *Symbol* | *Loss (dB)* |
|---|---|---|
| Transmission line loss, (transmitter to antenna) | $L_T$ | 1.0 |
| Beam pattern constant: | $L_n$ | 2.0 |
| Beam shape loss (2-D) | $L_p$ | 2.6 |
| Miscellaneous processing losses | $L_x$ | 3.0 |
| Receiver filter mismatch | $L_m$ | 0.8 |
| Integration loss (20 pulses) | $L_i$ | 3.0 |
| Fluctuation loss (Sw. I, $P_d$ 0.9, $P_{fa}$ $10^{-6}$) | $L_f$ | 8.4 |
| Atmospheric attenuation (two-way) | $L_a$ | 1.2 |
| Total | $L_s$ | 22.0 |

at any given frequency (i.e., $n_e = 10$). Then from [1], signal loss due to target fluctuation is

$$L_f(n_e)_{dB} = (1/n_e)L_f(1)_{dB} = (1/10)(8.4) = 0.84 \text{ dB} \tag{3.8}$$

We have now determined all of the parameters of (3.7) for the three types of pulsed radar except for $D_0(1)$, the detectability factor for a steady target. Figure 3.4 shows this to be approximately 13.2 dB for the stated probabilities of detection and false alarm.

Substituting the appropriate values for the radar parameters into (3.7), we have

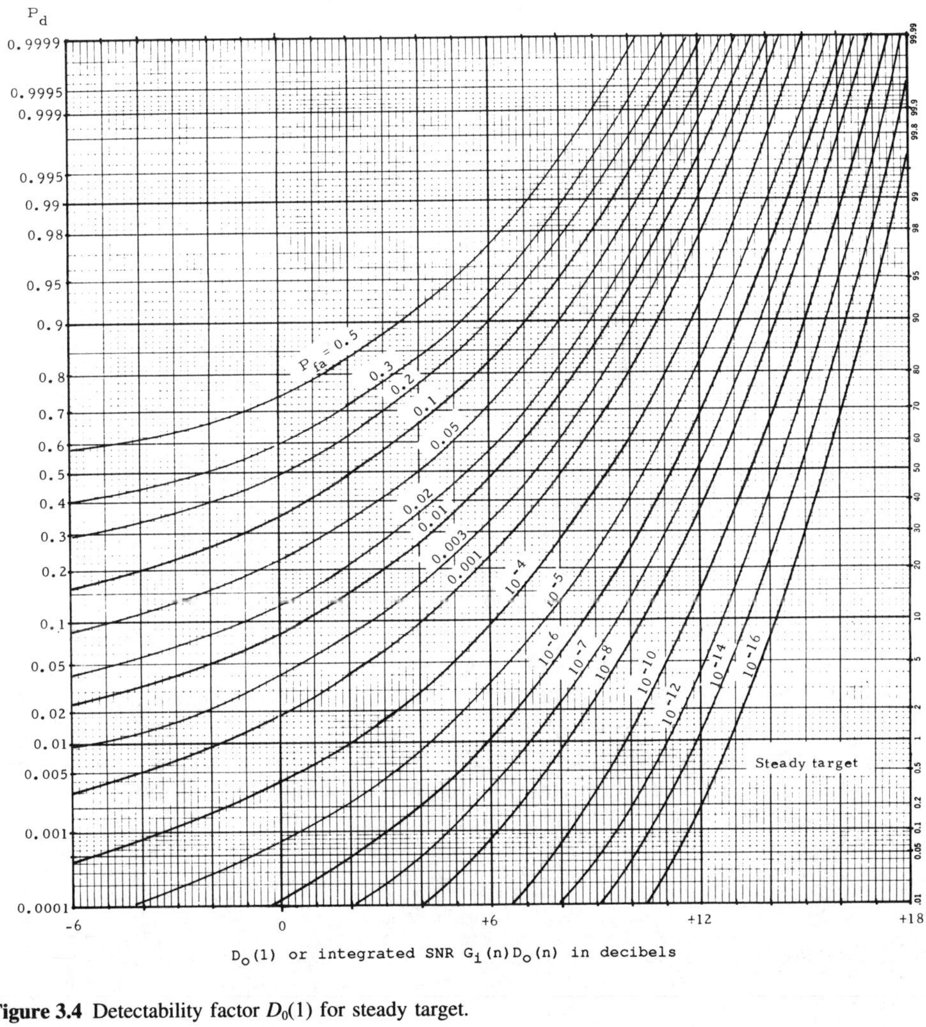


**Figure 3.4** Detectability factor $D_0(1)$ for steady target.

$$P_{avg} = \frac{(4\pi)(85{,}000)^4(0.55)(1.38 \times 10^{-23})(1000)(20.9)}{(5.9)(6)(1)} L_s = 2.94 L_s \tag{3.9}$$

Assuming all other losses remain the same except for the fluctuation loss $L_f$, the radar average and peak powers required for the three different waveforms in the order discussed are:

- Noncoherent pulse: $L_s$ = 22 dB = 158.5, $P_{avg}$ = 466W, and $P_t = P_{avg}/D_u$ = 466W/0.0012 = 388 kW;
- Coherent pulse: $L_s$ = 19 dB = 79.4, $P_{avg}$ = 233W, and $P_t$ = 194 kW;
- Noncoherent pulse with frequency diversity: $L_s$ = 14.4 dB = 27.8, $P_{avg}$ = 81.7W, and $P_t$ = 68 kW.

While we have covered some of the basic points involved in the preliminary design of a radar with this example, there are many additional issues to be considered in a preliminary design, even in the seemingly straightforward case of the 2-D search radar described. For example, use of a fan beam, assumed here for the sake of simplicity, may not represent the best approach to an actual search problem. If continuous coverage is required at high altitudes, at ranges closer to the radar than that covered by the half-power beamwidth of a fan beam, a csc$^2$-shaped elevation beam might be called for. Alternatively several stacked beams, each having a beamwidth of a few degrees or less, could be used to obtain the required coverage in elevation, while affording useful resolution in that dimension.

If Doppler processing were to be used, the additional simultaneous coverage provided by these alternatives would probably be required to ensure a sufficient time on target for adequate Doppler resolution. If the radar requirements include detection of small targets at low altitudes, the radar waveform and signal processor design must meet the clutter cancellation requirements dictated by the expected land or sea clutter characteristics, including clutter caused by rain or chaff. To allow cancellation of moving clutter caused by wind-swept rain and birds, the radar PRF of this S-band example would need to be increased to about two-and-a-half times that shown for the noncoherent waveform to accommodate a clutter notch filter while still permitting sufficient Doppler bandwidth for the range of expected target velocities. With this increase in PRF, the radar's unambiguous range will be reduced to about one-third its original value, and the use of multiple PRFs would be required to resolve targets at greater ranges.

## REFERENCES

[1] Barton, D. K., *Modern Radar System Analysis*, Norwood, MA: Artech House, 1988.

[2] Barton, D. K., C. E. Cook, and P. C. Hamilton, *Radar Evaluation Handbook*, Norwood, MA: Artech House, 1991.

[3] Barton, D. K., and W. F., Barton, *Modern Radar System Analysis Software*, Version 2.0, Norwood, MA: Artech House, 1993.

[4] Huizing, A. G., and A.Theil, *Computer-Aided Radar Performance Evaluation Tool (CARPET)*, Version 1.0, Norwood, MA: Artech House, 1993.

[5] Technology Service Corporation, *Radar Workstation*, Version 2.2, Dahlgren, VA: Technology Service Corporation, 1990.

# Chapter 4
# Foundations of Optical Engineering


**Edward N. Toughlian**
*Rome Laboratory*

**Henry Zmuda**
*Stevens Institute of Technology*

**Douglas Norton**
*Rome Laboratory*


## 4.1 INTRODUCTION

The goal of this chapter is to introduce the radar designer to some of the most essential and fundamental concepts used in optical system design. Both radar and optical engineering disciplines share the common language of Fourier analysis. For this reason, we open with a discussion of the two-dimensional Fourier transform. This then leads to the essential ideas of diffraction theory required for an understanding of optical information processing systems. We also discuss several common optical components, such as lenses, beam splitters, wave plates, isolators, and diffractive optics. These components are freely used in the implementation of optical systems, and a basic understanding of their operation is essential. The chapter closes with a discussion of optical heterodyne techniques.

## 4.2 TWO-DIMENSIONAL FOURIER TRANSFORM [1,2]

The Fourier integral is a valuable mathematical tool for the study of linear time-invariant dynamical systems. When applied to linear media (i.e., one whose permittivity, or refractive index, and permeability do not depend on the applied field), Maxwell's equation constitutes a set of linear differential equations. Hence, for optical systems, Fourier techniques are applicable. In the case of optics, the spatial coordinates are the independent variables used instead of time. The reader is assumed to be familiar with the mathematical development of the Fourier transform and the corresponding inverse transform. Theorems regarding the existence, uniqueness, and other mathematical properties of the transform can be found in the references cited at the end of the chapter.

Consider the function $f(x,y)$, where $x$ and $y$ are spatial coordinates. The 2-D Fourier transform (we adopt the notation $j = (-1)^{1/2}$) of $f(x,y)$ is

$$F(\omega_x, \omega_y) = \mathcal{F}[f(x,y)] = \int_{-\infty}^{\infty}\int_{-\infty}^{\infty} f(x,y)e^{-j(\omega_x x+\omega_y y)}dx\, dy \tag{4.1}$$

where $f_x = \omega_x/2\pi$ and $f_y = \omega_y/2\pi$ are *spatial frequency domain variables*. Figure 4.1 gives a geometrical interpretation of these spatial frequencies, showing how they are related to a plane wave propagating in some arbitrary direction.

Equation (4.1) is the successive application of two 1-D transforms, or

$$F(\omega_x, \omega_y) = \mathcal{F}_x[\mathcal{F}_y[f(x,y)]] = \mathcal{F}_y[\mathcal{F}_x[f(x,y)]] \tag{4.2}$$

where $\mathcal{F}_x$ and $\mathcal{F}_y$ represent the usual 1-D transforms with respect to $x$ and $y$, respectively. This allows the extension of many 1-D results directly to 2-D transforms. The inverse Fourier transform is given by

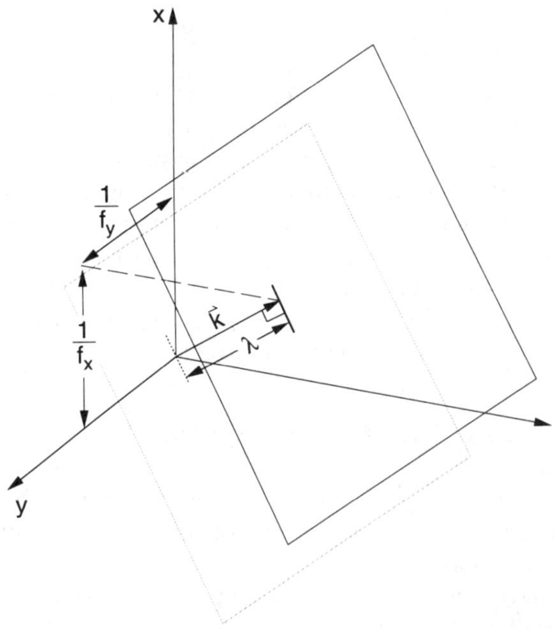


**Figure 4.1** Geometric interpretation of spatial frequencies.

$$f(x,y) = \mathcal{F}^{-1}[F(\omega_x, \omega_y)] = \frac{1}{4\pi^2}\int_{-\infty}^{\infty}\int_{-\infty}^{\infty} F(\omega_x, \omega_y)e^{j(\omega_x x+\omega_y y)}d\omega_x\, d\omega_y \tag{4.3}$$

which once again may be thought of as the successive application of two 1-D inverse transforms.

*Remark.* The location of the $1/4\pi^2$ factor varies in the literature.

Just as with its 1-D counterpart, the 2-D Fourier transform represents a spectral (plane wave) decomposition in terms of the spatial frequency components, which we now examine in greater detail.

Mathematically, (4.3) states that the function $f(x,y)$ may be constructed by a linear combination of complex exponential functions $\exp[j(\omega_x x + \omega_y y)]$, each with weight $F(\omega_x,\omega_y)d\omega_x d\omega_y$. The physical significance of these spatial spectral components is obtained as follows.

Consider the 3-D functional description of the (electric) field distribution $f(x,y,z)$, generated from a spatially distributed monochromatic source of (temporal) frequency $f = \omega/2\pi$. Let us form the 2-D function $f_0(x,y) = f(x,y,0)$ located at the plane $z = 0$, with Fourier transform $F_0(\omega_x,\omega_y)$ computed from (4.1). The function $f_0(x,y)$ can be expressed in terms of its inverse transform as

$$f_0(x,y) = \frac{1}{4\pi^2}\int_{-\infty}^{\infty}\int_{-\infty}^{\infty} F_0(\omega_x, \omega_y)e^{j(\omega_x x+\omega_y y)}d\omega_x d\omega_y \tag{4.4}$$

The unit normal $\mathbf{u}$ of an elemental plane wave is specified by its direction cosines $(\alpha, \beta, \gamma)$, which are the components of $\mathbf{u}$ along the cartesian axis, or $\mathbf{u} = \alpha\mathbf{a}_x + \beta\mathbf{a}_y + \gamma\mathbf{a}_z$, where $\mathbf{a}_x$, $\mathbf{a}_y$, and $\mathbf{a}_z$ are unit vectors, and $\mathbf{u}$ is of unit length $\alpha^2 + \beta^2 + \gamma^2 = 1$. Such a unit amplitude plane wave, shown in Figure 4.1, is described by

$$e^{jk(\alpha x+\beta y+\gamma z)} \tag{4.5}$$

where $k = \omega/c = 2\pi/\lambda$.

The integrand of (4.3) may be regarded as a plane wave, evaluated at $z = 0$, with amplitude $F_0(\omega_x,\omega_y)d\omega_x d\omega_y$, where $\omega_x = k\alpha$, and $\omega_y = k\beta$. An infinite number of these plane waves are summed (integrated) over all $\omega_x$ and $\omega_y$. The function $f_0(x,y)$ is thus (conceptually) constructed from an angular spectrum of plane waves whose directions are specified by the spatial frequencies $f_x = \omega_x/2\pi = \alpha/\lambda$ and $f_y = \omega_y/2\pi = \beta/\lambda$.

For $z \neq 0$, the angular spectrum of the function $f(x,y,z)$ is

$$F(\omega_x, \omega_y, z) = \int_{-\infty}^{\infty}\int_{-\infty}^{\infty} f(x,y,z)e^{-j(\omega_x x+\omega_y y)}dxdy \tag{4.6}$$

For sinusoidal steady-state time variation, the function $f(x,y,z)$ must satisfy the wave equation

$$\nabla^2 f(x,y,z) + k^2 f(x,y,z) = 0 \tag{4.7}$$

Expressing $f(x,y,z)$ by its inverse transform,

$$f(x,y,z) = \frac{1}{4\pi^2}\int_{-\infty}^{\infty}\int_{-\infty}^{\infty} F(\omega_x,\ \omega_y,\ z)e^{j(\omega_x x+\omega_y y)}d\omega_x d\omega_y \tag{4.8}$$

Substituting (4.8) into the wave equation (4.7) and solving the resulting second-order differential equation for $F(\omega_x,\omega_y,z)$ gives

$$F(\omega_x,\ \omega_y,\ z) = F_0(\omega_x,\ \omega_y)e^{\pm j\sqrt{k^2-\omega_x^2-\omega_y^2}z} \tag{4.9}$$

as independent solutions. Since $\omega_x^2 + \omega_y^2 + \omega_z^2 = k^2$, two distinct cases arise, namely

$$\omega_x^2 + \omega_y^2 < k^2 \qquad \omega_x^2 + \omega_y^2 > k^2 \tag{4.10}$$

When $\omega_x^2 + \omega_y^2 < k^2$, $\omega_z = -\gamma k$ is real and the wave propagates in the $z$ direction. When $\omega_x^2 + \omega_y^2 > k^2$, $\omega_z = -j|\gamma|k$ is purely imaginary and the wave decays exponentially or is evanescent in the $z$ direction. Because of this exponentially decaying nature, values of $\omega_x$ and $\omega_y$ satisfying $\omega_x^2 + \omega_y^2 > k^2$ are referred to as the *invisible* portion of the spectrum, whereas values of $\omega_x$ and $\omega_y$ satisfying $\omega_x^2 + \omega_y^2 < k^2$ are referred to as the *visible* portion of the spectrum.

Finally, the function $f(x,y,z)$ may be expressed in terms of its angular spectrum by inverse transforming the (positive $z$-propagating) angular spectrum $F_0(\omega_x,\omega_y)$ of the spatial field distribution function $f_0(x,y)$ given by (4.9), or

$$f(x,y,z) = \frac{1}{4\pi^2}\int_{-\infty}^{\infty}\int_{-\infty}^{\infty}\left(F_0(\omega_x,\omega_y)e^{-j\sqrt{k^2-\omega_x^2-\omega_y^2}z}\right)\cdot e^{j(\omega_x x+\omega_y y)}d\omega_x d\omega_y \tag{4.11}$$

With regard to the function $f(x,y,z)$ as the response of a system to an excitation function $f_0(x,y)$ with transform $F_0(\omega_x,\omega_y)$, (4.11) expresses the response as the inverse Fourier transform of the frequency domain function

$$F(\omega_x,\ \omega_y,\ z) = F_0(\omega_x,\ \omega_y)e^{-j\sqrt{k^2-\omega_x^2-\omega_y^2}z} \tag{4.12}$$

The exponential factor

$$H(\omega_x,\omega_y,z) = e^{-j\sqrt{k^2-\omega_x^2-\omega_y^2}z} \tag{4.13}$$

may be regarded as the transfer function description of how the excitation at $z=0$ continues for $z > 0$. This transfer function behavior characteristic of linear translation-invariant systems forms the fundamental basis of optical signal processing. (*Translational invariance* means that translation of the system input image results in pure (undistorted) translation of the output object. This is typically a reasonable approximation, at least over a finite portion of an optical system.)

It should be noted here that if the aperture distribution is not a steady-state monochromatic waveform, but some arbitrary function of time, then the field for $z \geq 0$ at time $t$ is represented by a spectrum which is a function of both direction and temporal frequency content. For this case, $f_0(x,y,t) \Leftrightarrow F_0(\omega_x,\omega_y,\omega)$, where $\omega$ is the usual Fourier (time-frequency) transform variable.

## 4.3 FUNDAMENTALS OF THE DIFFRACTION THEORY [3]

The light emanating from a diffracting aperture is usually classified in one of three regions: the evanescent-wave region, the Fresnel region, and the far-field or Fraunhofer region. The evanescent-wave region extends only several wavelengths from the aperture and is of no interest in the majority of optical signal processing applications. The region beyond this, where only the propagating portion of the angular spectrum contributes to the intensity, is known as the Fresnel region. Finally comes the far-field or Fraunhofer region, where, as we shall see, the field intensity at each observation point is determined by a single plane wave component of the angular spectrum. Propagation in the Fraunhofer region is governed by the rules of geometrical optics.

### *Fresnel Diffraction*

This section will show that the Fresnel diffraction integral gives the amplitude distribution of a wave at any position $z$ with respect to a boundary condition (aperture field distribution) at $z=0$. For sinusoidal steady-state angular frequency $\omega$, it is known that all electromagnetic field components satisfy the wave equation (4.7). In cartesian coordinates and in the absence of boundaries, the general solution is a plane wave,

$$e^{j(\omega_x x+\omega_y y+\omega_z z+\omega t)} \tag{4.14}$$

with

$$\omega_x^2 + \omega_y^2 + \omega_z^2 = k^2 = \frac{\omega^2}{c^2} = \frac{(2\pi)^2}{\lambda^2} \tag{4.15}$$

For many optical signal processing applications, it is usual to assume $\omega_x^2 + \omega_y^2 \ll k^2$, so that

$$\omega_z = \sqrt{k^2 - \omega_x^2 - \omega_y^2} \approx k - \frac{\omega_x^2 + \omega_y^2}{2k} \tag{4.16}$$

This is known as the *paraxial approximation* and assumes that the wave vector **u** makes a small angle with respect to the $z$-axis. The general situation is pictured in Figure 4.2. Note that the paraxial approximation implies that $F_0(\omega_x,\omega_y)$ must vanish for $\omega_x^2 + \omega_y^2 \ll k^2$ to satisfy the assumed paraxial approximation. Such an assumption is valid if the distance between the excitation aperture and plane of observation is very much greater than the greatest linear dimension of the input aperture. It further assumes that the plane of observation is of finite extent, with maximum linear dimension much less than that of the distance between observation point and excitation.

An arbitrary amplitude distribution $f(x,y,z)$ is thus constructed in the "Fourier sense" from a continuous superposition of elementary plane waves. Substituting the approximation of (4.16) into (4.11) gives

$$f(x,y,z) = \frac{1}{4\pi^2}\int_{-\infty}^{\infty}\int_{-\infty}^{\infty} F_0(\omega_x,\omega_y)\mathrm{e}^{j(\omega_x x+\omega_y y)}\mathrm{e}^{j\frac{(\omega_x^2+\omega_y^2)}{2k}z}\mathrm{d}\omega_x\mathrm{d}\omega_y \tag{4.17}$$

As is customary, we have suppressed both the $\mathrm{e}^{j\omega t}$ time dependence as well as the spatial phase factor $\exp\{-jkz\}$ variation in (4.17) for the sake of brevity. The Fourier integral theorem states that at points of continuity of $f(x,y)$, $f(x,y) = \mathcal{F}^{-1}\{\mathcal{F}[f(x,y)]\}$. At points of discontinuity, $f(x,y)$ converges to the average of its left- and right-hand limits. From the

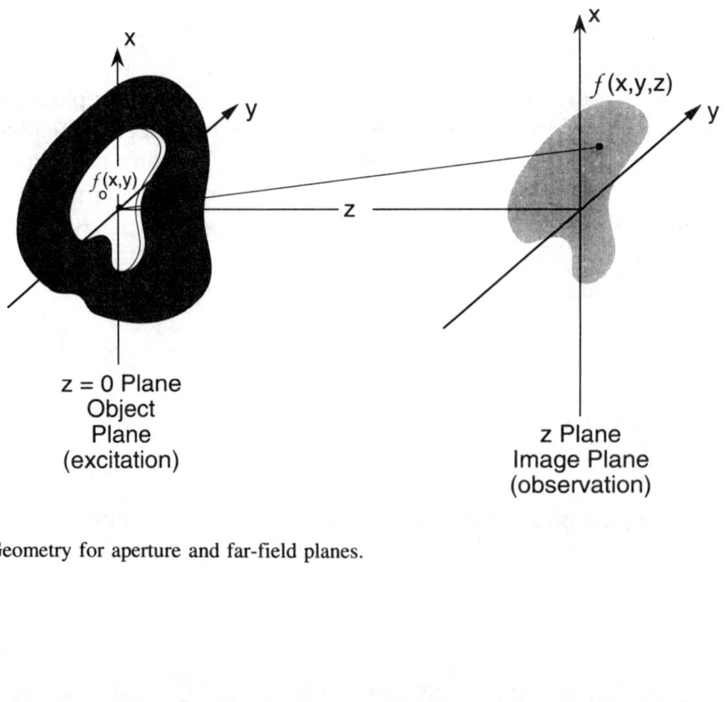


**Figure 4.2** Geometry for aperture and far-field planes.

Fourier integral theorem, under the paraxial approximation, we can express the field distribution $f(x,y,z)$ in terms of its source distribution. Expressing $F_0(\omega_x,\omega_y)$ in (4.17) in terms of $f_0(x,y)$ using the inverse transform gives

$$f(x,y,z) = \frac{1}{4\pi^2}\int_{-\infty}^{\infty}\int_{-\infty}^{\infty}\left[\int_{-\infty}^{\infty}\int_{-\infty}^{\infty} f_0(u,v)\mathrm{e}^{-j(\omega_x u+\omega_y v)}\mathrm{d}u\mathrm{d}v\right]\cdot \mathrm{e}^{j(\omega_x x+\omega_y y)}\mathrm{e}^{j\frac{\omega_x^2+\omega_y^2}{2k}z}\mathrm{d}\omega_x\mathrm{d}\omega_y \tag{4.18}$$

Reversing the order of integration between d$u$d$v$ and d$\omega_x$d$\omega_y$, we identify the *Fresnel kernel* $h(x,y,z)$ defined by

$$h(x,y,z) = \frac{1}{4\pi^2}\int_{-\infty}^{\infty}\int_{-\infty}^{\infty} \mathrm{e}^{j(\omega_x x+\omega_y y)}\mathrm{e}^{j\frac{\omega_x^2+\omega_y^2}{2k}z}\mathrm{d}\omega_x\mathrm{d}\omega_y \tag{4.19}$$

Equation (4.19) can be evaluated in closed form giving

$$h(x,y,z) = \frac{j}{\lambda z}\mathrm{e}^{-jk\frac{x^2+y^2}{2z}} \tag{4.20}$$

where we use the 1-D transform: $\mathcal{F}\{\exp[\alpha u^2]\} = (\pi/\alpha)^{1/2}\exp[-\omega^2/4\alpha]$. The Fresnel diffraction integral can thus be concisely expressed as the convolution

$$f(x,y,z) = f_0(x,y) ** h(x,y,z) = \int_{-\infty}^{\infty}\int_{-\infty}^{\infty} f_0(u,v)h(x-u,y-v,z)\mathrm{d}u\mathrm{d}v \tag{4.21}$$

where ** denotes 2-D convolution. Or, equivalently, since convolution in one domain corresponds to multiplication in the other, the Fresnel diffraction integral can be expressed in the spatial frequency domain by

$$F(\omega_x,\ \omega_y,\ z) = H(\omega_x,\ \omega_y,\ z)F_0(\omega_x,\ \omega_y) \tag{4.22}$$

where from (4.19) we see

$$H(\omega_x,\ \omega_y,\ z) = \mathcal{F}[h(x,y,z)] = \mathrm{e}^{j\frac{\omega_x^2+\omega_y^2}{2k}k} \tag{4.23}$$

Now, substituting (4.23) into (4.22) gives the result

$$F(\omega_x,\omega_y,z) = \mathcal{F}[f(x,y,z)] = F_0(\omega_x,\omega_y)\mathrm{e}^{j\frac{\omega_x^2+\omega_y^2}{2k}z} \tag{4.24}$$

We see that the Fresnel kernel $h(x,y,z)$ of (4.19) may be regarded as the impulse response of this free-space system with corresponding transfer function $H(\omega_x,\omega_y,z)$.

*Fraunhofer Diffraction*

If the distance between the plane of excitation at $z = 0$ (denoted by the subscript 0) and the observation plane at $z$ (denoted by the subscript 1) is large, or $z \gg k(x_0^2 + y_0^2)_{max}/2$, one generally approximates the exponential arguments with

$$(x_1 - x_0)^2 + (y_1 - y_0)^2 \approx (x_1^2 + y_1^2) - 2(x_0x_1 + y_0y_1) \tag{4.25}$$

and

$$\frac{x_0^2 + y_0^2}{z} \approx 0 \tag{4.26}$$

Under these approximations, the Fresnel diffraction integral of (4.21) becomes the *Fraunhofer* diffraction integral, or

$$f(x_1,y_1,z) = \frac{j}{\lambda z} e^{-jk\frac{(x_1^2+y_1^2)}{2z}} \int_{-\infty}^{\infty}\int_{-\infty}^{\infty} f_0(u,v)\, e^{j\frac{k}{z}(x_1u+y_1v)}\, du dv \tag{4.27}$$

where we recognize the integral in (4.27) has the functional form of the Fourier transform of $f_0(x,y)$, with frequency variables $-kx_1/z$ and $-ky_1/z$ replacing $\omega_x$ and $\omega_y$, respectively. Thus, we can write

$$f(x_1,y_1,z) = \frac{j}{\lambda z} e^{-jk\frac{x_1^2+y_1^2}{2z}} F_0\left(-k\frac{x_1}{z}, -k\frac{y_1}{z}\right) \tag{4.28}$$

Those familiar with antenna theory will recognize this result as a restatement of the far-field antenna theorem, which states that the far field of an aperture is asymptotically equal to its angular spectrum [4].

Figure 4.3 shows how a rectangular waveform (generated, say, by a plane wave incident on an opaque screen with a rectangular slit cut in it) would propagate into the Fresnel and Fraunhofer region.

## 4.4 TRANSFER FUNCTION OF A THIN LENS [5]

We have noted that the far field is related to the Fourier transform of the aperture distribution. A lens is capable of generating this far-field distribution a finite distance from the aperture. This fundamental and useful signal processing function is provided by the action of a thin lens, schematically shown in Figure 4.4.

To see how a lens can produce the spatial Fourier transform of the input excitation, we first note from Figure 4.4 that the portion of incident excitation $f(x,y)$ for values of

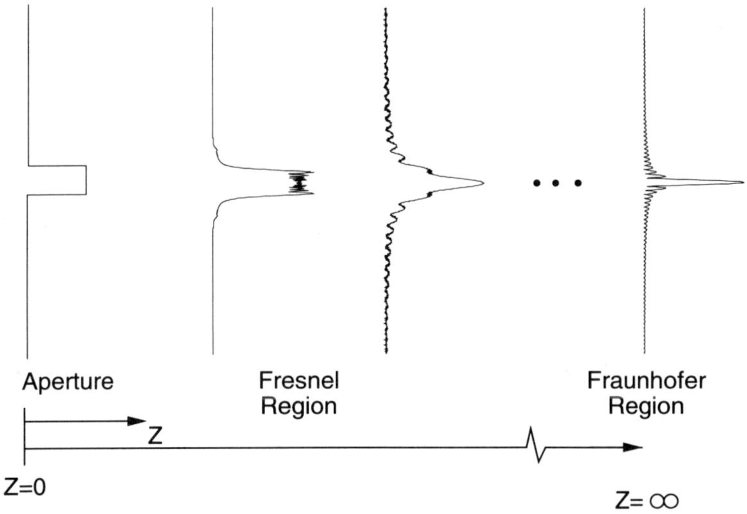


**Figure 4.3** Example of Fresnel and Fraunhofer diffraction.

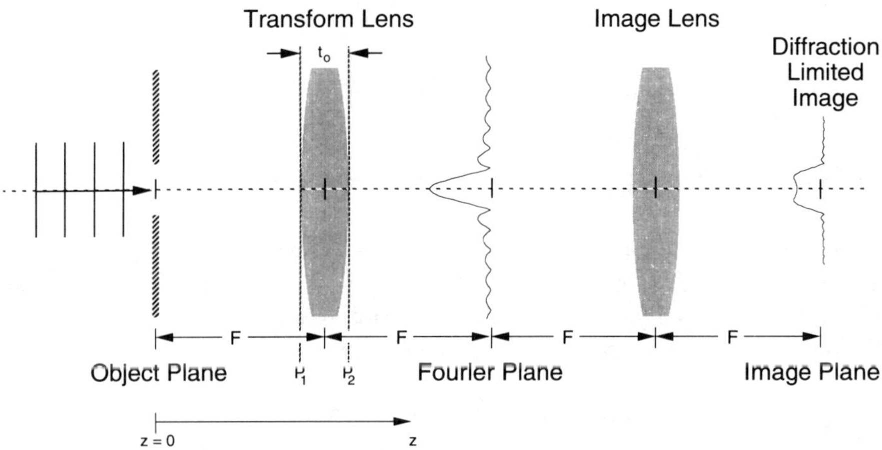


**Figure 4.4** Fourier transform property of a thin lens.

$x,y$ nearer the lens center will undergo a greater delay than will those nearer the outer edge. If $t_0$ is the maximum thickness of the lens, then the incident illumination will undergo a total phase delay passing from reference planes $P_1$ to $P_2$ of

$$\phi(x,y) = knt(x,y) + k[t_0 - t(x,y)] \tag{4.29}$$

where $n$ is the index of refraction of the lens and $t(x,y)$ is the thickness of the lens. Again making use of the paraxial approximation, this thickness function can be expressed as

$$t(x,y) \approx t_0 - \frac{x^2 + y^2}{2F(n-1)} \tag{4.30}$$

where $F$ is the focal length of the lens. The resulting phase delay through the lens is

$$h_\ell(x,y) = e^{j\phi} = e^{jknt_0} e^{-j(k/2F)(x^2+y^2)} \tag{4.31}$$

Consequently, the light distribution $g(x,y)$ emanating from the lens (at plane $P_2$) will be a phase-shifted version of the input illumination (plane $P_1$)

$$g(x,y) = h_\ell(x,y)f(x,y) \tag{4.32}$$

The thin lens approximation neglects the thickness of the glass and assumes that all rays emanate from approximately the same $(x,y)$ coordinate at which they entered.

In the spatial frequency domain, (4.32) can be expressed as the frequency domain convolution $G(\omega_x,\omega_y) = F(\omega_x,\omega_y) ** H_\ell(\omega_x,\omega_y)$, where

$$H_\ell(\omega_x,\omega_y) = \mathcal{F}[h_\ell(x,y)] = \frac{\lambda F}{j} e^{jknt_0} e^{(jF/2k)(\omega_x^2+\omega_y^2)} \tag{4.33}$$

*Transfer Function of a Lens System*

Referring again to Figure 4.4, an initial distribution $f(x,y,0) = f_0(x,y)$ at the front focal plane ($z = 0$) propagates a distance $F$ to the plane of the lens. The illumination on the lens (located at $z = F$) is given by the product of the free-space transfer function $H(\omega_x,\omega_y,F)$ and the transform of the aperture function $F_0(\omega_x,\omega_y,0)$.

$$F_{P_1}(\omega_x,\omega_y,F) = F_0(\omega_x,\omega_y,0)\,H(\omega_x,\omega_y,F) \tag{4.34}$$

Making use of (4.32), the corresponding output distribution

$$\begin{aligned} G_{P_2}(\omega_x,\omega_y,F) &= \mathcal{F}[f_{P_1}(x,y,F)\,h_\ell(x,y)] \\ &= [F_0(\omega_x,\omega_y,0)\,H(\omega_x,\omega_y,F)] ** H_\ell(\omega_x,\omega_y) \end{aligned} \tag{4.35}$$

This distribution again propagates at a distance $F$ to the image or back focal plane located at $z = 2F$, where the distribution $G_1(\omega_x,\omega_y,2F)$ is

$$G_1(\omega_x,\omega_y,2F) = \{[F_0(\omega_x,\omega_y,0)H(\omega_x,\omega_y,F)] ** H_\ell(\omega_x,\omega_y)\}H(\omega_x,\omega_y,F) \tag{4.36}$$

Evaluating the above, we find that

$$G_1(\omega_x,\omega_y,2F) \propto F_0(\omega_x,\omega_y) \tag{4.37}$$

which indeed shows that the resulting distribution at the back focal plane is proportional to the Fourier transform of the input distribution at the front focal plane.

Under ideal circumstances, a uniform plane wave incident on a lens will result in a spatial impulse (a delta function with zero spatial dimension) a focal length away from the lens. Since infinite-extent plane waves do not exist in nature, and since a lens has a finite diameter, a pupiling function is always associated with any optical system. This will limit the minimum spot size of the focussed beam. The effect of this pupiling function, known as *vignetting*, is minimized if the object is placed at the plane of the lens rather than at the front focal plane. This will introduce an additional phase factor which multiplies the Fourier transform of the input distribution at the back focal plane.

## 4.5 GAUSSIAN BEAM OPTICS

Uniform plane waves must be of infinite spatial extent to satisfy Maxwell's equations, and hence do not exist in nature. Although they are well approximated by the far field of a point source radiator, the near-field considerations, such as the degree to which a lens may focus an optical beam, often requires a more exact analytical description of the optical beam profile. A detailed analysis of the output light profile of most lasers shows that this profile is generally Gaussian in nature, and has intensity

$$I = I_0\ (z)\ e^{-(2r^2/w^2)} \tag{4.38}$$

where $I_0$ is the intensity at the center of the beam ($r = 0$) and $w$ is the $e^{-2}$ beam radius. Note that nearly 100% of the power lies within a radius $r = 2w$. This of course does not imply that an aperture may be used to filter out the small intensity tails, resulting in a more uniform field. Indeed, the result of such filtering would be, as seen in the last section, the convolution of the aperture function with the incident (Gaussian) illumination, resulting in a distinctly non-Gaussian far-field intensity function. The Gaussian function is in fact the only function whose Fourier transform is proportional to itself (i.e., also a Gaussian function). As a consequence, the field or intensity distribution anywhere in the (linear) optical system remains Gaussian if the effects of vignetting are neglected.

## 4.6 HOLOGRAPHY AND MATCHED FILTERING

Synthetic aperture radar and optical beam forming for phased-array antennas are examples of systems that make use of the concepts of coherent wavefront reconstruction or holography. Holography is the storage of both the amplitude and phase information of light reflected from an object.

*Holography*

The means of generating a hologram is illustrated in Figure 4.5(a). A light source (usually a plane wave) illuminates the object whose hologram is desired. In order to record phase information, a portion of this wave illumination must be used as a reference beam. This reference as well as the light reflected from the object are allowed to interfere on or within a storage medium. This material then bears the necessary information in the form of a transmittance function which can be processed to form the image of the object.

To view the image, one simply illuminates the hologram with a wave identical to the reference wave used to record the hologram and incident at the same angle. This is

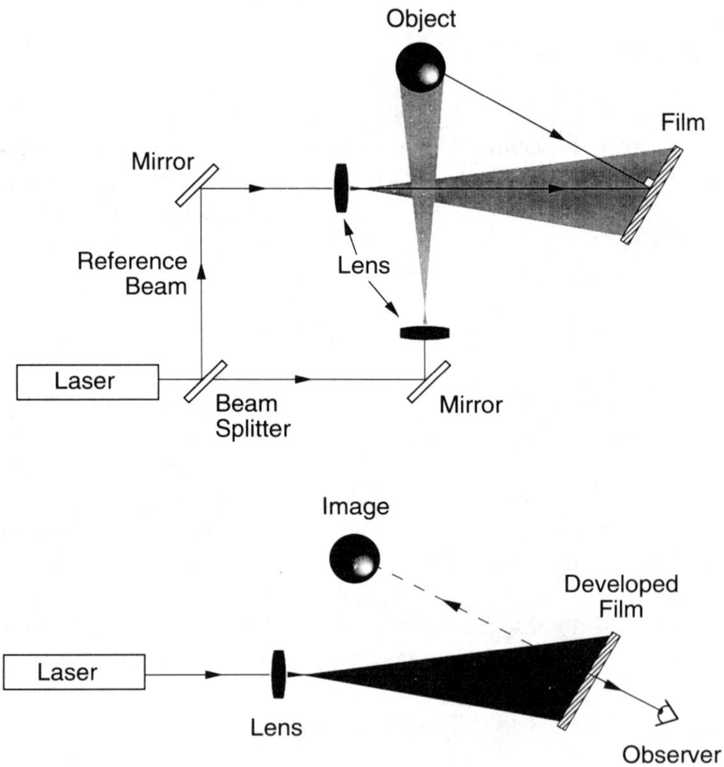


**Figure 4.5** (a) Hologram generation and recording; (b) hologram reproduction.

illustrated in Figure 4.5(b). Ideally, the viewer will see a 3-D image of the illuminated portion of the object.

The dependence on the angle of incidence for the recording and viewing of a hologram forms the basis for volume holographic storage. Different images are stored by using a different angle of incident illumination for each image.

*Matched Filtering [6]*

A matched filter is well known to both the radar and optics communities as a system that maximizes the signal-to-noise ratio. By definition, given an input signal $f(x,y)$, a filter is said to be *matched* to this signal if its impulse response is given by

$$h(x,y) = h_m^*(-x,-y) \tag{4.39}$$

The output $g(x,y)$ of a system with impulse response $h(x,y)$ and whose input is $f(x,y)$ is, by convolution,

$$g(x,y) = \int_{-\infty}^{\infty}\int_{-\infty}^{\infty} h(x-u, y-v) f(u,v)\,du\,dv \tag{4.40}$$

Substituting (4.39) into (4.40) gives

$$g(x,y) = \int_{-\infty}^{\infty}\int_{-\infty}^{\infty} h_m^*(u-x, v-y) f(u,v)\,du\,dv \tag{4.41}$$

We recognize (4.41) as the cross-correlation between $f$ and $h_m$. It is known that this function attains its maximum value when $f = h_m$.

An optical implementation of the cross-correlation operation is readily achieved and is illustrated in Figure 4.6(a). First, we recognize from (4.39) that the Fourier transform of $h(x,y)$ is

$$\mathcal{F}[h_m^*(-x,-y)] = H_m^*(\omega_x,\omega_y) \tag{4.42}$$

The input function $f(x,y)$ is somehow realized as a 2-D transmittance function and is illuminated by a plane wave as shown.† The resulting distribution is Fourier-transformed by lens $l_1$ to give $F(\omega_x,\omega_y)$, which is incident on the desired matched filter $H_m^*(\omega_x,\omega_y)$. The light emanating from the matched filter is the product $F(\omega_x,\omega_y)H_m^*(\omega_x,\omega_y)$. Lens $l_2$ (inverse) Fourier-transforms the product, with the result displayed on a screen (perhaps a detector

†One simple way of realizing a (fixed) transmittance function is to draw it on a a glass slide or transparency.

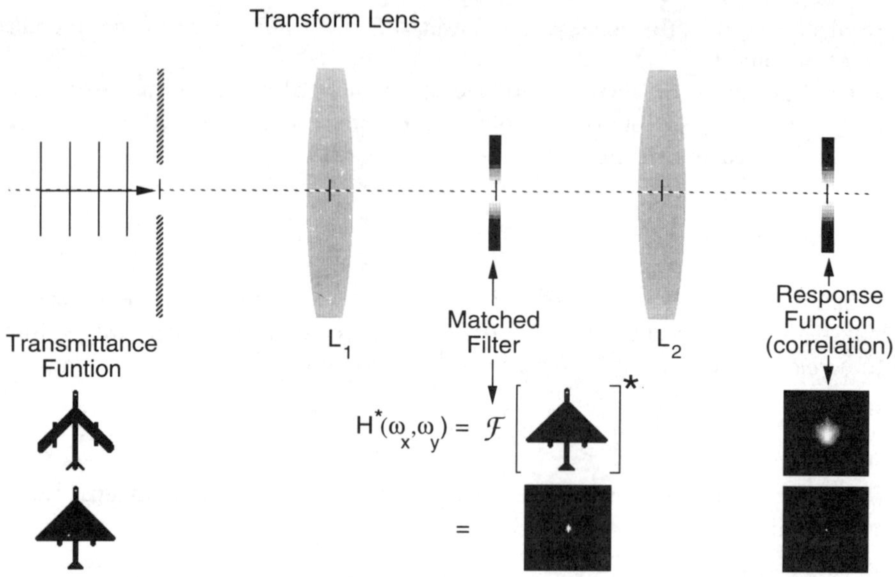


**Figure 4.6** Matched filtering operation.

array) a focal distance away. When $f(x,y)$ equals $h_m^*(-x,-y)$, the matched filter exactly "undoes" the distortion introduced by the first transmittance function $f(x,y)$, resulting once again in a plane wave incident on lens 2. Lens 2 will then focus this plane wave to a bright spot on the screen. Anything other than $f(x,y) = h_m^*(-x,-y)$ will not be "undone" by the matched filter and will yield something other than a plane wave incident on lens 2. The inverse transform will then disperse the energy to something other than a single bright spot on the screen. The energy contained in this spot is a measure of the probability of a match between input and output. A concrete example of the matched filtering operation is shown in Figure 4.6(b). This simple yet effective scheme for pattern recognition could be used for image analysis. Current research interests are concerned with being able to dynamically alter the matched filter using various classes of spatial light modulators (SLM) [7]. (An SLM can be thought of as an "dynamically addressable" transparency. It is discussed further in Chapter 6.)

## 4.7 PASSIVE COMPONENTS [8]

Many optical systems and subsystems use some rather standard optical components. Although these are quite familiar to the optical engineer, a brief summary of some of these components and their applications is presented here for the benefit of those new to optical processing concepts. Components such as attenuators, mirrors, prisms, filters (e.g., colored glass), polarizers, wedges, and gratings have obvious effects and are not discussed.

Many optical components are coated to minimize surface reflections, thus maximizing the light transmitted into the optical system. Such a coating is referred to as an *antireflection* (AR) coating. Microwave matching structures often use a quarter-wave section of transmission line to match the impedances at a dielectric boundary. The same concepts apply at optical frequencies. Since a section of dielectric is a quarter-wave thick at only one frequency, multiple dielectric coatings are used to match over a band of optical frequencies.

*Beam Splitters.* The need to split (or combine) two separate beams occurs frequently in optical systems, and a beam splitter is a device used to accomplish this. Additionally, a polarization beam splitter, constructed from a pair of dielectric-coated prisms cemented together to form a cube, can be used to decompose polarized light into two orthogonal beams, as shown in Figure 4.7. Extinction ratios of better than 1,000:1 are readily achieved over an optical band as broad as a few hundred nanometers.

*Wave Plates.* Wave plates are used to alter the polarization of light. They operate on the principle of birefringence, where the index of refraction of the material varies spatially over the cross section of the crystal so that different polarizations of the electric field experience different refractive indices. Retardation plates or wave plates exploit this effect to modify the polarization of light in a desired way. Specifically, a properly cut crystal will exhibit a minimum refractive index (slow axis) for one polarization and a maximum (fast axis) 90 deg away. This in turn implies that the phase velocities will differ for each polarization. Since for a refractive index $n$ the phase constant is simply $n\omega/c$, the retardation (in radians) for a length $\ell$ of crystal is

$$\Gamma = \omega(\ell/c)(n_{\text{fast}} - n_{\text{slow}}) \tag{4.43}$$

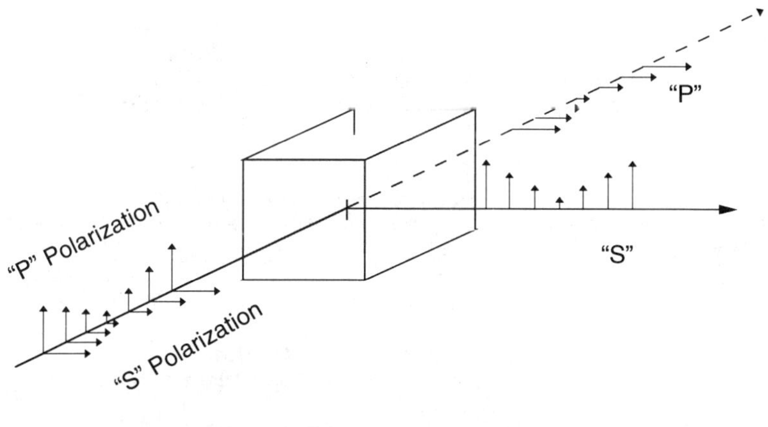


**Figure 4.7** Polarizing beam splitter cube operation.

where $n_{fast}$ and $n_{slow}$ are the minimum and maximum refractive indices, respectively. Normally, the retardation is expressed in wavelengths. The two most commonly found wave plates are half-wave and quarter-wave, for which $\Gamma = \pi$ and $\Gamma = \pi/2$, respectively. Half-wave plates are used to rotate the plane of polarization to any value. Quarter-wave plates are usually used to convert linear polarization to circular and vice versa.

Note that at the design wavelength, a retardation of $\Gamma$ is identical to $\Gamma + 2m\pi$ for $m = 0, 1, 2, \ldots$, because of the periodic nature of wave propagation. This avoids the necessity of cleaving very thin crystals, and is called a *multiple-order wave plate*. For this case, however, the sensitivity of retardation $\Gamma$ with respect to wavelength is greater than for the case of $m = 0$, called a *zero-order wave plate*. An equivalent zero-order wave plate can be realized by cascading two multiple-order wave plates whose retardations differ by half a wave. In this way, a true half- or quarter-wave plate can be constructed.

*Isolators.* An optical isolator allows light to propagate through it only in one direction and results in severe attenuation for reverse propagation. Commonly, isolators are used to prevent laser instabilities due to reflections back into a laser cavity. Such isolation is usually essential for semiconductor lasers. A simple yet effective isolator can be made from a polarizing beam splitter cube and quarter-wave plate, as previously discussed. Light in one direction is polarized by the cube and then circularly polarized by the quarter-wave plate. For the other direction, however, the back-reflections will be changed by the quarter-wave plate from circular to linear polarization and reflected away from the source by the beam splitter cube, as shown in Figure 4.8. Backward attenuation of roughly 30 dB can be achieved at the design wavelength, but only over a relatively narrow optical band and only against the particular state of return polarization generated by the signal laser. Broadband devices using Faraday rotation, on the other hand, are effective against arbitrary polarization states of back-propagating light. Broadband devices use Faraday

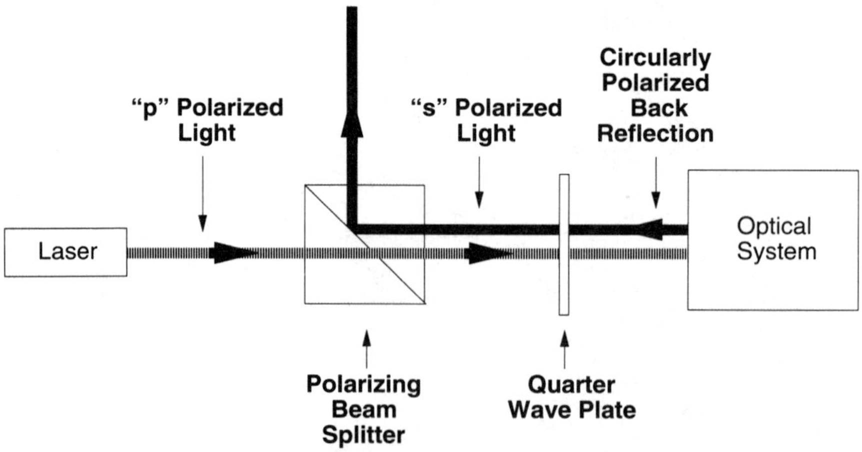


**Figure 4.8** An optical isolator.

rotation effects in much the same way as their microwave frequency counterparts. For example, an yttrium-iron-garnet (YIG) or terbium-gallium-garnet (TGG) crystal, when in the presence of a fixed (dc) magnetic field, will rotate the polarization of light propagating through it in the same direction regardless of the direction of the light. Hence, with polarizers fore and aft, effective isolation is achieved. Isolators may also be cascaded to enhance the degree of isolation.

For fiber-based applications, fiber-coupled (known as *pigtailed*) isolators exist. Since ordinary optical fiber does not preserve polarization over any appreciable length, the fiber-based isolators must be able to handle both polarizations. One way to accomplish this is by using a polarizing beam splitter to separate the light into two polarizations and then using two Faraday rotation–type isolators, as described above, one for each polarization. The output of each isolator is then recombined with a polarizing cube and coupled into an output fiber.

### *Diffractive Optical Elements [9,10]*

Recent advances in very-large-scale integration (VLSI) integrated circuit techniques have made possible the efficient manufacture of a class of optical component known as diffractive optics or binary optics. Utilizing the ideas of an optical transfer function, any arbitrary spatial domain response function, generally computer-generated, can be specified. A diffractive optical surface which will produce this response is then approximated on an optical substrate in a discrete or quantized manner. In one approach a microlithographic mask pattern is transferred to a photoresist coating and then transferred to the optical substrate by ion milling or plasma etching. This process can be repeated to produce a multilevel phase relief structure that approximates the idealized diffractive surface. A single etching produces a two-level phase relief structure, hence the term *binary optic*. Two primary categories of diffractive optics using specialized periodic structures have evolved and have experienced rapid development. The first is analogous to a 2-D diffraction grating based on a periodic pixelated structure. The grid pattern possesses translational symmetry with respect to the transverse axes. Proper computer-optimized design of the pixel pattern phase depth and shape permits great flexibility in the resulting energy distribution of the diffracted orders. The familiar blazed diffraction grating could be viewed as a 1-D version of such a structure, having energy concentration in only one selected order. Improvements in the technology continue to improve performance by reducing the minimum feature size beyond the present limit on the order of tenths of a micron. Figure 4.9 shows an example of how a binary optic device may be used to efficiently split a single laser beam into 25 equal intensity beams. The same device could be used to efficiently *and coherently* sum 25 input beams. The pattern shown in the figure is lithographically etched onto a glass slide. A laser beam (plane wave) incident on this pattern produces the far-field pattern shown. The particular device shown is being used to deliver the modulated RF information to a 25-element phased-array radar antenna system [11].

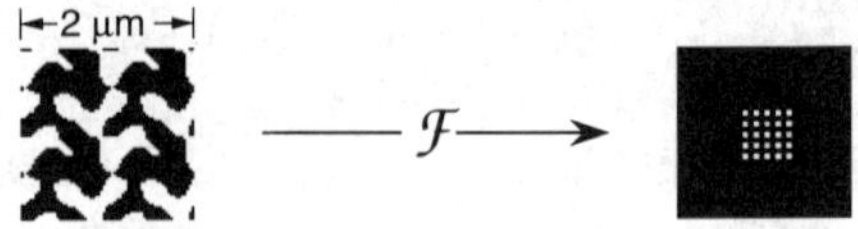


**Figure 4.9** A diffractive optical element.

The second major category involves circular pattern structures having radial symmetry about the optic axis. Many of these are referred to as *zone plates* and possess focal properties resembling those of a lens. Furthermore, the focusing performance is free of all aberrations at the design wavelength. The revolutionary impact results from the ability to fabricate such components (or arrays of them) onto almost any substrate material. Micron dimensions are not a problem, and further integration in micro-optic systems enables many system designs that are simply not possible with conventional optics. The usual planar fabrication permits further weight reduction, as well as cost-effective molding techniques where appropriate.

For many applications where the illumination is monochromatic, efficient (better than 90%) devices can be made with a single binary layer. This requires the relief structure detail to be smaller than or at least comparable to an optical wavelength, with typically a half wave in depth. High diffraction efficiency requires a multilevel relief structure. With every step of the etching process, doubling the number of phase levels, highly efficient (>99%) devices are possible.

## 4.8 OPTICAL HETERODYNE PROCESS [11,12]

Optical heterodyning is a method of achieving RF/microwave phase shift using optical processing and has extensive application in the coherent photonic signal processing systems considered in later chapters. The heterodyne process is the coherent detection of the interference between two wavefronts of differing temporal radian frequencies $\omega_0$ and $\omega_1$. Figure 4.10(a) illustrates two such (plane) wavefronts impinging on a large-area photodetector. Specifically, these plane waves may be mathematically described by

$$a_0(x,z,t) = A_0 e^{j(\omega_0 t + 2\pi\alpha_0 x + \beta_0 z + \phi_0)} \tag{4.44}$$

and

$$a_1(x,z,t) = A_1 e^{j(\omega_1 t + 2\pi\alpha_1 x + \beta_1 z + \phi_1)} \tag{4.45}$$

where $A_0$ and $A_1$ are, respectively, the amplitudes of waves of frequency $\omega_0 = \beta_0 c$ and $\omega_1 = \beta_1 c$ with arbitrary phase shifts $f_0$ and $f_1$. In most radar applications, the difference

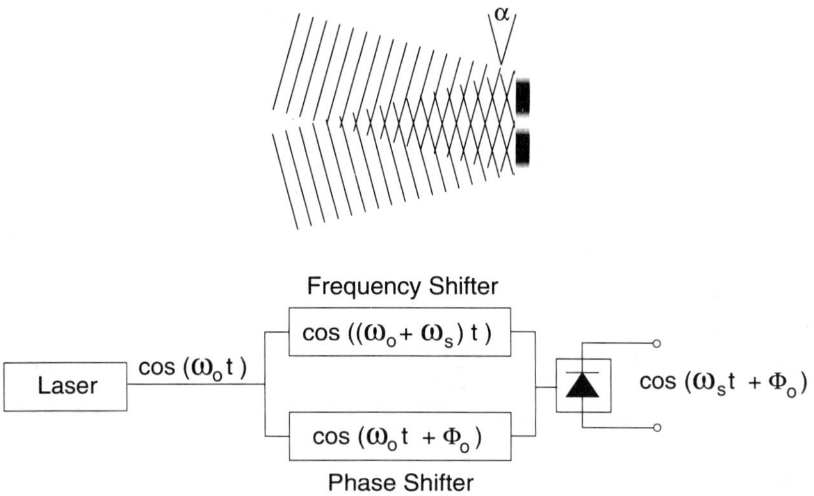


**Figure 4.10** (a) Heterodyne detection; (b) electrical equivalent circuit.

$\Delta\omega = |\omega_0 - \omega_1|$ is a microwave radian frequency. Note also that each beam strikes the photodetector (located in the $z = 0$ plane) at angles $\theta_0 = \alpha_0\lambda$ and $\theta_1 = \alpha_1\lambda$ measured normal to the detector. First, let us see that these waves will produce a detectable beat signal only when they are of the same angle (spatial frequency.) The photodetector integrates the intensity over all space of the summed signal and reference plane waves producing the electrical signal $i(t)$ proportional to the time-averaged intensity

$$i(t) = \frac{1}{2}\int_{-\infty}^{\infty} |A_0 e^{j(\omega_0 t + 2\pi\alpha_0 x + \beta_0 z + \phi_0)} + A_1 e^{j[(\omega_1 t + 2\pi\alpha_1 x + \beta_1 z + \phi_1)]}|^2 \mathrm{d}x \tag{4.46}$$

Using an identity from Fourier analysis, the impulse or delta function $\delta(\cdot)$ can be expressed as

$$\delta(u) = \int_{-\infty}^{\infty} e^{j2\pi ux}\mathrm{d}x \tag{4.47}$$

The integral of (4.46) is readily evaluated. Neglecting the dc and higher order terms, we find that

$$i(t) = A_0A_1 \cdot \cos[\Delta\omega t + \phi_0 - \phi_1] \cdot \delta(\alpha_0 - \alpha - \alpha_1) \tag{4.48}$$

Clearly, a nonzero output will occur only for $\alpha_0 = \alpha_1$. Of course, in a nonideal system, plane waves are not realizable, and therefore diffraction-limiting effects give the system finite resolution.

As seen in the equivalent circuit of Figure 4.10(b), the resulting electrical signal at the output of the photodetector contains the time-varying term that is proportional to $\cos[\Delta\omega t + (f_0 - f_1)]$. We see that the resulting beat frequency term $\Delta\omega$ is phase-shifted by the same value $f_0 - f_1$ as were the optical beams. Consequently, the optical heterodyne process serves as an optical-to-RF phase converter. In this way, substantial RF phase shift is readily obtained and forms the basis for several of the phased-array antenna controllers presented in later chapters.

Signal-to-noise and dynamic range are important issues in the design of microwave systems. For optically based radar systems, the fundamental concepts are identical to RF/microwave system design, and these basics can be found in [13,14]. Specific concerns such as laser/detector noise, optical-microwave modulation noise, and bandwidth considerations with regard to particular systems are dealt with in later chapters on these topics.

## REFERENCES

[1] Goodman, J. W., *Introduction to Fourier Optics*, New York: McGraw-Hill, 1968.

[2] Papoulis, A., *Systems and Transforms With Applications in Optics*, New York: McGraw-Hill, 1970.

[3] Born, M., and E. Wolf, *Principles of Optics*, New York: Macmillan, 1964.

[4] Clark, R. H., and J. Brown, *Diffraction Theory and Antennas*, New York: Halsted Press, 1980.

[5] Hecht, E., *Optics*, 2nd edition, Reading, MA: Addison-Wesley, 1987.

[6] Vander Lugt, A., *Optical Signal Processing*, New York: John Wiley, 1992.

[7] Bartlet, H., "Unconventional Correlators," in *Optical Signal Processing*, J. Horner, ed., New York: Academic Press, 1987, pp. 97–127.

[8] Newport Research Corporation, *Precision Laser and Optics Products Catalog*, 1991.

[9] Leger, J. R., M. Holz, G. J. Swanson, and W. B. Veldkamp, "Coherent Laser Beam Addition: An Application of Binary-Optics Technology," *The Lincoln Laboratory Journal*, Vol. 1, No. 2, 1988, pp. 225–246.

[10] Kathman, A., and E. Johnson, "Binary Optics: New Diffractive Elements for the Designer's Tool Kit," *Photonic Spectra*, Sept., 1992, pp. 125–132.

[11] Zmuda, H., and E. N. Toughlian, "Adaptive Microwave Signal Processing: A Photonic Solution," *Microwave J.*, Vol. 35, No. 2, Feb. 1992, pp. 58–71.

[12] Whitman, R., A. Korpel, and S. Lotsoff, "Applications of Acoustic Bragg Diffraction to Optical Processing Techniques," *Modern Optics*, New York: Polytechnic Press, 1967.

[13] Yariv, A., *Optical Electronics*, 3rd edition, New York: Holt, Rinehart and Winston, 1985.

[14] Haus, H. A., *Fields and Waves in Optoelectronics*, Englewood Cliffs, NJ: Prentice Hall, 1984.

# Chapter 5
# High-Speed Semiconductor Laser Sources and Detectors for Microwave Fiber-Optic Applications

**K. Y. Lau**
*University of California at Berkeley*

**J. Paslaski**
*Ortel Corporation*

## 5.1 LASERS

### 5.1.1 Introduction

The field of high-speed semiconductor lasers has undergone many substantial advances since the first demonstration of a laser with modulation bandwidth beyond 10 GHz [1]. State-of-the-art high-speed lasers offer bandwidths up to 27 GHz [2]. This type of capability makes possible many new applications in microwave signal transmission and processing.

Much of the present understanding of the high-speed modulation of semiconductor lasers comes from a small-signal analysis of the laser rate equations, which is basically a bookkeeping of the rate of supply, annihilation, and creation of carriers and photons inside the laser cavity, and which describes laser dynamics in a basic manner. The resulting small-signal bandwidth is given by the now standard formula [1]

$$2\pi B \cong 2\pi f_r = \sqrt{\frac{G'S_0}{\tau_p}} \tag{5.1}$$

where $f_r$ is the relaxation oscillation frequency of the laser, which is commonly taken to be the nominal bandwidth $B$, $G'$ is the differential optical gain, $S_0$ is the average photon density in the laser cavity, and $\tau_p$ is the photon lifetime. Noting that the rate of loss of carriers to stimulated emission is $G'SN$, where $N$ is the carrier density and $S$ is the photon density, the term $1/G'S_0$ is actually the stimulated lifetime of the carriers, so that the modulation bandwidth is the geometric mean of the stimulated carrier lifetime ($\tau_{\text{stim}}$) and the photon lifetime

$$2\pi B = \sqrt{\frac{1}{\tau_p \tau_{\text{stim}}}} \tag{5.2}$$

This is the nominal bandwidth in the sense that in the absence of parasitics or other undesirable effects, the modulation response of the laser takes the form of a second-order low-pass network response with corner frequency at $B$, falling off rapidly at the rate of 40 dB/decade beyond $B$. Figure 5.1 shows a plot of the theoretical intrinsic small-signal modulation response of a semiconductor laser. The height of the resonance at $B$ varies among laser structures and is related to intrinsic laser parameters by the different spontaneous emission factors and gain compression [3] in various structures, and to extrinsic parameters such as parasitic elements. To maximize modulation bandwidth, (5.1) gives the following guidelines: maximize the differential gain and photon density, and minimize photon lifetime. As one approaches bandwidths beyond 10 GHz, special structures must be developed that minimize parasitics as well.

### 5.1.2 Quantum Well and Strained-Layer Quantum Well Lasers

*High-Speed Quantum Well Lasers*

Quantum well (QW) lasers are predicted to possess superior modulation bandwidth as compared to common semiconductor lasers. It turns out that there are other factors which may limit the full potential of these lasers. We shall revisit these issues later, but in the meantime consider the potential advantages offered by QW lasers due to a high differential gain. A high differential gain should lead to a high-modulation bandwidth, as the simple formula (5.1) implies. This is the basis on which predictions of high-modulation bandwidths in QW lasers are made. Sample results of these predictions are shown in Figures 5.2 [4] and 5.3 [5].

Results such as those in Figure 5.3 [5] inspire high hope for strained-layer QW lasers as high-speed optical sources. It is based again on the large $G'$ predicted for these lasers due to a reduced valence band density of states as a result of a reduced valence band effective mass [6–9]. We recognize that the maximum optical gain available from transition between two quantized states (such as the first quantized states of the electrons and holes) is finite and first-order analysis does not depend strongly on the density of states. It follows that if one considers transitions involving the first quantized states only (optical gain below 100 $cm^{-1}$) [10], a high differential gain cannot be maintained over an arbitrarily wide range of electron density, since at high electron densities the gain must necessarily saturate (as a function of electron density). The advantage of a strained-layer QW laser as compared to an unstrained QW laser is that the former can attain a higher

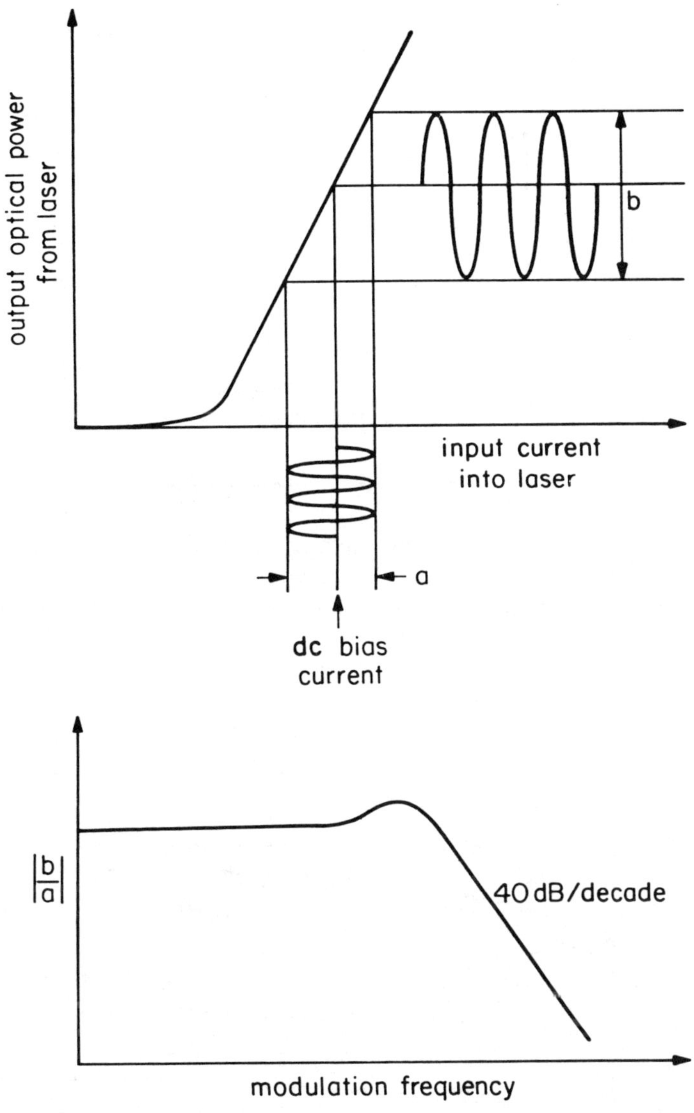


**Figure 5.1** Theoretical modulation response of laser diodes.

differential gain at a nominal range of optical gain than is usually encountered. To quantify these advantages, we use an expression for relaxation oscillation frequency given by (5.1), where the photon lifetime is related to the optical gain by

$$\frac{1}{\tau_p} = \Gamma G \equiv g_m \tag{5.3}$$

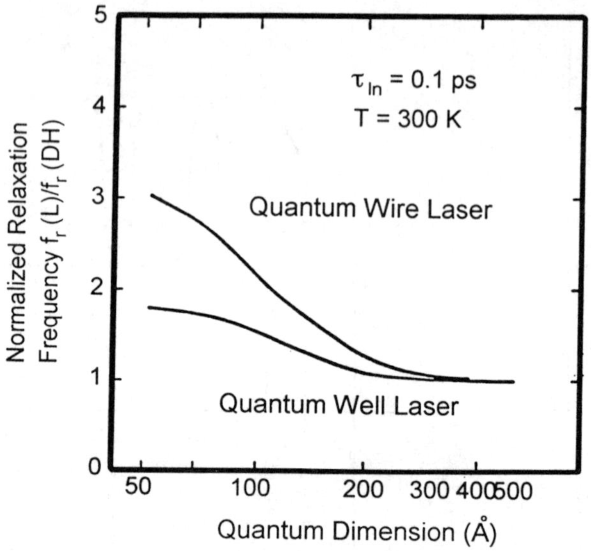


**Figure 5.2** Dependence of the relaxation oscillation frequency on well dimension for QW and quantum wire (from [4]).

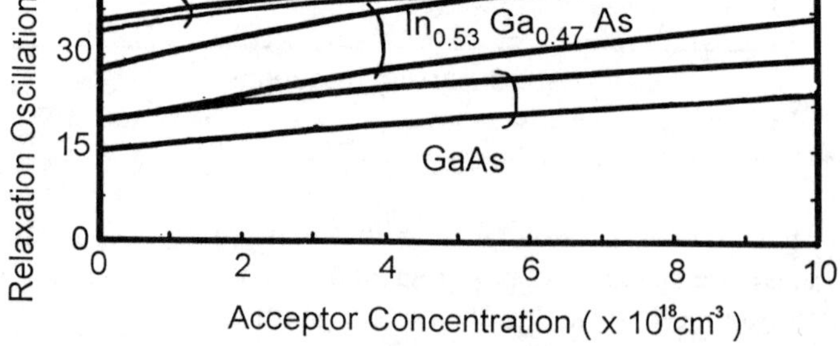


**Figure 5.3** Relaxation oscillation frequency of QW and strained QW laser diodes (from [5]).

where $\Gamma$ is the optical confinement factor. The relaxation oscillation frequency is

$$f_r = \frac{1}{2\pi}\sqrt{\Gamma G G' S_0} \tag{5.4}$$

The important quantity is thus the product $GG'$. This product is a strong function of the band structure of the QW material, which can be represented by a single parameter $D$, the ratio of the electron mass to the hole mass, at the simple level of approximation. A common unstrained QW corresponds to $D = 10$ and for a strained QW, $D = 1$. The results have been reported in [11], in which the mode gain $g_m$ is plotted versus injection current density, as shown in Figure 5.4, for various values of $D$. The current density where $g_m$ crosses zero is the transparency level and decreases with decreasing $D$. The maximum achievable gain is also slightly reduced for small $D$. Using these results, we can compute the product $gg'$ and hence the relaxation oscillation frequency $f_r$, which is plotted versus $g_m$, as shown in Figure 5.5, for different values of $D$ at the same photon density $P_0$. One striking feature of this plot is that the relaxation oscillation frequency of a regular QW laser ($D = 10$) is quite independent of $g_m$ while that of a strained layer ($D = 2$ or 1) shows a clear maximum. Thus, by fitting actual modulation data to theory, one can obtain an estimation of the value $D$, the ratio of the hole to electron density of states, and hence the amount of strain in the QW. The plot also shows that the optimum value of $D$ is around 2, at which point the relaxation oscillation frequency can be enhanced by a factor of approximately 1.8 by applying a suitable amount of strain.

The general behavior described above can be observed by using devices of various lengths, and the data points are superimposed on the theoretical plots of Figure 5.5 for both single QW lasers and strained-layer single QW lasers operated at an identical (linear) optical power density of 3 mW/μm. One observes that the measured relaxation oscillation frequency of standard single QW lasers is relatively independent of gain, while that for strained-layer QW shows an obvious peak at an optical gain of around 60 $cm^{-1}$, consistent with the theoretical predictions described above. In fact, the peak of the observed data matches the case for $D \cong 2$. The improvement is about a factor of two and can be sustained only over a narrow range of optical gain. To take advantage of this improvement, it is necessary to pay some attention to the device design. These conclusions are confirmed by recent detailed measurements of the optical gain [12] and modulation responses [13] of strained QW lasers. Other results of modulation response measurements of QW lasers can be found in [14,15].

Since quantum confinement generally leads to a higher differential gain, it is logical to expect higher order quantum-confined structures (quantum wires and dots) to exhibit even higher differential gains than QWs. Theoretical calculations showed that this is indeed the case [16,17]. Avoiding the difficulty of actually fabricating these structures, their dynamic properties have been studied through a series of experiments in which QW lasers are placed in a high-magnitude field which produces quantization of carriers moving perpendicular to the field [18,19]. The problem of fabrication is indeed a severe one,

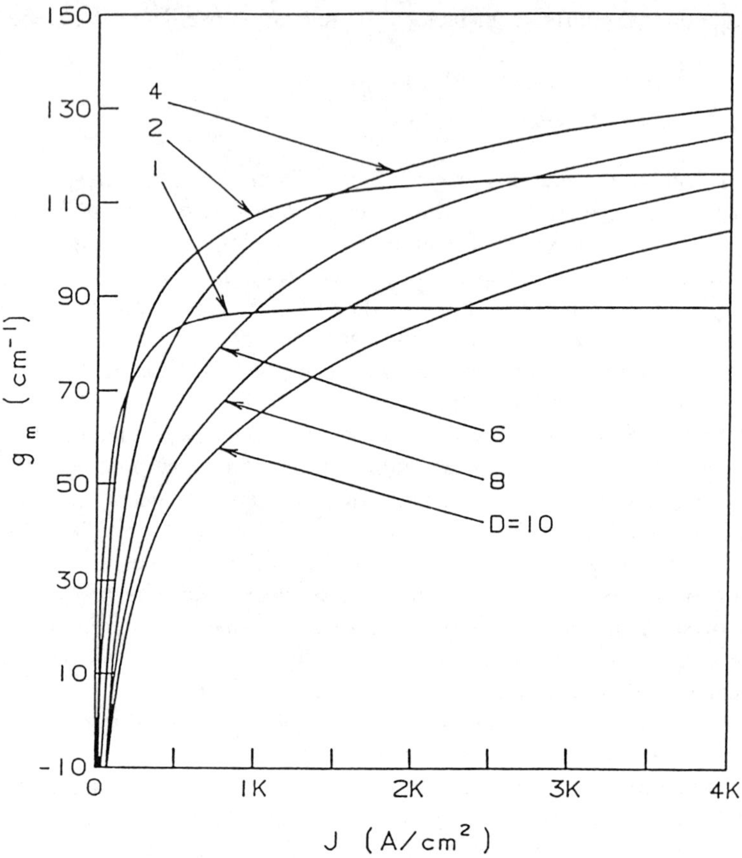


**Figure 5.4** Computed mode gain versus current density for a single QW with the electron density of states equal to that of a regular single QW, while the hole density of states is $D$ times that of the electron density of states.

since even a small degree of physical inhomogeneity can lead to the ''smearing'' of quantum-confined effects to the degree that they become irrelevant [20,21]. It may thus be premature to place a judgment on the potential of these lower dimension structures.

*Gain Compression*

One issue that needs to be considered when one attempts to achieve very wide bandwidth lasers is gain compression. This issue is best described by noting that, since the electron density never fluctuates too far from its steady-state value under modulation, the gain $g(N)$ can be expressed as a linear function of electron density as

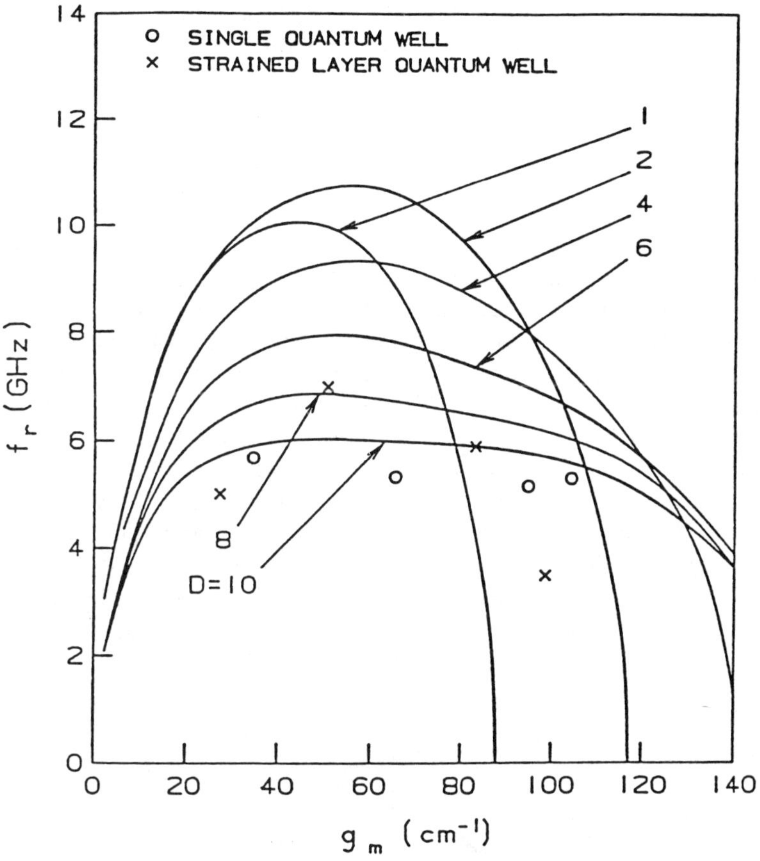


**Figure 5.5** Plot of $f_r$ versus $g_m$ for various values of $D$ for a strained QW laser.

$$G(N) = \frac{G_0 + G'_n}{1 + \epsilon S} \tag{5.5}$$

where $G_0$ is the average optical gain, $G'$ is the differential gain, $n$ is the small-signal deviation of the electron density from the steady-state value, and $\epsilon$ is the gain compression parameter, whose origin is still a matter of active investigation and will not be discussed here. The relative amount of gain compression, $\epsilon S$, is very small compared to 1, even at very high optical power, so that its effect on the dc lasing characteristics is only secondary. Its effect on the dynamic characteristic is, however, profound. To put it simply, this is due to the fact that the dynamics of the laser depend not so much on the absolute value of the gain, but on the difference between the gain and the cavity loss. This difference is no more than a few percent of the gain itself, and thus even a very small gain compression translates into a substantial effect. Quantitatively, a small-signal analysis of the rate equations gives the following modulation response function for the photon density:

$$s(\omega) = \Gamma G' S_0 (j/ed)/f(\omega) \tag{5.6}$$

$$f(\omega) = \left(\frac{i\omega + \beta R_{sp}}{(S_0 + \epsilon S_0)/\tau_p}\right)\left(\frac{i\omega + 1}{\tau_s + G' S_0}\right) + \frac{G' S_0}{\tau_p(1 + \epsilon S_0)} \tag{5.7}$$

where $s(\omega)$ is the small-signal photon density $S_0$ is the average photon density, $j$ is the small-signal modulation current. Equation (5.6) represents a conjugate pole-pair (second-order low-pass filter) type of response, and under the assumption of a small $\epsilon$, the resonance frequency is approximately given by (5.7). The locations of the poles of this response function are shown in Figure 5.6. As $\epsilon$ increases the poles move in the direction as shown. Eventually they reach the real axis and split. At this point the response makes the transition from being critically damped to overdamed, thus reducing the −3-dB modulation bandwidth [22–24]. The reduction is more severe at high optical power because of a stronger compression effect, thus negating the advantage of a higher modulation bandwidth at high optical power as (5.1) implies. The effect of gain compression on the modulation response and the −3-dB modulation bandwidth as a function of $S_0$ is shown in Figure 5.7(a,b) for different compression parameters.

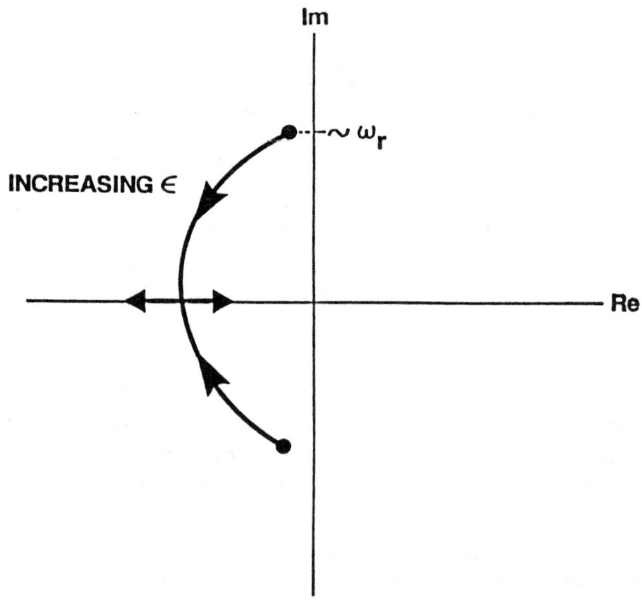


**Figure 5.6** Location of the poles of the modulation response function as gain compression is increased.

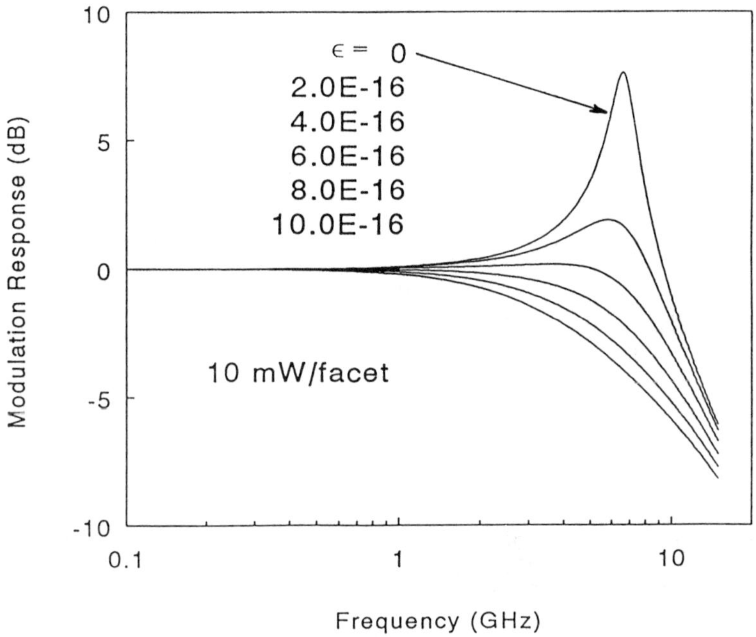


**Figure 5.7(a)** Damping in the modulation response due to gain compression.

### 5.1.3 Narrowband Modulation Beyond Bandwidth Cutoff by Mode-Coupling Enhancement

Beyond the relaxation oscillation frequency of a laser, the modulation efficiency drops rapidly at a theoretical rate of 40 dB/decade. However, the modulation efficiency can be enhanced in a narrow frequency band by coupling the laser to an external optical resonator with a round-trip delay equal to the reciprocal of the modulation frequency (Figure 5.8). This is actually the configuration of performing active mode locking at a high repetition rate [25]. The principle of operation can be described simply in terms of resonant modes introduced by the presence of the external optical cavity, which are equally spaced in the frequency domain with a spacing equal to

$$\Delta f = c/2L \tag{5.8}$$

where c is the velocity of light and $2L$ is the round-trip optical path length of the external resonator. When the laser is modulated at $\Delta f$ or integral multiples of it, coupling occurs between these resonant modes, which serves to enhance the modulation response of the laser over a small frequency band. When a large number of such modes are coupled by a large applied modulation signal, the response of the laser is highly nonlinear and is in

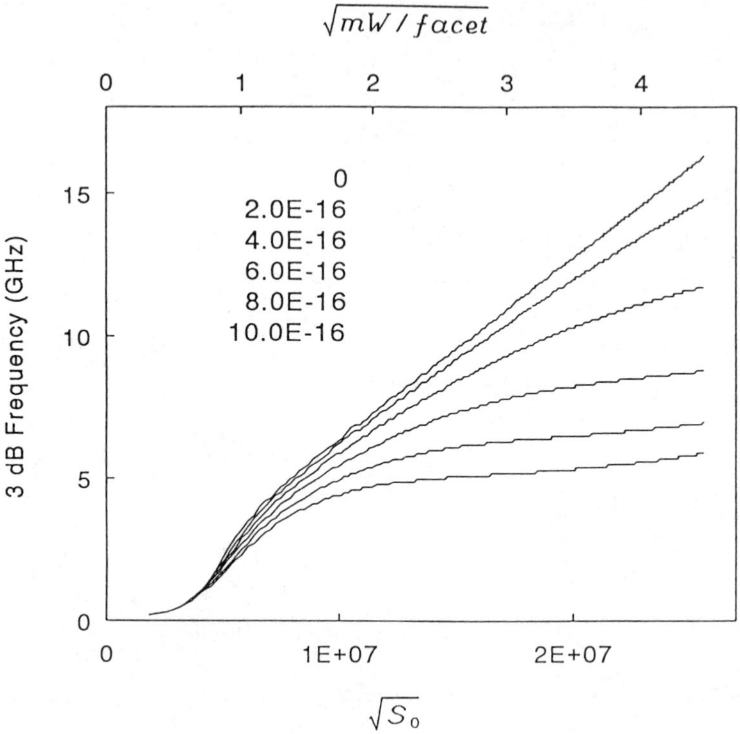


**Figure 5.7(b)** The dependence of the –3-dB bandwidth on output optical power for various values of gain compression.

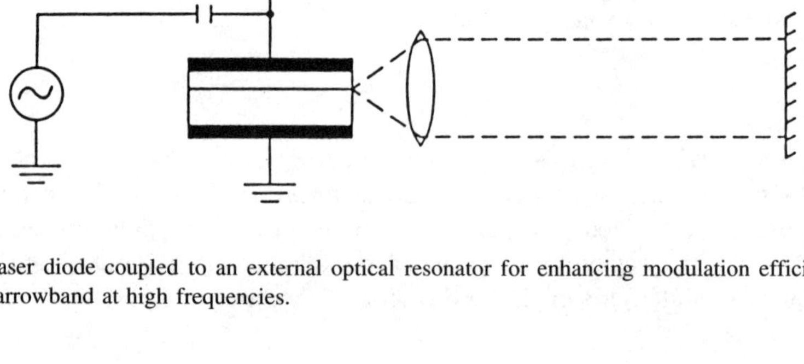

**Figure 5.8** Laser diode coupled to an external optical resonator for enhancing modulation efficiency over a narrowband at high frequencies.

the form of a periodic train of pulses of 100% optical modulation depth—an effect known as *mode locking*. When the applied modulation signal is kept small such that the laser response is kept in the linear regime, one can use this arrangement as a narrowband optical transmitter centered at a high frequency.

While the first demonstration of this narrowband enhancement technique was done at relatively low (<20 GHz) frequencies, it was recently shown that this technique can be extended to beyond 100 GHz [25]. At these millimeter-wave frequencies, the length of the ''external'' cavity is only a few hundred micrometers, so that for all practical purposes, a fully monolithically integrated laser device is used [25], which, from a practical viewpoint, is actually more desirable.

Here we describe, for the purpose of illustration, one of the first demonstrations of this narrowband modulation technique at 17 GHz [25]. A high-speed GaAs window-type buried heterostructure laser was used. The small-signal modulation response at an output power of 10 mW is shown by the heavy curve in Figure 5.9. The –3-dB bandwidth is 10.3 GHz, and the response drops to –20 dB at 18 GHz. When coupled to a length of fiber 6.3 mm long with a cleaved end on the far end and lensed end on the near end, a broad resonance is induced in the frequency response at about 16 GHz—the round trip frequency of the fiber cavity—as shown by the dashed curve in Figure 5.9. When the far end of the external fiber is coated with gold, the induced resonance becomes very sharp, as shown by the light curve in Figure 5.9. The modulation efficiency at resonance is enhanced by more than 15 dB and is at the same level as that within the modulation bandwidth. When the laser is driven on resonance by a microwave source whose power output is <6 dBm, the optical output is not fully modulated and the laser is operating in

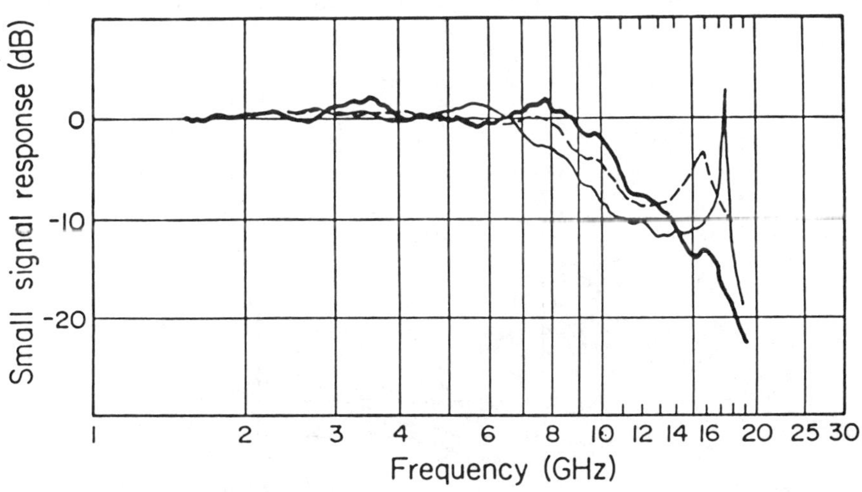


**Figure 5.9** Measured small-signal modulation response of a laser: intrinsic laser response (dark solid curve); weakly coupled to an external fiber cavity (dotted curve); with increased coupling (light solid curve).

the small-signal regime. As the microwave drive power is increased to >10 dBm, the optical modulation depth approaches unity and the optical waveform becomes pulselike. In the linear regime, the modulation response can be maintained to within 3 dB over a modulation bandwidth of 1 GHz, as shown in Figure 5.10. These techniques can be extended, in principle, to beyond 100 GHz [26], provided that the parasitic elements of the laser diode itself would not impede the application modulation signal to the laser diode.

### 5.1.4 Intermodulation

In many microwave applications, such as satellite transmission or radar systems, the signal being transmitted or processed usually occupies less than an octave bandwidth centered at a certain high frequency. High-fidelity transmission of this type of signal relies not so much on low harmonic distortions as on low third-order intermodulation (IM) products. Most active microwave components specify a third-order intercept, which is the input drive power at which the magnitude of the third-order IM products equals the signal. Of course, in most devices, including laser diodes, the third-order intercept level occurs (well) beyond the input rating as determined by other considerations. The third-order intercept, based on the assumption that the third-order product varies as the third power of the input, is a specification used primarily to determine the dynamic range in any specific application. The third-power dependence of third-order intermodulation has been verified in laser diodes [27].

Measurements of third-order IM products on high-speed laser diodes show that the IM product is strongly frequency-dependent, but at each frequency it does vary as the

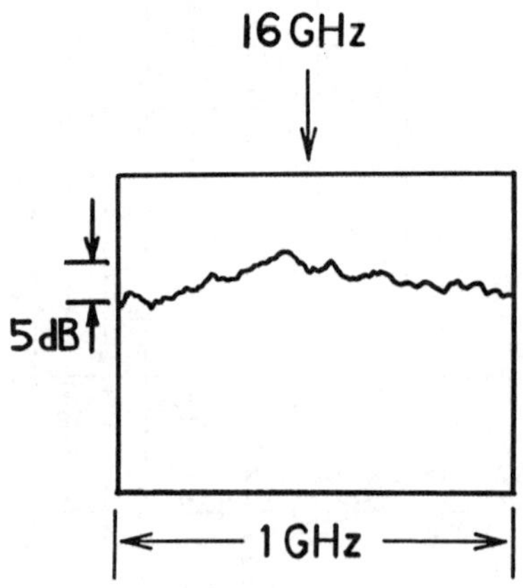


**Figure 5.10** Sweep-frequency modulation response of the resonantly enhanced laser at 16 GHz, over a 1-GHz bandwidth.

third power of the input drive. Figure 5.11 shows the measured IM product of a window GaAs laser as a function of frequency at a fixed optical modulation depth of 70% for two different bias levels. Quaternary lasers at 1.3 μm behave quite similarly. The relaxation oscillation frequency occurs at 5 GHz and 10 GHz for the two bias levels shown. One observation immediately apparent is that there is a continous rise in the third-order IM as the frequency increases, which is quite independent of the bias level *as long as* one maintains a *constant* optical modulation depth. The rise levels out at around one-half of the relaxation oscillation frequency.

To explain the IM behavior, we can use the perturbative analysis [27] of the rate equations to give the following expression for the ratio of the third-order IM product to the signal:

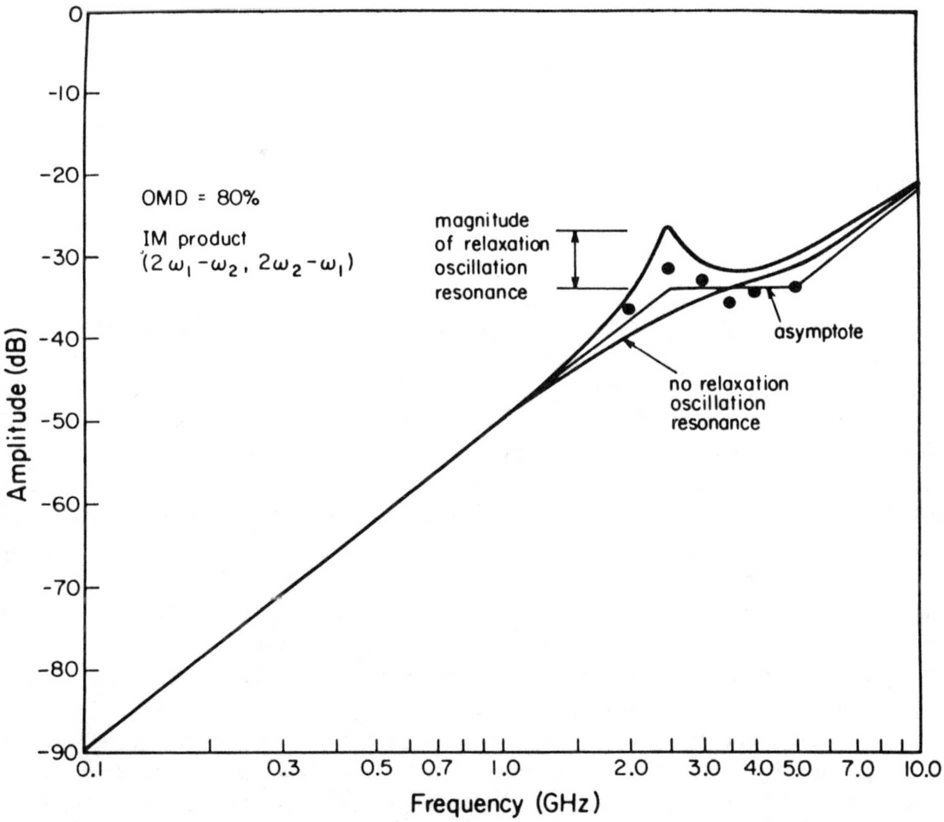


**Figure 5.11** Measured and calculated third-order intermodulation distortion product at various optical bias levels.

$$\frac{s(IM)}{s(\omega)} = -\frac{im^2}{4}\omega\tau_p\left(\frac{1+\left(\frac{\omega}{\omega_0}\right)^2}{f\left(\frac{2\omega}{\omega_0}\right)}\right) \tag{5.9}$$

where $m$ is the optical modulation depth, defined here as the half-wave amplitude of the sinusoidal modulation divided by the dc optical power, $\tau_p$ is the photon lifetime, $\omega$ is the modulation frequency, and the relaxation oscillation frequency is $\omega_0$. The function $f(x) = 1 + i\omega/\omega_0 Q + (i\omega/\omega_0)^2$ is the inverse of the intrinsic modulation response of the laser diode. Equation (5.6) is plotted in Figure 5.11 for $\omega_0 = 5$ GHz, 10 GHz, and 20 GHz. One interpretation that can be drawn from these results is that high-speed lasers operating at high optical bias are expected to exhibit a higher IM product at higher frequencies when the comparison is done at a fixed optical modulation depth. However, if a constant modulation *signal* is maintained, then increasing the optical bias level will lead to a lower IM product as well as a higher bandwidth, which is due to the $m^2$ dependence of (5.9).

### 5.1.5 Intensity Noise

*High-Frequency Intensity Noise*

The dynamic range of a microwave fiber-optic system is determined at the high end by the intermodulation distortions discussed above, and at the low end by laser intensity noise. Many microwave systems have quite stringent dynamic range requirements that far exceed those of digital transmission. Laser noise usually dominates receiver noise in high-frequency systems because the transmission distance is limited mainly by fiber bandwidth instead of attenuation, and, as a result, the optical power at the receiver is usually quite high. Laser noise is commonly quantified by the relative intensity noise (RIN), which is the rms value of the intensity fluctuation normalized by the average intensity [28]. The overall intrinsic laser intensity noise behavior has been quite well characterized, with the noise spectral density peaking at the relaxation oscillation frequency, and the overall noise level dropping as the optical power is increased [29–31]. This general behavior can be explained quite satisfactorily by a model using the laser field equations with Langevin source terms.

Laser diodes coupled to fiber systems show noise behaviors that deviate, sometimes quite substantially, from the general behavior described above. These excess noises are mostly induced by external elements. The existence of these excess noises limits the variety of microwave systems that can possibly benefit from fiber-optic techniques, since they cannot tolerate SNRs even marginally above the intrinsic laser noise level (RIN approximately −150 dB/Hz). The most noticeable of such effects is the occurrence of periodic noise spikes in the noise spectrum, which is caused by optical feedback from a distant (anywhere from a few centimeters to kilometers) interface such as connectors or the photodiode surface at the receiver end. The frequency spacing of the noise spikes is approximately proportional to the inverse of the round-trip time to the point of reflection. This type of noise spike is very severe and can rise 30 to 40 dB above the intrinsic noise level, particularly around the relaxation oscillation resonance. The effect starts to appear with very small amounts of feedback, on the order of $10^{-5}$. (Fig. 5.12 shows a typical example of a reflection-induced noise spike.) For this reason, particular care must be taken in assembling the fiber-optic system to avoid reflection, such as using reflection-

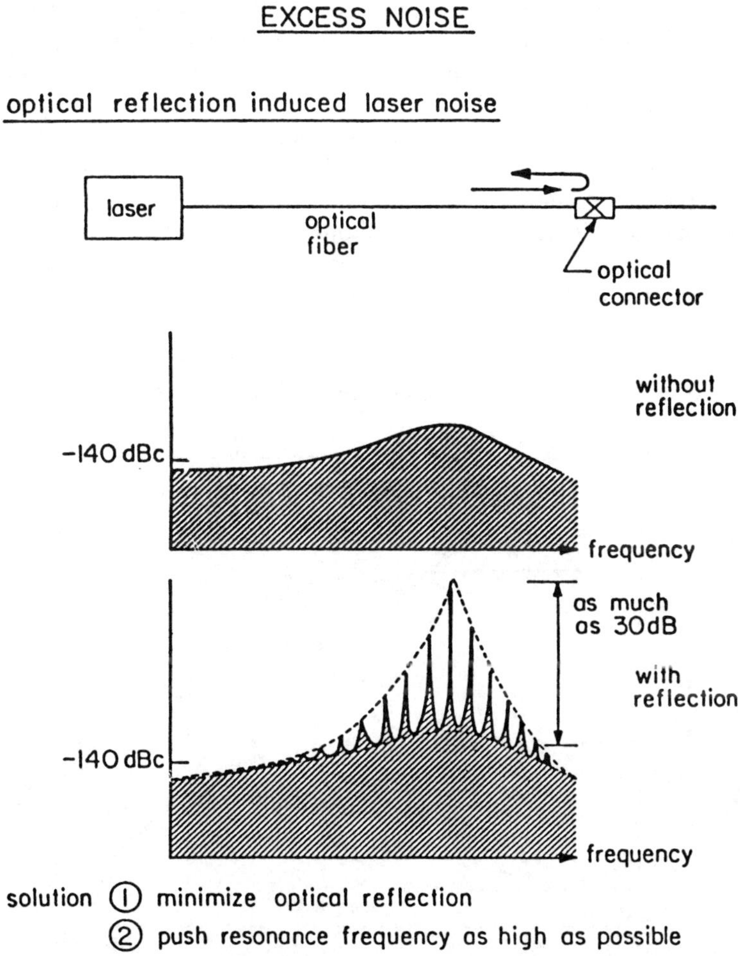


**Figure 5.12** Example of noise spike induced by reflection from an optical connector without proper index matching. The peak occurs at the relaxation oscillation frequency.

free fiber connectors or, preferably, fusion splicing, and geometrically arranging the receiver end to prevent reflection from the photodiode surface from reentering the fiber.

*Low-Frequency Mode-Partition Noise*

Mode partition noise is associated with mode competition in a multimode laser. This effect has been analyzed in detail, and the general characteristics are shown in Figure 5.13. In short, gain saturation tends to stabilize the total power fluctuation in the laser output, but there is little active mechanism to stablize the distribution of energy between the modes in a multimode laser. The result is a large power fluctuation in each mode taken alone in a multimode laser. The power fluctuation spectrum of an isolated mode in a multimode laser, as shown in Figure 5.13, has a much enhanced low-frequency part, with enhancement factors as large as 40 dB. As long as all the modes are launched with equal efficiency into an optical-fiber system, the negative correlations of fluctuations between the modes cancel out one another and the resultant noise spectrum is not too different from that of a single-mode laser emitting at the same total power. Even a slight frequency discrimination in the transmission characteristics of the fiber system (either in group velocity or amplitude) upsets the delicate balance and produces enhanced noise at low frequencies (<1 GHz).

This section summarizes the derivation of mode partition noise. The approach is a very standard one using multimode rate equations [32]. We simplify the results using a

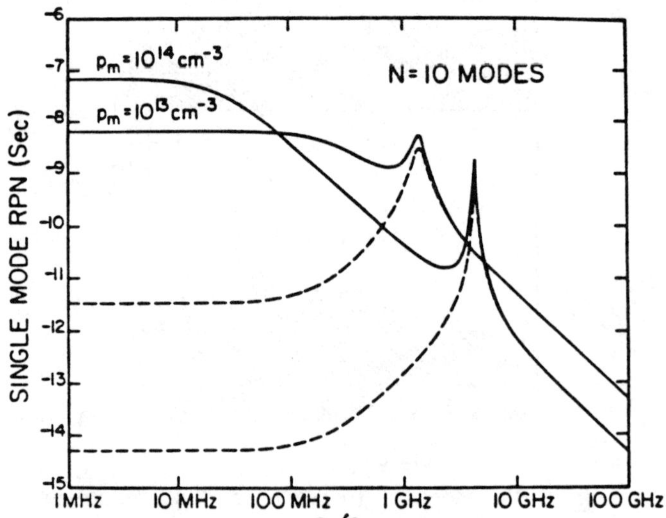


**Figure 5.13** Calculated mode-partition noise with 10 modes (from [29]).

two-mode approximation, with one mode dominant similar to that in [33]. Let $S_i$ be the photon density in the $i$th longitudinal mode. The multimode rate equations are [32]

$$\frac{dN}{dt} = \frac{J}{ed} - \frac{N}{\tau_s} - v\sum_i g_i(N)S_i + F_N(t) \tag{5.10a}$$

$$\frac{dS_i}{dt} = v\Gamma g_i(N)S_i - \frac{S_i}{\tau_p} + R_{sp} + F_{si}(t) \tag{5.10b}$$

where $N$ is the carrier density, $\Gamma$ is the optical confinement factor, $J$ is the pump current density, $d$ is the thickness of the active region, $\tau_s$ is the recombination lifetime (radiative and nonradiative) of the carriers, $\tau_p$ is the photon lifetime, $g_i(N)$ is the optical gain per unit length of the $i$th mode as a function of the carrier density, $v$ is the group velocity, $R_{sp}$ is the spontaneous emission rate into each mode, and $e$ the electronic charge. $F_N(t)$ and $F_{si}(t)$ are Langevin noise sources driving the electrons and the modes; their correlation characteristics have been derived in detail [34–36]. The form of optical gain is assumed to be

$$g_i(N) = g'_{i0}(N - N_0)(1 - v\epsilon S_i) \tag{5.11}$$

where $\epsilon$ is the gain compression parameter, $N_0$ is the transparency electron density, and $g'_{i0}$ assumes a parabolic gain profile near the gain peak. Note that we have neglected the cross-compression terms between different longitudinal modes. Their effects were studied [33] and shown to produce a low-frequency "mode-hopping noise" in situations where two or more longitudinal modes have almost equal power. The noise spectra are obtained by a small-signal solution of (5.10), using the proper Langevin correlation characteristics. In the case of a nearly single mode laser ($S_1 \gg S_2$), one can obtain the noise spectra in closed (albeit approximate) form in an approach similar to that used in [33]:

$$s_1(\omega) = F_{s1}(\omega)A(\omega) - F_{s2}(\omega)B(\omega)) \tag{5.12a}$$

$$s_2(\omega) = F_{s2}(\omega)B(\omega)) \tag{5.12b}$$

where $s_1(\omega)$ and $s_2(\omega)$ are Fourier transforms of the small-signal modal fluctuations. $A(\omega)$ and $B(\omega)$ are given by

$$A(\omega) = \frac{i\omega + 1/\tau_R}{(i\omega)^2 + i\omega\gamma_1 + \omega_r^2} \tag{5.13a}$$

$$B(\omega) = \frac{i\omega + 1/\tau_R}{(i\omega)^2 + i\omega\gamma_2 + \delta\omega_r^2} \tag{5.13b}$$

where expressions for the corner frequencies $\omega_r$, $\delta\omega_r$, and the damping constants $\gamma_1$, and $\gamma_2$ are

$$\gamma_{1,2} = \frac{R_{sp}}{S_{1,2}} + \frac{1}{\tau_R} + v\epsilon S_{1,2} \tag{5.13c}$$

$$\omega_r^2 \sim g_1' S_1/\tau_p \tag{5.13d}$$

$$\delta\omega_r^2 = \left[\frac{R_{sp}}{S_2} + v\epsilon S_2\right]\frac{1}{\tau_R} \tag{5.13e}$$

where $\tau_R$ is the effective carrier lifetime (stimulated and spontaneous), and $F_{s1}(\omega)$ and $F_{s2}(\omega)$ are Fourier transforms of the Langevin nðises driving the respective modes with the following correlation relations [36]:

$$<F_{si}(\omega)F_{sj}^*(\omega)> = 2R_{sp}S_i\delta_{ij} \tag{5.14}$$

It has been assumed, as is customarily the case, that the Langevine force driving the electron reservoir ($F_N(t)$) is negligible.

The small-signal fluctuation of the total power after propagating through a dispersive fiber is

$$s(\omega) = s_1(\omega) + ks_2(\omega)e^{i\omega dL} \tag{5.15}$$

where *d*, *k*, and *L* are the differential delay between the two modes per unit length, relative coupling coefficient of the two modes into the fiber, and fiber length, respectively. The RIN is given by

$$\text{RIN} \cdot (S_1 + S_2)^2 = <s(\omega)s^*(\omega)> \tag{5.16}$$

which can be evaluated using (5.15). The result is given by

$$\text{RIN} \cdot S^2 = 2R_{sp}(S_1\,|A(\omega)|^2 + S_2|(1 - ke^{i\omega dL})B(\omega)|^2) \tag{5.17}$$

where $S = S_1 + S_2$ is the total optical power. The significance of these results will be further discussed in Section 5.1.5.3 when we consider the effect of these noises on signal transmission.

The periodic variation in the low-frequency noise as a function of the distance between the fiber pigtail and the laser leads to speculations that the Fabry-Perot resonator formed by the laser facet and the fiber end results in selective filtering of the laser modes, thus leading to the enhanced low-frequency noise.

### *Translation of Low-Frequency Noise to High Frequency by Intermodulation*

To evaluate the effect, we assume that there are a few dominant lasing modes of roughly equal amplitudes and very nearly equal model gain. We then use the perturbative approach developed for calculating IM products [27] to calculate the beat noise. Let $S_i$ represent the photon number in the $i$th mode and $N$ be the electron density. The variables are then split into the following parts:

$$S_i = S_{0i} + S_{1i}e^{i\omega_0 t} + s_{2i}(\Omega) + S_{3i}(\omega = \omega_0 + \Omega) \tag{5.18a}$$

$$N = N_0 + n_1 e^{i\omega_0 t} + n_2(\Omega) + n_3(\omega = \omega_0 + \Omega) \tag{5.18b}$$

where the variables with subscripts 0, 1, 2, and 3 denote respectively the steady state, modulation signal, low-frequency fluctuation part, and the beat noise; $\Omega$ denotes the low-frequency noise; and $\omega_0$ is the signal frequency. The three components are shown schematically in Figure 5.14. Assuming $s_{1i} \gg s_{2i} \gg s_{3i}$, $s_{1i}(n_1)$ can be determined independently using the small-signal rate equations with a current modulation signal. The beat noise $s_{3i}(\omega)$ is then found by writing the small-signal rate equations with drive terms derived from the products $n_1 s_{2i}(s_{1i} n_2)$.

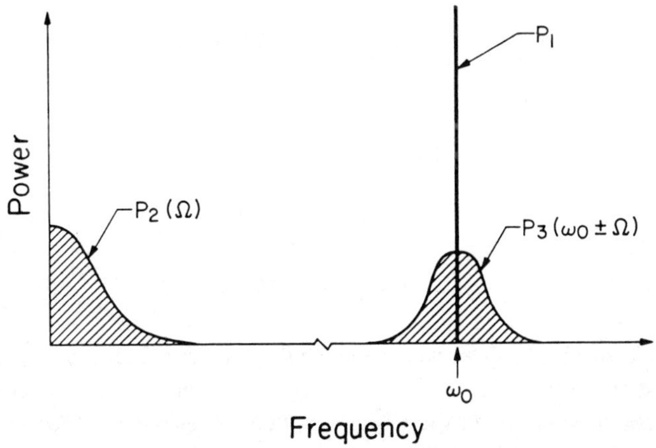


**Figure 5.14** Schematic representation of the various noise spectra used in calculating the translation factor of low-frequency noise to high frequencies by intermodulation.

$$i\omega n_3 = -\left[\sum_i g_i(N_0 s_{3i} + S_{0i} n_3) + n_3 + D(\omega)\right] \quad (5.19a)$$

$$i\omega s_{3i} = \gamma[g_i N_0 s_{3i} + g_i S_0 n_3 - s_{3i} + \beta n_3 + G(\omega)] \quad (5.19b)$$

where $\gamma$ is the ratio of carrier spontaneous to photon lifetime, $\beta$ is the spontaneous emission factor, and $g_i$ is the modal gain. The drive terms are given by

$$G(\omega) = \frac{g_i}{2}[n_2(\Omega)s_{1i}(\omega_0) + s_{2i}(\Omega)n_1(\omega_0)] \quad (5.20a)$$

$$D(\omega) = \sum_i G(\omega) \quad (5.20b)$$

At this point, two approximations are made: (1) the low-frequency fluctuation in photon number in each mode is much greater than the fluctuation in electron number due to the gain saturation stabilization effect, so that only the second term in (5.20a) is retained; and (2) the modal gain in the dominant modes differ from each other by only parts in $10^4$ and are thus all taken to be equal in this calculation. The result for the beat noise $s_3$ (normalized to the modulation signal $s_1$) is given in the following:

$$\frac{s_3(\omega = \omega_0 + \Omega)}{s_1(\omega_0)} = \frac{1}{2}\frac{\sum_i s_{2i}(\Omega)}{S_0 + \beta}\left\{\frac{g(\omega)(S_0 + \beta + h(\omega))}{f(\omega)}\right\} \quad (5.21a)$$

where

$$s_3(\omega) = \sum_i s_{3i} \quad (5.21b)$$

$$g(\omega) = i\omega + \gamma(1 - N_0) \quad (5.21c)$$

$$h(\omega) = i\omega + 1 + S_0 \quad (5.21d)$$

$$f(\omega) = h(\omega)g(\omega) + \gamma N_0(S_0 + \beta) \quad (5.21e)$$

The left-hand side of (5.21a) represents the SNR. The first factor on the right-hand side of (5.21a) is just the RIN value of the total low-frequency intensity noise. Figure 5.15 plots the factor contained in braces in (5.21a), which represents the translation factor from the low-frequency noise to high frequencies due to the presence of a modulation signal. At frequencies approaching the relaxation oscillation frequency, there is almost a one-to-one translation of the low-frequency RIN to the maximum achievable high-frequency

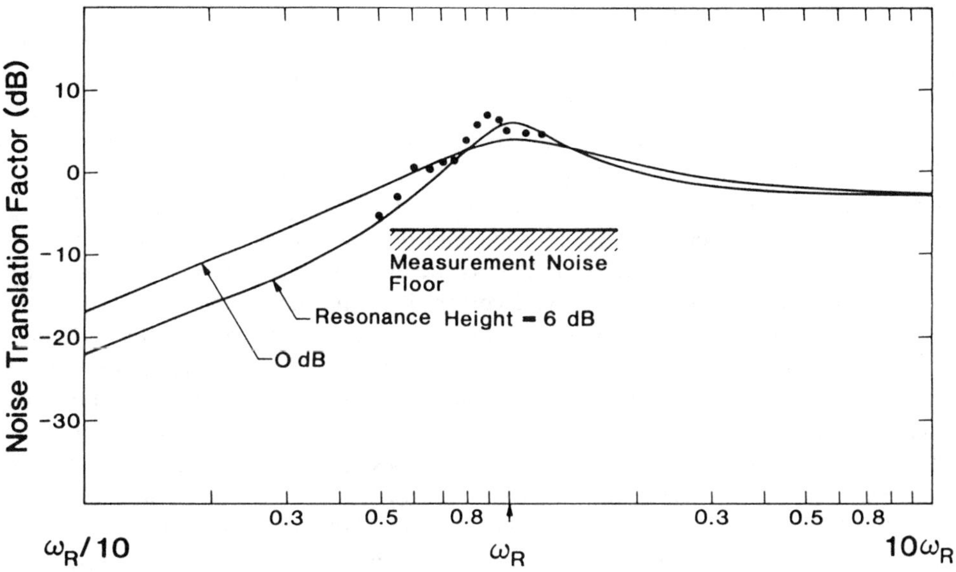


**Figure 5.15** Low-frequency noise translation factor.

SNR. The measured data points show a reasonable match to the theoretical plot of Figure 5.15.

### 5.1.6 Modulation Efficiency and Signal Throughput

*Impedance Matching*

A primary reason for using fiber optics for microwave transmission is the very low loss of optical fibers (below 1 dB/km) as compared to coaxial cables (dB/100 ft). This type of comparison does not take into account the losses encountered in the electrical-to-optical-to-electrical conversion and in launching the laser radiation into and out of the optical fiber. If we take some typical numbers for the efficiencies of state-of-the-art lasers and detectors (0.25 mW/mA for 1.3-μm laser and 0.65 mA/mW for pin detectors) and assuming 30% laser-to-fiber coupling and 95% fiber-to-detector coupling, then the end-to-end electrical signal throughput suffers as loss of 27 dB. Note that each additional kilometer of fiber adds only a fraction of a decibel to the signal loss, so that for any appreciable distance (>100m) fiber links are almost always superior to coaxial links. Nonetheless, if this throughput loss can be made to be minimal or even zero, it would eliminate the need for an active amplifier following the photodiode and make fiber optics attractive even to applications requiring short distances.

To reduce the throughput loss, one must improve on the efficiencies of all the components involved. It would probably be quite difficult to improve significantly on the intrinsic efficiencies of the laser and detector without major breakthroughs in device technology. However, the external efficiency of the laser can be improved by roughly a factor of two by applying AR coating to the back facet. The laser-fiber coupling is primarily a package problem made quite difficult by thermal and shock requirements and other military specifications. Nevertheless, one can expect coupling efficiencies of around 50% to be achieved routinely with a production line process that conforms to military specifications. These improvements will bring the throughput loss down to about 16 dB.

The next area for throughput enhancement is in impedance matching of the laser and photodiode. At microwave frequencies, one is expected to work in a 50Ω system. A forward-biased diode presents an impedance of about 5Ω, and a 45Ω chip resistor is customarily placed in series with the laser diode for broadband impedance matching. The photodiode is basically a high-impedance current source and is launched directly into a 50Ω line without any special matching. These schemes provide very simple, yet effective, broadband matching, but are not optimized for efficiency. Consider the laser-matching problem: if a lossless impedance-matching network is placed between the 50Ω source driver and the laser diode as in Figure 5.16, such that the combined matching network and the laser diode presents a perfect 50Ω to the source, all the power delivered from the source must be absorbed by the resistive part of the laser diode, in which case the current gain over that of broadband resistive matching is $(50/R^{1/2}$. For $R = 5\Omega$, a gain of 10 dB is achieved for a perfect match, in which case the throughput loss will be reduced to only 6 dB. The procedure of synthesizing the matching network can be found in many standard textbooks [37]. The actual circuit design will not be discussed here; rather,

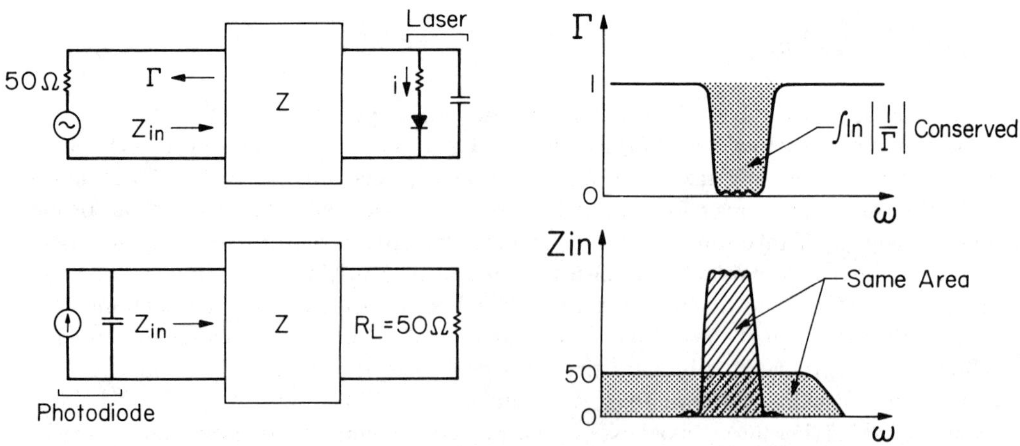


**Figure 5.16** Lossless impedance matching of laser diode and photodiodes.

we shall discuss some of the fundamental limitations as far as matching bandwidth is concerned.

The type of lossless matching described above must necessarily be narrowband, from a network theorem first derived by Bode [38] for a simple resistance-capacitance (RC) network and later expanded by Fano and others to any circuit. For an RC network, the following gain-bandwidth relationship applies:

$$\int_0^\infty \ln\left|\frac{1}{\Gamma}\right| d\omega = \frac{\pi}{\mathrm{RC}} \tag{5.22}$$

where $\Gamma$ is the reflection coefficient of the combined impedance-matching network and the load (the laser). For a perfect match, $\Gamma = 0$ and throughput efficiency will attain the theoretical gain described above. If good matching is desired for the frequency range of $f_a$ to $f_b$, the best result will be obtained when $\Gamma = 1$ (i.e., perfect mismatch) at all frequencies outside the band. In this case, assuming $\Gamma$ to be constant inside the band, (5.22) gives

$$|\Gamma| = \mathrm{e}^{-\pi f_0 \Delta\omega} \tag{5.23}$$

where $\Delta\omega = 2\pi(f_a - f_b)$ and $f_0 = 1/\mathrm{RC}$. This represents a gain in throughout efficiency (over that of wideband resistive matching) of

$$G = 10 \log \frac{50}{R}(1 - \mathrm{e}^{-\pi f_0/\Delta f}) \tag{5.24}$$

The ideal ''square'' passband described above can be approached by using a higher order network of various types, the choice of which will depend on the design complexity and ripple requirements within the passband. Figure 5.17 plots $G$ against $\mathrm{RC}\Delta f$, which is the matching bandwidth as a fraction of the RC bandwidth of the laser, according to (5.24), and assuming $R = 5\Omega$, typical of a laser diode. A significant observation from this plot is that near-perfect matching can be accomplished for $\mathrm{RC}\Delta f \rightarrow 1$, which means that matching can be accomplished without much sacrifice in the laser bandwidth as limited by the parasitic RC. It is obvious that further gain can be obtained if the laser resistance can be lowered.

Although photodiodes are the subject of a later section, it is appropriate to point out here that significant gain can be obtained by proper impedance matching of the photodiode. The photodiode can be modeled as a high-impedance current source shunted by a parallel capacitor. A simple receiver can be implemented by launching the photodiode current directly into a $50\Omega$ load. An impedance-matching network placed between the photodiode and the load is in Figure 5.16 can lead to a voltage gain $F$ with the following gain-bandwidth product limitation:

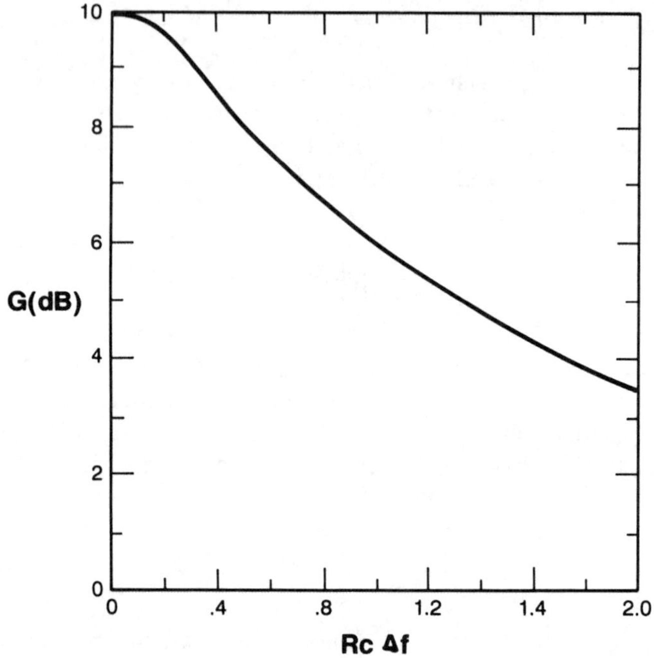


**Figure 5.17** Current gain in a narrowband matched laser diode as a function of matching bandwidth $\Delta f$ normalized to the RC bandwidth of the laser.

$$\int_0^\infty F(\omega)\mathrm{d}\omega = \frac{\pi}{2CR_L} \tag{5.25}$$

The impedance matching synthesizes a network, with the parasitic capacitance and the load resistance as part of the circuit components, which presents a high impedance to the current source over a specific frequency band. By reducing the detector bandwidth fourfold, the remaining 6 dB of throughput loss as discussed in the previous part of this section can be overcome. Further reduction in bandwidth can lead to overall throughput gain.

*Quantum Well Gain-Lever Effect*

Gain lever is an effect that ultilizes the highly sublinear nature of the QW gain characteristic to accomplish very high modulation efficiencies in intensity modulation and frequency modulation, as well as broad wavelength tunability [39, 40]. To appreciate the gain-lever effect, consider a single QW laser, which has a typical gain characteristic as shown in Figure 5.18 and which is inhomogeneously pumped in a tandem contact geometry. The

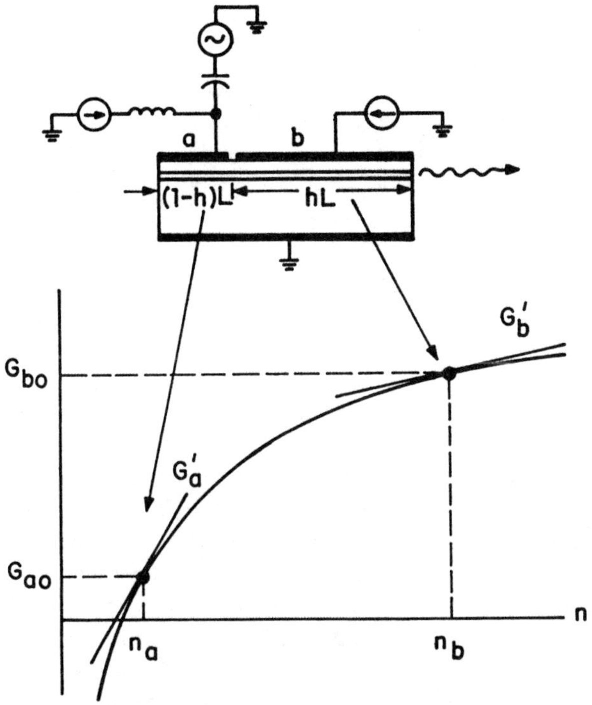


**Figure 5.18** Schematic diagram of a QW gain-levered laser.

principle of operation is as follows: when the device is above lasing threshold, the sum of the optical gain of the sections is clamped at a constant value (equal to the cavity loss). If one section increases in optical gain by, say, injection of extra electrons, then the other section must automatically reduce its gain by the same amount by ejecting the extra electrons. If each section is pumped by a constant current source, then the extra electrons cannot be ejected electrically but will do so optically through the emission of photons. According to Figure 5.18, when the sections are biased as shown, then a small modulation applied to section a will result in a large modulation in the optical output by the "levering" effect.

Quantitatively, the modulation performance of this device is described by the rate equations

$$\dot{S} = S[\Gamma G_a(1 - h) + \Gamma G_b h - (1/\tau_p)] \tag{5.26a}$$

$$\dot{N}_a = (J_a/ed) - BN_a^2 - G_a S \tag{5.26b}$$

$$\dot{N}_b = (J_b/ed) - BN_b^2 - G_b S \tag{5.26c}$$

where $S$, $N_{a,b}$, $J_{a,b}$, and $G_{a,b}$ are the photon density, carrier densities, injection currents, and optical gain in the respective sections; $\Gamma$ is the optical confinement factor; $h$ is the fractional length of the gain section; and a bimolecular recombination is used which is shown experimentally to be appropriate for QW structures [18]. A small-signal analysis yields the desired expression for the photon density modulation amplitude as shown in Figure 5.19. It can be shown that [39,40] the relative modulation efficiency at frequencies below relaxation oscillation is

$$\eta = \frac{\gamma_b}{\gamma_a} k \tag{5.27}$$

valid for $h$ approximately equal to 1. At low optical power, the spontaneous lifetime dominates and $\gamma_b \cong \gamma_a$. The enhancement in modulation efficiency is thus approximately equal to the ratio of the slopes of the differential gain. At high optical power, the stimulated lifetime dominates and $\gamma_b / \gamma_a \cong 1/k$. Because of a shorter stimulated lifetime in the modulation section, a larger modulation current is needed to produce the same amount of modulation in the electron density, thus reducing the effectiveness of the levering effect. In fact, at a sufficiently high optical power, the levering effect is completely neutralized and the modulation efficiency is no different from that of a uniformly pumped device. This implies a limited modulation bandwidth for gain-lever lasers, except for those with a very short photon lifetime (cavity length).

The gain-lever effect has been experimentally demonstrated first by optical modulation [39] and then by electrical modulation [40]. An example of the latter is shown in

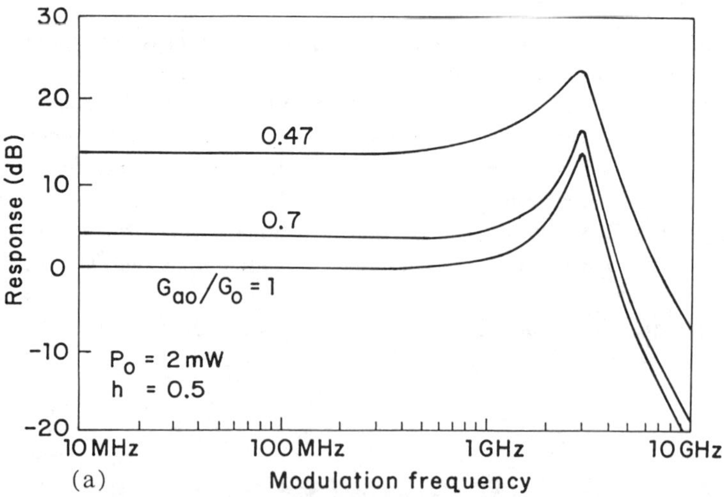


**Figure 5.19** Theoretical modulation response for a split contact QW gain-levered laser, as compared to that of a uniformly pumped laser.

Figure 5.20. The laser bandwidth is between 3 and 6 GHz at output power levels up to 4 mW for this particular device under uniform injection, as shown by curves labeled *uniform* in Figure 5.20(a) for a 400-μm laser and in 5.20(b) for a 220-μm device. The two sections are then biased separately, with the bias current into the modulation section reduced and that into the gain section increased correspondingly to keep the output power constant. The relaxation oscillation frequency remains essentially unchanged. Microwave modulation is applied to the modulation section whose gain is estimated to be 0.2 that of threshold gain. The modulation responses are plotted in Figure 5.20 as shown by curves labeled *tandem*. The improvement for lower $f_r$ and for shorter devices is evident. The largest improvement observed was about 23 dB, with the 220-μm device at $f_r \cong 3$ GHz.

### 5.1.7 Microwave Phase Noise

*Intrinsic Phase Noise*

Many applications of fiber optics in microwave systems require transmitting stable microwave phase information, an example of which is in the transmission of feed and/or return signals to and from the individual elements of a large-scale phased-array antenna. Another example that was mentioned in Section 5.1.1 involves synchronizing an antenna network. One can distinguish microwave phase noise into the following categories: (1) long-term phase drift, (2) short-term (high frequency) phase noise, and (3) phase variation due to physical disturbances.

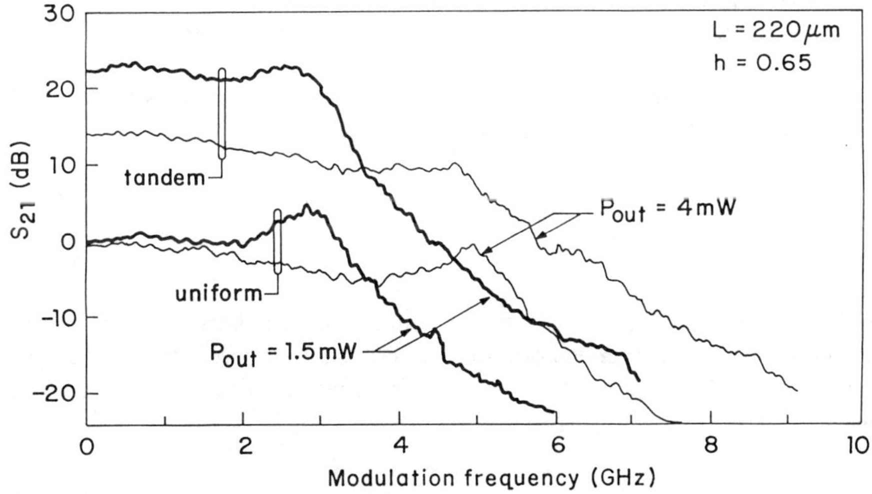


**Figure 5.20** Measured modulation response at output powers of 1.5 mW and 4 mW. Thin curves: uniformly pumped; thick curves: tandem pumped with $G_{a0}/G_0 \cong 0.2$ and a 220-μm cavity with $h = 0.65$.

Category 1 can result from a drift in the operating bias point of the laser, the wavelength of the laser (if it is not exactly at the zero dispersion point of the fiber), or a gradual change in the propagating properties of the fiber itself (for example, as a result of gradual relief of stresses induced during fiber installation). High-frequency phase noise originates almost exclusively from the laser itself. Here we are not concerned about the *optical* phase noise of the laser field, since intensity detection is used. Rather, it is the quadrature part of the additive intensity noise in the laser output that contributes to phase noise in the modulation signal.

Physical disturbance of the optical-fiber cable causes phase variation in the transmitted microwave signal through two mechanisms: change in the fiber characteristic, and change in the amount of reflection back to the laser. Physical disturbances are sometimes inadvertent, but quite often they are part of the nature of the system. Examples of this are signal-feed to the horn of a rotating antenna or systems on an airborn platform. Studies have shown that [41] phase shift caused by physical movement of the optical fiber consitutes by far the most severe handicap. It should be noted that movement of a coaxial cable also causes phase disturbance in the transmitted RF signal, and this problem is not unique in optical-fiber systems. Multimode fiber systems have been ruled out for this type of application, because it was observed that movement of multimode fibers cause as much and sometimes even bigger RF phase disturbance in the transmitted signal than coaxial cables do, for reasons not yet understood. Use of single-mode fiber systems does bring about an order of magnitude improvement in this respect, but the amount of RF phase shift observed was still much higher than predicted from the bending of the fiber alone [42]. The excess phase disturbance disappears when an optical isolator is incorporated in the laser assembly, leading to the conclusion that most of the phase disturbance is due to a varying amount of optical reflection back into the laser as the fiber is physically moved.

We will first analyze the microwave phase noise resulting from laser intensity noise. The output intensity from the laser can be written as

$$I = I_0 + I_1 e^{i\omega_1 t} + \rho(t) \tag{5.28}$$

where $I_0$ is the average intensity, $I_1$ is the amplitude of the modulated waveform, and $\rho(t)$ is the intensity noise. $\rho(t)$ is delta-correlated in time and has a spectral density distribution given by the solution to the small-signal noise equations. Separating $\rho$ into real and imaginery parts,

$$\begin{aligned} I &= I_0 + \{I_1 + i\ \mathrm{Im}[\rho(t)]e^{i\omega_1 t}e^{i\omega_1 t}\} + \mathrm{Re}[\rho(t)] \\ &= I_0 + I_1' e^{i\omega_1 t + \phi(t)} + \mathrm{Re}[\rho(t)] \end{aligned} \tag{5.29}$$

where

$$\phi(t) = \mathrm{Im}\left[\frac{\rho(t)e^{-i\omega_1 t}}{I_1}\right] \tag{5.30}$$

and the spectral density of the phase fluctuation is

$$W_\phi(\omega) = F_T\left[\frac{\mathrm{Im}(\rho)e^{-i\omega_1 t}}{I_1}\right] \tag{5.31}$$

Since $\rho(t)$ represents a real fluctuation in the time domain and is analytic, the real and imaginary parts of $\rho(t)$ are Hilbert transforms of each other [43], which leads to the following:

$$F_T[\mathrm{Im}(\rho)] = -iF_T[\mathrm{Re}(\rho)] = \frac{1}{2}F_T(\rho) \tag{5.32}$$

The spectral density of the phase fluctuation as given by (5.31) and therefore has exactly the same form for the laser intensity noise, except for a shift in the frequency scale by $\omega_1$. If $I_1 = I_0$ (i.e., 100% modulation), then the microwave phase noise density spectrum, measured in rad$^2$/Hz, is 3 dB below the RIN of the laser. The microwave frequency noise spectrum is $(\omega/2\pi)^2$ multiplied by the phase noise spectrum. Another quantity frequently used in the microwave literature is the "SSB (single side-band) phase noise to carrier ratio" [38], which is the actual power density of the ratio of phase noise to carrier, as measured on a microwave spectrum analyzer. In the case of small phase deviations ($\ll 1$ rad), the SSB ratio is $W_\phi(\omega) - 3$ dB. For a fiber-coupled laser free from severe reflection-induced noise effects, the RIN is roughly in the −140-dB/Hz level, and the SSB ratio for a signal at 100% optical modulation depth is 6 dB below the level. These numbers are quite comparable with a typical synthesized signal generator source, which has an SSB ratio of −120 dB/Hz at an offset of 1 kHz from the signal, falling to −160 dB/Hz at 100-kHz offset [44].

*Microwave Phase Fluctuation Due to Optical Reflection*

Phase fluctuation in the microwave-modulated output of a laser diode can result from the physical movement or bending of the fiber, which varies the amount of feedback into the laser diode. Since feedback can induce intensity noise, which translates into microwave phase noise as discussed above, the amount of high-frequency phase noise can vary as a result. However, in addition to this, a rather sizable phase shift, on the order of a few degrees, occurs as a result of bending the fiber to a radius of a few centimeters. This effect disappears when reflection is minimized by using an optical isolator in the laser assembly, leading to the conclusion that the phenomenon is definitely related to optical reflection into the laser. To analyze this effect, we use the small-signal laser field equations with the Langevin terms replaced by a current modulation term, $j\exp(i\omega t)$, in the carrier equation. The reflection causes undulation in both the amplitude and phase of the modulation response. Solution of the modified equation (5.15) gives the following result for the optical modulation:

$$\rho = \frac{(1/2)G'E_0 j}{[i\omega + \epsilon + \kappa(1 - e^{-i\omega\tau})][S + (1/\tau_R)] + GG'E_0^2} \tag{5.33}$$

where $\kappa$ is the reflection parameter

$$\kappa = \frac{(1 - R)R_0^{1/2}}{R^{1/2}\tau} \tag{5.34}$$

where $R$ is the facet reflectivity, $R_0$ is the reflectivity of the external reflector, $\epsilon = 2\beta/E_0^2$ is the laser linewidth, and the phase difference between the returned light and the laser field is taken to be zero. The phase of this response function is

$$\phi = \tan^{-1}\frac{\omega[\kappa(1 - \cos\omega\tau) + \epsilon + (1/\tau_R)]}{\omega_R^2 - \omega^2 - \kappa\omega\sin\omega\tau} \tag{5.35}$$

The quantity of interest here is the differential change in phase due to a fractional change in the feedback, $\Delta = \delta F/F$:

$$\frac{d\phi}{d\Delta} = \sqrt{\frac{R_0}{R}}\frac{(1 - R)}{\tau_d}\frac{d\phi}{d\kappa} \tag{5.36a}$$

$$\frac{d\phi}{d\kappa} = \frac{\omega(\omega_R^2 - \omega^2 - \kappa\omega S) + \omega^2 S[\kappa(1 - C) + (1/\tau_R) + 2\epsilon]}{(\omega_0^2 - \omega^2 - \kappa\omega S)^2 + \omega^2[\kappa(1 - C) + (1/\tau_R) + 2\epsilon]^2} \tag{5.36b}$$

where $S = \sin\omega\tau$ and $C = \cos\omega\tau$. For a distant reflector, the sinusoids oscillate rapidly as a function of modulation frequency, and $d\phi/d\Delta$ does likewise. Figure 5.21(a) plots $d\phi/d\Delta$ as a function of $\omega$ for all possible values of $\omega\tau$ from 0 to $2\pi$. The value of $R_0$ is taken to be $10^{-3}$, which is the approximate value for reflection from a fiber connector, assuming a 10% laser-to-fiber coupling. At frequencies close to the relaxation oscillation resonance, a 1% change in feedback causes ~3.5-deg change in phase, which is quite substantial. Since $d\phi/d\Delta$ scales with $\rho(R_0)^{1/2}$, a 40-dB optical isolator can reduce the phase drift by two orders of magnitude. Figure 5.21(b) shows experimentally measured results which closely match that of Figure 5.21(a).

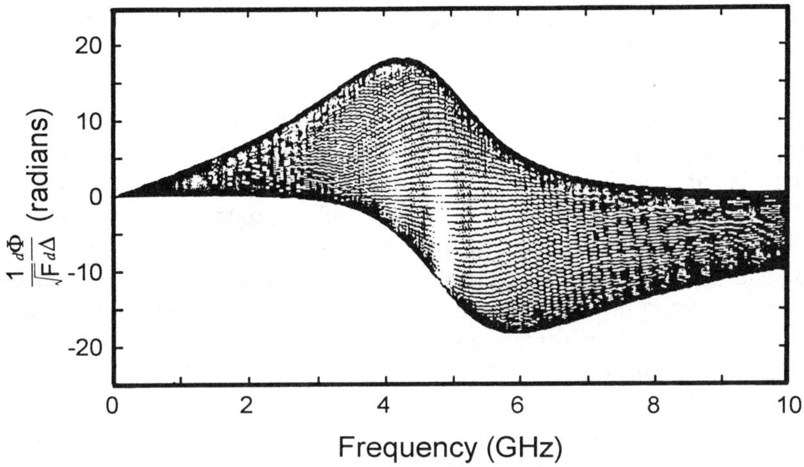


(a)

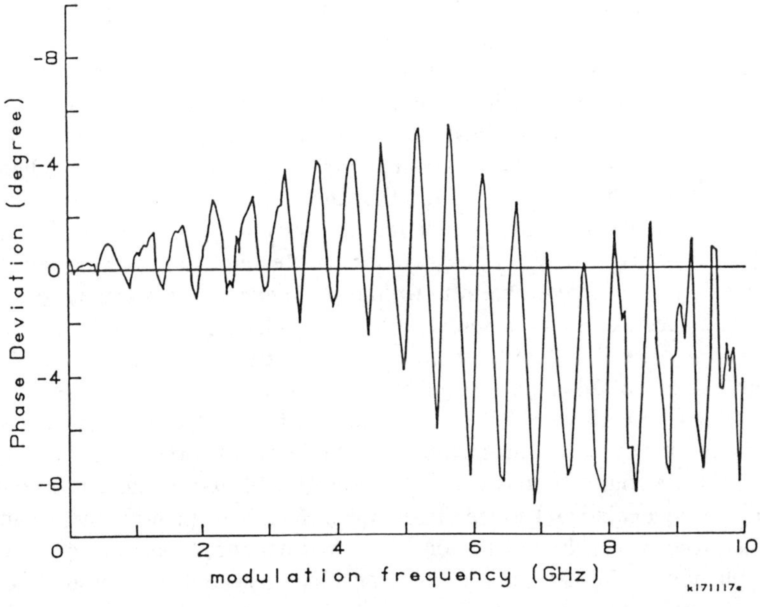


(b)

**Figure 5.21** (a) Sensitivity of the microwave phase to fractional change in the amount of optical feedback; (b) experimental observations.

### 5.1.8 SNR of Fiber Links Consisting of Fabry-Perot and Distributed Feedback Lasers

This section is concerned with a quantitative evaluation of the various sources of signal-induced noise in a high-frequency analog, single-mode fiber-optic link using directly modulated multimode (Fabry-Perot (FP)) and single-frequency (distributed feedback (DFB)) lasers. The SNR in a typical fiber-optic link is commonly evaluated by treating the various sources of noises, such as RIN, mode partition noise, and shot and thermal noises, as additive quantities independent of the modulation signal. However, there are sources of noise which become prominent only in the presence of a modulation signal. This section presents experimental and theoretical studies on this type of noise, which arises from mode-partitioning in FP lasers and interferometric phase-to-intensity noise conversion in DFP lasers, the former incited by fiber dispersion and the latter by fiber reflections such as Rayleigh scattering. Both of these effects, which are well known, increase with fiber length, and so does the signal-induced noise instigated by these effects.

Historically, the development of high-speed semiconductor lasers for microwave/analog applications took place with conventional FP lasers due to the simplicity in their construction and analysis. FP lasers lend themselves to convenient demonstrations of ideas related to speed alone. To date, the highest bandwidth semiconductor laser has been demonstrated in FP lasers [45,46], although single-frequency lasers also show exceptional performances [47]. The multimode lasing spectrum of FP lasers will be an issue in any wideband system due to fiber dispersion, even for wavelengths near (but not exactly at) the dispersion minimum. This problem may not be significant for many microwave applications, since most of these systems operate in a relatively narrow subgigahertz band, albeit centered at a high frequency of multigigahertz. A more serious concern is the manifestation of mode-partition noise due to fiber dispersion, a subject studied in considerable detail [32]. This type of noise usually does not extend to beyond a few hundred megahertz, and is not considered harmful to high-frequency microwave systems; hence, they need not be specified in such applications. A similar type of low-frequency noise, the *mode-hopping* noise [32,48,49], which can manifest itself in the laser diode output alone without the aid of fiber dispersion, was similarly considered not significant for microwave applications. However, it has been shown that in a directly modulated laser diode, the low-frequency noise can be translated to the neighborhood of the modulation carrier [50]. This is a source of signal-induced noise for FP lasers that can become quite serious at high frequencies and longer fiber links. We shall quantify these parameters later. It is therefore a mistake not to consider low-frequency noise in lasers for high-frequency applications. Furthermore, it would be very misleading to measure the system noise level at high frequencies without any applied modulation to the laser, and then to calculate the anticipated SNR based on these measurements, as if the signal and the noise are independent entities.

It should be noted that a similar translation of low-frequency noise onto the high-frequency modulation carrier also exists for systems using diode-pumped YAG lasers with external modulators. The low-frequency noise from the diode-pumped YAG laser, typically below tens of megahertz, originates from relaxation oscillation and from beating between longitudinal modes, the latter being very significant even with an excellent side-mode rejection. These low-frequency noises must be eliminated by feedback using an additional modulator.

For single-frequency lasers such as DFB lasers, neither mode-partition nor mode-hopping noises becomes much of an issue. The dominant source of low-frequency noise comes from double reflections along the fiber length, which serve to convert the laser phase noise to intensity noise [51,52]. With the use of good optical connectors, reflections along the link can be minimized except for intrinsic Rayleigh backscattering [53]. The resulting intensity noise spectrum is basically that of the Lorentzian laser line shape centered at dc, with a typical linewidth of a few tens of megahertz. On applying high-frequency modulation, this noise will be translated along with the modulation carrier. In general, this effect is much less than that of FP lasers, provided proper optical connectors and splices are employed. These parameters will be quantified again later.

To illustrate the effects discussed above, we first show measured results for links consisting of a 1.3-μm (a) FP laser and 1.3 μm (b) DFB laser, modulated at frequencies of 6.5 and 10 GHz, propagating through a distance of 1 km, 6 km, and 20 km of single-mode fibers. Both of these are high-speed lasers with a 3-dB modulation bandwidth well beyond 10 GHz, as shown in Figure 5.22(a). A single high-speed pin photoreceiver with a 12-GHz 3-dB bandwidth is used for all of the measurements. The spectra of the lasers

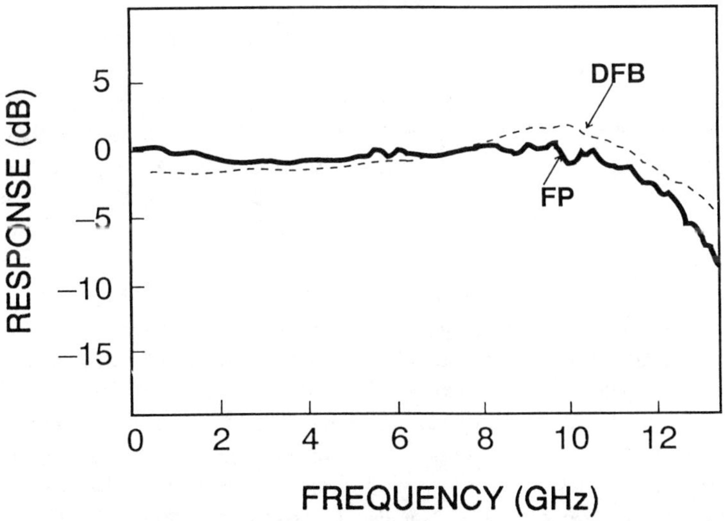


**Figure 5.22(a)** Direct modulation response of the high-speed FP and DFB lasers used in this experiment. Both lasers have an estimated relaxation oscillation frequency at slightly below 10 GHz, although the resonances of both lasers are strongly damped.

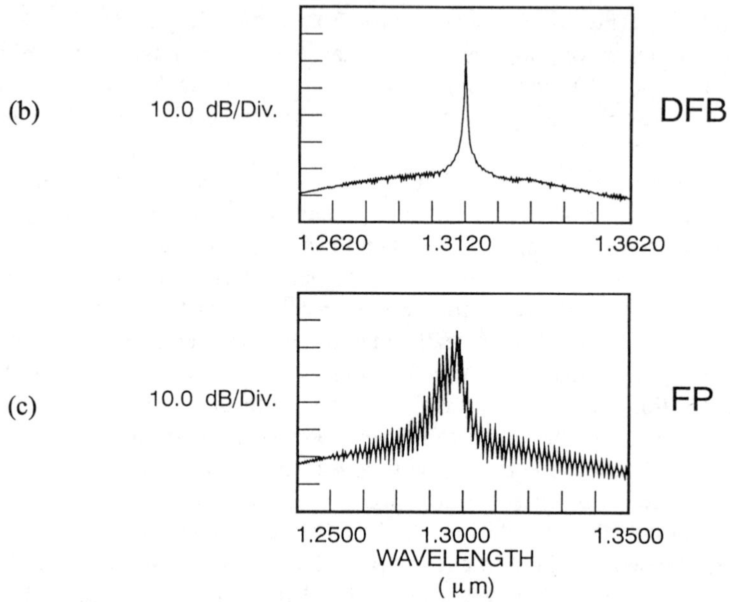


**Figure 5.22(b,c)** Lasing spectrum of the DFP and FB lasers, respectively.

are shown in Figure 5.22(b,c). The measurements are done with an input RF drive level into the lasers at +10 dBm. As a point of reference, the 1-dB compression level of the laser is +15 dBm. The optical input into the photoreceiver is adjusted to give 1 mA of dc photocurrent in all cases, except where noted in a few instances. Angled polished optical connectors (APC) are used whenever a connection is required, which amounts to between two and four pairs of connectors for the links used in this experiment.

For the FP laser, low-frequency mode-partition noise can be clearly observed on propagation through only a few kilometers of fibers, as shown in Figure 5.23(a,b). The highest of these noise levels corresponds to RIN figures of: <−145 dB/Hz (this number is receiver noise limited); with no fiber −132 dB/Hz after propagating through 6 km; and −115 dB/Hz after 20 km. In contrast, for the DFB laser the noise spectrum is at the receiver noise limit of <145 dB/Hz at 6 km (Figure 5.24(a), and remains so after 20 km (Figure 5.24(b)). Nevertheless, at 20 km one can already observe a hint of the interferometric noise protruding above the receiver noise level at low frequencies (Figure 5.24(b)). This low level of interferometric noise is due to Rayleigh scattering of a long piece of fiber and reflection from the APCs. The nature of the interferometric noise can be more easily observed when bad splices or connectors exist in the link, as illustrated in Figure 5.24(c)—a situation clearly to be avoided.

The kinds of low-frequency noises described above are quite commonly observed in typical fiber links. In the following, we illustrate how the low-frequency noises are

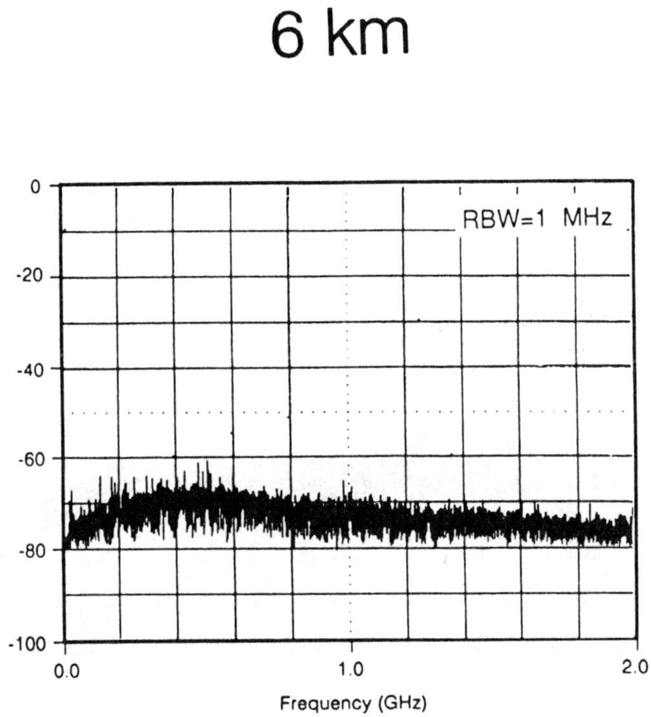


**Figure 5.23(a)** Mode partition noise in FP lasers after transmission through 6 km.

translated to the neighborhood of the modulation carrier upon applying a direct high-frequency modulation to the laser. Figure 5.25 shows the results of applying a 6.5-GHz modulation carrier to the FP (Fig. 5.25(a–c)) and the DFB lasers (Fig. 5.25(d–f)), and measured after transmission through 1 km, 6 km and 20 km. The blanks in the traces for FP lasers in Fig. 5.25(a–c) establish the RIN level with the RF drive to the laser disconnected (i.e., without noise translation). These plots illustrate that the low-frequency noise and its translation are very significant. The drop in the RF signal level at longer fiber lengths for the FP laser (Fig. 5.25(c)) is due to fiber dispersion, but not attenuation (recall that all measurements, except where noted, are done with 1 mA of photocurrent from the receiver). One should contrast the above results with those obtained with the DFB laser (Fig. 5.25(d–f)). Only at 20 km does the translated noise become slightly visible, consistent with the low-frequency noise observations of Figure 5.24.

All of the above observations hold for modulation at 10 GHz, as shown in Figure 5.26(a–f), except for the even more striking difference between the FP laser (Fig. 5(a–c)) and the DFB laser (Fig. 5.26(d–f)). It is worthwhile to note that for a DFB laser, only a very slight degradation is observed, even at 10 GHz and at 20 km, while the FP laser is

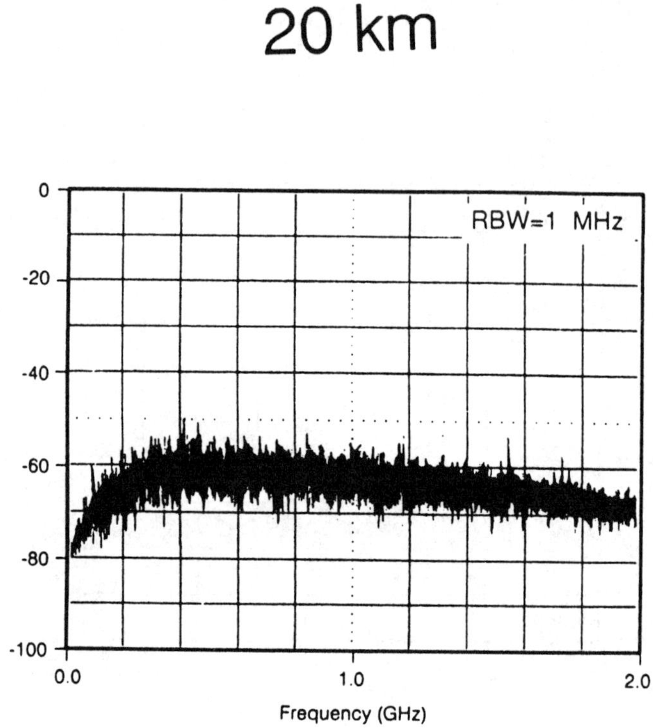


**Figure 5.23(b)** Mode partition noise in FP lasers after transmission through 20 km of single-mode fibers.

all but nonfunctional at these frequencies and distances. Note also from Figures 5.25(b) and 5.26(b) that even at a relatively short distance of 6 km, the actual SNR of the high-speed FP laser is, depending on the modulation frequency, between 10 and 20 dB worse than that predicted from a standard RIN measurement alone (without modulation applied).

### 5.1.9 Conclusion

We have examined some of the topics of high-speed characteristics of semiconductor lasers that are relevant to their applications in microwave fiber-optic systems. We have purposely omitted much of the detail in laser and device physics, including those of QW lasers and transport dynamics, since this chapter is directed primarily towards applications. To this end, many of the key formulas are simply written down without detailed derivations. The interested reader should consult the reference list for detailed descriptions of these topics. From an applications viewpoint, the issues of parasitics, intermodulation, and noise are particularly important. Furthermore, a topic quite specific to microwave fiber-optic systems concerns modulation efficiency and signal throughput, which is seldom discussed

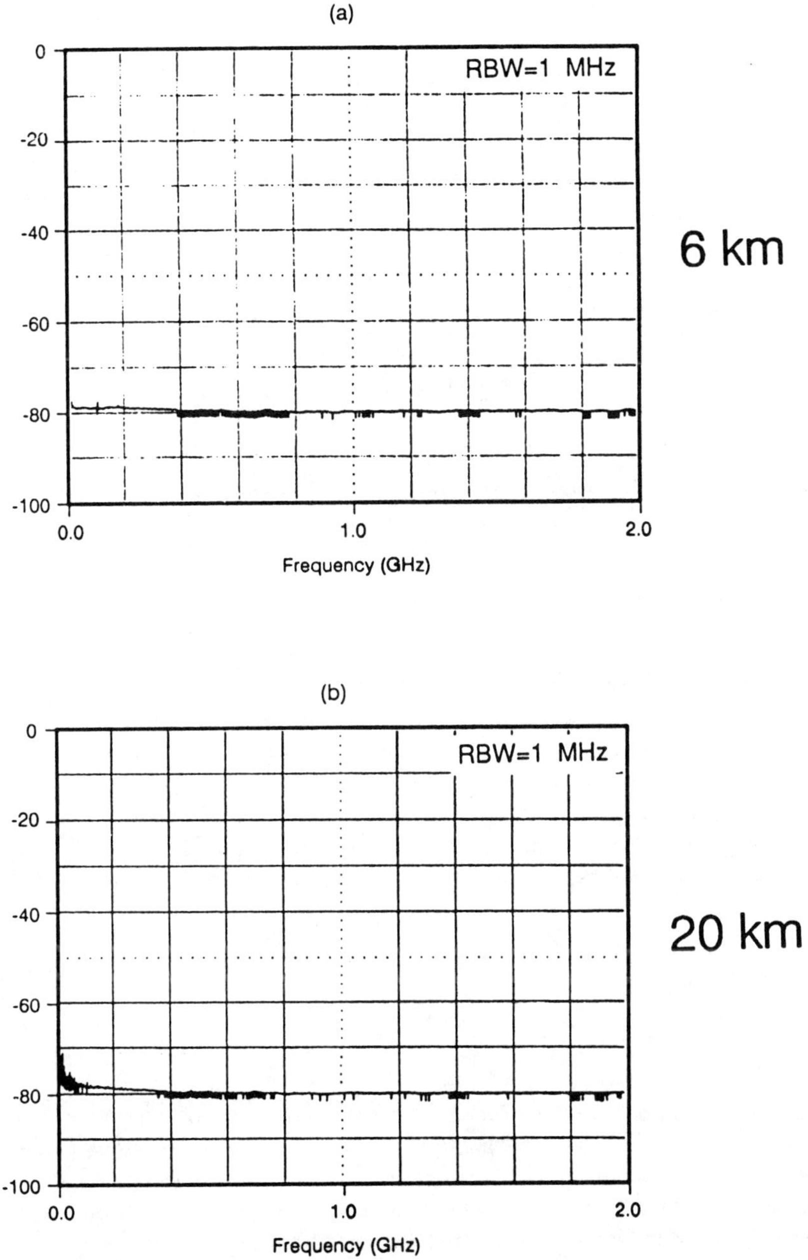


**Figure 5.24** Interferometric phase intensity–converted noise for DFB lasers due to double Rayleigh scattering for transmission through (a) 6 km and (b) 20 km.

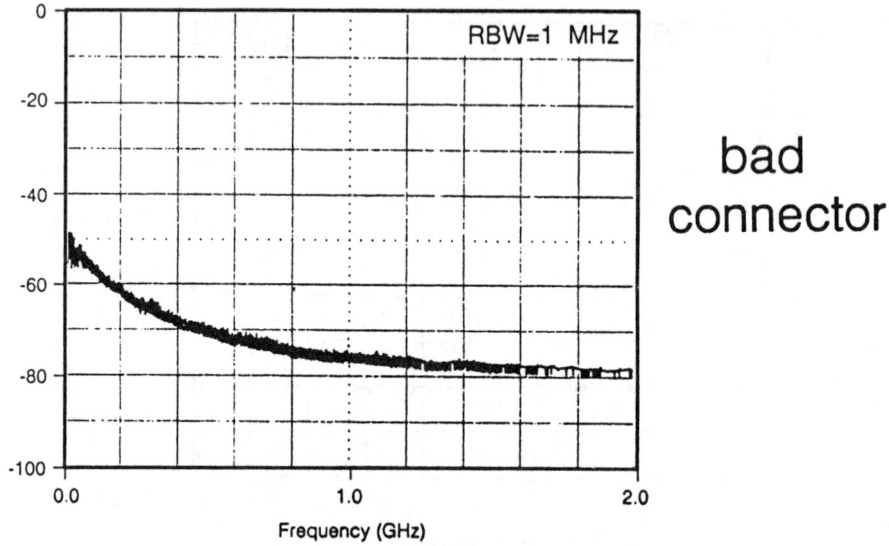


**Figure 5.24(c)** Interferometric phase intensity–converted noise for DFB lasers due to double Rayleigh scattering for transmission through result of a bad splice in the fiber link.

in the literature but which is included in this chapter. A similar situation exists in the microwave phase noise of microwave fiber-optic systems as related to fluctuations in the laser itself. Finally, although this chapter is intended to describe the laser devices, it would not be complete without a description of how these lasers perform, especially in terms of noise characteristics, when they are employed in a real-life fiber-optic microwave transmission system. A substantial portion of the latter part of Section 5.1 is devoted to this topic, which establishes the importance of using single-frequency lasers in these type of systems.

## 5.2 DETECTORS

### 5.2.1 Introduction

The three most commonly used detectors for high-speed optical detection are pin detectors, avalanche photodiodes (APD), and photoconductors. Fundamentally (and perhaps overly simplistically), they can all be described by an absorbing semiconductor region under an applied electric field which extracts the photogenerated carriers to the outside circuit. In pin detectors and APDs, the high electric field in the depletion region of a pn junction is utilized, the latter being operated under breakdown condition, whereas in a photoconductor, the field is applied through an external ohmic contact. The main difference between a pin detector (or Schottky barrier detector) and a photoconductor is that in the former,

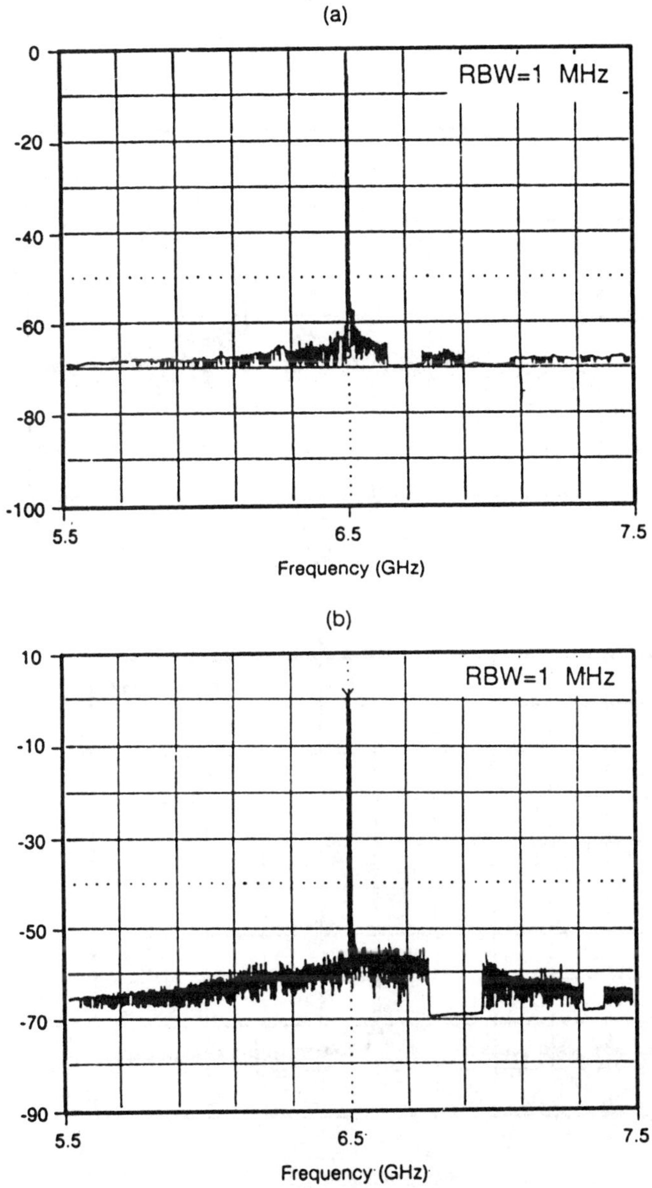


**Figure 5.25** Noise spectra under an applied modulation signal at +10 dBm at 6.5 GHz: (a–c) FP laser, (d–f) DFB laser. Three cases are shown for each laser: transmission through 1 km (a,c), 6 km (b,e), and 20 km (c,f) of single-mode fibers. The dc photocurrent is adjusted to be 1 mA in all the measurements except (c), which is at 0.43 mA. The RF output from the photodiode is amplified by a 20-dB amplifier in these measurements.

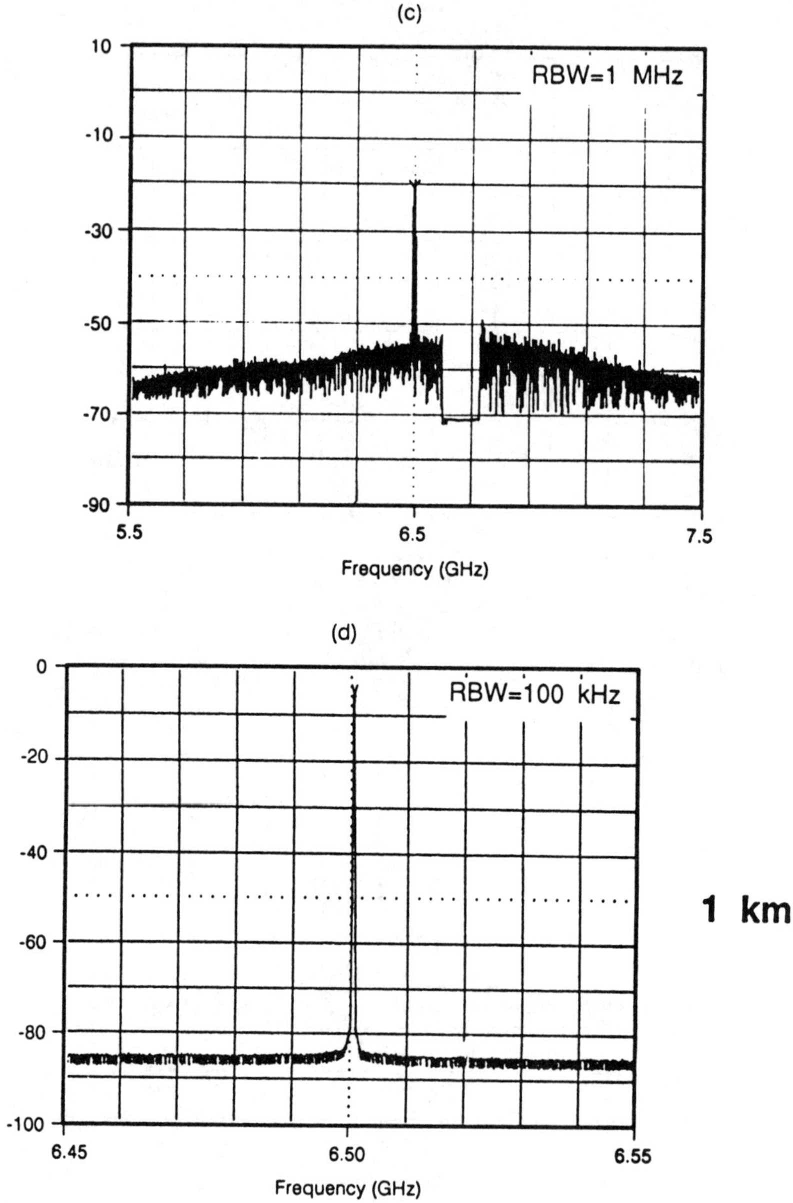


**Figure 5.25** Noise spectra under an applied modulation signal at +10 dBm at 6.5 GHz: (a–c) FP laser, (d–f) DFB laser. Three cases are shown for each laser: transmission through 1 km (a,c), 6 km (b,e), and 20 km (c,f) of single-mode fibers. The dc photocurrent is adjusted to be 1 mA in all the measurements except (c), which is at 0.43 mA. The RF output from the photodiode is amplified by a 20-dB amplifier in these measurements.

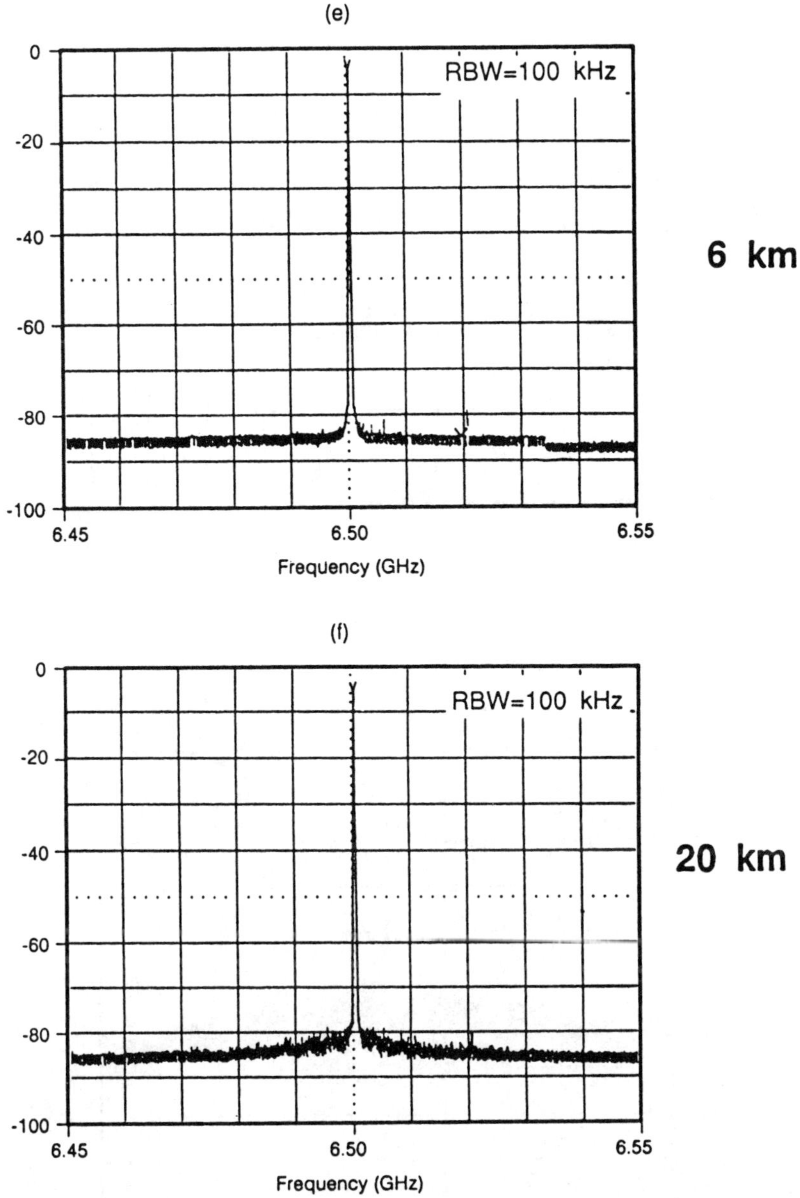


**Figure 5.25** Noise spectra under an applied modulation signal at +10 dBm at 6.5 GHz: (a–c) FP laser, (d–f) DFB laser. Three cases are shown for each laser: transmission through 1 km (a,c), 6 km (b,e), and 20 km (c,f) of single-mode fibers. The dc photocurrent is adjusted to be 1 mA in all the measurements except (c), which is at 0.43 mA. The RF output from the photodiode is amplified by a 20-dB amplifier in these measurements.

(a)

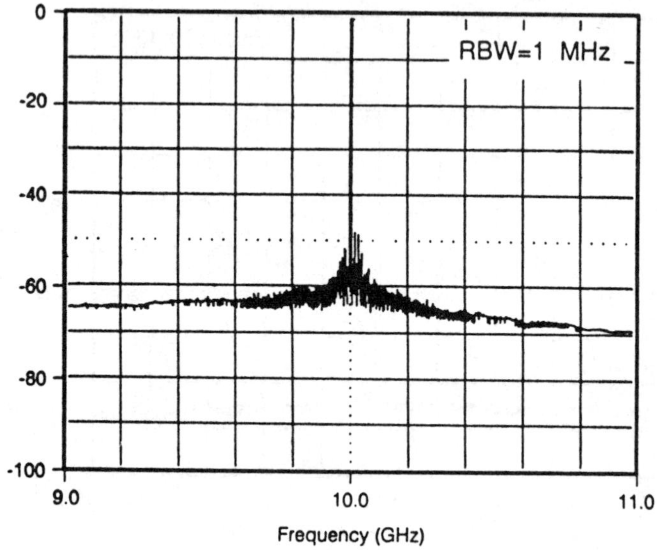


(b)

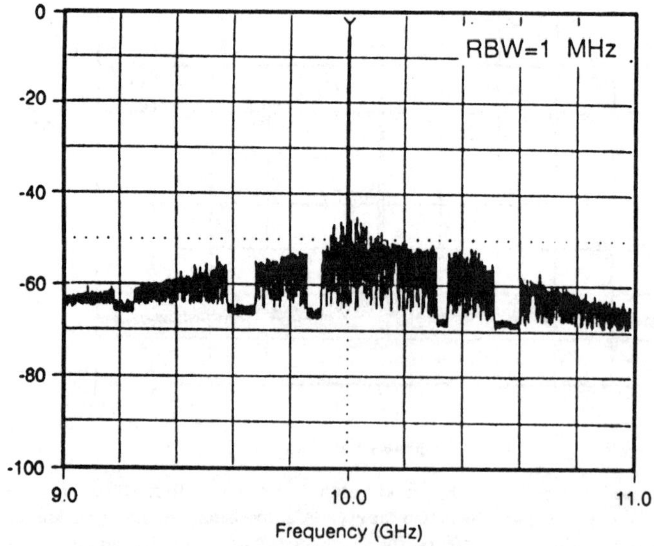


**Figure 5.26(a,b)** Similar to Figure 5.25, but at 10-GHz modulation frequency.

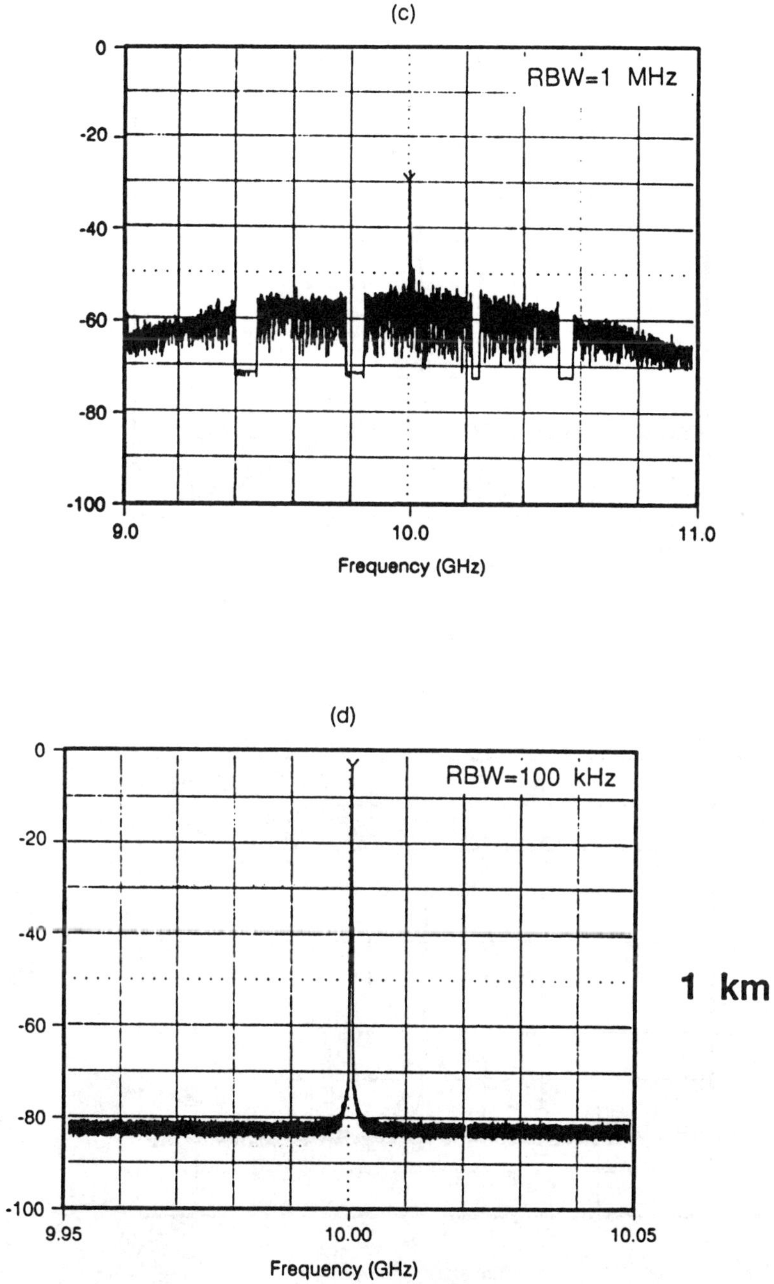


**Figure 5.26(c,d)** Similar to Figure 5.25, but at 10-GHz modulation frequency.

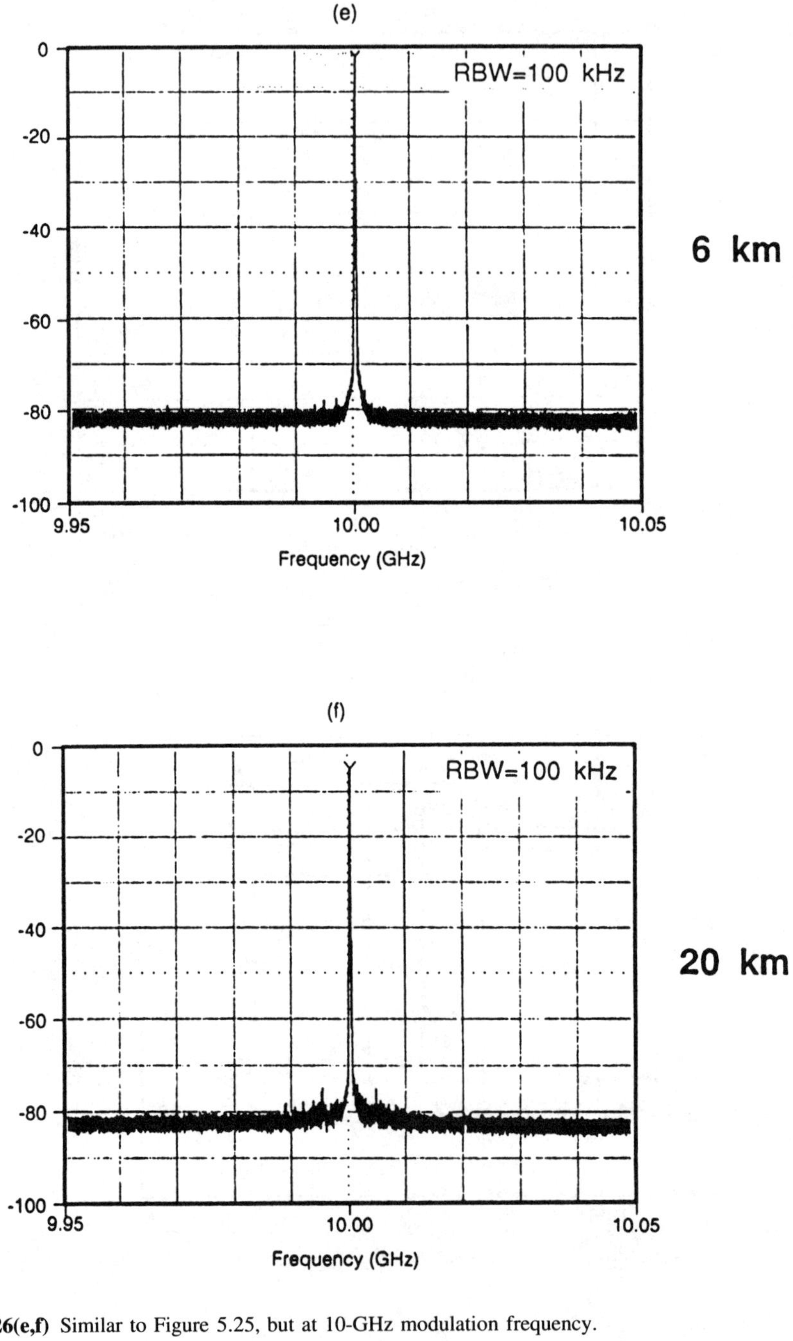


**Figure 5.26(e,f)** Similar to Figure 5.25, but at 10-GHz modulation frequency.

the contacts *collect* but do not inject carriers, whereas in the latter, the ohmic contact permits a free flow of carriers in and out of the external circuit. The consequence is that in pin or Schottky barrier detectors, once all the photogenerated carriers are collected, the photocurrent will cease, but in the photoconductor, a photocurrent keeps on flowing until the carriers dissipate by recombination. A single incident photon can therefore generate a flow of many electrons through the external circuit, and therefore exhibits an apparent "detector gain." The drawback is that the speed is limited by the recombination lifetime, which can be rather slow. APDs exhibit gain by generating a large number of carriers through impact ionization, which slows down the detection process somewhat and exhibits higher noise. Of the three types of detectors, the pin or Schottky barrier detector is the most commonly used in high-frequency microwave systems, due to their inherent speed and quantum-limited noise performance. We shall, in this section, concentrate on pin detectors. We shall also consider a different type of detector specific to microwave systems: a *down-converting* detector, which simultaneously performs the function of detecting a microwave-modulated optical beam and down-mixing the detected signal to an intermediate frequency (IF) band. The detection/mixing process is performed with a photoconductor, making use of the fact that, unlike photodiodes, the responsivity of a photoconductor varies with bias voltage.

### 5.2.2 High-Speed PIN Photodiodes

*Speed Considerations*

In conventional photodiodes, there is a well-known trade-off between speed and quantum efficiency above a certain frequency. This is shown in the plots of Figure 5.27, where a 20-GHz photodiode at 1.3 μm can have at most a quantum efficiency of 82%, and at 1.55 μm it is only 70% due to the lower absorption coefficient [54]. This is because higher speed demands thin depletion layers, which absorb less light once they are less than a few absorption lengths. The proposed solution is a waveguide structure with absorption and carrier transport in orthogonal directions so that they are not constrained to the same layer thickness. A number of devices have been in the literature primarily aimed at frequencies of 40 to 60 GHz [55–57] with less attention to high efficiency. As such, efficiencies have been only modest at 30% to 70%. This section addresses the issue of high-efficiency coupling to such waveguide photodiodes while still maintaining a reasonable frequency goal.

As mentioned above, in a properly designed pin photodiode, the frequency response is limited by two major contributions: the transit time of carriers out of the depletion region and the RC time associated with charging and discharging the junction capacitance plus any parasitic capacitance. Modeling of the frequency response starts with a current source model of the photodiode whose frequency dependence is determined by the transit time limitation. This then drives a circuit consisting mainly of the intrinsic diode capaci-

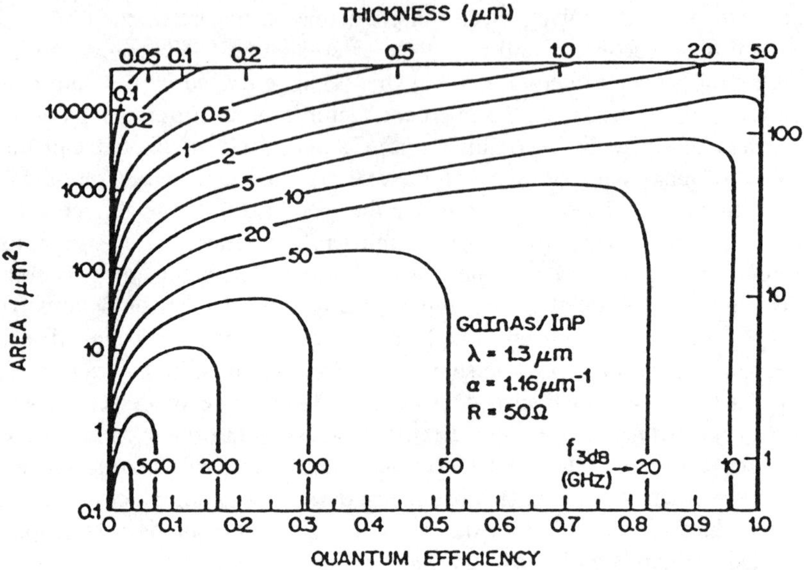


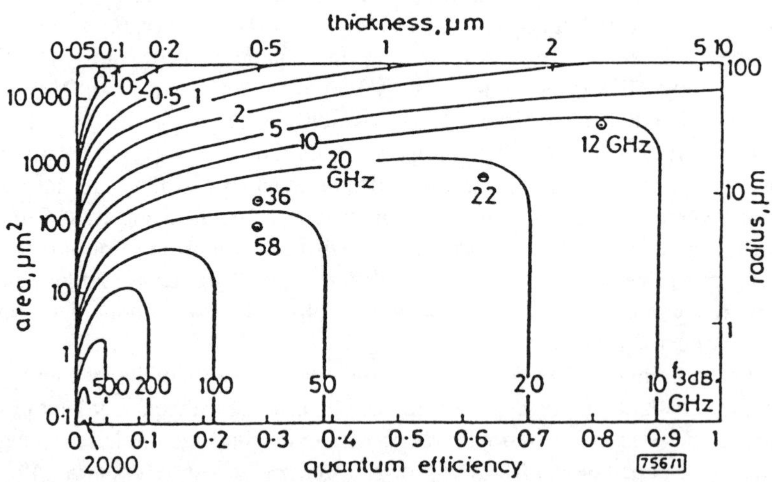


**Figure 5.27** Theoretical 3-dB bandwidth contours for conventional InGaP pin photodiodes: $\lambda$ = 1.3 μm (top); $\lambda$ = 1.5 μm (bottom) (from [58,59]).

tance and any parasitic capacitance in parallel with the load resistance. This simple model can be easily expanded to include additional parasitics such as bond wire inductance and contact resistance, as well as any external matching circuit. Denoting the current source response with $F_i(\omega)$ and the circuit response as $F_{RC}(\omega)$, the total response of the photodiode $F_{PD}(\omega)$ is

$$F_{PD}(\omega) = F_i(\omega)F_{RC}(\omega) \tag{5.37}$$

The current response term $F_i(\omega)$ is governed by the transit time of carriers out of the depletion region and is most easily derived by Fourier-transforming the time domain impulse response for removing photocarriers. Typically, pin diodes have a uniform electric field in the depletion layer and are usually biased such that the electron and hole velocities are constant at their saturation values, which we denote with $v_n$ and $v_p$, respectively. In general, the impulse response depends on the transit times

$$\tau_n = L/v_n \qquad \tau_p = L/v_p \tag{5.38}$$

where $L$ is the depletion layer thickness and the "flat-response" regime of $F_i(\omega)$ is often roughly estimated from the inverse of these times. Beyond this, one must determine the shape of the impulse response which, in the constant velocity approximation, is governed by the distribution of photogenerated carriers in the depletion layer. For illumination normal to the depletion layer, the distribution is given by the absorption coefficient $\alpha$ of light in the depletion layer, and the expression for $F_i(\omega)$ depends on the parameter $\alpha L$ as well as the direction of the illumination [64]. For the waveguide photodiode, we assume the illumination will be symmetric and uniform across the depletion layer, so the impulse response for each carrier type is a single sawtooth given by

$$i(t) = \frac{2}{\tau}(1 - t/\tau) \qquad 0 < t < \tau \tag{5.39}$$

where $\tau(= L/v)$ is the relevant transit time. The total frequency response is then half the sum of the electron and hole contributions, which are

$$\begin{aligned} F_i(\omega) = {} & \frac{1}{j\omega\tau_n}\left(1 - \exp(-j\omega\tau_n)\frac{\sin(\omega\tau_n/2)}{\omega\tau_n/2}\right) \\ & + \frac{1}{j\omega\tau_p}\left(1 - \exp(-j\omega\tau_p)\frac{\sin(\omega\tau_p/2)}{\omega\tau_p/2}\right) \end{aligned} \tag{5.40}$$

The circuit response is given by

$$F_{RC}(\omega) = \frac{1}{1 + j\omega RC} \tag{5.41}$$

where the total capacitance is a combination of diode and bond pad parts

$$C = C_{\text{diode}} + C_{\text{bond pad}} \tag{5.42}$$

and the load resistance is the line impedance or half this if there is a matching resistance at the diode:

$$R = Z_0 \tag{5.43a}$$

with no matching and

$$R = Z_0/2 \tag{5.43b}$$

with matching.

In Figure 5.28 is plotted the expected response for both the impedance-matched and nonmatched cases. For a typical photodiode. The diode and bond pad area dimensions are fairly fixed by practical considerations. Allowing variation of only the i-layer thickness, we optimize for the maximum −3-dB frequency, which occurs for a thickness of 1.0 μm. The corresponding total capacitance is 90 fF. As seen, the matched case has a −3-dB point of 27 GHz, since the load on the photodiode is only 25Ω.

Other limitations on the speed of pin photodiodes include diffusion tails and charge trapping at heterojunctions. Both of these effects mandate careful attention to the design of the vertical layer structure, since the appropriate choice of band gaps and doping can make or break the desired performance.

Diffusion tails are caused by absorption of light outside of the depletion region, which results in minority carriers that can contribute to the photocurrent until they either diffuse to the depletion region or recombine. Since recombination is unacceptably slow (typically 3 to 10 ns), absorption outside the depletion layer must either be eliminated or restricted to a thin enough layer that diffusion out of this layer is fast enough to be insignificant. This is accomplished by restricting the absorption layer (InGaAs) to lie within or be slightly larger than the depletion layer and using a larger band gap material outside this which is transparent to the signal wavelength. Carriers generated outside the depletion layer but within a distance $d$ will diffuse with an exponential time constant $\tau_{\text{diff}}$ given by

$$\tau_{\text{diff}} = \frac{4d^2}{\pi^2 D} = \frac{4d^2}{\pi^2 \mu V_T} \tag{5.44}$$

where $D$ and $\mu$ are the appropriate carrier diffusion and mobility, respectively, and $V_T$ is the thermal voltage (= 25 mV at 300K). Holes are the primary concern, since their mobility of about 150 cm$^2$/V · s is about 60 times less than that of electrons in InGaAs [65]. For

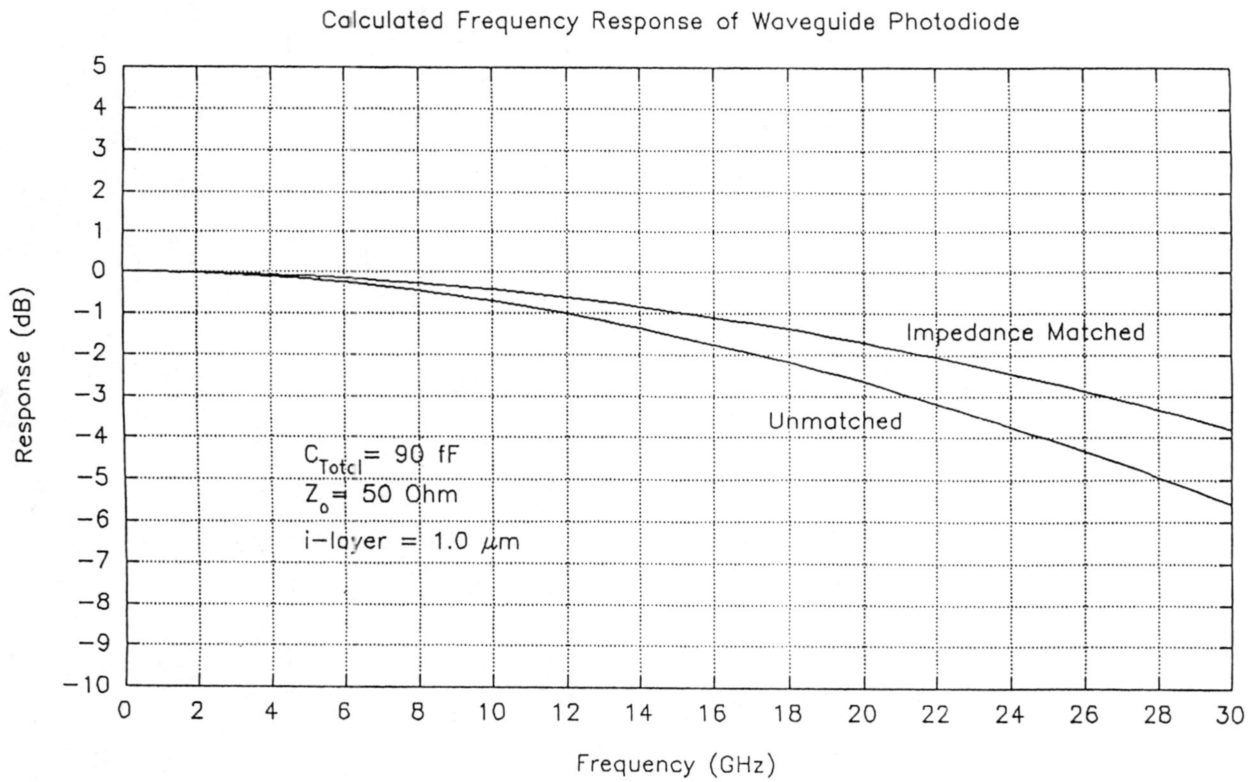


**Figure 5.28** Calculated frequency response of a waveguide photodiode.

a thickness $d = 0.1$ μm, the decay time is about 10 ps, which corresponds to a −3-dB frequency of 15 GHz. Since this frequency varies as $d^{-2}$, the layer does not have to be much thinner before it will be well beyond 20 GHz. Furthermore, the diffusion tail effect is also reduced by the thin layer in that the number of carriers contributing to the tail is small compared with that of the main signal (i.e., in the depletion layer). Thus, for a layer of 0.1 μm outside of a 1.0-μm depletion layer, the diffusion tail will only amount to about 10% of the main signal. Thus, the effect of this diffusion tail will be only a 0.8-dB total drop going from below to above the related corner frequency.

While the use of a large band gap material outside the depletion layer can be used to avoid diffusion tail problems, it can also result in another problem, which is charge trapping at the heterojunction. Since a higher band gap material is used for the transparent layers surrounding the absorption layer; both electrons and holes will encounter an energy barrier as they are swept out of the depletion layer. If this barrier is large enough, the charge will be trapped there and only slowly be removed by thermionic emission over the barrier. This will result in a slow component of the impulse response and consequent degradation of the frequency response. It is generally believed that holes are the major problem due to their higher mass and the larger portioning of the band gap difference to the valence band. According to [66], the conduction band discontinuity $\Delta E_c$ and band gap difference $\Delta E_g$ for InGaAsP on InP are related by

$$\Delta E_c = 0.39\Delta E_g \tag{5.45}$$

meaning that the valence band barrier is 60% of the band gap difference.

These effects are well documented in investigations of APDs [67]. Suggested solutions include grading of the band gap to avoid an abrupt heterojunction or insertion of an intermediate layer to split the energy barrier into two smaller barriers [68] (which can be viewed as a first approximation to grading). While grading appears to be the ideal approach, it complicates the wafer growth by requiring a continuous change of quaternary material composition while still maintaining correct proportions for lattice matching. An alternative approach, which avoids even a single intermediate quaternary layer, uses a graded gap superlattice to simulate band gap grading [69]. Such grading schemes also involve a trade-off between grading outside the depletion layer, which allows absorption outside of there, and grading inside, which makes the field nonuniform in the depletion layer.

The effect of an abrupt junction can be reduced by keeping it as small as possible and as thin as possible to enhance tunneling. Thin barriers can be achieved by placing the barrier in a highly doped region so that there is enough available charge for rapid build-up of the barrier potential. As a rough approximation, the barrier thickness $W$ is overestimated by

$$W = \sqrt{\frac{2\epsilon_r\epsilon_0 V_b}{qN_b}} \tag{5.46}$$

where $V_b$ and $N_b$ are the band edge potential difference and doping level at the barrier, and $\epsilon_r\epsilon_0$ is the dielectric constant. Taking typical values of 0.2 eV for the barrier height and a doping of $3 \times 10^{18}$ cm$^{-3}$, the barrier thickness is only about 100Å, which should enable tunneling. Thus, we can achieve both low heterojunction effects and low diffusion tails by extending the absorption layer into the doped regions by a distance of between 100Å and 1000Å.

*Collection Efficiency*

In designing the waveguide photodiode for high collection efficiency, several aspects of the collection process must be considered. First, the effective collection aperture must be large enough to collect the optical spot size, which will depend on the fiber used and the use of any optics. Also, if the absorption layer is thinner than the optical guiding layer, the effective absorption coefficient will be proportionately reduced and a longer distance will be required to fully absorb the light. One must be careful that the design is still an improvement over the traditional photodiode design when making such structures. Finally, an AR coating should be used to reduce the 30% reflection that occurs at a bare InP surface.

In designing the collection aperture, we start first with the anticipated spot size of the optical beam. Standard single-mode fiber for use at 1.3 μm has a core diameter of 9 μm and a numerical aperture (NA) of 0.11. The corresponding Gaussian beam waist $\omega_0$ (radius) is 5 μm. In our devices, the width of the aperture is about 20 μm, so we can consider that all of the light will be collected in this direction if butt coupling or reducing optics is used. In the vertical direction, speed considerations dictate that the absorption layer be about 1 μm for a 20-GHz device. This is obviously not large enough to catch much of the fiber spot size, and some combination of optics and a larger aperture are needed to achieve high efficiency. In order to evaluate this, we need a quantitative measure of how much light will be collected by a structure.

For an exact calculation, what should be done is to find the modes of the waveguide and their absorption coefficients. Then the incident field should be expanded in terms of these modes and the net absorption of each mode over the length of the guide evaluated to determine the portion of the mode power that is absorbed. These powers would then be summed with the proper modal weighting to determine the total fraction of the incident power that is absorbed. In most cases, the device is long enough so that all modes are completely absorbed. This case has been treated by expanding the incident field over the guided modes of the device structure and summing the squares of the expansion coefficients [70]. Comparison of this theoretical treatment with measurements showed good results and the theory actually underestimated the collection efficiency by a few percent. This is actually to be expected since nonguided portions of the incident light can also be partially absorbed. As an extreme example of this, the light traveling normal to the active layer in a conventional photodiode is not guided but is nonetheless absorbed. An exact treatment is obviously very complicated, and it is not clear that such an exercise is warranted for the design of a good device. Even the expansion over the guided modes is probably not necessary, especially for multimode guides, which is usually the case. We develop instead a very simple theory for estimating the collection efficiency of the aperture, which slightly underestimates the true value and is therefore a conservative indicator of a good design.

In this simple theory, we assume that the collected power is that portion which falls within the waveguide aperture and within the acceptance angle of the structure. Such an approximation works well for multimode waveguides, where the wave nature of the light becomes less apparent and ray optics is a good approximation. In the structures we are considering, there are usually at least three modes and as many as ten in the large optical cavity structure. The acceptance angle $\theta_a$ is determined by the refractive index step of the structure and is often expressed in terms of the NA as

$$\sin\theta_a = \mathrm{NA} = (n_1^2 - n_2^2)^{1/2} \tag{5.47}$$

where $n_1$, and $n_2$ are the refractive indices inside and outside the guide layer, respectively. For the structures used here, the indices were about 3.4 and 3.2, which gives an NA of 1.15, so the acceptance angle constraint is ignored. Then the coupling efficiency in the vertical direction $\eta_y$ is given by the integral of the Gaussian beam profile from the fiber over the aperture of the photodiode as

$$\eta_y = \frac{\int_0^{L/2} \exp(-2_y^2/\omega^2)\mathrm{d}y}{\int_0^{\infty} \exp(-2_y^2/\omega^2)\mathrm{d}y} = \mathrm{erf}\left(\frac{L}{\sqrt{2}\omega}\right) \tag{5.48}$$

where $L$ is the thickness of the waveguide collecting aperture and $\omega$ is the Gaussian beam waist of the incident optical beam at the waveguide facet (i.e., the optical electric field behaves as $\exp(-y^2/\omega^2)$). The error function is plotted in Figure 5.29, and for an efficiency of 80%, it is seen that $L/\sqrt{2}\omega$ must be >0.9. Very near 100% is reached at a value of 2.0, and it would seem that this is a good design objective.

Next, we have to be sure there is enough distance along the waveguide for complete absorption of the light, especially if an external waveguide is used. In a conventional photodiode, the quantum efficiency varies as $1 - e^{-\alpha L}$, while in the waveguide photodiode the efficiency will be approximately $1 - e^{-\Gamma\alpha z}$, where $z$ is the length of absorption along the waveguide. The factor $\Gamma$ is due to the incomplete confinement of the optical power

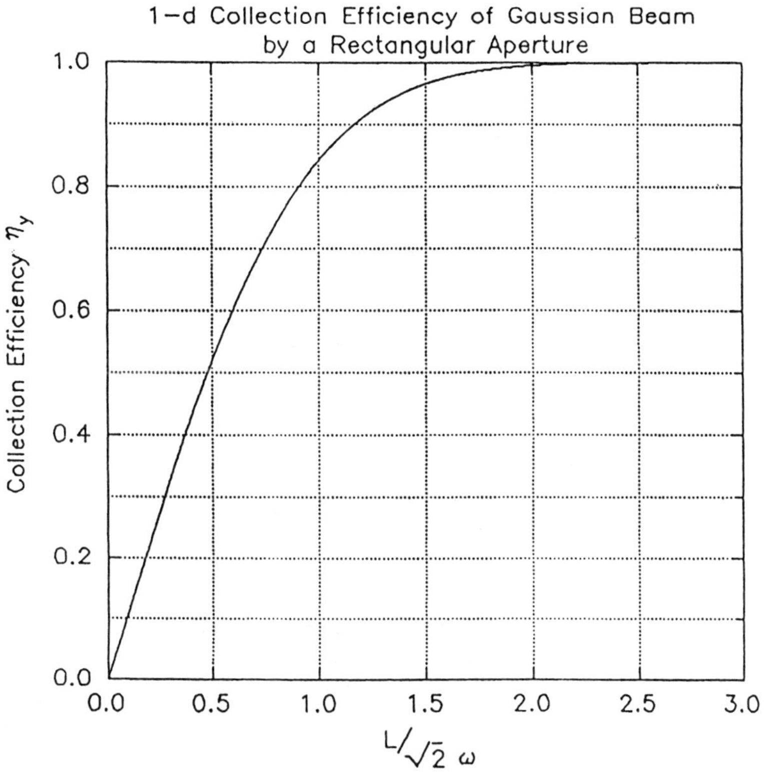


**Figure 5.29** 1-D collection efficiency of Gaussian beam by a rectangular aperture.

in the absorbing layer and is given approximately by $\Gamma \approx L/L_c$, where $L_c$ is the total thickness of the optical waveguide and absorption layer. In order to gain an improvement over a conventional photodiode with the same absorption thickness (set by speed considerations), then we must have $z/L_c > 1$. This will be easily achievable, since we expect optical cavity thickness to be 3 to 4 μm, while the length $z$ will be 20 to 50 μm.

Finally, the reflection of incident light at the device facet must be minimized, since reflected light is lost and causes potential return loss problems in the system. An AR coating is the most practical means to achieve this, and we routinely use such coatings on photodiodes and lasers. The ideal AR coating is a quarter wavelength thick and has a refractive index which is the geometric mean of those of the incident and device media (e.g., air and semiconductor). We do not require an extremely low reflectivity, but are only seeking a few percent, so the ideal index is not critical. In Figure 5.30(a,b) are shown the reflectivities versus coating thickness for alumina and zirconia coatings, respectively. These are two materials with an appropriate refractive index, which we commonly use for coating and for protection of the exposed facet. One added consideration with the

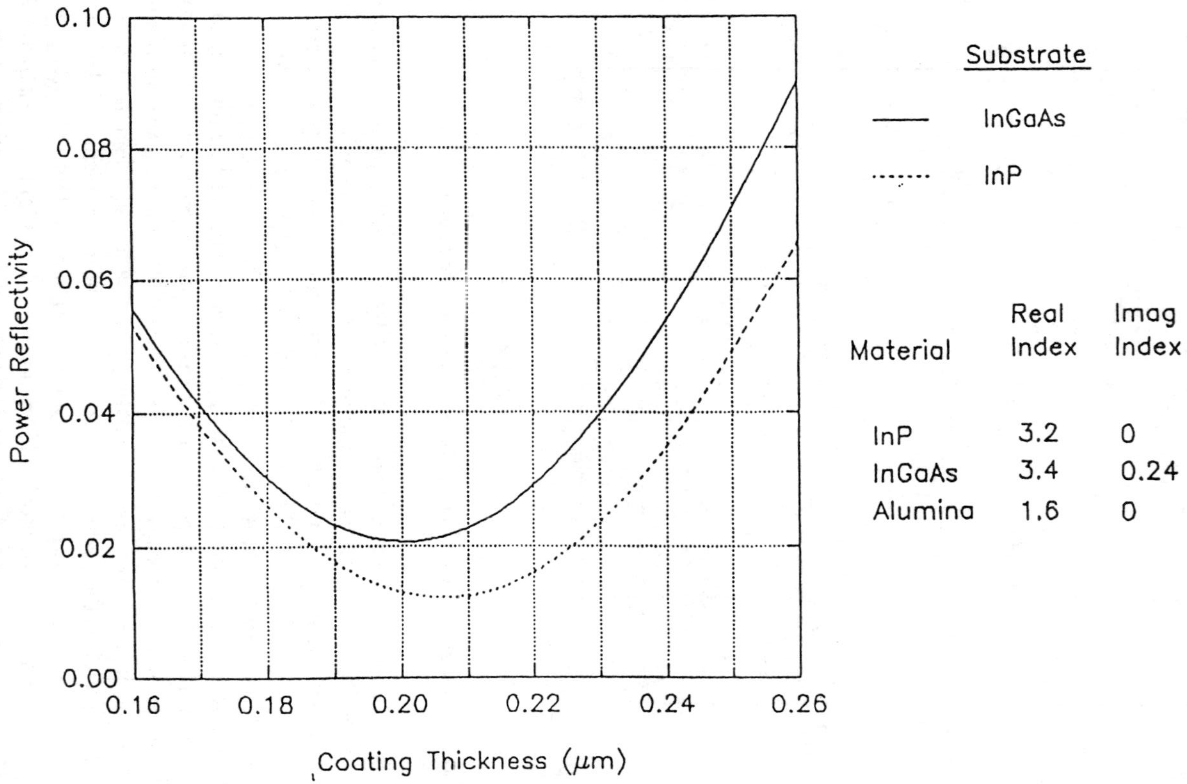


**Figure 5.30(a)** Reflectivity of: alumina ($A1_2O3$) AR coating at $\lambda = 1.3\mu m$.

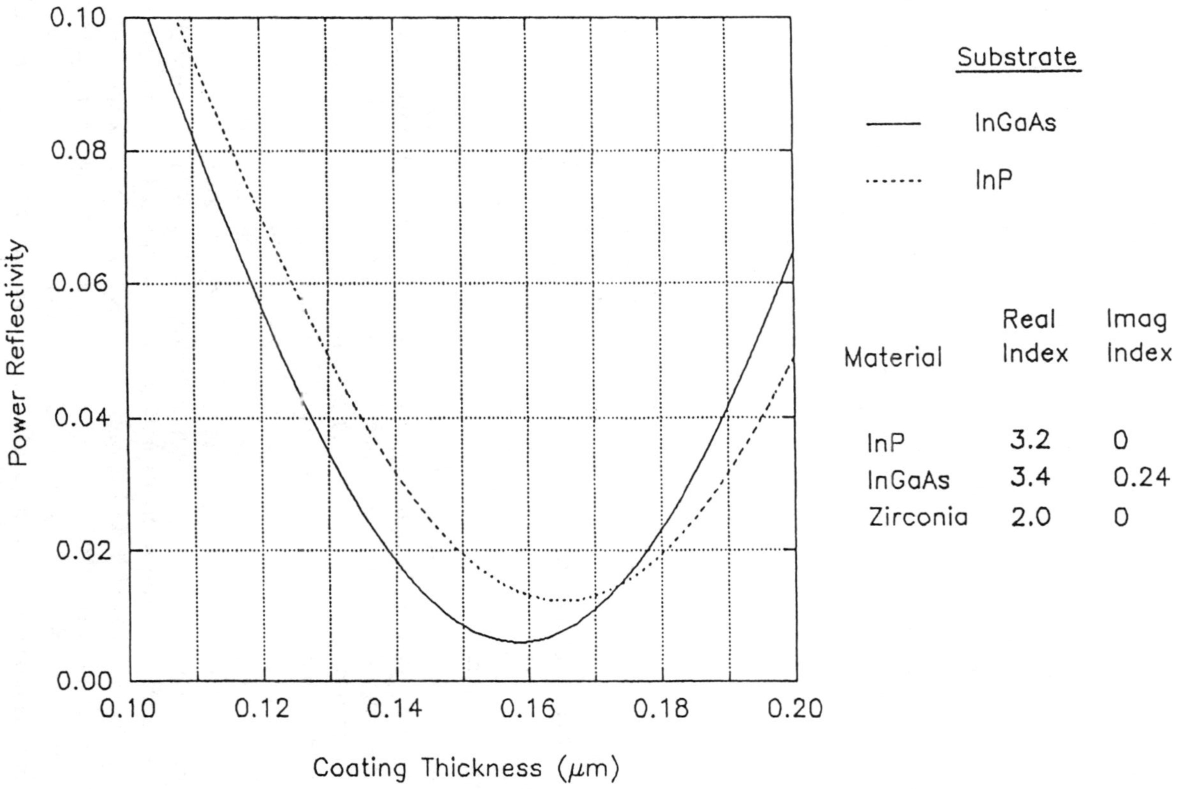


**Figure 5.30(b)** Ziconia (ZrO2) at $\lambda = 1.32$ μm.

waveguide is that the incident beam sees several different materials (e.g., an InGaAs absorption layer, InP cladding layers, and possibly InGaAsP guide layers), and it is not clear that a single coating can give low reflection for all of these interfaces. Not only do these materials present different refractive indices, but the absorbing layer has an imaginary component to the refractive index which changes the quarter-wave thickness requirement on the coating. In order to evaluate these concerns, we calculated the reflectivity curves for coatings on both InP and InGaAs, including the effect of absorption in the InGaAs. The index of the InGaAsP guide layers would be between these two extremes and is transparent, so we can expect its curve would lie between the two and with a minimum reflectivity at the same thickness as InP. While the absorption in InGaAs does shift the optimum thickness from that of InP, the shift is small enough that coatings can be made which have reflectivities which are <2% for alumina and <1.5% for zirconia. The small effect of the InGaAs absorption can be attributed to the small imaginary part of the refractive index (0.24) compared to the real part (3.4).

*Device Structure*

Two structures described in this section will be referred to as device 1 and device 2. Both structures are a pin doping sequence grown on an $n^+$ InP substrate. The primary difference between the two is that device 2 has guide layers outside the absorption layer to form a large optical cavity to enhance light collection, while in device 1 the absorption and guide layers are the same.

The structure of device 1 is shown in Figure 5.31 and consists of an $n^+$ InP layer (1 μm) grown as a buffer, the undoped InGaAs absorbing layer (1.2 μm), a $p^+$ InP guiding layer (0.5 μm), and a $p^+$ InGaAs layer (0.2 μm) for ohmic contact. In this structure the absorbing layer is the same as the optical waveguide layer, so optical confinement is high, but the vertical dimension of the waveguide is rather small (1.2 μm). This resulted in lower coupling than desired and so the next structure was modified to include a large optical cavity for improved collection of the focused light.

The layer structure of device 2 is shown schematically in Figure 5.32 and includes several improvements over the design of device 1. The growth starts with a buffer layer of $n^+$ InP(0.5 μm) as before. Next there is an optical waveguide layer of $n^+$ InGaAsP (1.1 μm). The composition of this layer is chosen so that the band gap energy corresponds to a wavelength of 1.15 μm and so is transparent to the 1.3-μm and 1.55-μm signal wavelengths for which it is designed. Together with the corresponding $p^+$ InGaAsP guide layer of the same composition and thickness and the InGaAs absorption layer between, an optical waveguide is formed that has an overall thickness of 3.4 μm. As discussed later, this significantly improves the collection efficiency of the device, since the focused light spot is about 3 μm is diameter.

The thin (0.1 μm) doped layers of InGaAs on either side of the undoped i-layer are intended to improved transport at the heterojunctions by providing the charge needed

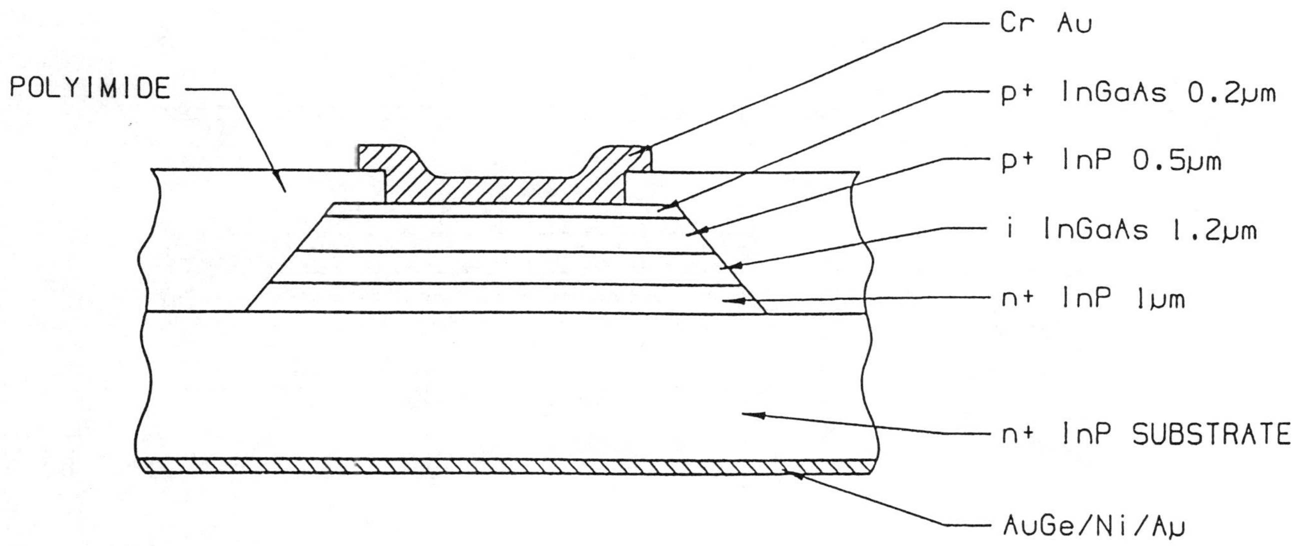


**Figure 5.31** Structure of waveguide photodiode.

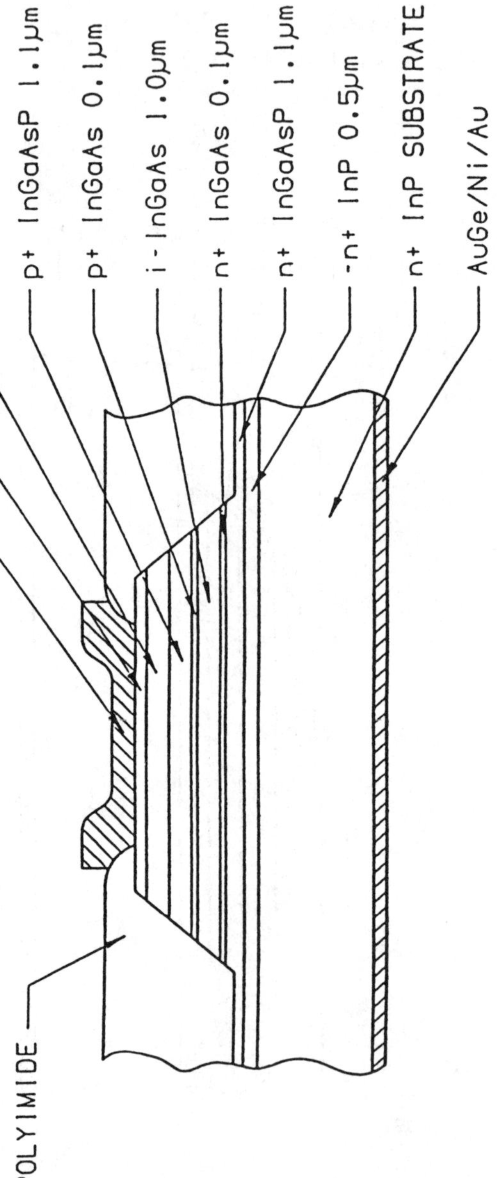


**Figure 5.32** Structure of waveguide photodiode with large optical cavity.

for a narrow barrier which should enable tunneling. If transport is poor at these junctions, charge trapping can occur which results in long tails in the temporal impulse response of the photodiode. However, the doped InGaAs layers must not be too wide, since absorption will occur in these layers, resulting in diffusion tails. With the 0.1-μm width used here, the diffusion time should be <10 ps, and the relative contribution to the total signal should be <10%. The use of the intermediate InGaAsP layers also helps to reduce the effect of charge trapping by splitting the overall barrier due to the band gap difference of InP and InGaAs into two smaller barriers.

The i-layer thickness was also reduced from 1.2 μm in device 1 to 1.0 μm in device 2. Calculations indicated that the first thickness resulted in a response which was primarily transit time–limited, and by reducing the thickness, we achieved a better balance between the transit time and RC limits. This also improved the expected 3-dB response point from 20 GHz to about 21.5 GHz to provide a little margin for the target.

### *dc Collection Efficiency*

Coupling efficiency was evaluated by measuring the photocurrent from the photodiode when illuminated by a known power from a single-mode fiber. The coupling is expressed in terms of photocurrent electrons per photon emitted from the fiber. A gradient-index (GRIN) lens was usually used for maximum coupling, and the magnification was optimized by varying the lens position and aligning the fiber for best position and focus. The single-mode fiber used has a mode field diameter ($2\omega$) of 10 μm and the optical power used was typically 0.6 mW. Measurements were done with no bias on the photodiode, so there is no dark current or possible amplification, and a low-impedance ammeter was used to measure the photocurrent. The results of these measurements are summarized Table 5.1 along with the relevant parameters of the devices and the estimated coupling efficiency based on the simple theory of Section 5.1.2.2.

The estimate of the simple theory is seen to be a very good indicator for designing the aperture size, as shown by the good agreement between the theory and experiment

**Table 5.1**
Collection Efficiency Results

| *Measurement* | *Device 1* | *Device 2* | *Device 2* |
|---|---|---|---|
| Device number | 1 | 2 | 2 |
| Aperture height $L$(μm) | 1.2 | 3.4 | 3.4 |
| Coupling scheme | GRIN | GRIN | Direct |
| Magnification | 1/2.65 | 1/2.6 | 1 |
| $L/\sqrt{2}\omega$ | 0.45 | 1.25 | 0.48 |
| Simple theory coupling | 0.47 | 0.92 | 0.50 |
| Experimental coupling | 0.53 | 0.93 | 0.74 |

of the first two columns. In fact, the theory tends to slightly underestimate the coupling, which is to be expected since it ignores light outside the guide which can be partially absorbed. We also see that the use of external waveguide of device 2 has a very significant impact on the coupling efficiency, especially when considering that the absorption layer of device 2 is actually thinner (1.0 μm) than that of device 1 (1.2 μm). This efficiency of 93% is the highest reported to date—the next highest is 68% [71]—for a waveguide photodiode and can be attributed to a combination of the use of the GRIN lens and a larger waveguide thickness. Also shown in the table above is a result for direct butt-coupling of the fiber to the photodiode facet, which is more desirable in terms of packaging cost. The result here is also quite good, although there is significant discrepancy with the theory we have not been able to explain.

A combination of a larger waveguide (4 to 5 μm) and some lensing on the end of the fiber would probably be the most cost-effective means to get efficiencies of 80% to 90% in a packaged device. However, for the top efficiency, the GRIN lens is probably best, since there are practical limits on the waveguide thickness and the GRIN lens is not very expensive.

*Frequency Response*

The measured frequency response of a device 2 device is shown in Figure 5.33 for several values of the reverse-bias voltage. In these data, one should neglect the low-frequency portion of the response, since the roll-off is due to the specific bias circuit used in the package. Based on dc measurements of the photocurrents from the individual lasers, the ideal 100% response should be at −37.9 dB, which we take as the reference value. At a bias of −5V, the average response at 20 GHz is at about −40.5 dB or down about 2.5 dB. Thus, the 3-dB response frequency is believed to be well above the target of 20 GHz.

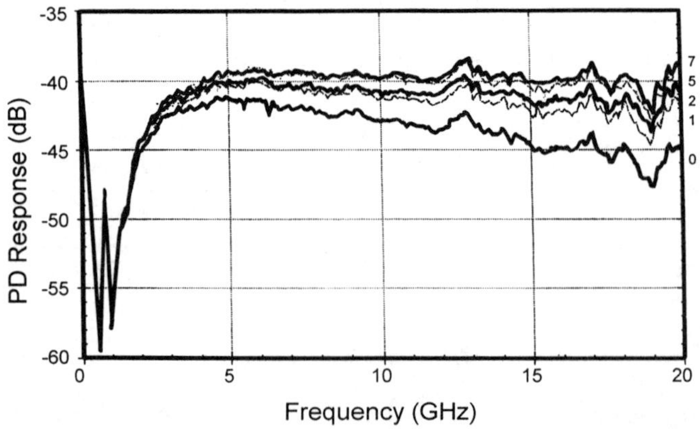


**Figure 5.33** Measured frequency response of waveguide photodiode.

The response is also very flat except for a few glitches at 12.5 GHz and 17 to 19 GHz, which we believe are due to the package. Even if one includes the package artifacts, the 3-dB frequency is about 19 GHz.

The variation in response with bias voltage shows that these devices require about 5V of bias in order to reach their optimum high-speed performance. This is common with most 20-GHz photodiodes, and is consistent with the field needed for velocity saturation of the holes (~ 50 kV/cm) [72] times the typical i-layer thickness of about 1 μm. If the structure were modified so that the response is dominated by electron transit, then the required voltage should be reduced by a factor of at least five [73, 74] to about 1V or so. Furthermore, most of this would be provided by the built-in voltage of the pn junction, so there is potential for a very low or even zero-bias high-speed photodiode with this structure. Note that there is no evidence of charge trapping or diffusion tails in the measured frequency response.

### 5.2.3 Down-Converting Photodetector by Photoconductive Mixing

*Introduction*

Recent developments in high-speed optoelectronic technology have created new areas of applications of optical techniques in microwave systems. High-speed lasers and detectors operating at frequencies up to 20 GHz made possible distribution of microwave signals using low-loss optical fibers. The kilometer distances attainable with fiber-optic systems offer improvement over conventional coaxial and waveguide systems by many orders of magnitude. In fact, a combination of optoelectronics and monolithic microwave integrated circuit (MMIC) technologies has been cited as a must for implementation of future phased-array radar systems, where tens of thousands of radiating and receiving elements are involved. One of the most common operations in a microwave system is frequency up- and down-conversion using diode mixers. If optics were to play a dominant role of signal transport in a microwave system, it would be highly desirable to perform the function of RF mixing directly in optical form without first converting it into electrical form. Wideband photoconductive mixers [76, 77] can be used to perform the function of optoelectronic mixing. It will be shown that such optoelectronic mixers offer not only a much simpler (and hence more reliable) component design, but compare very favorably in terms of performance specifications to the arrangement where the optical signal is first demodulated using a photodiode and then frequency up- or down-converted using microwave mixers. In addition, by suitably modifying certain characteristics of the photoconductor, the opto-electronic mixer can potentially perform over a wide frequency range (over tens of gigahertz) that would be very difficult to achieve with conventional electrical diode mixers.

In the following, an analysis of the basic operation of the optoelectronic mixer is shown. The performance of the optoelectronic mixer is then compared with the conventional photodiode/mixer approach. Actual devices were fabricated with epitaxially grown

InGaAs layers on InP substrate, which operates in the 1.3-μm range. The device is examined in detail, and good agreement is obtained between the measured results and the theoretical predictions. Based on these theoretical and experimental findings, means to further improve the device will also be discussed.

### *Background Analysis*

*General Microwave Frequency Mixing.* We start with a general analysis of mixing in an ideal case to establish some limitations and reveal important parameters of the mixer. The basic idea is to modulate either the transmission or reflection of a microwave signal by modulating the conductance of an element in the circuit. For a signal-element mixer, the peak conversion using transmission is 6 dB lower than that of a reflection configuration, and so we only consider the latter.

The circuit configuration is shown in Figure 5.34 in which a time-varying conductance $G(t)$ terminates a transmission line of impedance $Z_0$. Denoting incident and reflected voltage waveforms with $v_i(t)$ and $v_r(t)$, respectively,

$$v_r(t) = \rho(t) v_i(t) \tag{5.49}$$

For now it is assumed that the parasitic elements $L$ and $C_g$ can be neglected, and thus the reflection coefficient $\rho$ is simply

$$\rho(t) = \frac{1/G(t) - Z_0}{1/G(t) + Z_0} = \frac{1 - Z_0 G(t)}{1 + Z_0 G(t)} \tag{5.50}$$

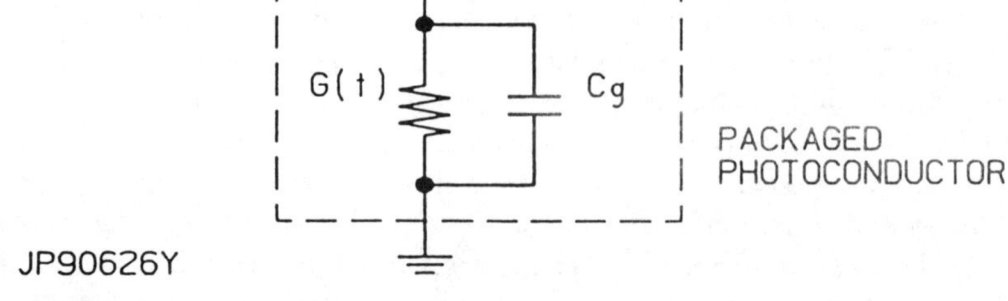


**Figure 5.34** Schematic operation of photoconductive mixer.

If $G(t)$ is periodic with a frequency $\omega_{LO}$ and symmetric about $t = 0$, then we can write

$$\rho(t) = \sum_{n=0}^{\infty} \rho(n\omega_{LO}) \cos(n\omega_{LO}t) \tag{5.51}$$

Likewise, if there is a microwave source in the circuit at a frequency $\omega_{RF}$, then the incident and reflected waveforms are, in general,

$$v_{i,r}(t) = \sum_{n} v_{i,r}(\omega_{RF} \pm n\omega_{LO}) \exp(i(\omega_{RF} \pm n\omega_{LO})t) \tag{5.52}$$

The frequency $|\omega_{RF} - \omega_{LO}|$ will be referred to as $\omega_{IF}$ and is typically the desired mixing component for down-mixing. In order to proceed, the external circuit must be specified in terms of the impedance seen by signals originating at the mixer. For the case here, it is assumed that all frequencies are impedance-matched (e.g., $Z_0$), except for the IF impedance, which is some arbitrary $Z_s(\omega_{IF})$. Then the only signals incident on the mixer are the source RF and possibly some IF. Inserting (5.51) and (5.52) into (5.49) and taking the IF component gives

$$v_r(\omega_{IF}) = v_i(\omega_{IF})\rho(0) + \frac{1}{2}\rho(\omega_{LO})v_i(\omega_{RF}) \tag{5.53}$$

This is just equivalent to an IF voltage source with impedance $Z_m(\omega_{IF})$ and amplitude $V_m(\omega_{IF})$ given by

$$Z_m(\omega_{IF}) = \frac{1 + \rho(0)}{1 - \rho(0)} Z_0 \tag{5.54a}$$

$$V_m(\omega_{IF}) = [1 + Z_m(\omega_{IF})/Z_0]\frac{1}{2}\rho(\omega_{LO})v_i(\omega_{RF}) \tag{5.54b}$$

From these expressions it is straightforward to calculate the efficiency of the mixer in converting RF power to IF power. The efficiencies for external IF impedances of $Z_0$ (line-matched) and $Z_m(\omega_{IF})$ (source-matched) are

$$\eta = \frac{1}{4}\rho^2(\omega_{LO}) \tag{5.55}$$

$$\eta = \frac{1}{4}\rho^2(\omega_{LO})/[1 - \rho^2(0)] \qquad Z_{ext} = Z_m(\omega_{IF}) \tag{5.56}$$

Typically, a photoconductive mixer has a linear response with respect to optical power and will be illuminated by a sinusoidally modulated light source. Thus, we consider the case in which the conductance is given by

$$G(t) = G_0[1 + m \cos(\omega_{LO}t)] \tag{5.57}$$

The theoretical conversion efficiencies of (5.55) and (5.56) versus the parameter $Z_0G_0$ are plotted in Figures 5.35 and 5.36 for various values of the modulation index $m$. Also plotted are curves based on analytic approximations valid for small $m$ or $Z_0G_0$, which are

$$\eta \approx m^2(Z_0G_0)^2/(1 + Z_0G_0)^4 \tag{5.58}$$

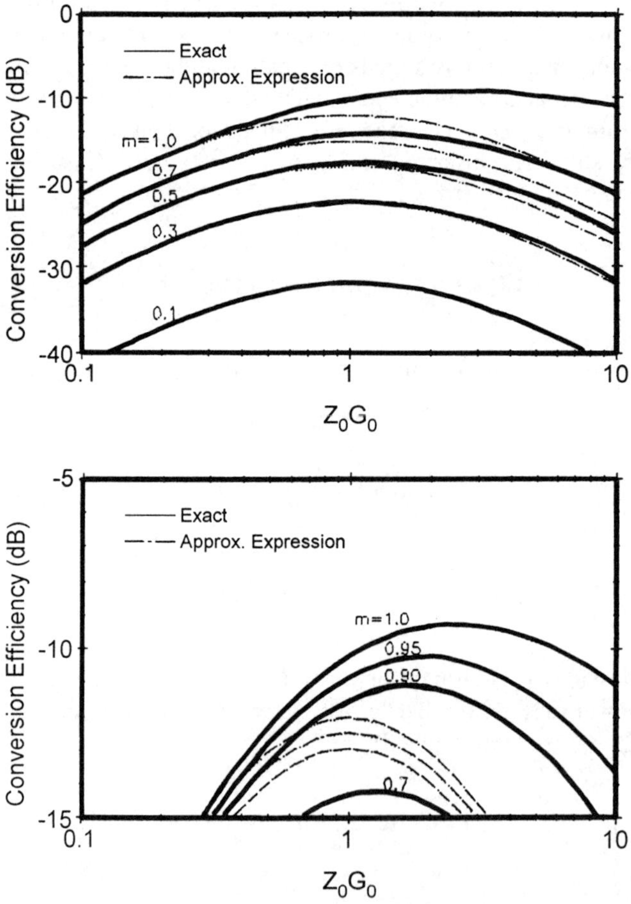


**Figure 5.35** Theoretical efficiency of mixer with line-matched IF load = $Z_0$.

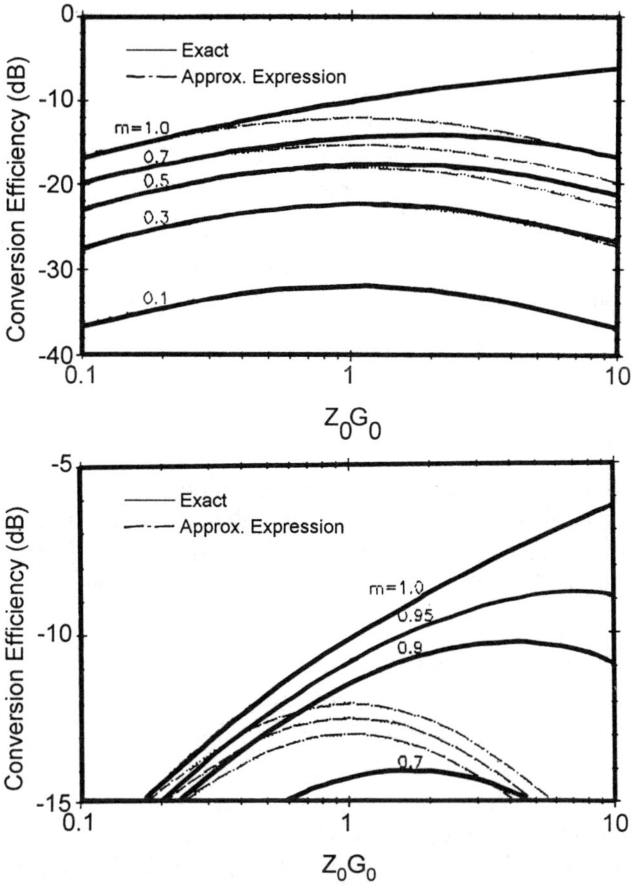


**Figure 5.36** Theoretical mixer efficiency with source-matched load = $Z_m(\omega_{IF})$.

$$\eta \approx m^2 Z_0 G_0 / 4(1 + Z_0 G_0)^2 \tag{5.59}$$

For an IF load of $Z_0$, the peak conversion is –9.3 dB and occurs at a $Z_0G_0$ of 2.3. For an IF load matched to the mixer, higher conversion efficiences ae possible, though at rather large values of $Z_0G_0$, which would require more optical power. The $m = 1$ curve indicates that significantly better performance can be obtained; however, it must be remembered that full 100% modulation may be difficult to obtain. The $m = 0.95$ curve indicates that only a slight deviation from the ideal $m = 1$ significantly reduces the peak efficiency obtainable.

Finally, it is possible that performance could be improved by modifying the assumption that all reflected waves are absorbed by a matched impedance of the external circuit. Specifically, signals reflected by the external circuit (other than the RF and IF) could be remixed and, if properly phased, generate additional IF power. With conventional diode mixers, this can result in a 0.9-dB improvement in conversion efficiency. If the conductance varies sinusoidally as given by (5.57), then another useful result can be obtained by making the external impedance zero (as seen at the mixer) to all but the RF. In this case, the mixing components will be limited to the incident RF and two sidebands ($\pm\omega_{LO}$) as opposed to the "comb" of frequencies implied in (5.51).

*Photoconductor Properties.* The conductance of the photoconductor varies through the number of carriers in the semiconductor gap between two contacts. If carriers are distributed uniformly between the contacts, then the conductance is given by

$$G = \frac{\eta q \mu A}{L} = \frac{N q \mu}{L^2} \tag{5.60}$$

where $q$ is the electronic charge, $\mu$ is the relevent mobility, $n$ is the carrier density, $N$ is the total number of carriers in the gap, $A$ is the effective cross-sectional area, and $L$ is the length of the gap. A dark conductivity $G_d$ is given by the background doping level of the photoconductor. For linear recombination, the excess carrier number $N$ is governed by

$$\frac{dN}{dt} = \frac{\eta_d P(t)\lambda}{hc} - \frac{N}{\tau_r} \tag{5.61}$$

where $\eta_d$ reflects the collection efficiency due in part to surface reflection and exposed active area, $P(t)$ is the incident optical power, $\lambda$ is the optical wavelength, $h$ is Planck's constant, and $\tau_r$ is the carrier lifetime. A more detailed model is required when nonlinear recombination becomes significant, which is typically at carrier densities greater than $10^{18}$ $cm^{-3}$. For the devices used here, carrier densities are estimated to be only $10^{15}$ $cm^{-3}$, so the linear recombination model is quite good. If the optical power is modulated as

$$P(t) = P[1 + m_{\text{opt}} \exp(i\omega_{\text{LO}}t)]$$

then solving (5.61) and using (5.60) gives

$$G(t) = \frac{\eta_d P \lambda q \mu \tau_r}{hcL^2}\left[1 + \frac{m_{\text{opt}}}{1 + i\omega_{\text{LO}}\tau_r}\exp(i\omega_{\text{LO}}t)\right] \tag{5.62}$$

With the inclusion of the dark conductivity and neglecting the phase shift, the parameters of the conductivity model (5.57) are

$$G_0 = G_{\text{opt}} + G_d \tag{5.63a}$$

$$m = \frac{m_{\text{opt}}}{\sqrt{1 + (\omega_{\text{LO}}\tau_r)^2}}\frac{1}{1 + G_d/G_{\text{opt}}} \tag{5.63b}$$

$$G_{\text{opt}} = \frac{\eta_d P \lambda q \mu \tau_r}{hcL^2} \tag{5.63c}$$

For good mixing efficiency, $m$ must be large so that the lifetime $\tau_r$ is small enough to keep $\omega_{\text{LO}}\tau_r < 1$ at the frequency of interest. The dark conductivity should also be significantly lower than the light-induced conductivity. Since the desired $G_0$ is typically $(50\Omega)^{-1}$, the dark resistance need only be greater than a few $\kappa\Omega$. This is in sharp constrast to detector applications, where low light level considerations favor dark resistances of >10 M$\Omega$, which is difficult to achieve with low band gap, liquid-phase epitancy (LPE)-grown material. Finally, the optical modulation index should obviously be as high as possible for efficient mixing.

While we have modeled the photoconductor as a light-controlled conductance, this model breaks down at large voltages due to sweep-out of the excess carriers. Specifically, if the carriers are swept out before they recombine, then the maximum current available is obtained and increasing the voltage does not increase the current. This saturation can take two variations depending on the nature of the contacts at the edge of the photoconductor gap. If the contacts are of very low resistance to electrons, then charge can be injected to maintain neutrality in the gap, and saturation will be governed by sweep-out of the low mobility holes. This is also the mechanism which leads to the well-known photoconductive gain. The voltage at which the hole transit time becomes less than the recombination time is

$$L^2/\mu_h\tau_r \tag{5.64}$$

Below this voltage, the conductance will still be given by the expression $G_{\text{opt}}$, with the mobility being the sum of the electron and hole mobilities. If the contacts are not injecting, then there are two cases, depending on the amount of free charge in the gap.

If the charge is small, then the fields in the gap will not be disturbed by the movement of charge, and both electrons and holes will be swept out independently. This would then result in a current-voltage curve with a low-voltage ''knee'' corresponding to sweep-out of electrons, given by (5.64) with $\mu_e$, and a high voltage saturation with the sweep-out of holes. On the other hand, if there are a lot of free carriers, then as electrons and holes move in opposite directions, leaving excess holes at one contact and electrons at the other, the electric field in the gap will be altered. This shift will tend to increase the field at the holes and decrease it at the electrons such that the current is uniform across the gap. Thus, the saturation behavior will be governed solely by the sweep-out of holes. Furthermore, the small-signal conductance will be dominated by the movement of holes and will consequently be less than would be expected using the electron mobility in the expression for $G_{opt}$. A good measure of whether the free charge in the gap is "small" or "large" is to compare it to the charge on the gap capacitance at a typical voltage.

The gap capacitance is also important to the mixing performance in that it must not short the conductance $G(t)$ as was assumed in the derivation of the previous section. The capacitance of interdigitated structures as are used here has been theoretically evaluated in the literature [78]. In summary, it is essentially proportional to the effective width of the straightened-out gap and a nonlinear increasing function of the ratio of the finger width to the space between fingers. This will be further discussed in the device description section. In general, the planar nature of the device leads to very low capacitance while maintaining a reasonably large active area.

### *Microwave Mixing*

The microwave mixing measurements were done as illustrated in Figure 5.37. A microwave modulation—labeled RF1—is applied to the laser diode source illuminating the photoconductive mixer. A second microwave signal—RF2—is applied as a bias to the mixer and adjusted to give a swept (± 1 MHz) IF output around 5 MHz. This IF is filtered out through the dc port of a bias tee and measured with a spectrum analyzer. All sources, bias tees, and the spectrum analyzer were carefully calibrated with a separate microwave power meter over all frequencies of interest.

The linearity of the mixer is shown in Figure 5.38 with a plot of IF output power versus RF input power to the mixer at various RF frequencies. For ideal linear operation, we expect to get 1 dBm of IF per dBm of RF. At the lower frequencies, there is a noticeable compression of the mixing at RF power levels of about –5 dBm and above. This level corresponds to a voltage amplitude of about 0.2V and is explained by the saturation of the conductance due to carrier sweep-out as seen in the dc measurements. At higher frequencies, the carriers are not swept out as they would be by a dc field, and the mixing is more linear even at the higher RF levels.

In Figure 5.39, the measured conversion efficiency relative to that expected from theory is plotted versus the RF frequency. The *theoretical expected value* is calculated using (5.58) and neglecting the effects due to carrier lifetime. The value used for $G_0$ is the dc measured value of $(320\Omega)^{-1}$ and the optical modulation $m_{opt}$ is calculated based on the RF power applied to the laser diode relative to the dc current above threshold. At low frequencies, measurements were made with very low RF levels applied to the mixer to avoid the nonlinear problems just discussed. At higher frequencies, higher RF power had to be used to enable detection with the spectrum analyzer, but this is probably not serious, as was also discussed. Also plotted is the photoconductor detector response from Figure

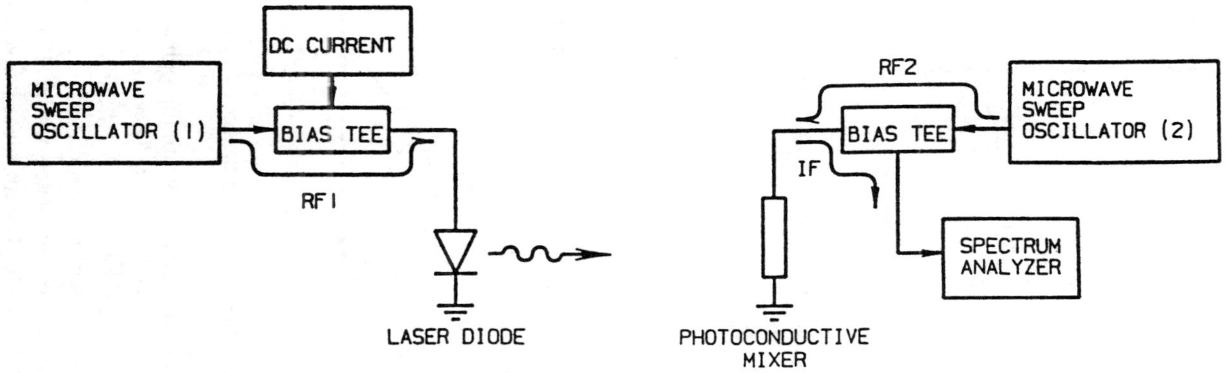


**Figure 5.37** Microwave optoelectronic mixing measurement setup.

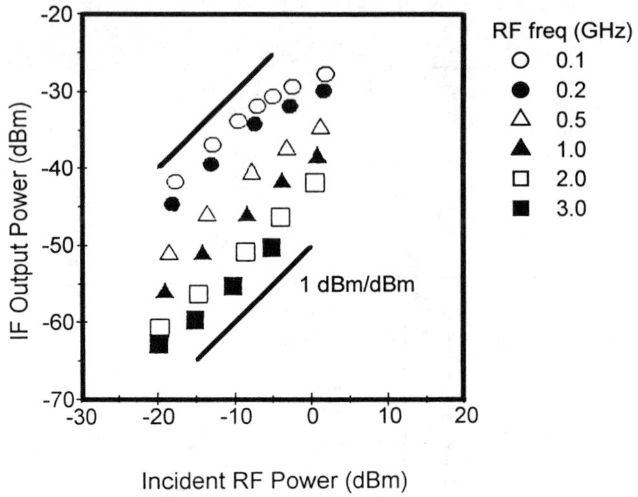


**Figure 5.38** Linearity of mixing versus RF power.

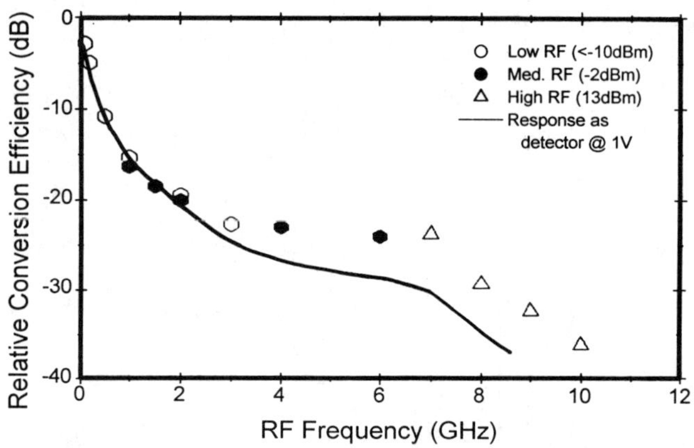


**Figure 5.39** Mixer efficiency relative to theory illustrating lifetime effects.

5.37, situated so as to match the low-frequency mixing behavior. It can be seen that the detector mode response curve fits the mixing data very well, and thus the carrier lifetime similarly influences the mixing performance.

At a frequency of 100 MHz, the conversion efficiency is only 3 dB below that expected theoretically, and this can be attributed to the lifetime roll-off, which has an estimated corner frequency of 80 MHz. The absolute conversion at this frequency was –24 dB, with an optical modulation of about 0.73 (2.7 dB below $m$ = 1). The 14.7-dB difference between this and the theoretical maximum of –9.3 dB can be broken down as follows: 3 dB due to lifetime effects, 2.7 dB due to incomplete optical modulation, 6 dB due to the nonoptimum $Z_0G_0$, and a 3-dB difference between the actual curve and the approximate expression at the peak position in Figure 5.35. Thus, the operation of the mixer is well described by the theory developed above.

At higher frequencies, the relative mixing appears to be 3 to 4 dB better than the detector modulation response. This can be explained by recalling the $S_{11}$ scattering measurements, which indicate that the resistance of the gap is dropping with frequency. As seen in Figure 5.35, the peak mixer efficiency occurs when the average gap resistance is about $Z_0$. Thus, other than the lifetime effects, the mixer performs better at higher frequencies as the effective dc resistance is reduced toward optimum. This suggests that these devices would work very well as frequency up-converters, where an RF signal in the 5 to 10-GHz range would be mixed with <100-MHz modulation on an optical signal.

*Carrier Lifetime Reduction*

An obvious way to improve the bandwidth is to reduce the carrier lifetime. This, however, invariably leads to a reduction in the detection sensitivity; that is, a high optical power is needed to bring the photoconductance up to a certain level. On the other hand, considering the fact that the mixing efficiency peaks at around $Z = 50\Omega$, it is not important that the resistivity be brought to a very low value. Optical power requirements for an efficient mixing operation can still be maintained at the milliwatt level. The most promising means to reducing the carrier lifetime is by some sort of ion bombardment. This has frequently been shown to be effective in reducing recombination lifetimes from the nanosecond range down to 10 ps [79] or less [80]. In addition to improving the frequency response, a reduction in lifetime will increase the voltage level at which current saturation occurs, thereby increasing dynamic range. One concern with ion bombardment is that the mobility can be significantly reduced [81]. This, together with the lifetime reduction, will greatly increase the optical power needed to achieve the desired values of $G_0$ as seen in (5.63c). This can be at least partially offset by better contacts and a feasible reduction in the finger spacing by a factor of 3. An alternative scheme may be to use a surface-depleted junction as the conducting region and use a vertical sweep-out to "turn off" the conductivity [82]. Such a scheme is attractive in that it can maintain a high mobility.

### 5.2.4 Conclusions

To date, photodetetectors (including high-speed photodectors) available on the market are those with vertical coupling, and hence they are limited by the speed-efficiency trade-off as dicussed in the above sections. A commerical grade edge-couple photodiode, such as the waveguide photodiode described in the above sections, has yet to be developed and qualified in terms of its long-term reliability and reproducibility. We have shown in this section a particular device design which gives a –2.5-dB response at 20 GHz and a quantum efficiency of 93%. A simple theory is also presented for designing an appropriate aperture size for efficient light collection. This theory was also successfully used to design a structure with a large optical waveguide outside the thin absorption layer. This structure resulted in improvement of the efficiency from 53% to 93% over the device with only the thin absorption layer for light collection. This result demonstrates that the waveguide structure does allow freedom in choosing the absorption layer thickness without sacrificing quantum efficiency as with conventional photodiodes. This has immediate consequences in the potential for efficient, higher speed devices, as well as for other design objectives, such as low-bias photodiodes.

To develop these devices into a usable product, further work is needed in the following areas: reduction of leakage currents, lower cost package development, improvement of yields, and assessment of lifetime and reliability. For higher speed operation at 30 to 40 GHz, the contacts should be designed for flip chip mounting, and the mesa should be made narrower to reduce capacitance. Investigations should also be continued into other performance criteria such as linearity and power handling, which are becoming increasingly important in high-fidelity microwave links.

The down-converting photoconductive detector described in this section performs a function that is very common in microwave systems. However, one should compare the performance of this device with a more conventional approach, namely, detecting the microwave-modulated optical signal with a conventional photodetector and then using a conventional microwave mixer to perform the down-mixing function. Amplifiers must be used to boost the detected signal following the photodetector, and this invariably limits the operating bandwidth of this approach. A properly designed photoconductive mixer can be expected to achieve wideband operation with a conversion efficiency comparable to that of a conventional microwave mixer.

## ACKNOWLEDGMENTS

Much of the material described in this chapter represent results obtained over a number of years by the authors in collaboration with numerous colleagues. Of these, particular thanks go to Dr. H. Blauvelt, N. Bar-Chaim, I. Ury, C. Gee, and T. R. Chen of Ortel Corporation, Alhambra, California, and Dr. A. Yariv of California Institute of Technology, Pasadena, California. Much of the work on high-speed lasers and detectors was performed over the years with support from DARPA and Rome Laboratories. The work on the photoconductive mixer was performed at Ortel Corporation with support from Harry Diamond Laboratories.

## REFERENCES

[1] K. Y., Lau, and A. Yariv, "Ultra High Speed Semiconductor Lasers," *IEEE J. Quantum Electron.*, Vol. QE-21, 1985, pp. 121–138.

[2] Lester, L. F., S. D. Offsey, B. K. Ridley, W. J. Schaff, B. A. Foreman, and L. F. Eastman, *Appl. Phys. Lett.*, Vol. 59, 1991, p. 1162.

[3] Bowers, J. E., B. R. Hemmenway, A. H. Gnauck and D. P. Wilt, ''High Speed InGaAsP Constricted Mesa Lasers,'' *IEEE J. Quantum Electron.*, Vol. QE-22, 1986, pp. 833–844.

[4] Arakawa, Y., and A. Yariv, *IEEE J. Quantum Electron.*, Vol. QE-22, 1986, p. 1887.

[5] Suemune, I., L. A. Coldren, M. Yamaniski, and Y. Kan, *Appl. Phys. Lett.*, Vol. 53, 1988, p. 1378.

[6] Adams, M. J., *Electron. Lett.*, Vol. 22, 1986, p. 250.

[7] Yablonovitch, E., and E. O. Kane, *IEEE J. Lightwave Tech.*, Vol. LT-4, 1986, p. 504.

[8] Houng, M. P., Y. C. Chang, and W. I. Wang, *J. Appl. Phys.*, Vol. 64, 1988, p. 4609.

[9] Iye, Y., E. E. Mendez, W. I. Wang, and L. Esaki, *Phys. Rev. B*, Vol. 33, 1986, p. 5854.

[10] Mittelstein, M., Y. Arakawa, A. Larsson, and A. Yariv, *Appl. Phys. Lett.*, Vol. 49, 1986, p. 1689.

[11] Lau, K. Y., W. Wang, S. Xin, M. Mittlestein, and N. Bar-Chaim, *Appl. Phys. Lett.*, Vol. 55, 1989, p. 1173.

[12] Rideout, W., B. Yu, J. LaCourse, P. K. York, K. J. Beernink, and J. J. Coleman, *Appl. Phys. Lett.*, Vol. 56, 1990, p. 706.

[13] Fukushima, T., J. G. Wasserbauer, J. E. Bowers, R. A. Logan, T. Tanbun-Ek, and H. Temkin, *IEEE Int. Semiconductor Laser Conf.*, Davos 1990, paper PD-18.

[14] Uomi, K., M. Aoki, T. Tsuchiya, M. Suzuki, and N. Chinone, *IEEE Photonic Tech. Lett.*, Vol. 3, 1991, p. 493.

[15] Sakakibara, Y., A. Yakemoto, Y. Nakajima, M. Fujiwara, N. Yoshida, and S. Kakimoto, *Electron. Lett.*, Vol. 25 1989, p. 1530.

[16] Arakawa, Y., K. Vahala, and A. Yariv, *Appl. Phys. Lett.*, Vol. 45, 1990, p. 950.

[17] Asada, M., Y. Miyamoto, and Y. Suematsu, *IEEE J. Quantum Electron.*, Vol. QE-22, 1986, p. 1915.

[18] Arakawa, Y., H. Sakaki, M. Nishioka, J. Yoshino, and T. Kamiya, *Appl. Phys. Lett.*, Vol. 46, 1985, p. 519.

[19] Vahala, K. J., Y. Arakawa, and A. Yariv, *Appl. Phys. Lett.*, Vol. 50, 1987, p. 365.

[20] Vahala, K. J., *IEEE J. Quantum Electron.*, Vol. QE-24, 1988, p. 523.

[21] Zarem, H., K. Vahala, and A. Yariv, *IEEE J. Quantum Electron.*, Vol. QE-25, 1989, p. 705.

[22] Olshansky, R., D. M. Fye, J. Manning, and C. B. Su, *Electron. Lett.*, Vol. 21, 1985, p. 893.

[23] Bowers, J. E., and M. A. Pollack, Ch. Bin *Optical Fiber Telecommunications*, S. Miller and I. Kaminow, eds., Academic Press, 1988,

[24] Eom, J., C. B. Su, J. LaCourse, and R. B. Lauer, *Appl. Phys. Lett.*, Vol. 56, 1990, p. 518.

[25] Lau, K. Y., and A. Yariv, ''Direct modulation and active mode-locking of ultra-high speed GaA1As lasers at frequencies up to 18GHz'', *Appl. Phys. Lett.*, Vol. 46, 1985, p. 326.

[26] Lau, K. Y., *IEEE J. Quantum Electron.*, Vol. QE-26, 1990, p. 250.

[27] Lau, K. Y., and A. Yariv, *Appl. Phys. Lett.*, Vol. 45, 1984, pp. 1034–1036.

[28] Sato, K., *IEEE J. Quantum Electron.*, Vol. QE-19, 1983, pp. 1380–1391.

[29] Vahala, K., and A. Yariv, *Phys. Rev.*, 1986,

[30] Vahala, K., and A. Yariv, *IEEE J. Quantum Electron.*, Vol. QE-19, 1983, pp. 1096–1101, 1102–1109.

[31] Henry, C.H. *IEEE J. Quantum Electron.*, Vol. QE-18, 1982, pp. 259–264.

[32] Petermann, K., *Laser Diode Modulation and Noise*, Kluwer Academic Publishers, 1988.

[33] Su, C. B., J. Schlafer, and R. B. Lauer, *Appl. Phys. Lett.*, Vol. 57, 1990, pp. 849–851.

[34] Vahala K. J., and A. Yariv, *IEEE J. Quantum Electron.*, Vol. QE-19, 1983, pp. 1096–1109.

[35] Harder, Ch., J. Katz, S. Margalit, J. Schacham, and A. Yariv, *IEEE J. Quantum Electron.*, Vol. QE-18, 1982, pp. 333–225.

[36] Henry, C. H., *J. Lightwave Tech.*, Vol. LT-4, 1986, p. 288.

[37] Matthaei, G., L. Young, and E. M. T. Jones, *Microwave Filters, Impedance Matching Networks and Coupling Structures*, Norwood, MA: Artech House, 1980,

[38] Bode, H. W., *Network Analysis and Feedback Amplifier Design*, Van Nostrand, 1947, p. 306.

[39] Vahala, K. J., M. A. Newkirk, and T. R. Chen, *Appl. Phys. Lett.*, Vol. 54, 1989, p. 2506.

[40] Moore, N., and K. Y. Lau, *Appl. Phys. Lett.*, Vol. 55, 1989, p. 936.

[41] George Lutes

[42] Lau, K., ''Telecomm. and Data Acquisition Pro. Report,'' Jet Propulsion Laboratory, Vols. 42–36, 1981, pp. 26–32.

[43] Born, M., and E. Woff, *Principles of Optics*, Pergamon Press, 1975, p. 494.

[44] Hewlett-Packard Application Note 11729B-1, ''Phase Noise Characteristics of Microwave Oscillators.''

[45] Lester, L. F., S. S. Keefe, W. J. Schaff, and L. Eastman, *Electron. Lett.*, Vol. 28, 1992, pp. 383–385.

[46] Nagarajan, R., T. Fukushima, J. Bowers, R. Geels, and L. Coldren, *Appl. Phys. Lett.*, Vol. 58, 1991, pp. 2326–2328.

[47] Uomi, K., H. Nakano, and N. Chinone, *Electron. Lett.*, Vol. 25, 1989, 668–669.

[48] Yamada, M., *IEEE J. Quantum Electron.*, Vol. QE-22, 1986, pp. 1052–1059.

[49] Meslener, G. J., *Photonic Tech. Letters.*

[50] Lau, K. Y., and H. Blauvelt, *Appl. Phys. Lett.*, Vol. 52, 1988, pp. 694–696.

[51] Choy, M. M., J. L. Gimlett, R. Welter, L. G. Kazovsky, and N. K. Cheung, *Electron. Lett.*, Vol. 23, 1987, pp. 1151–1152.

[52] Gimlett, J. L., and N. K. Cheung, *IEEE J. Lightwave Tech.*, Vol. 7, 1989, pp. 888–895.

[53] Wu, S. W., A. Yariv, H. Blauvelt, and N. Kwong, *Appl. Phys. Lett.*, Vol. 59, 1991, pp. 1156–1158.

[54] Wentworth, R. H. G. E. Bodeep, and T. E. Darcie, *IEEE J. Lightwave Tech.*, Vol. 10, 1992, pp. 84–89.

[55] Lau, K. Y., and A. Yariv, *Appl. Phys. Lett.*, Vol. 47, 1985, p. 84.

[56] Pepeljugoski, P. K., and K. Y. Lau, *IEEE J. Lightwave Tech.*, Vol. 7, 1992, pp. 957–963.

[57] Yariv, A., H. Blauvelt, and S. W. Wu, *IEEE J. Lightwave Tech.* Vol. 7, 1992, pp. 978–981.

[58] Bowers, J. E., and C. A. Burrus, *J. Lightwave Technol.*, Vol. LT-5, 1987, p. 1339.

[59] Bowers, J. E., C. A. Burrus, and F. Mitschke, *Electron. Lett.*, Vol. 22, 1986, p. 633.

[60] Humphreys, D. A., R. J. King, D. Jenkins, and A. J. Moseley, *Electron. Lett.*, Vol. 21, 1985, p. 1187.

[61] Kato, K., S. Hata, A. Kozen, and J.-I. Yoshida, *Tech. Dig. OFC'91*, San Diego, CA, paper ThO4, 1991,

[62] Bowers, J. E., and C. A. Burrus, *Proc. SPIE*, Vol. 716, 1986, p. 109.

[63] Wake, D., T. P. Spooner, S. D. Perrin, and M. J. Harlow, *Electron. Lett.*, Vol. 27, 1991, p. 1073.

[64] Bowers, J. E., C. A. Burrus, and R. J. McCoy, *Electron. Lett.*, Vol. 21, 1985, p. 812.

[65] Hayes, J. R., A. R. Adams, and P. D. Greene, *GaInAsP Alloy Semiconductors*, T. P. Pearsall, New York: John Wiley & Sons, pp. 189–211.

[66] Forrest, S. R., P. H. Schmidt, R. B. Wilson, and M. L. Kaplan, *Appl. Phys. Lett.*, Vol. 45, 1984, p. 1199.

[67] Forrest, S. R., O. K. Kim, and R. G. Smith, *Appl. Phys. Lett.*, Vol. 41, 1982, p. 95.

[68] Campbell, J., A. G. Dentai, W. S. Holden, and B. L. Kasper, *Electron. Lett.*, Vol. 19, 1983, p. 818.

[69] Capasso, F., H. M. Cox, A. L. Hutchinson, N. A. Olsson, and S. G. Hummel, *Appl. Phys. Lett.*, Vol. 45, 1984, p. 1193.

[70] Kato, K., S. Hata, A. Kozen, J.-I. Yoshida, and K. Kawano, *IEEE Photonic Tech. Lett.*, Vol. 3, 1991, p. 820.

[71] Kato, K., S. Hata, A. Kozen, J.-I. Yoshida, and K. Kawano, *IEEE Photonic Tech. Lett.*, Vol. 3, 1991, p. 473.

[72] Hill, P., J. Schlafer, W. Powazinik, M. Urban, E. Eichen, and R. Olshansky, *Appl. Phys. Lett.*, Vol. 50, 1987, p. 1260.

[73] Windhorn, T. H., L. W. Cook, and G. E. Stillman, *IEEE Electron. Dev. Lett.*, Vol. EDL-3, 1982, p. 18.

[74] Windhorn, T. H., L. W. Cook, and G. E. Stillman, *J. Electron. Mat.*, Vol. 11, 1982, p. 1065.

[75] Bowers, J. E., and C. A. Burrus, *Electron. Lett.*, Vol. 22, 1986, p. 905.

[76] Foyt, A. G., and F. J. Leonberger, Ch. 9 in *Picosecond Optoelectronic Devices*, C.H. Lee, ed., Academic Press, 1984.

[77] Foyt, A. G., F. J. Leonberger, and R. C. Williamson, *Appl. Phys. Lett.*, Vol. 40, 1982, p. 447.

[78] Engan, H., *IEEE Trans. Electron. Dev.*, Vol. ED-16, 1969, p. 1014.

[79] Downey, P. M., R. J. Martin, R. E. Nahory, and O. G. Lorimor, *Appl. Phys. Lett.*, Vol. 46, 1985, p. 396.

[80] Loepfe, R., A. Schaelin, H. Melchior, and M. Blaser, *Appl. Phys. Lett.*, Vol. 52, 1988, p. 2130.
[81] Tell, B., K. F. Brown-Goebeler, T. J. Bridges, and E. G. Burkhardt, *J. Appl. Phys.*, Vol. 60, 1986, p. 665.
[82] Chen, C. Y., A. G. Dentai, B. L. Kasper, and P. A. Garbinski, *Appl. Phys. Lett.*, Vol. 46, 1985, p. 1164.

# Chapter 6

# Spatial Light Modulators for Optically Controlled Phased-Array Radar Signal Processors

**Cardinal Warde**

*Massachusetts Institute of Technology*

It is widely believed that optics will play an increasingly important role in computers and signal and image processors of the future. These processors will exploit and combine the flexible numerical processing strengths of electronics with the communication and parallel processing strengths of optics to accomplish computationally intensive tasks with high processing speed. Such tasks include radar and sonar signal processing, pattern recognition, machine vision, industrial inspection, automatic target recognition, autonomous vehicle guidance, sensor fusion, symbolic inference, and very large data/knowledge base operations.

Spatial light modulators (SLM) [1–15], or light valves as they are often called, are the building blocks of optical information processors. These devices modify the polarization, phase, and/or amplitude of a readout light beam as a function of space and time in response to an electrical or optical drive signal which may also be 2-D.

Of particular interest here are SLMs suited for use in optically controlled phased-array radar (OCPAR) systems [16–35]. OCPAR is important for applications involving a large number of antenna elements and in applications where light weight and processing speed are important. Such is the case for space and airborne radar, as well as modern radar systems, which are moving in the direction of dense (hundreds of) lightweight antenna arrays.

While most research in this field has been focused on optically generating the desired phase and amplitude of the signals for the transmitter array (outgoing wave problem), it is recognized that optical signal processing needs to simultaneously solve the receive problem so that the system is also endowed with the capability to extract the angular position, range, and velocity of targets, or excise jammers.

Like other high-performance special-purpose optical processors, OCPAR systems require SLMs that simultaneously offer high framing speed (<1 ms), large gray-scale dynamic range (>40 dB), and high spatial resolution (>20 line pairs/mm). Thus, the full potential of optics to solve the OCPAR problem has not yet been realized because of the shortcomings of existing SLMs.

This chapter reviews the principles of operation and recent advances in those 2-D SLMs that are most likely to find applications in OCPAR. The list includes devices based on nematic and ferroelectric liquid crystals (FLC), electro-optic crystals, photorefractive materials, deformable membrane mirrors, and multiple-QW (MQW) layers.

## 6.1 TYPICAL OCPAR SIGNAL-PROCESSING ARCHITECTURES

A generic transmitter architecture for an OCPAR system is shown in Figure 6.1. The signal from a microwave generator operating at the radar frequency $\omega_r$ is mixed with an optical signal of frequency $\omega_o$ from a laser to produce an up-shifted optical signal at $\omega_o + \omega_r$. The up-shifted optical signal is then distributed to an array of channels, one for each antenna in the array. Each optical signal is delayed (phase-shifted) and amplitude-modulated before being heterodyned with the unshifted reference optical beam. The resulting optical signal, which is modulated at the beat (radar) frequency, is detected, amplified, and used to drive the transmitters. The optical frequency shifter is typically an acoustic modulator operating in the Bragg regime. Phase-only and amplitude-only modulators are used for optical delay and amplitude modulation, respectively. Several

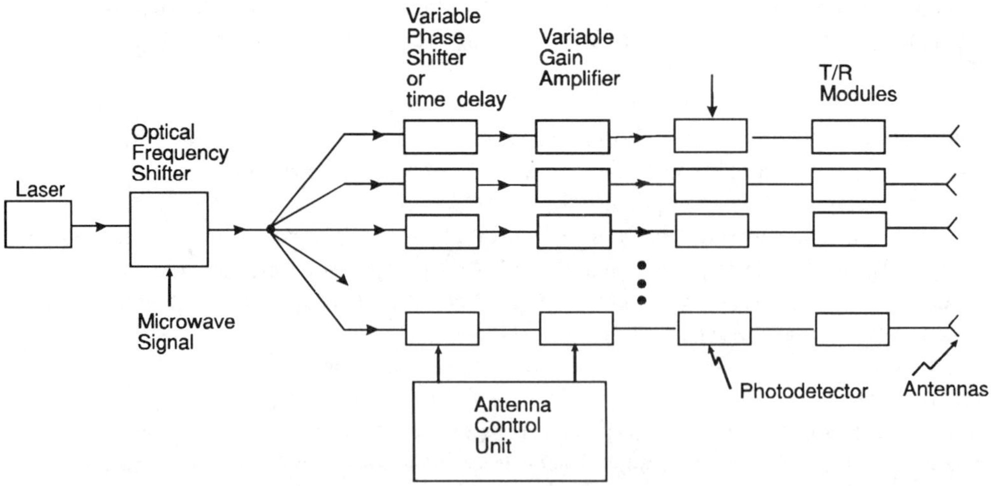


**Figure 6.1** Generic transmitter architecture of a phased-array radar system.

means of optically controlling the phase and amplitude have been proposed [16–35], but those that employ SLMs typically fall into three basic classes: phase-amplitude distribution systems, Fourier-beam-forming systems, and true time-delay systems.

Figure 6.1 is also the basic architecture of a phase-amplitude distribution system. In this system, discrete electro-optic elements are used to control the amplitude and phase of the optical beams of each channel. The amplitude-only and phase-only devices needed may be electrically or optically addressed, depending on the specifics of the architecture.

The typical Fourier-beam-forming transmitter system [21–25] shown in Figure 6.2 is based on the fact that the microwave field (phase and amplitude distribution) at the antennas is the Fourier transform of the desired far-field pattern and that a simple lens performs a Fourier transformation on an optical field passing through it. Thus, in this system, one constructs an optical replica of the desired far-field microwave pattern on an SLM and uses a lens to Fourier-transform this pattern. This Fourier-transformed optical pattern is then heterodyned detected using a plane wave reference beam, which is up-shifted by the radar frequency. The resulting spatially distributed optical signal component at the beat (radar) frequency is detected, amplified, and applied to the microwave transmitters. SLMs for these systems require high spatial resolution and very large gray-scale dynamic range (>40 dB) to accurately encode the far-field microwave patterns. True time-delay systems [26–35] are ideal for broadband radar, since in these systems the steered angle is not frequency-dependent and squint is eliminated. They involve generating the exact time delay between the antenna elements. Unfortunately, the typical phase delays needed are so large that they are difficult to implement on existing SLMs. Most systems, therefore, use SLMs to route light between one or more delay lines, which are typically free-space links or optical fibers of different lengths. A block diagram of such a true time-delay OCPAR system is shown in Figure 6.3.

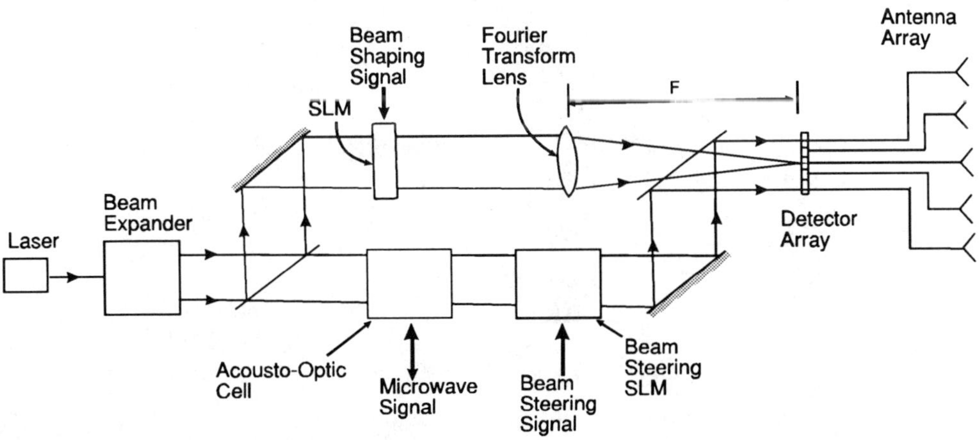


**Figure 6.2** Typical Fourier-beam-forming OCPAR transmitter architecture.

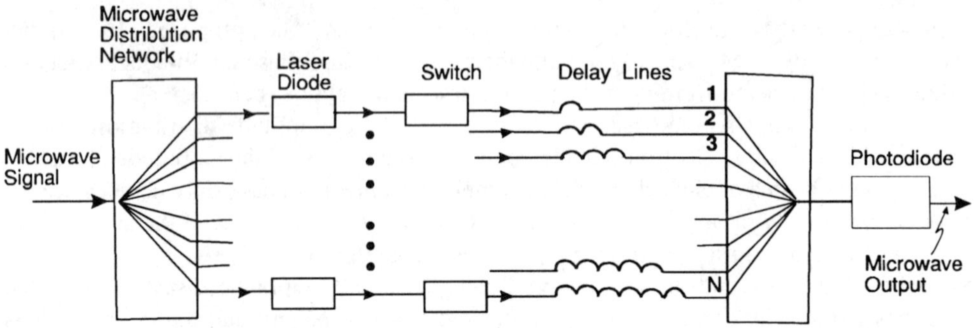


**Figure 6.3** One channel of a true time-delay OCPAR system employing electro-optic switching of laser diodes through any of $N$ lengths of optical fiber.

## 6.2 SLMs: STRUCTURES, PRINCIPLES, AND LIMITATIONS

The structure of a typical SLM is shown in Figure 6.4. It consists of a charge-generating element, an optional gain element, and a light-modulating element. For optically addressed SLMs, charge-generating elements include, photoconductors, photodiodes and photodiode layers, phototransistors, and photocathodes. For electrically addressed SLMs, the charge-generating element may be a hard-wired electrode array, an electron gun, a VLSI chip

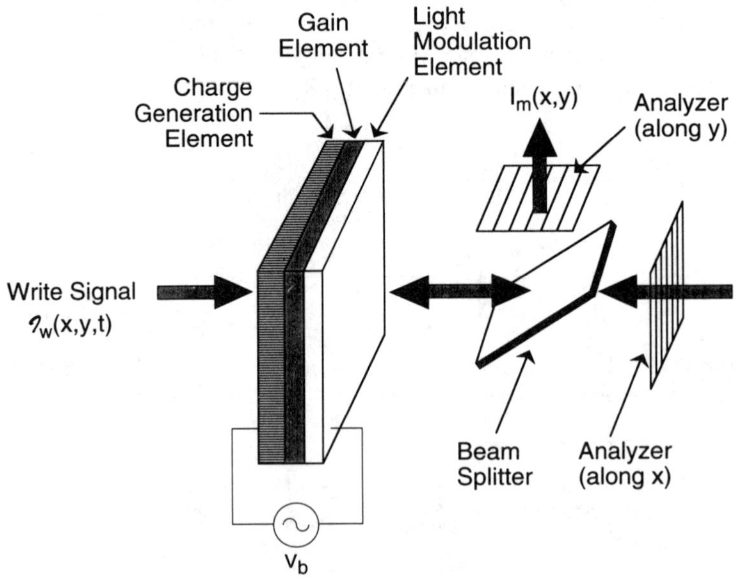


**Figure 6.4** Structure of a generic SLM.

or other electrical circuit. Examples of gain elements include microchannel plates (as in the microchannel SLM) [36–40] and transistor amplifier gain (as in some VLSI-driven SLMs). Gain may also be built into the charge-generating element (avalanche gain in a photodiode, for example). A wide variety of physical phenomena and materials have been used to realize light-modulating elements. These include:

- The electro-optic effect in inorganic materials such as $KH_2PO_4$, $Bi_{12}SiO_{20}$, and $LiNbO_3$, and in organic materials such as MNA and DAST;
- The photorefractive effect in $BaTiO_3$, $LiNbO_3$, and GaAs, for example;
- Molecular alignment and polarization switching via applied electric fields in liquid crystals;
- The Faraday effect in garnets;
- Electroabsorption in semiconductors and insulators (e.g., GaAs, GaAs MQW structures, and $KTaO_3$);
- Electrostatic deformation of membranes, gels, and oil films;
- The acousto-optic effect in materials such as $LiNbO_3$ and $TeO_2$; and
- Emission from light-emitting diode and laser diode arrays and from electroluminescent and cathodoluminescent phosphors.

In the context of Figure 6.4, the readout optical wave $\epsilon_r(x,y,t)$ can be written as

$$\epsilon_r(x,y,t) = \mathrm{Re}\{E_r(x,y)e^{-jwt}\} \tag{6.1}$$

where the spatial part of the wave is given by

$$E_r(x,y) = \hat{e}_r(x,y)A_r(x,y)e^{j\phi_r(x,y)} \tag{6.2}$$

Here $\hat{e}_r$ is the polarization vector, $A_r(x,y)$ is the amplitude, and $\phi_r(x,y)$ is the phase of the readout beam at the face of the modulator. Similar expressions hold for the modulated beam $\epsilon_m(x,y,t)$. For brevity, we will suppress the temporal (harmonic) component and represent the optical waves only by their spatial parts. Also, we will generally represent the write signal as $\varsigma_w(x,y,t)$, whether that signal is an optical exposure or intensity, or an electrical voltage or current.

SLMs may be classified as amplitude-only, phase-only, polarization-only, or intensity modulators. An amplitude-only modulator is defined by

$$A_m(x,y,t) = T_a(\varsigma_w(x,y,t))A_r(x,y) \tag{6.3}$$

$$\phi_m(x,y) = \phi_r(x,y) + \phi_{mo}(x,y) \tag{6.4}$$

$$\hat{e}_m(x,y) = [M_{ij}]\hat{e}_r(x,y) \tag{6.5}$$

where $T_a(\varsigma_w(x,y))$ is the amplitude transmittance of the modulator, $\phi_{mo}$ is the static additive phase owing to the light-modulating element, and $M_{ij}$ is the static Jones matrix [41] characterizing the polarization properties of the light-modulating element. Note that for an amplitude-only modulator, only the amplitude of the modulated beam is affected by the write signal. Thus, during operation, amplitude modulators follow trajectories that lie along the real axis of Figure 6.5.

In a similar vein, a phase-only modulator is defined by

$$A_m(x,y) = A_r(x,y)[1 - \alpha(x,y)]^{1/2} \tag{6.6}$$

$$\phi_m(x,y,t) = \phi(\varsigma_w(x,y,t)) \tag{6.7}$$

$$\hat{e}_m(x,y) = \hat{e}_m(x,y) \tag{6.8}$$

where $\alpha(x,y)$ is the static absorption loss in the modulator, and $\phi(\varsigma_w(x,y,t))$ is the phase modulation function. Note also that for phase-only modulators, we generally preserve the input polarization as indicated by (6.8). Phase-only modulators therefore follow trajectories that lie on or inside the unit circle of Figure 6.5.

A polarization-only modulator is defined by

$$A_m(x,y) = A_r(x,y)[1 - a(x,y)]^{1/2} \tag{6.9}$$

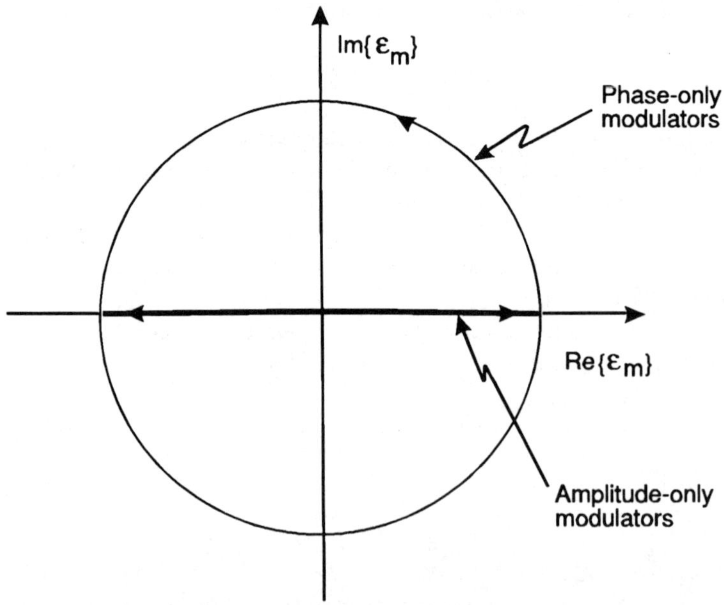


**Figure 6.5** Phase-space diagram for the modulated wave of an SLM.

$$\phi_m(x,y) = \phi_r(x,y) + \phi_{mo}(x,y) \tag{6.10}$$

$$\hat{e}_m(x,y,t) = [M_{ij}(\varsigma_w(x,y,t))]\hat{e}_r(x,y) \tag{6.11}$$

Here we see that only the Jones matrix $M_{ij}$ is a function of the write signal $\varsigma_w(x,y,t)$. Finally, an intensity modulator is defined by

$$I_m(x,y,t) = T_i(\varsigma_w(x,y,t))I_r(x,y) \tag{6.12}$$

$$\hat{e}_m(x,y) = [M_{ij}]\hat{e}_r(x,y) \tag{6.13}$$

where $T_i(\varsigma_w(x,y,t))$ is the intensity transmittance function of the modulator. Clearly, $T_i = T_a^2$.

Phase-only modulators are the most versatile of the four types. Indeed, a phase-only SLM offering very high resolution, high framing speed, and large phase dynamic range would permit the real-time synthesis of almost any conceivable optical element or transformation, since it could function as either a refractive or a diffractive, or both (e.g., holographic) element. The list of exotic active optical elements that, in principle, could be achieved with high-performance phase-only SLMs includes holographic optical elements, phase-only spatial filters, dynamic interconnection elements, and new optical transforms.

## 6.3 CHARACTERIZATION OF SLMs

The parameters by which SLMs are characterized and compared are defined in this section. In choosing an SLM for OCPAR systems or any general application, careful attention must be paid to these device parameters. The SLM responsivity characteristics are the functions $T_a(\varsigma_w(x,y))$, $T_i(\varsigma_w(x,y))$, $\phi(\varsigma_w(x,y))$, and $M_{ij}(\varsigma_w(x,y))$ in (6.3) through (6.13). Each is a measure of the sensitivity of that specific device characteristic to the write signal. Each characteristic embodies all the nonlinearities in the charge-generating and light modulating elements. For optically addressed devices, each characteristic is further a function of the spectral content of the write light. The spectral dependence of the responsivity is, therefore, an important characteristic, since it defines the write light wavelength range; that is, the range of wavelengths over which the charge-generating element is responsive. For electrically addressed devices, write signal current or voltage levels, rather than exposure, are the appropriate write signal characteristics.

For optically addressed SLMs, there is a specific value of $\varsigma_w(x,y)$, called the *write light exposure* (J/cm$^2$), which is the energy per unit area that must be received on the charge-generating element to change the amplitude of the modulated light from OFF to ON. The 10% to 90% value is often quoted, since most responsivity characteristics tend to be nonlinear. In a phase-only modulator, the half-wave exposure may be defined as the write light exposure that alters the phase of the modulated light by $\pi$ radians.

The switching energy per pixel (joules) is the total write energy (electrical and optical) required to switch a pixel from the OFF to the ON state. For phase-only light modulators, it is the total energy required to change the phase of the modulated light by $\pi$ radians.

The framing speed of an SLM, measured in hertz, is a measure of how fast it can update the modulated light in response to changes in the write signal. The framing speed is the reciprocal of the sum of the write and erase periods of the modulator. It is a critical parameter in OCPAR systems because it often determines how fast the antenna weights can be recomputed to steer the radar beam or to track moving targets. The framing speed of most SLMs is limited by either the response time of the light-modulating element or the ability of the charge-generating element to either supply current or establish the desired voltage.

A nematic liquid crystal SLM with an array of electrodes driven by a set of amplifiers is an example of a device whose framing rate is limited by the light-modulating material. The speed of this SLM is limited to about 50 Hz because of the slow rotational rate of the nematic liquid crystal molecules in response to the driving electric field. On the other hand, the microchannel SLM (MSLM) [36–40] is an example of a device whose framing speed is limited by the charge-generating element. Specifically, the response time of the MSLM is limited by the inability of the photocathode/microchannel-plate assembly to produce enough current to drive the light-modulating $LiNbO_3$ crystal at kilohertz rates.

The storage time of an SLM is the time that an image persists in the SLM after the write light is turned off. Storage time tends to depend on the capacitance and resistance of the light-modulating and charge-generating elements and, therefore, this characteristic varies widely from device to device. For example, nematic liquid crystal devices tend to have no intrinsic storage capability. The liquid crystal molecules relax back to the OFF state when the drive signal is turned off. On the other hand, the molecules in FLC SLMs switch between two stable orientations during operation and will remain in one of these states after the driving voltage is turned off. The MSLM employs a $LiNbO_3$ crystal, which has a very large resistivity, and it can store images as charge patterns on the surface of the crystal for several days.

The spatial resolution (cycles/mm or pixels/mm) is a measure of the smallest resolvable dimension within the active area of the SLM. For modulators with spatially discrete pixels, it is simply the number of pixels per unit length. For modulators with spatially continuous pixels, it is defined as the reciprocal of the width at half height of the spatial point spread function for amplitude modulation. It can also be measured by exciting the SLM with spatially sinusoidal write light of varying spatial frequency $f$ and plotting the roll-off of the response (MTF of intensity or phase) as a function of spatial frequency. The modulation transfer function (MTF) is defined by

$$\mathrm{MTF} = \frac{1}{m} \frac{I_{\max}(f) - I_{\min}(f)}{I_{\max}(f) + I_{\min}(f)} \tag{6.14}$$

where $I_{max}$ and $I_{min}$ are the peak and minimum intensities in the modulated beam, and $m$ is the modulation index of the write signal. The spatial resolution is usually quoted at the 50% point on the MTF curve. The spatial resolution is typically limited by either the charge-generating element, or the light-modulating element, depending on the geometry and materials parameters of the SLM.

The space-bandwidth product is simply the number of pixels. The number of pixels is determined by the device active area and the spatial resolution. For OCPAR Fourier-beam-forming transmitter systems, for example, the required number of pixels in the SLM can far exceed the number of antenna elements. A typical electrically addressed SLM today may offer 256 × 256 pixels with 100-Hz framing rate, and this performance falls short of what is often needed in a high-performance OCPAR system.

The contrast ratio of an amplitude-only or intensity SLM is defined as the ratio of the modulated-light ON-state intensity to the OFF-state intensity. To get a meaningful estimate of contrast ratio, this measurement should be made over the entire device and not at a single point or pixel. Thus, the contrast ratio (CR) of an SLM is defined by

$$CR = \frac{\langle I_{max} \rangle}{\langle I_{min} \rangle} \tag{6.15}$$

where the angle brackets denote spatial averaging over the active area of the SLM.

The gray-scale dynamic range is simply the number of resolvable gray levels. It is typically specified in bits. To be meaningful, this measurement should be made as an average over the entire active area of the device. It is defined by

$$DR = \frac{\langle I_{max} \rangle - \langle I_{min} \rangle}{\langle \Delta I \rangle} \tag{6.16}$$

where quantity $I_{min}$ is the intensity of the modulated beam with the write signal turned off, and $I_{max}$ is the maximum (but nonsaturated) value of the modulated output light. The quantity $\langle \Delta I \rangle$ is the rms value of the spatial fluctuations in the modulated light owing to a uniform write signal that yields a modulated output signal that is nominally midway between $I_{max}$ and $I_{min}$. A similar expression can be written for the gray-scale phase dynamic range.

There are two principal types of spatial nonuniformities. Static spatial nonuniformities are those which are write-signal-independent. These are represented by the terms $\alpha(x,y)$ and $\phi_{mo}(x,y)$ in (6.3) through (6.13). They can generally be compensated out with amplitude and phase masks, and often the masks can be made with holographic recording techniques. The more serious type of spatial nonuniformities are dynamic (write-signal-dependent) nonuniformities; that is, spatial nonuniformities in the responsivity characteristics $T_a(\varsigma_w(x,y))$, $T_i(\varsigma_w(x,y))$, $f(\varsigma_w(x,y))$ and $M_{ij}(x,y)$. Look-up tables are often needed to perform the correction. Spatial nonuniformities act to reduce the gray-scale dynamic range of the modulator.

The readout wavelength range is the range of readout wavelengths over which the light-modulating element can perform useful modulation. The readout wavelength range can be limited by the spectral transmission or reflectivity of the light-modulating element, or by the fall-off of the modulation sensitivity with readout wavelength. Readout-wavelength-dependent sensitivity is common in electro-optic and in some diffractive-light-modulating elements where phase modulation is employed. Thus, in the context of Figure 6.4, the readout optical efficiency for an SLM is defined as

$$\eta(\lambda) = \frac{I_{\max}(\lambda_r)}{I_r(\lambda_r)} \tag{6.17}$$

where $I_{\max}(\lambda_r)$ is the maximum value of the modulated light, $I_r(\lambda_r)$ is the corresponding readout light intensity, and $\lambda_r$ is the wavelength of the readout light.

If the write and readout wavelength ranges overlap, then the SLM is said to be cascadable. That is, cascadability is the property of a modulator to be written on and read out with light of the same wavelength.

Some SLMs, because of the peculiarities of their design, offer unique image processing functions. For example, the PRIZ (a Russian acronym) [42,43] operates normally in an edge enhancement mode, and the variable grating mode device [44–47] naturally performs conversion of intensity to spatial frequency. Such special-purpose devices can be quite useful and should be considered for exploitation in OCPAR systems.

Other parameters by which SLMs are characterized include optical power handling capacity, which is the maximum readout optical power that can be tolerated without damage to the SLM; SLM voltage, current, and/or electrical power requirements; operating temperature and thermal stability; and device lifetime, ruggedness, and cost.

## 6.4 SPECIFIC SLM ARCHITECTURES

In this section, we select and discuss a few of the most promising approaches being pursued for realizing high-performance SLMs that could find applications in OCPAR systems.

### 6.4.1 Electro-Optic SLMs

This class of devices employs materials that change their refractive index in response to an applied electric field. As an example, let us consider an optically addressed longitudinal electro-optic SLM with the architecture shown in Figure 6.4. The write light of intensity $\varsigma_w$ and wavelength $\lambda_w$ incident on the charge-generating element generates charge with a quantum efficiency $\eta$. This charge is amplified with a gain $G$ by the gain element and deposited at the interface of the light-modulating element. Thus, after an exposure time $t_w$, the accumulated charge density at the light-modulating material interface is

$$\sigma = \frac{\varsigma_w t_w \eta e G \lambda_w}{\text{hc}} \text{ Coul/cm}^2 \tag{6.18}$$

and thus the voltage developed across the material is

$$V = \frac{\sigma L_c}{\epsilon_{\|}\epsilon_0} \tag{6.19}$$

where $e$ is the electron charge, h is Planck's constant, c is the speed of light, $L_c$ is the thickness of the electro-optic material, $\epsilon_{\|}$ is the longitudinal dielectric constant of the material, and $\epsilon_0$ is the permittivity of free space.

For a material exhibiting the linear electro-optic effect, this voltage induces a refractive index change of the form [48]:

$$\Delta n_i = \pm(1/2)n_0^3 r_{ij} E_j \tag{6.20}$$

where $E_j$ is the applied electric field ($E_j = V/L_c$), $n$ is the refractive index of the material, and $r_{ij}$ the appropriate electro-optic coefficient. Here, $i = 1, 2, \ldots, 6$, $j = 1, 2, 3$, with $1 = x$, $2 = y$, $3 = z$, $4 = yz$, $5 = xz$, and $6 = xy$. For example, an applied electric field along the z-axis of an electro-optic uniaxial crystal could induce a phase retardation of $\Gamma = \phi_x - \phi_y = (2\pi/\lambda)(\Delta n_x - \Delta n_y)$, where $x$ and $y$ are two electric-field-induced principal axes.

For amplitude modulation, such an SLM is read out between crossed polarizers with the polarizer axes bisecting the induced principal axes. It is straightforward to show that the modulated light intensity is given by [48]

$$I_m = I_r \sin^2\left(\frac{\Gamma}{2}\right) = I_r \sin^2\left[\left(\frac{\pi}{2}\right)\frac{V}{V_\pi}\right] \tag{6.21}$$

where the quantity $V_\pi$ (corresponding to $\Gamma = \pi$) is a materials characteristic called the *half-wave voltage*. For a device operated in the reflection mode with a double pass through the electro-optic material, $V_\pi$ is of the form [48]

$$V_\pi = \frac{\lambda}{4n_0^3 r_{ij}} \tag{6.22}$$

For phase-only modulation, both polarizers are oriented along one of the induced principal axes, although in principle the second polarizer (analyzer) can be omitted. For polarization modulation, the readout light is polarized along the bisector of the induced axes, and no analyzer is used. This device will generate all possible states of polarization as the electric field across the light-modulating element is varied.

The spatial resolution of longitudinal electro-optic devices is generally limited by field fringing in the electro-optic material. Field fringing is a function of the thickness and dielectric constants of the material. Theoretically, the crystal-thickness-limited resolution (owing to the fringing of the electric fields in the light-modulating element) is [49]:

$$f^* = \frac{\epsilon_{\parallel}/L_c + \epsilon_d/L_d}{2\pi[(\epsilon_{\perp}\epsilon_{\parallel})^{1/2} + \epsilon_d]} \tag{6.23}$$

and for $\epsilon_d < \epsilon_{\parallel}, \epsilon_{\perp}; L_d > L_c$:

$$f^* \approx \frac{1}{2\pi L_c}\left(\frac{\epsilon_{\parallel}}{\epsilon_{\perp}}\right)^{1/2} \tag{6.24}$$

Here $\epsilon_{\perp}$ is the transverse dielectric constant of crystal, and $\epsilon_d$ and $L_d$ are the dielectric constant and thickness, respectively, of either the charge-generating element itself or the layer (if any) that shares the applied voltage with the light-modulating element.

For optimum performance, the electro-optic light-modulating material for a longitudinal SLM with the architecture shown in Figure 6.4 should have large electro-optic coefficients, large refractive index, large bulk and surface resistivity, large dielectric strength, large thermal conductivity, high crystal symmetry, wideband spectral transmission, low optical dispersion, low optical absorption and scattering losses, small dielectric constants, small temperature coefficients, low vapor pressure, no optical activity, no photoconductivity, no photovoltaic effect, and no pyroelectric or piezoelectric effects. The most significant of these requirements are not met simultaneously in any known material, and therein lies the difficulty.

Electrically addressed electro-optic SLMs typically employ a hard-wired matrix of discrete electrodes, high-voltage integrated-circuit arrays, or electron beams [50–56] as the charge-generating element. Optically addressed electro-optic crystal SLMs that have been realized include: the MSLM [36–40], which employs a photocathode and a microchannel plate addressing a lithium niobate crystal; a deuterated potassium dihydrogen phosphate (DKDP) light valve employing amorphous silicon as the charge-generating element [55–58]; the Pockels readout optical modulator (PROM) [59–62]; and the PRIZ [42,43]. The latter devices both employ a single photorefractive material that functions as both the charge-generating and light-modulating element (see Section 4.2.4). Liquid crystals also exhibit the electro-optic effect, but these are a special case, which will be treated in greater detail in the next section.

### 6.4.2 Photorefractive Materials and Devices

A material is said to be photorefractive if its refractive index changes upon exposure to optical radiation [63–67]. Materials exhibiting this effect include $Bi_{12}SiO_{20}$, $BaTiO_3$, $Sr_xBa_{1-x}Nb_2O_6$, $LiNbO_3$:Fe, $KNbO_3$:Fe, GaAs:Fe, and InP. Photorefractive devices often consist simply of a rectangular block of photorefractive material. Electrodes are optional in some devices. Photorefractive devices can be classified as either volume holographic or nonholographic, and the volume holographic devices are capable of real-time holographic recording. Photorefractive materials are finding applications in optical memory, optical interconnects, and real-time wavefront phase conjugation.

In volume holographic devices, information is written to the material using holographic recording techniques as depicted in Figure 6.6. In this case, the nonuniform write light ionizes impurity ions and/or lattice defects, creating spatially varying internal charge distributions. This charge is subsequently transported by drift and/or diffusion from the brighter to the darker regions of the material. The resulting space-charge electric fields modulate the refractive index of the material via the electro-optic effect.

Photorefractive materials, therefore, are essentially electro-optic photoconductors operating at photon energies below the band gap of the material. The best performance is generally obtained by operating in the regime where the spatial structure in the input image is comparable to the drift and diffusion lengths in the material. Spatial resolutions on the order of 2,000 cycles/mm are usually achieved.

By employing two-, three-, or four-wave mixing, the volume holographic devices can perform multiplication, amplification, and incoherent-to-coherent image conversion [68–73]. However, because readout is often destructive and the writing process is interferometric, applications of these devices are usually constrained by the need for a source of polarized coherent light, a mechanically stable platform suitable for holographic recording,

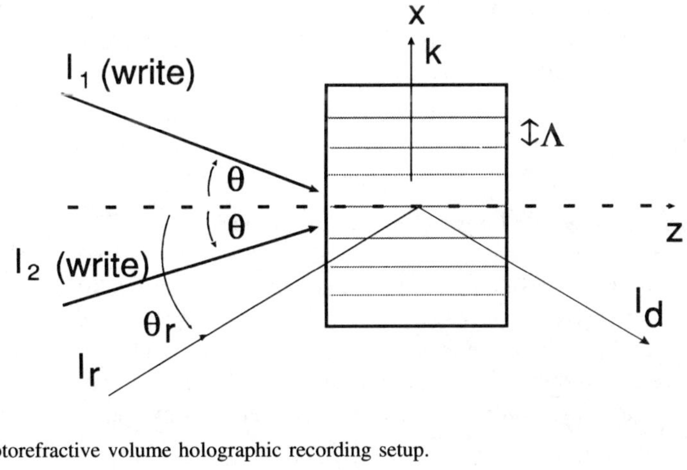


**Figure 6.6** Photorefractive volume holographic recording setup.

and auxiliary light beams with specific geometric, polarization, and/or intensity characteristics relative to the input information-carrying beams.

The framing speed of a photorefractive device must include both the write and erase times. The write time is exposure-dependent and hence can be decreased by increasing the intensity of the writing beams. On the other hand, the erase time tends to be strongly limited by the recombination time and the mobility of the carriers. High carrier mobilities and low dielectric constants lead to fast response times.

The ideal photorefractive material for high-speed information processing should exhibit:

- Excellent spatial uniformity (>97%) in its optical and electrical characteristics across large dimensions (~25 mm);
- Large electro-optic coefficients;
- Large dark resistivity;
- Large photoconductivity;
- Large mobility-lifetime product;
- Large carrier drift or diffusion lengths;
- Low residual strain birefringence;
- Large saturation diffraction efficiency (>50%) without the need for an externally applied electric field;
- Large phase conjugate reflectivity;
- Large four-wave mixing gain (>10 $cm^{-1}$);
- Low dielectric constant;
- No photovoltaic effect;
- No optical activity;
- Efficient operation at near infrared write wavelengths, making it compatible with semiconductor lasers;
- Operation with very low-power (<100 $\mu W/cm^2$) write beams and;
- Operation in an anisotropic diffraction mode such that the output signal beams can be isolated from the write beams through the use of polarizers for high SNR, and in addition, it should be simple to produce and process.

The above list implies certain conflicting materials properties, so trade-offs have to be made when choosing a material (e.g., low-level intensity recording for speed).

*Self-Pumped Photorefractive Phase-Conjugating Mirrors*

Optical phase conjugation is the process of the reversal of the direction of wave propagation. A system that performs phase conjugation, therefore, behaves like a special kind of mirror that causes an incident beam to retrace its path upon reflection. A self-pumped phase conjugator is a system in which the beam to be phase-conjugated serves as its own pump [74–77]. This unique property has been observed in some photorefractive materials (e.g., $BaTiO_3$).

In $BaTiO_3$, for example, it is believed that the mechanism is as follows. A strong pump beam entering an appropriately oriented crystal scatters from imperfections within the crystal. The beams scattered from one location within the pump beam interfere with the spatially coherent pump beam at other locations to set up internal space-charge gratings. However, the particular set of beams which are scattered in a direction such as to suffer internal corner-cube reflections are special. These are retroreflected into the gratings and thereby generate a counterpropagating pump beam which exits the crystal as a phase-conjugated beam. This process is illustrated schematically in Figure 6.7. Self-pumped phase-conjugated systems have already been proposed for use in OCPAR systems for self-alignment and aberration correction [24].

*Photorefraction-Based 3-D Optical Memories*

Because information can be recorded holographically in photorefractive materials, these systems are also being investigated for optical memory applications [78–87]. Both angle- and wavelength-multiplexed schemes are being vigorously investigated. The simultaneous storage of several hundred holograms has already been demonstrated in iron-doped lithium niobate [78]. Techniques to fix the holograms so that they are resistant to erasure by light are being explored in a number of materials. These 3-D page-oriented memories may possibly be useful for lookup table operations in an OCPAR system.

*Photorefractive Incoherent-to-Coherent Optical Converter*

This SLM consists of a slab of photorefractive material with the write/readout setup illustrated in Figure 6.8 [88–90]. In the work reported on the photorefractive incoherent-to-coherent optical converter (PICOC), a slab of bismuth silicon oxide was used as the

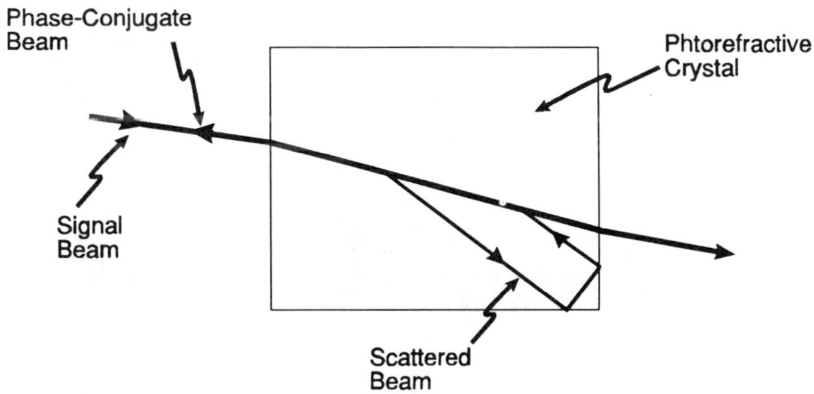


**Figure 6.7** Self-pumped phase-conjugating mirror system.

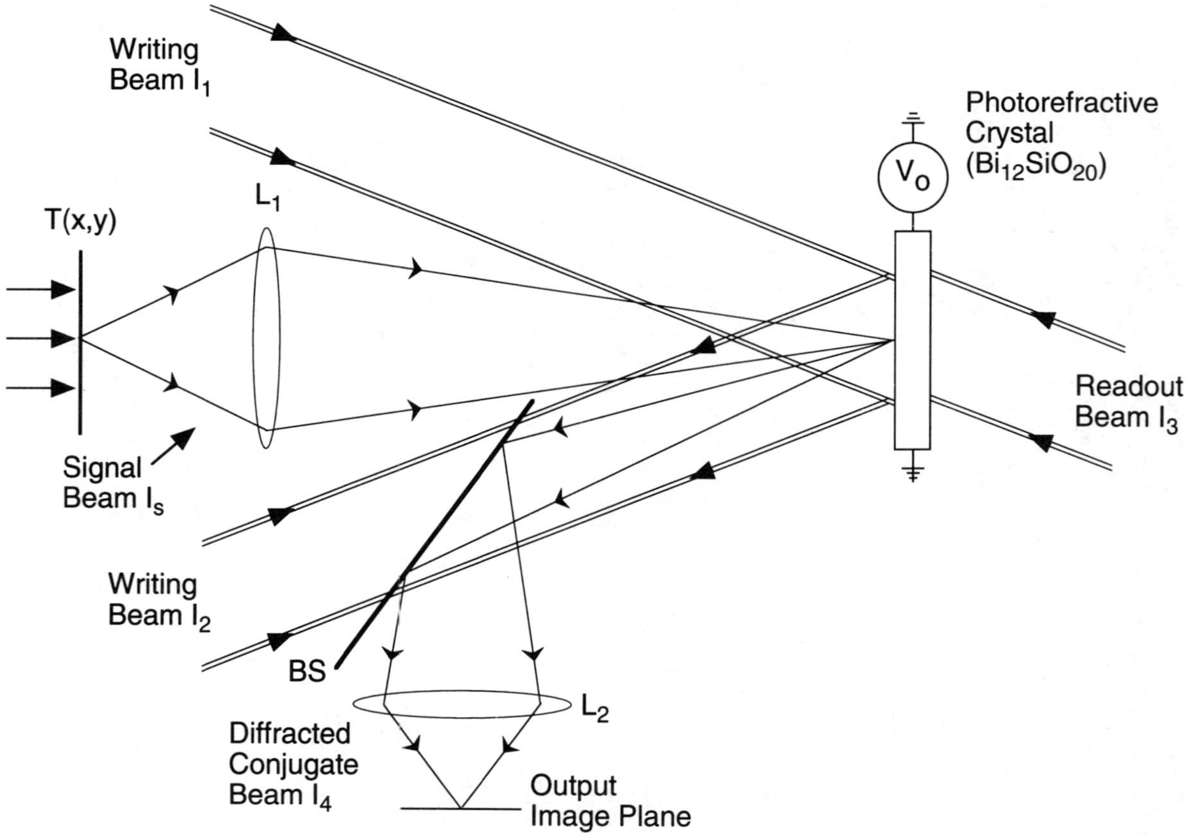


**Figure 6.8** PICOC recording and readout scheme.

photorefractive material. Drift was the dominant mode of charge transport in this device, with a transverse electric field applied across the slab.

In operation, a grating is first written in the material with blue or green coherent light of wavelength $\lambda_w$. The incoherent image to be processed is then imaged into the crystal so that its intensity selectively erases the grating. A third coherent readout beam of red or infrared light of wavelength $\lambda_w$ reads out the grating at the Bragg angle to produce a diffracted image, which is a negative replica of the incoherent image. The applied electric field in the $Bi_{12}SiO_{20}$ device [89,90] was on the order of 4 kV/cm, the limiting spatial resolution of the device was observed to be ~500 line pairs/mm, the framing rate was approximately 1 Hz, and the storage time was about 1 hour in the dark. Its applications include incoherent-to-coherent image conversion, Fourier-plane spatial filtering, and programmable mask generation.

*Pockels Readout Optical Modulator*

The PROM [59–62] is a nonholographic longitudinal photorefractive device. It uses a thin (~500 $\mu$m) wafer of photorefractive $Bi_{12}SiO_{20}$ with a pair of transparent electrodes that are separated from the crystal by 5-$\mu$m-thick blocking layers of paralene. The architecture of the PROM is actually a special case of the generic SLM architecture of Figure 6.4, in which the charge-generating element and the light-modulating element are one and the same.

Blue or near-ultraviolet light incident on the device generates an image-wise distribution of hole-election pairs within the bulk of the crystal. The mobile electrons drift through the bulk to the electrodes under the influence of an applied longitudinal electric field. The space-charge field set up by the immobile hole pattern and the transparent electrodes modulates the refractive index of the crystal and thereby the phase (or amplitude, when polarizers are employed) of the readout light. To avoid destructive readout, red readout light is employed. This device is thus not cascadable.

Erasure of the stored image is accomplished by flooding the crystal with light containing blue or near-ultraviolet components. Current devices exhibit a spatial resolution of about 5 to 10 cycles/mm at 50% MTF, framing rates of 10 to 30 Hz, 1/e exposures in the neighborhood of 10 $\mu J/cm^2$, contrast ratios of about 1,000:1, and operating bias voltages of about 4 kV [59,60]. The storage time depends on the ambient illumination. Storage times of about one hour in the dark have been reported.

### 6.4.3 Liquid Crystal SLMs

Liquid crystal SLMs are considered to be among the most promising for low-voltage, high-speed, high-resolution applications. These devices can be optimized for phase-only or amplitude-only operation. They typically operate on about 5V to 20V and at low currents. Many types of liquid crystals have been employed in SLMs, most notably the nematics and the ferroelectrics.

Liquid crystal light-modulating cells are usually fabricated by sandwiching a liquid crystal layer between two glass plates coated with molecular alignment layers. Figure 6.9 shows a typical optically addressed nematic liquid crystal SLM employing the traditional sandwich structure of Figure 6.4. This structure includes a photoconductive layer, a light-blocking layer, a dielectric mirror stack, two molecular alignment layers, transparent electrodes, and the liquid crystal layer. These devices are usually read out by placing polarizers in both the readout and modulated beams as shown in Figure 6.9. For phase-only modulators, the alignment layers are oriented parallel to each other and so are the input and output polarizers.

For intensity modulators, the angle between the alignment layers may be 900 or 2,700 for transmission modulators and 450 for reflection modulators. The nematic liquid crystal molecules are pinned along the alignment layer directions at the walls of the cell, and so within the cell the molecular axes twist in going from one alignment layer to the other. When linearly polarized light is passed through such a cell, its plane of polarization

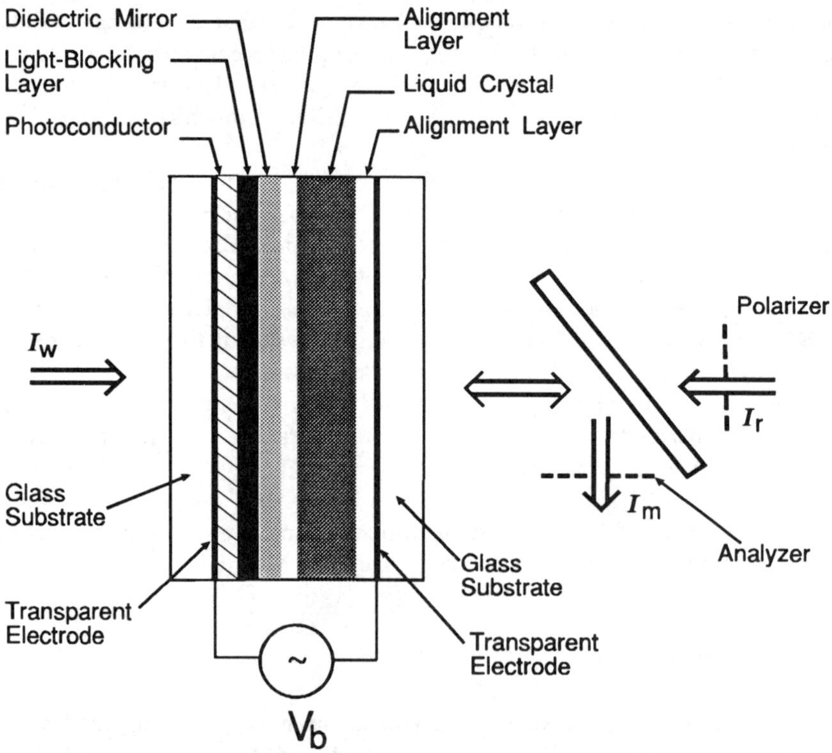


**Figure 6.9** Architecture of an optically addressed nematic liquid crystal light valve.

rotates because of coupling to the molecular axis (twisted nematic effect). In these geometries, the input and output polarizers are aligned for extinction or crossed for maximum transmission when the applied electric field is zero. A spatially varying longitudinal electric field applied across the nematic liquid crystal layer causes the molecular axes to reorient along the field lines. This alters the average birefringence of the liquid crystal layer. The (output polarizer) analyzer converts this birefringence to intensity modulation.

A rich variety of nematic liquid crystal SLMs have been built to date [91–105]. These include optically addressed devices driven by a variety of photoconductor charge-generating elements, such as cadmium sulfide, GaAs, bismuth silicon oxide, and amorphous and crystalline silicon photoconductive layers [91–97]. Electrically addressed nematic liquid crystal devices have been driven by hard-wire electrode arrays, VLSI chips, and electron beams [98–105]. Even pocket televisions driven by thin-film transistors employing nematic liquid crystals have been modified for use as SLMs [106–114].

Nematic liquid crystal SLMs offer gray-scale intensity modulation but slow framing speeds (~40 Hz) because the molecular rotation is sluggish. Intensity-modulation devices tend to have very thin liquid crystal layers (on the order of 2 to 10 $\mu$m) and consequently can offer spatial resolutions up to 50 line-pairs per millimeter. However, the typical intensity-modulating nematic liquid crystal SLM offers low contrast ratio (on the order of 40:1), and the ON-OFF exposure of photoconductor-based optically addressed devices typically falls in the 1- to 5-$\mu$J/cm$^2$ range. Phase-only modulators with nematic liquid crystal layers as thick as 15 $\mu$m have been built and have yielded phase dynamic ranges as large as $30\pi$ radians [98], but at low framing speeds. Such large phase shifts could be useful for OCPAR system concept verification.

FLC SLMs [115–127] are optical polarization switching devices in which the liquid crystal molecules rotate between two stable states of equal energy. The switching between these two states is quite rapid, and this accounts for the high switching speeds (~100 kHz) of these materials. Binary intensity modulation is achieved by reading out the device between crossed polarizers with the readout light polarizer along one of the two stable molecular orientations. In this case, the modulated light intensity is given by [127]

$$I_m = I_r \sin^2(4\theta) \sin^2(\pi \Delta n L_c / \lambda) \tag{6.25}$$

where $\theta$ is the optical tilt angle of the molecules, $\Delta n$ is the birefringence, $L_c$ is the liquid crystal thickness, and $\lambda$ is the wavelength of the light. Thus, in contrast to the nematics, FLCs offer high switching speeds but no gray scale. FLCs driven by VLSI chips [115–121] and by photodiode layers [122–126] have been demonstrated.

Several of the liquid crystal SLMs referred to above (both nematics and ferroelectrics) have been proposed for use in OCPAR systems [23,26,28,31,32].

### *E-Beam-Addressed Liquid Crystal CTP SLMs*

The structure of the electron-beam-addressed charge-transfer plate (CTP) liquid crystal light modulator (e-LCLM) is shown in Figure 6.10. It consists of an electron gun that

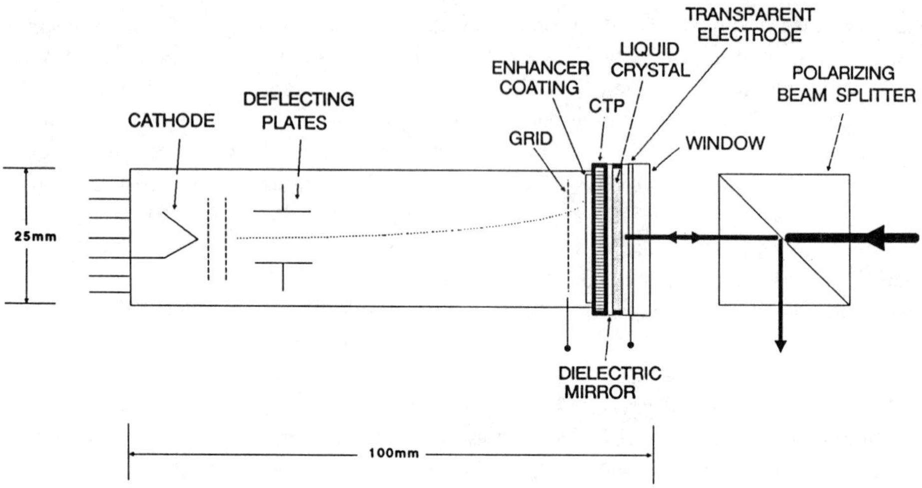


**Figure 6.10** E-beam-addressed liquid crystal SLM.

scans a focused electron beam across a special CTP liquid crystal anode, which also acts as the face plate of the electron gun [98].

As illustrated in Figure 6.11, a CTP is a dense array of longitudinally oriented conducting pins embedded in a wafer of electrically insulating material. CTPs of 25 mm in diameter with conductors 10 $\mu$m in diameter on 13-$\mu$m centers in hexagonal close-packed patterns have been built in thicknesses up to 4 mm. The surfaces of the CTP are highly polished to an optical flatness of $\lambda/10$ before one surface is overcoated with the alignment layer.

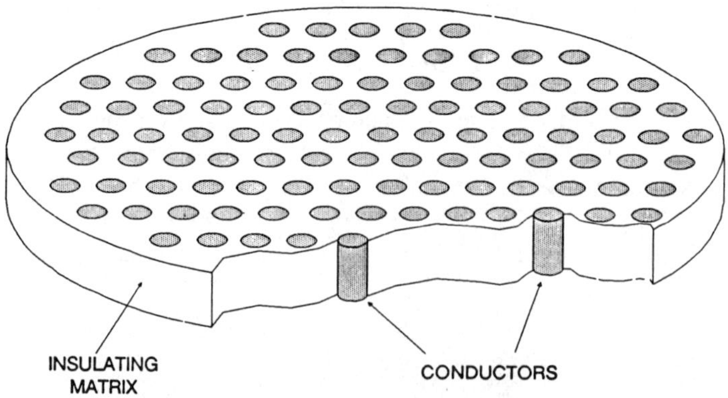


**Figure 6.11** CTP consisting of a dense array of conducting pins embedded in an insulating medium.

A novel addressing scheme that exploits the secondary electron emission characteristics of the CTP is employed to ensure that the time-averaged addressing voltage across the CTP is zero (liquid crystals undergo electrochemical degradation when exposed to long-term dc fields). In a recently built phase-only e-LCLM, for example, dynamic lenses of varying focal lengths were demonstrated by writing the appropriate charge distributions on the CTP surface [98]. Without some kind of interface component, such as a CTP, e-beam addressing of a liquid crystal layer would be impossible. Details on other e-beam-addressed liquid crystal light modulators are discussed elsewhere [99–105].

Possible applications of the phase-only e-LCLM in OCPAR systems include time delay, Fourier beam forming, beam steering, and programmable general-purpose optical elements. Other applications of these devices include nearly real-time computer-generated holographic displays, phase-only spatial filters, wavefront phase correction for communication and imaging through turbulence, and correction of aberrations in optical instruments. The amplitude device has been applied to large-screen projection displays [99–104]. It can also be used as an input-plane transducer or as a Fourier-plane filter in optical processors.

*Electrically Addressed Silicon VLSI Liquid Crystal Light Modulators*

These silicon VLSI circuit liquid crystal SLMs consist of an electrically addressed silicon VLSI chip driving a liquid crystal light-modulating element [115–119]. As such, they combine the computation strengths of integrated circuits (IC) with the processing and communication strengths of optics. Their advantages also include (1) low cost and rapid fabrication, since they use the mature silicon VLSI technology for leverage, (2) flexible device design and functionality, (3) low-voltage operation (signal levels for nematics are the same as for typical VLSI chips) and compact structures, and (4) easy integration into existing electronic processors.

The disadvantages of devices built thus far include (1) a compromise of the response time for devices employing ferroelectrics at low voltages, (2) a loss of contrast ratio due to nonflat chip surface topology, and (3) currently low fill factors.

Electrically addressed silicon VLSI liquid crystal SLMs have been demonstrated using ferroelectrics, nematics, and nematic guest-host liquid crystal structures. Static random-access memory (SRAM) chips and dynamic random-access memory (DRAM) chips have been used as drivers. The ferroelectrics, although they typically lack gray-scale light-modulating capability, are preferred over nematics, because they offer higher modulation speed.

One of the major fabrication difficulties of silicon VLSI liquid crystal SLMs is that unless special processes are employed to planarize the surface of the chip and protect the circuits, alignment layers on the chip are impractical. By using a CTP as an interface element, this problem is overcome [98,99]. Electrically addressed silicon VLSI CTP liquid crystal SLMs are under development [99]. These devices have the general structure shown in Figure 6.12, in which VLSI chips are bump bonded to CTP liquid-crystal anode assemblies to form SLMs.

The specific advantages of using a CTP are: no possible damage to the chip from alignment layer deposition and processing; the CTP permits the use of two liquid crystal alignment layers and so there is no loss of contrast ratio as compared with the early VLSI liquid crystal devices that employed only one alignment layer; and excellent spatial uniformity can be achieved in the liquid crystal cell because the CTP can be polished flat to $\lambda/10$ in the visible.

### *Optoelectronic IC Liquid Crystal SLMs*

Optoelectronic IC (OEIC) SLMs [119–121,128–144] are an extremely powerful class of compact optically addressed hybrid processors. Both silicon-based and GaAs-based devices are being investigated. These smart-pixel SLMs combine an optical detector, electronic circuitry, and either a light-generating or light-modulating element in each pixel (see Figure 6.13). Recent advances in such devices are bringing us closer to optimally combining the long-range off-chip communication strengths of optics with the computational and short-range on-chip interconnection strengths of IC technology.

Among the important devices being investigated are the optically addressed silicon VLSI circuits driving on-chip liquid crystal light-modulator cells [119–121]. For example, a 32 × 32 optically addressed silicon VLSI FLC SLM has been fabricated using a 2-$\mu$m, n-well, low-noise, analog complementary metal oxide semiconductors (CMOS) process [119,120]. The circuit at each pixel consisted of a phototransistor, an amplifier, and a CMOS inverter that drove the liquid crystal modulator pads. The electrical circuitry in

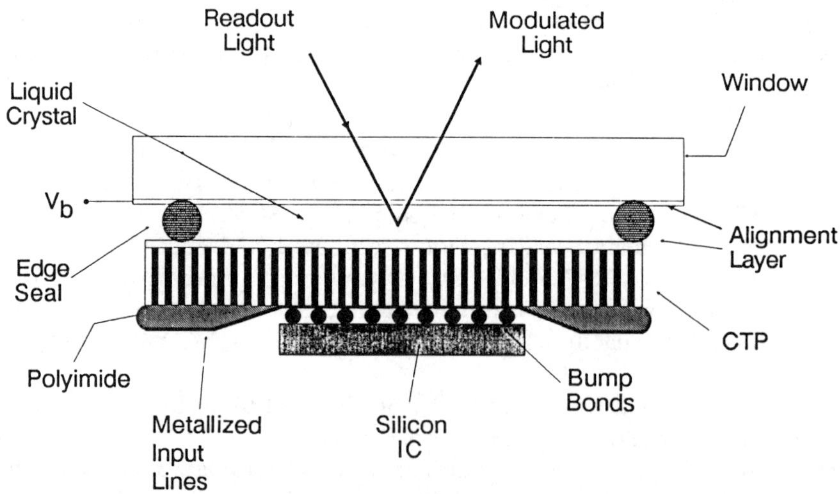


**Figure 6.12** VLSI liquid crystal SLM employing a CTP as a planarizing and support substrate.

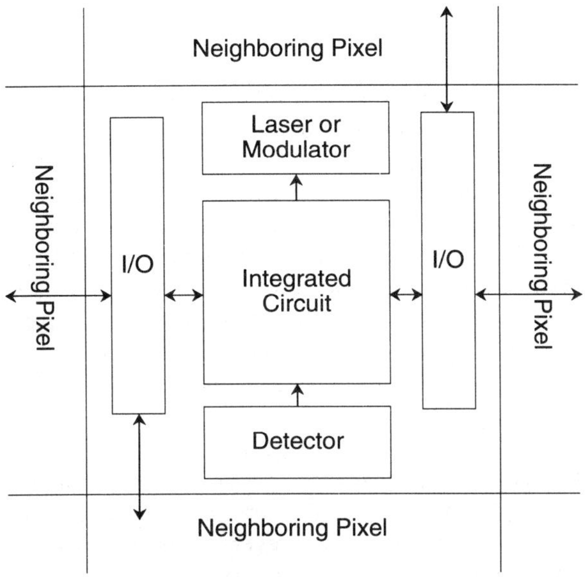


**Figure 6.13** Pixel structure of an optoelectronic IC SLM.

each pixel performed either thresholding or logarithmic amplification. The pixels were located on $100 \times 50$-$\mu$m centers, the modulating pad was $46 \times 50$ $\mu$m, and the n-well region of the phototransistor was $40 \times 36$ $\mu$m. A major disadvantage of this approach was the undesirable operating constraints arising from having the readout and write beams incident from the same side of the device.

Two-dimensional OEIC SLMs, when cascaded with holographic interconnection elements, will find applications in neural network processors, optical inference engines, cellular automaton networks, and pattern recognition processors. Because of their versatility, these smart-pixel SLMs are expected to find future applications in OCPAR systems.

### 6.4.4 Optoelectronic Bistable Optical Devices

These are a special class of optoelectronic SLMs. Basically, a bistable optical device is one that will switch the readout light into one of two stable states, depending on either the intensity or the exposure of the write light. Both two-port and three-port devices exist. To observe bistability, an optical feedback mechanism must exist in conjunction with an optically induced change in an optical property of the device material. Fabry-Perot cavities are often used to provide the feedback. The most popular technologies employed involve optically addressed GaAs VLSI chips driving arrays of MQW modulators, light-emitting diodes, and laser diodes.

*GaAs MQW Self-Electro-Optic Device*

The self-electro-optic device (SEED) [129–138] is a bistable optical switch that employs a pin diode structure with GaAs/GaAlAs MQW layers embedded in the insulating region of the diode. A series resistor and a constant-voltage power supply provide the biasing network for the device. In operation, the wavelength of the incident light is chosen near the exciton resonance for zero voltage bias across the diode. With low optical power, nearly all the supply voltage is dropped across the diode. This voltage shifts the exciton absorption to longer wavelengths and lower absorption results. The operation of the device is based on the quantum-confined Stark effect, whereby increasing the optical power reduces the voltage across the diode and gives rise to increased absorption at the bistable wavelength, leading to regenerative feedback and switching. Switching times of about 1 ns with 7.5 fJ/$\mu m^2$ are typical.

To overcome problems associated with the absolute optical power required to switch on the device, a two-component symmetric SEED has been built, in which bistable switching is based on the ratio of two input beams rather than on the absolute power in a single beam. This device consists of two reverse-biased MQW pin diodes placed in series and often grown on top of a mirror structure.

Very large arrays of symmetric SEEDs (up to 2,048 in 1991) have been built. Recently, FET SEEDS with about 20 transistors per pixel have been built and operated with less than 100 fJ and run at hundreds of megahertz. Their applications include optical logic elements and optical computing, but so far these devices appear not to have found applications in OCPAR.

*Thyristor-Type Devices*

The vertical-to-surface transmission electro-photonic (VSTEP) device is essentially a GaAs/AlGaAs double heterostructure optical thyristor [139,140]. The inner multilayer structure, consisting of p-AlGaAs/n-GaAs/p-GaAs/n-GaAs, acts as a signal absorber in the OFF state and as a light emitter in the ON state. With an applied bias below the thyristor turn-on voltage, the device can be switched on with the optical input, and it maintains this state until switched off with a negative voltage pulse. Optical switching energies below 26 fJ/pixel and a power dissipation of 2 $\mu$W per pixel have been achieved, along with integration of these devices in a 1k array. Other related devices are discussed in detail elsewhere [141–144].

### 6.4.5 Deformable Mirror Devices

The family of deformable mirror devices (DMD) [145–154] consists of a silicon IC chip electrostatically driving an array of deformable metal mirrors suspended above the circuitry. These electromechanical devices are fabricated entirely by microelectronic techniques, and as such are expected to be low in cost. Three basic pixel mirror structures are used: cantilever-beam, torsion-beam, and flexure-beam elements [145]. The set of DMDs includes the inverted clover leaf DMD (both line- and frame-addressed), and the torsion beam DMD. The line-addressed DMD is a monolithic silicon integrated device consisting of a 2-D array of inverted clover leaf mirrors with a fill factor of about 25%. Each pixel ($51 \times 51$ $\mu$m) consists of four cantilever hinged mirrors that are electrostatically addressed by electrodes connected to the floating sources of an array of metal oxide semiconductors (MOS) transistors. Addressing circuitry consists of a charge-coupled serial-to-parallel analog data converter, 128 analog drive amplifiers, and a 7- to 128-line digital demultiplexer. The mirrors are deflected by the electrostatic forces between the mirrors and the underlying electrodes when the electrodes are addressed via the transistors (one line at a time from a single analog input of 10V to 15V, up to a 20-MHz rate). The framing rate of this device is circuit-limited to 180 Hz.

In the frame-addressed devices, four mirrors in each pixel (with an 80% fill factor) are hinged at a central support post 2 to 3 $\mu$m high. The addressing signal is applied as charge to the post in a framed mode using serial-to-parallel charge-coupled devices. These devices offer kilohertz framing rates [146].

In the torsion beam (bistable) device [147], each pixel consists of a square, light-reflecting torsion beam made from a layer of thick metal that is connected to a surrounding conducting support layer by two thin metal torsion rods at diagonal corners. Beneath each torsion beam are two landing electrodes and two addressing electrodes. When addressed, the mirrors rotate about 9 deg until their tips touch the landing electrodes. The device needs about 10V for addressing and a 60V, 10-$\mu$s reset pulse. Pixel response times of 6 $\mu$s at atmospheric pressure are typical, and large arrays for high-definition projection television display applications are planned.

Applications of the DMDs include incoherent-to-coherent image conversion, Fourier-plane spatial filtering, programmable mask generation, optical correlators, large-screen projection displays, printer technology, fiber-optic switching, and bistability. DMDs have been proposed for use in OCPAR systems [17,33].

### 6.4.6 Membrane Mirror Light Modulators

Membrane mirror SLMs (MLM) [155–161] employ a monolithic deformable membrane mirror that is supported over a substrate containing an array of pixel wells, as illustrated in Figure 6.14. A driving electrode at the bottom of each well deforms the membrane into the well by means of electrostatic forces. The substrate is typically either a glass plate, a CTP, a VLSI chip, or other insulating plate. The wells are created by overcoating the substrate with a layer of insulating material and etching away unwanted material.

Most devices employ a polymeric membrane that is bonded via the van der Waals forces to the insulating layer containing the wells. The exposed surface of the membrane is metallized to convert it to a flexible mirror and to provide electrical biasing during operation.

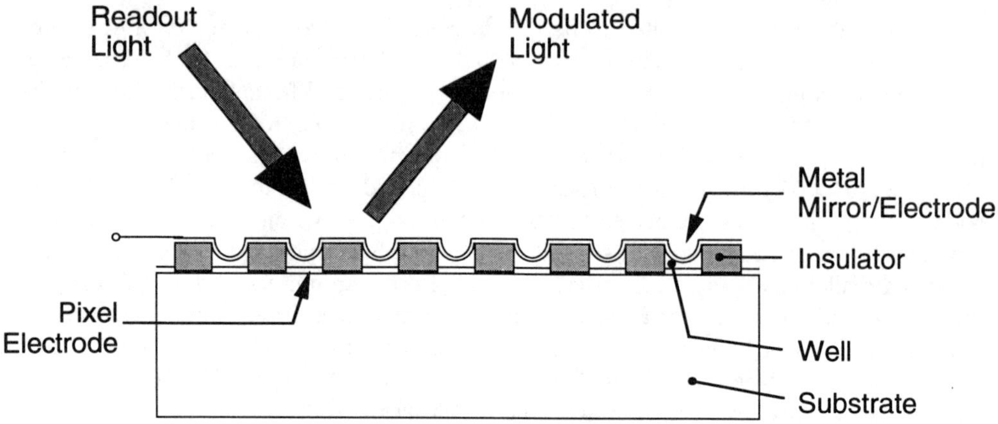


**Figure 6.14** General structure of a membrane mirror SLM.

The etched-well approach offers tremendous flexibility in the design and performance of the device. This is because: the photolithographic techniques used to etch the wells yield extremely high spatial uniformity, thereby leading to SLMs with low fixed-pattern noise; (2) the photolithographic techniques also allow complete freedom to choose well diameter, pitch, depth, and pattern (e.g., circular wells on hexagonal or square centers); and (3) the membrane tension is determined almost exclusively by the materials used for the membrane and the insulating well layer.

The basic operation of the MLMs is such that in the unaddressed state, the membrane mirror is flat and all of the readout light is specularly reflected from the mirror and focused by the Fourier transform lens of Figure 6.15 into a single zero-order spot. In the addressed state where the membrane is deformed into the wells, as illustrated in Figure 6.14, the readout light diffracts from the membrane mirror into a pattern that consists of the central zero order surrounded by a set of higher order spots that reflect the packing symmetry of the membrane wells. As charge is added to a well electrode, the membrane deforms into that well until the restoring force exerted by the tension on the membrane exactly cancels the attractive electrostatic force. Thus, gray-scale phase modulation is achieved. Intensity modulation is achieved by spatially filtering the diffracted light and reimaging the membrane surface as shown in Figure 6.15. For example, with a Fourier-plane, zero-order-stop spatial filter, a positive of the write image is formed.

Altering device parameters such as well diameter $a$ and well depth $s$ has a profound effect on the voltage required to deflect the membrane. It has been shown that the voltage required for a membrane deflection $d$ is given by [155]:

$$V = 4\frac{s}{a}\left[\frac{2Td}{\epsilon_o}\right]^{1/2} \tag{6.26}$$

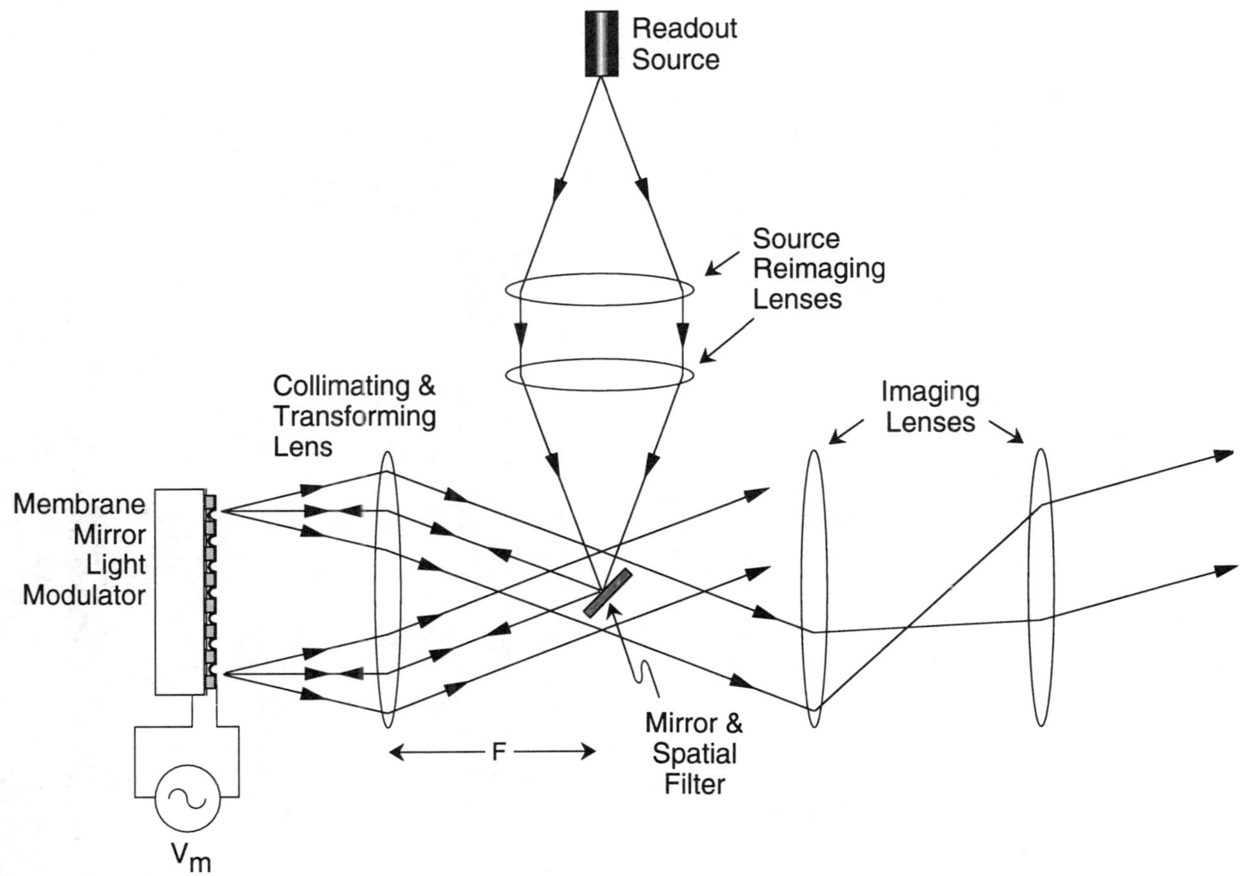


**Figure 6.15** Fourier-plane spatial filtering readout system for membrane mirror SLMs.

where $T$ is the membrane tension and $\epsilon_0$ is the permittivity of free space. Thus, it can be seen that for a voltage-driven device, the operating voltage scales as the ratio of well depth to well diameter. For a current-driven device, the required charge per pixel $q$ for $d \ll s$ is given by

$$q = \pi a[2Td\epsilon_o]^{1/2} \tag{6.27}$$

Thus, for $d \ll s$, $q$ scales as the well diameter $a$ and is independent of the well depth.

The various types of membrane mirror SLMs include: a hard-wired electrode-matrix-addressed device in which an array of electrodes is deposited on an electroded glass or CTP substrate and driven by a bank of voltage amplifiers; an e-beam-addressed device in which a high-resolution electron gun is used to address a CTP-based membrane mirror anode [99,156]; an optically addressed device employing a photocathode/microchannel-plate assembly driving a CTP-based membrane mirror anode [157–159]; and an IC-addressed device in which a high-voltage VLSI chip bump-bonded to a CTP drives the membrane. The general architecture of the CTP-based membrane mirror SLMs is shown in Figure 6.16.

Membrane mirror SLMs are designed to operate with about 40V to 200V for maximum deflection by suitably scaling the well diameter and well depth with wavelength. They also offer reasonably high (kilohertz) framing speeds. Assuming the drive circuits

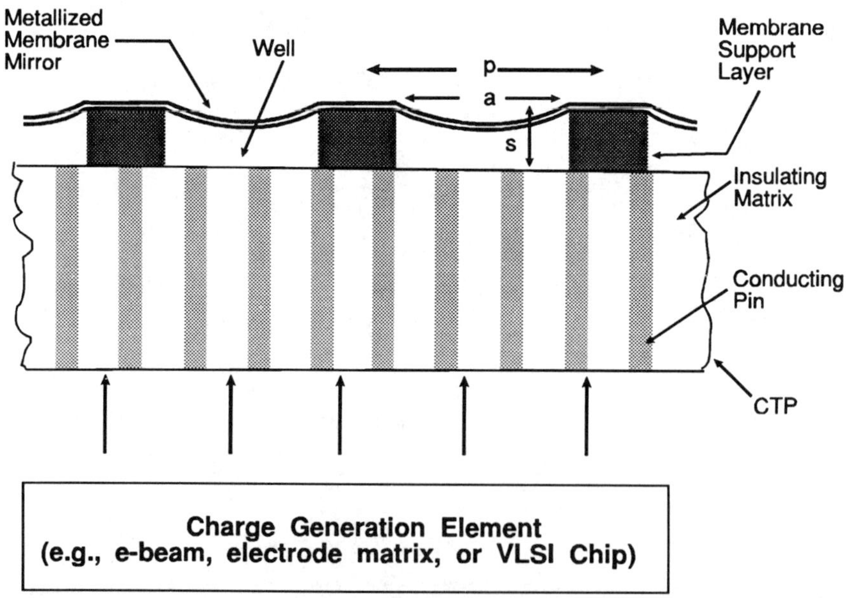


**Figure 6.16** (a) CTP-based membrane mirror anode in the unaddressed state; (b) parabolic deformation of the membrane in the addressed state.

are not current-limited, the device can be driven up to the ultimate practical speed of response of the membrane, which is limited only by the first resonant frequency of the membrane (typically about 1 MHz for pixels 50 $\mu$m in diameter [99]).

Because the membrane deflects into the wells with a parabolic profile, membrane mirror SLMs can also function as an array of electrically controlled dynamic focusing mirrors. Other applications of these devices include real-time wavefront phase correction for communication and imaging through turbulence, correction of aberrations in optical instruments, and phase-only Fourier-plane spatial filtering. For OCPAR processors, these devices can be used as phase retarders.

### 6.4.7 Magneto-Optic SLMs

The most common magneto-optic SLM (MOSLM) utilizes the Faraday magneto-optic effect in an epitaxial garnet film [162–166]. The film is a thulium iron garnet that is heavily doped with bismuth to increase its Faraday rotation. The film is divided into an array of square mesas on centers of about 75 $\mu$m, as shown in Figure 6.17. The magnetization direction of each mesa (pixel) can be individually flipped through the combined magnetic field from an $x$-$y$ matrix of conductors running between the pixels.

The two magnetization orientations produce opposite rotations of the polarization of a light beam transmitted through the garnet film. Therefore, an amplitude-modulated image is obtained by reading out the device between polarizers. The Faraday rotation and absorption of the garnet film vary with film thickness and readout wavelength. Thus,

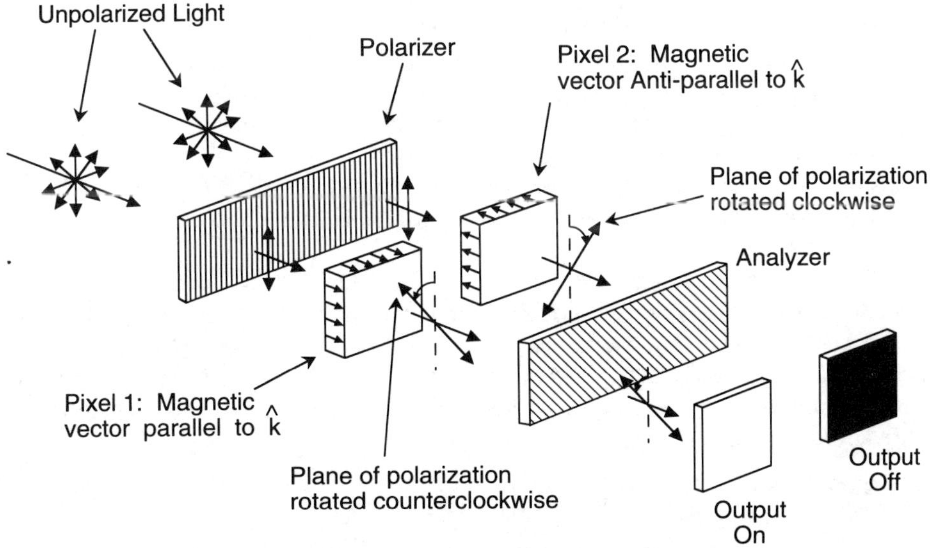


**Figure 6.17** Architecture and principle of operation of the MOSLM showing polarization rotation in two pixels.

when the device is operated with white light, the transmission of the device becomes color-dependent. This device is capable of very high contrast ratios (>10,000:1 for monochromatic light, and approximately 20:1 for white light) and framing rates of approximately 100 Hz. It offers long storage times (months), but suffers from high absorption for readout light in the visible. The MOSLM has been proposed for use in an OCPAR architecture [23].

## 6.5 FUTURE DIRECTIONS

The state of development of SLMs suitable for OCPAR is still quite primitive. Devices with large phase modulation depth, large gray-scale dynamic range, high framing speed, and high spatial resolution are still needed for beam steering and beam shaping, and for searching/tracking receivers (an area that has received very little attention).

Framing speeds rarely exceed 1 kHz for most devices available. The exposure sensitivities of existing optically addressed SLMs range from pJ/cm$^2$ to hundreds of $\mu$J/cm$^2$. Spatial resolutions of up to about 50 line-pairs/mm have been achieved in liquid crystal light valves, and this is adequate for some applications. Clearly, high-speed, high-resolution optically addressed SLMs with built-in gain are needed. The MSLM and some of the opto-electronic SLMs are among the few devices with gain elements. Energy dissipation considerations reveal that novel high-performance materials may be needed to realize a megapixel SLM that frames at megahertz while offering $\mu$J/cm$^2$ sensitivity. Additionally, operation at near-infrared wavelength is desirable, since semiconductor laser diodes can be used, thus allowing the realization of compact OCPAR and other optical computing systems.

Future developments of SLMs for OCPAR systems could be enhanced by the discovery of more efficient thin-film photoconductors and light-modulating materials or structures that offer very low-voltage operation along with very high charge sensitivity. Ongoing research on polymers and organic crystals does offer some hope and so does the promise of artificially layered MQW structures. Also, research on paraelectric liquid crystals promises gray-scale operation simultaneous with high switching speeds. Vigorous research also continues on the development of highly sensitive, high-speed photorefractive materials and devices.

The ongoing research and development of GaAs OEIC SLMs promises a new class of versatile devices that exploit and combine the digital computation strengths of electronics with the parallel processing and communication strengths of optics. Also, there is wide appeal and acceptance for the incorporation of this technology into computers and other electronic processors. These trends, we hope, will lead to breakthroughs that could have significant impact on the development of OCPAR systems.

## REFERENCES

[1] Casasent, D., "Spatial Light Modulators," *Proc. IEEE,* Vol. 65, 1977, p. 143
[2] Casasent, D., ed., *Opt. Eng.,* Special Issue on Spatial Light Modulators, Vol. 17, 1978, p. 307.

[3] Knight, G. R., ''Interface Devices and Memory Materials,'' in *Optical Information Processing: Fundamentals, Topics in Applied Physics,* Vol. 48, S. H. Lee, ed., New York: Springer-Verlag, 1981, pp. 119–121.

[4] Tanguay, A. R., ed., *Opt. Eng.,* Special Issue on SLMs: Critical Issues, Vol. 22, 1983, p. 663.

[5] Efron, U., ed., ''Spatial Light Modulators and Applications,'' *Proc. SPIE* Vol. 465, 1984.

[6] Tanguay, A. R., ''Materials Requirements for Optical Processing and Computing,'' *Opt. Eng.,* Vol. 24, 1985, p. 2.

[7] Tanguay, A. R., and C. Warde, eds., *Opt. Eng.,* Special Issue on Optical Information Processing Components, Vol. 24, 1985, p. 91.

[8] Fisher, A. D., ''A Review of Spatial Light Modulators,'' *Technical Digest of the OSA '85 Topical Meeting on Optical Computing,* TuC1, 1985.

[9] Warde, C., and U. Efron, eds., *Opt. Eng.,* Special Issue on Materials and Devices for Optical Information Processing, Vol. 25, 1986, p. 197.

[10] Fisher, A. D., and J. N. Lee, *Proc. SPIE,* Vol. 634, 1987.

[11] Warde, C., and A. D. Fisher, ''Spatial Light Modulators: Applications and Functional Capabilities,'' in *Optical Signal Processing,* J. Horner, ed., Academic Press, 1987.

[12] *Spatial Light Modulators and Applications,* 1988 Technical Digest Series, Vol. 8, Optical Society of America, June 1988.

[13] Efron, U., A. D. Fisher, and C. Warde, eds., *Opt. Eng.,* Special Issue of *Applied Optics* on Spatial Light Modulators for Optical Information Processing, Vol. 28, 1989.

[14] Esener, S., J. L. Horner, and K. M. Johnson, eds., *Opt. Eng.,* Special Issue of *Applied Optics* on Spatial Light Modulators and Their Applications, Vol. 31, 1992.

[15] Spatial Light Modulators and Applications, 1993 Technical Digest Series, Vol. 6, Optical Society of America, 1993.

[16] Birkmayer, W., C. Schaffer, and B. Hosselbarth, ''Scenarios and System Architectures Advantageous for Optical Technologies in Phased Array Antennas,'' *Proc. SPIE,* Vol. 1217, 1990, pp. 14–25.

[17] Zmuda, H., and E. N. Toughlian, ''Adaptive Microwave Signal Processing: A Photonic Solution,'' *Microwave Journal,* Vol. 35, 1992, pp. 58–71.

[18] Kunath, R. R., and K. B. Shasin, ''Optically Controlled Phased-Array Technology for Space Communication Systems,'' *Proc. SPIE,* Vol. 886, 1988, pp. 234–238.

[19] Seeds, A., ''Optical Beamforming Techniques for Phased-Array Antennas,'' *Microwave Journal,* July 1992, p. 74.

[20] Wale M. J., and W. S. Birkmayer, ''Coherent Optical Techniques in Antenna Beamforming for Satellite Communications,'' *Optical Intersatellite Links and On-Board Techniques Conference,* Digest #1990/008, IEE, Jan. 1990.

[21] Koepf, G. A., ''Optical Processor for Phased-Array Antenna Beam Formation,'' *Proc. SPIE,* Vol. 477, 1984, pp. 75–81.

[22] Koepf G. A., and J. L. Molz, ''Compact Optical Beam-Forming System for Large Phased-Array Antennas,'' MILCOM 84, IEEE Military Communications Conference, 1984, pp. 335–337.

[23] Nickerson, K. A., P. E. Jessop, and S. Haykin, ''Optical Processor for Array Antenna Beam Shaping and Steering,'' *Opt. Eng.,* Vol. 30, 1991, pp. 1497–1502.

[24] Hong J. H., and I. McMichael, ''Novel Optical Technique for Phased-Array Processing,'' *Opt. Eng.,* Vol. 30, 1991, p. 1976.

[25] Konishi, Y., W. Chujo, I. Iwasaki, and K. Yasukawa, ''Techniques for Antenna Sidelobe Suppression and Laser Power Reduction in Optically Controlled Array Antenna,'' *Proc. SPIE,* Vol. 1217, 1990, pp. 26–34.

[26] Dolfi, D., J. P. Huignard, and M. Baril, ''Optically Controlled True Time Delays for Phased Array Antenna,'' *Proc. SPIE,* Vol. 1102, 1989, pp. 152–161.

[27] Dolfi, D., F. Michel-gabriel, S. Bann, and J. P. Huignard, ''Two-Dimensional Optical Architecture for Time-Delay Beam Forming in a Phased Array Antenna,'' *Opt. Lett.,* Vol. 16, 1991, pp. 255–257.

[28] Riza, N. A., "Transmit/Receive Time-Delay Beam-Forming Optical Architecture for Phased-Array Antennas," *Applied Optics,* Vol. 30, 1991, pp. 4594–4595.

[29] Curtis, D. D., "Frequency Domain Analysis and Performance of a True Time Delay Fiber Optic Beamforming Network for Array Antennas," *Proc. SPIE,* Vol. 1217, 1990, pp. 226–234.

[30] Newberg, I. L., A. A. Walston, J. J. Lee, W. W. Ng, and G. L. Tangonan, "Electro-Optical Switch for Antenna Beamsteering," *Proc. SPIE,* Vol. 1217, 1990, pp. 126–131.

[31] Anderson, L. P., F. Boldissar, and D. C. D. Chang, "Antenna Beamforming Using Optical Processing," *Proc. SPIE,* Vol. 886, 1988, pp. 228–232.

[32] Guggenmos, J. L., R. L. Johnson, and B. D. Seery, "Optical Distribution Manifolds for Phased Array Antennas," *Proc. SPIE,* Vol. 886, 1988, pp. 214–217.

[33] Toughlian, E. N., H. Zmuda, and P. Kornreich, "A Deformable Mirror-Based Optical Beamforming System for Phased Array Antennas," *IEEE Photonics Tech. Lett.,* Vol. 2, 1990, pp. 444–446.

[34] Goutzoulis A., and K. Davies, "Compressive 2-D Delay Line Architecture for Time-Steering of Phased-Array Antennas," *Proc. SPIE,* Vol. 1217, 1990, pp. 270–283.

[35] Goutzoulis, A. P., D. K. Davies, and J. M. Zomp, "Hybrid Electronic Fiber Optic Wavelength-Multiplexed System for True Time-Delay Steering of Phased Array Antennas," *Opt. Eng.,* Vol. 31, 1992, pp. 2312–2322.

[36] Warde, C., A. M. Weiss, A. D. Fisher, and J. I. Thackara, "Optical Information Processing Characteristics of the Microchannel Spatial Light Modulator," *Applied Optics,* Vol. 20, 1981, pp. 2066–2074.

[37] Warde, C., and J. I. Thackara, "Oblique-Cut LiNbO3 Microchannel Spatial Light Modulator," *Opt. Lett.,* Vol. 7, 1982, pp. 344–346.

[38] Warde, C., and J. I. Thackara, "Operating Modes of the Microchannel Spatial Light Modulator," *Opt. Eng.,* Vol. 22, 1983, pp. 295–702.

[39] Hara, T., K. Shinoda, T. Kato, M. Sugiyama, and Y. Suzuki, "Microchannel Spatial Light Modulator Having the Functions of Image Zooming, Shifting, and Rotating," *Applied Optics,* Vol. 25, 1986, p. 2306.

[40] Hara, T., Y. Ooi, Y. Suzuki, and M. H. Wu, "Transfer Characteristics of the Microchannel Spatial Light Modulator," *Applied Optics,* Vol. 28, 1989, p. 4781.

[41] Jones, R. C., "A New Calculus for the Treatment of Optical Systems," *J. Opt. Soc. Am.,* Vol. 32, 1942, pp. 486–493.

[42] Casasent, D., F. Caimi, and A. Khomenko, "Test and Evaluation of the Soviet PROM and PRIZ Spatial Light Modulators," *Applied Optics,* Vol. 20, 1981, p. 4215.

[43] Petrov, M. P., A. V. Khomenko, M. V. Krasin'kova, V. I. Marakhonov, and M. G. Shiyagin, "The PRIZ Image Converter and Its Use in Optical Data Processing Systems," *Sov. Phys. Tech. Phys.,* Vol. 26, 1981, p. 816.

[44] Chavel, P., A. A. Sawchuk, T.C. Strand, A. R. Tanguay, Jr., and B. H. Soffer, "Optical Logic With Variable-Grating Mode Liquid Crystal Devices," *Opt. Lett.,* Vol. 5, 1980, p. 398.

[45] Soffer, B. H., J. D. Margerum, A. M. Lackner, D. Boswell, A. R. Tanguay, Jr., A. A. Sawchuk. T. C. Strand, and P. Chavel, "Variable Grating Mode Liquid Crystal Device for Optical Processing and Computing, *Mol. Cryst. Liq. Cryst.,* Vol. 70, 1981, p. 145.

[46] Tanguay, A. R., Jr., C. S. Wu, P. Cheval, T. C. Strand, A. A. Sawchuk, and B. H. Soffer, "Physical Characterization of the Variable Grating Mode Liquid Crystal Device," *Opt. Eng.,* Vol. 22, 1983, p. 687.

[47] Tanguay, A. R., Jr., P. Chavel, T. C. Strand, C. S. Wu, and B. H. Soffer, "Polarization Properties of the Variable-Grating-Mode Liquid-Crystal Device," *Opt. Lett.,* Vol. 9, 1984, p. 174.

[48] Yariv, A., Ch. 9 in *Introduction to Optical Electronics,* 3rd edition, Holt Rinehart and Winston.

[49] Roach, W. R., "Resolution of Electrooptic Light Valves," *IEEE Trans. Electron. Devices,* Vol. ED21, 1974, pp. 453–459.

[50] Schwartz, A., X.-Y. Wang, and C. Warde, "Electron-Beam-Addressed Microchannel Spatial Light Modulator," *Proc. SPIE,* Vol. 465, 1984, pp. 23–28.

[51] Schwartz, A., X.-Y. Wang, and C. Warde, ''Electron-Beam-Addressed Microchannel Spatial Light Modulator,'' *Opt. Eng.,* Vol. 24, 1985, p. 119.
[52] Poon, T. C., B. D. Duncan, M. H. Wu, K. Shinoda, and Y. Suzuki, ''Real Time Optical Holography Using a Spatial Light Modulator,'' *Jap. J. Appl. Phys.,* Vol. 29, 1990, pp. L1840–L1842.
[53] Shinoda K., and Y. Suzuki, ''Electron Beam Addressed Spatial Light Modulator,'' *Proc. SPIE,* Vol. 613, 1986, p. 158.
[54] Hillman, R. L., G. A. Melnik, T. N. Tsakiris, F. Leard, R. Jurgilewicz, and C. Warde, ''Electron-Beam-Addressed Lithium Niobate Spatial Light Modulator,'' *SPIE Proc.,* Vol. 1562, 1991, pp. 136–142.
[55] Marie, G., and Donjon, J. ''Single Crystal Ferroelectrics and Their Applications in Light Valve Display Devices,'' *Proc. IEEE,* Vol. 61, 1973, p. 942.
[56] Marie, G., ''Light Valves Using DKDP Operated Near Its Curie Point: Titus and Phototitus,'' *Ferroelectrics,* Vol. 10, 1976, p. 9.
[57] Donjon, J., F. Dumont, M. Grenot, J. P. Hazan, G. Marie, and J. Pergrale, ''A Pockels-Effect Light Valve: PHOTOTITUS. Applications to Optical Image Processing,'' *IEEE Trans. Electron. Devices,* Vol. ED-20, 1973, p. 1037.
[58] Casasent, D., ''Photo DKDP Light Valve: A Review,'' *Opt. Eng.,* Vol. 17, 1978, pp. 365–370. Also, D. Casasent, ''e-beam DKDP Light Valves: A Review,'' *Opt. Eng.,* Vol. 17, 1978, pp. 344–252.
[59] Lipson, S. G., and P. Nisenson, ''Imaging Characteristics of the Itek PROM,'' *Applied Optics,* Vol. 13, 1974, p. 2052.
[60] Horwitz B. A., and F. J. Corbett, ''The PROM--Theory and Applications for the Pockels Readout Optical Modulator,'' *Opt. Eng.,* Vol. 17, 1978, pp. 353–364.
[61] Nagao, Y., H. Sakata, and Y. Mimura, ''Bi12SiO20 Thin-Film Spatial Light Modulator,'' *Applied Optics,* Vol. 31, 1992, p. 3966.
[62] Osugi, Y., H. Abe, A. Honda, A. Hamajima, and S. Toyoda, ''A Method for Making High Resolution PROM,'' *Conf. Record of 1990 Int. Topical Meeting on Optical Computing,* Kobe, Japan, 1990, pp. 21,22.
[63] Chen, F. S., ''Optically Induced Change of Refractive Indices in LiNbO3 and LiTaO3,'' *J. Appl. Phys.,* Vol. 40, 1969, pp. 3389–3396.
[64] Glass, A. M., ''The Photorefractive Effect,'' *Opt. Eng.,* Vol. 17, 1978, pp. 470–479.
[65] Kukharev, N. V., ''Kinetics of Hologram Writing and Erasure in Electro-Optic Crystals,'' *Sov. Tech. Phys. Lett.,* Vol. 2, 1976, pp. 438–440.
[66] Kukhtarev, N. V., V. B. Markov, S. G. Odulov, M. S. Soskin, and V. L. Vinetskii, ''Holographic Storage in Electrooptic Crystals I: Steady State,'' *Ferroelectrics,* Vol. 22, 1979, pp. 949–960.
[67] Feinberg, J., D. Heiman, A. R. Tanguay, Jr., and R. W. Hellwarth, ''Photorefractive Effects and Light-Induced Charge Migration in Barium Titanate,'' *J. Appl. Phys.,* Vol. 51, 1980, pp. 1297–1305; Vol. 52, 1981, p. 537.
[68] Hermann, J., J. P. Herriau, and J. P. Huignard, ''Nanosecond Four-wave Mixing and Holography in BSO Crystals,'' *Applied Optics,* Vol. 20, 1981, pp. 2173–2175.
[69] Huignard, J. P., and A. Marrakchi, ''Coherent Signal Beam Amplification in Two-Wave Mixing Experiments With Photorefractive Bi12SiO20 Crystals,'' *Opt. Comm.,* Vol. 38, 1981, pp. 249–254.
[70] Huignard J. P., and A. Marrackchi, ''Two Wave Mixing and Energy Transfer in Bi12SiO20 Crystals: Application to Image Amplification and Vibration Analysis,'' *Opt. Lett.,* Vol. 6, 1981, pp. 622–624.
[71] Pepper, D. M., ed., *Opt. Eng.,* Special Issue on Nonlinear Optical Phase Conjugation, Vol. 21, 1982, p. 156.
[72] Cronin-Golomb, M., B. Fischer, J. O. White, and A. Yariv, ''Theory and Applications of Four-Wave Mixing in Photorefractive Media,'' *J. Quantum Elec.,* Vol. QE-20, 1984, pp. 12–29.
[73] Strait J., and A. M. Glass, ''Time Resolved Photorefractive Four-Wave Mixing in Semiconductor Materials,'' *J. Opt. Soc. Am. B,* Vol. 3, 1986, p. 342.
[74] Feinberg, J., ''Self-Pumped, Continuous-Wave Phase Conjugator Using Internal Reflection,'' *Opt. Lett.,* Vol. 7, 1982, p. 486.

[75] Cronin-Golomb, M., B. Fischer, J. O. White, and A. Yariv, "Passive (Self-Pumped) Phase Conjugate Mirror: Theoretical and Experimental Investigation," *Appl. Phys. Lett.*, Vol. 41, 1982, pp. 689–691.

[76] Millerd, J. E., E. M. Garmire, and M. B. Klein, "Investigation of Photorefractive Self-Pumped Phase-Conjugate Mirrors IN The Presence of Loss and High Modulation Depth," *J. Opt. Soc. Am. B*, 1992, p. 1499.

[77] Shih, Y. H., X. H. He, and H. Y. Zang, "High Reflectivity Broadband Self-Pumped Phase Conjugator in Near-Infrared Wavelengths," *Appl. Phys. Lett.*, Vol. 61, 1992, p. 1910.

[78] Mok, F. H., M. C. Tackitt, and H. M. Stoll, "Storage of 500 High-Resolution Holograms in a LiNbO3 Crystal, *Opt. Lett.*, Vol. 16, 1991, pp. 605–607.

[79] d'Auria, L., J. P. Huignard, C. Slezak, and E. Spitz, "Experimental Holographic Read-Write Memory Using 3-D Storage," *Applied Optics*, Vol. 13, 1974, pp. 808–817.

[80] Staebler, D. L., W. J. Burke, W. Phillips, and J. J. Amodei, *Appl. Phys. Lett.*, Vol. 26, 1975, p. 182.

[81] Kurz, H., "Lithium Niobate as a Material for Holographic Information Storage," *Philips Tech Rev.*, Vol. 37, 1977, pp. 109–120.

[82] Huignard, J. P., and F. Micheron, "Optical Storage in LiNbO3:Fe With Selective Erasure Capability," *Rev. Phys. Appl.*, Vol. 10, 1975, pp. 417–423.

[83] Rakuljic, G. A., V. Leyva, and A. Yariv, "Optical Data Storage by Using Orthogonal Wavelength-Multiplexed Volume Holograms," *Opt. Lett.*, Vol. 17, 1992, p. 1471.

[84] Sasaki, H., J. Ma, Y. Fainman, S. H. Lee, and Y. Taketomi, "Fast Update of Dynamic Photorefractive Optical Memory," *Opt. Lett.*, Vol. 17, 1992, p. 1468.

[85] Segev, M., A. Yariv, and J. Shamir, "Limitation on Holographic Storage in Photorefractive Waveguides," *Opt. Lett.*, Vol. 17, 1992, p. 1076.

[86] Qiao, Y., and D. Psaltis, "Sampled Dynamic Holographic Memory," *Opt. Lett.*, Vol. 17, 1992, p. 1376.

[87] Ford, J. E., J. Ma, Y. Fainman, S. H. Lee, Y. Taketomi, D. Bize, and R. R. Neurgaonkar, "Multiplex Holography in Strontium Barium Niobate With Applied Field," *J. Opt. Soc. Am. A*, Vol. 9, 1992, pp. 1183–1192.

[88] Kamshilin, A. A., and M. P. Petrov, "Holographic Image Conversion in a Bi12SiO20 Crystal," *Sov. Tech. Phys. Lett.*, Vol. 144, 1980.

[89] Shi, Y., D. Psaltis, A. Marrakchi, and A. R. Tanguay, Jr., "Photorefractive Incoherent-to-Coherent Optical Converter," *Applied Optics*, Vol. 22, 1983, pp. 3665–3667.

[90] Marrakchi, A., A. R. Tanguay, Jr., J. Yu, and D. Psaltis, "Physical Characterization of the Photorefractive Incoherent-to-Coherent Optical Converter," *Opt. Eng.*, Vol. 24, 1985, p. 124.

[91] Bleha, W. P., L. T. Lipton, E. Wiener-Avnear, J. Grinberg, P. G. Reif, D. Casasent, H. B. Brown, and B. V. Markevitch, "Application of the Liquid Crystal Light Valve to Real-Time Optical Data Processing," *Opt. Eng.*, Vol. 17, 1978, pp. 371–384.

[92] Braatz, P. O., K.Chow, U. Efron, J. Grinberg, and M. J. Little, "A Fast Silicon Photoconductor-Based Liquid Crystal Light Valve," *Proc. Int. Electron. Devices Mtg.*, 1979, p. 540.

[93] Efron, U., P. O. Braatz, M. J. Little, R. N. Schwartz, and J. Grinberg, "Silicon Liquid Crystal Light Valves: Status and Issues," *Opt. Eng.*, Vol. 22, 1983, p. 682.

[94] Kompanets, I. N., A. V. Parfenov, and Yu. M. Popov, "Spatial Modulation of Light in a Photosensitive Structure Composed of a Liquid Crystal and an Insulated Gallium Arsenide Crystal," *Sov. J. Quantum Electron.*, Vol. 9, 1979, p. 1070.

[95] Parfenov, A. V., I. N. Kompanets, and Yu. M. Popov, "Spatial Modulation of Light in Photosensitive High-Resolution Metal-Insulator-Semiconductor Structures With Liquid Crystals," *Sov. J. Quantum Electron.*, Vol. 10, 1980, p. 167.

[96] Vasiliev, A. A., I. N. Kompanets, and A. V. Parfenov, "Advances in Development and Applications of Optically Controlled Liquid Crystal Spatial Light Modulators," *Optik*, Vol. 67, 1984, p. 223.

[97] Loiseaux, B., G. Illiaguer, and J. P. Huignard, "Dynamic Optical Cross-Correlator Using a Liquid Crystal Light Valve and a BSO Crystal in the Fourier Plane," *Opt. Eng.*, Vol. 24, 1984, p. 144.

[98] Melnik, G. A., G. C. Gilbreath, T. N. Tsakiris, R. P. Jurgilewicz, D. M. O'Mara, T. N. Horsky, and C.

Warde. "Operating Modes of a Charge-Transfer-Plate Liquid-Crystal Phase Modulator," *Applied Optics,* Vol. 31, 1992, p. 3892.

[99] Warde, C., C. M. Schiller, J. Bounds, T. N. Horsky, G. Melnik, and R. Dillon, "Charge-Transfer-Plate Spatial Light Modulators," *Applied Optics,* Vol. 31, 1992, pp. 3871–3879.

[100] Buzak, T. S., R. S. Vatne, G. A. Nelson, and D. E. Whitlow, "Electron-Beam-Addressed Light Valve Operating Modes and Performance," *SID Int. Symp. Digest of Technical Papers,* Vol. 18, 1987, pp 64–67.

[101] Haven, D. A., "Electron-Beam Addressed Liquid-Crystal Light Valve," *IEEE Trans. on Electron. Devices,* Vol. ED-30, 1983, pp. 489–492.

[102] Zhao, J-A., S-S Tung, and L. Ruan, "A Study of Liquid Crystal Light Valve for Large-Screen Projection TV," *Disp. Technol.,* Vol. 1, 1986, pp. 151–158.

[103] Van Raalte, J. A., "Reflective Liquid Crystal Television Display," *Proc. IEEE,* Vol. 56, 1968, pp. 2146–2149.

[104] Nixon, W. C., and C. E. Bryan, "Light Valve," International Patent Publication No. WO85/01362, March 1985.

[105] Hanson, J. R., and R. J. Schneeberger, "Liquid Crystal Media for Electron Beam Recording," *IEEE Trans. Electron. Devices,* Vol. Ed-15, No. 11, 1968, pp. 896–906.

[106] Liu, H.-K., J. A. Davis, and R. A. Lilly, "Optical-Data-Processing Properties of a Liquid-Crystal Television Spatial Light Modulator," *Opt. Lett.,* Vol. 10, 1985, p. 635.

[107] M., Young, "Low-Cost LCD Video Display for Optical Processing," *Applied Optics,* Vol. 25, 1986, p. 1024.

[108] Liu, H.-K., and T.-H. Chao, "Liquid Crystal Television Spatial Light Modulators," *Applied Optics,* Vol. 28, 1989, p. 4772.

[109] Barnes, T. H., T. Eiju, K. Matusda, and N. Ooyama, "Phase-Only Modulation Using a Twisted Nematic Liquid Crystal Television," *Applied Optics,* Vol. 28, 1989, p. 4845.

[110] Li, Y., A. Kostrzewski, D. H. Kim, and G. Eichmann, "Liquid-Crystal-TV Based White Light Optical Tracking Novelty Filter," *Applied Optics,* Vol. 28, 1989, p. 4861.

[111] Jutamulia, S., G. M. Storti, A. M. Seiderman, J. Lindmayer, and D. A. Gregory, "Infrared Signal Processing Using a Liquid Crystal Television," *Opt. Eng.,* Vol. 30, 1991, p. 178.

[112] Kirsch, J. C., D. A. Gregory, M. W. Thie, B. K. Jones, "Modulation Characteristics of the Epson Liquid Crystal Television," *Opt. Eng.,* Vol. 31, 1992, p. 963.

[113] Sakai, H., and J. Ohtsubo, "Image Subtraction Using Polarization Modulation of Liquid-Crystal Television," *Applied Optics,* Vol. 31, 1992, p. 6852.

[114] Yang, X., T. Lu, and F.T.S. Yu, "Compact Optical Neural Network Using Cascaded Liquid Crystal Television," *Applied Optics,* Vol. 29, 1990, p. 5223.

[115] Collings, N., W. A. Crossland, P. J. Ayliffe, D. G. Vass, and I. Underwood, "Evolutionary Development of Advanced Liquid Crystal Spatial Light Modulators," *Applied Optics,* Vol. 28, 1989, p. 4740.

[116] Jared, D. A., R. Turner, and K. M. Johnson, "Electrically Addressed Spatial Light Modulator That Uses a Dynamic Memory," *Opt. Lett.,* Vol. 16, 1991, pp. 1785–1787.

[117] Cotter, L. K., T. J. Drabik, R. J. Dillon, and M. A. Handschy, "Ferroelectric-Liquid-Crystal/Silicon-Integrated Circuit Spatial Light Modulator," *Opt. Lett.,* Vol. 15, 1990, pp. 291–293.

[118] McKnight, D. J., D. G. Vass, and R. M. Sillitto, "Development of Spatial Light Modulator: A Randomly Addressed Liquid-Crystal-Over-nMOS Array," *Applied Optics,* Vol. 28, 1989, p. 4757.

[119] Johnson, K. M., D. J. McKnight, and I. Underwood, "Smart Spatial Light Modulators Using Liquid Crystals on Silicon," *IEEE J. Quantum Electron.,* Vol. 29, 1993, pp. 699–714.

[120] Jared, D. A., and K. M. Johnson, "Optically Addressed Thresholding Very-Large-Scale Integration/Liquid-Crystal Spatial Light Modulators," *Opt. Lett.,* Vol. 16, 1991, pp. 967–969.

[121] Drabik, T. J., and M. A. Handschy, "Silicon VLSI VLSI/Ferroelectric Liquid Crystal Technology for Micropower Optoelectronic Computing Devices" *Applied Optics,* Vol. 29, 1990, pp. 5220–5225.

[122] Barbier, P. R., and G. Moddel, "Hydrogenated Amorphous Silicon Photodiodes for Optical Addressing of Spatial Light Modulators," *Applied Optics,* Vol. 31, 1992, p. 3898.

[123] Moddell, G., K. M. Johnson, W. Li, and R. A. Rice, "High-Speed Binary Optically Addressed Spatial Light Modulator," *Appl. Phys. Lett.,* Vol. 55, 1989, pp. 537–539.

[124] Landreth, B., and G. Moddell, "Gray Scale Response From Optically Addressed Spatial Light Modulators Incorporating Surface-Stabilized Ferroelectric Liquid Crystals," *Applied Optics,* Vol. 31, 1992, p. 3937.

[125] Cambon, P., M. Killinger, J. L. de Bougrenet de la Tocnaye, "Optically Addressed Spatial Light Modulator With Nipin aSi:H Layers and Bistable Ferroelectric Liquid Crystal," *SPIE Proc.,* 1991.

[126] Fukushima, S., T. Kurokawa, and M. Ohno, "Ferroelectric Liquid-Crystal Spatial Light Modulator Achieving Bipolar Image Operation and Cascadability," *Applied Optics,* Vol. 31, 1992, p. 6859.

[127] Patel, J. S., and J. W. Goodby, "Properties and Applications of Ferroelectric Liquid Crystals," *Opt. Eng.,* Vol. 26, 1987, pp. 373–384.

[128] Brown, J. J., J. T. Gardner, and S. R. Forrest, "An Integrated, Optically Powered, Optoelectronic 'Smart' Logic Pixel for Interconnection and Computing Applications," *IEEE J. Quantum Electron.,* Vol. 29, 1993, pp. 715–726.

[129] Miller, D.A.B., D. S. Chelma, T. C. Damen, A. C. Gossard, W. Weigmann, T. H. Wood, and C. A. Burrus, "Novel Hybrid Bistable Switch: The Quantum Well Self-Electro-Optic Effect Device," *Appl. Phys. Lett.,* Vol. 45, 1984, p. 13.

[130] Livescu, G., D.A.B. Miller, J. E. Henry, A. C. Gossard, and J. H. English, "Spatial Light Modulator and Optical Dynamic Memory Using Integrated Self-Electrooptic-Effect Devices," *Opt. Lett.,* Vol. 13, 1988, pp. 297–299.

[131] Lentine, A. L., H. S. Hinton, D.A.B. Miller, J. E. Henry, J. E. Cunningham, and L.M.F. Chirovsky, "Symmetric Self-Electrooptic Device: Optical Reset Latch, Differential Logic Gate, and Differential Modulator Detector," *IEEE J. Quantum. Electron.,* Vol. 25, 1989, pp. 1928–1936.

[132] Miller, D.A.B., "Quantum Well Self-Electrooptic-Effect Devices," *Optical and Quantum Electron.,* Vol. 22, 1990, pp. S61-S98.

[133] Boyd, G. D., A. M. Fox, and D.A.B. Miller, L.M.F. Chirovsky, L. A. D'Asaro, J. M. Kuo, R. F. Kopf, and A. L. Lentine, "33 ps Optical Switching of Symmetric Self-Electro-Optic Effect Devices," *Appl. Phys. Lett.,* Vol. 57, 1990, pp. 1843–1845.

[134] Giugni, S., K. Kawashima, K. Fujiwara, and N. Sano, "New Self-Electro-Optic Effect Device Using Two Wavelengths in InGaAs/AIGaAs Multiple Quantum Wells," *Appl. Phys. Lett.,* Vol. 61, 1992, p. 376.

[135] Lentine, A. L., and D.A.B. Miller, "Evolution of the SEED Technology: Bistable Logic Gates to Optoelectronic Smart Pixels," *IEEE J. Quantum Electron.* Vol. 29, 1993, pp. 655–669.

[136] Miller, D.A.B., "Novel Analog Self-Electrooptic-Effect Devices," *IEEE J. Quantum Electron.,* Vol. 29, 1993, pp. 678–698.

[137] Lentine, A. L., F.A.P. Tooley, S. L. Walker, F. B. McCormick, R. L. Morrison, L.M.F. Chirovsky, M. W. Focht, J. M. Freund, G. D. Guth, R. E. Leibenguth, G. J. Przybylek, L. E. Smith, L. A. D'Asaro, and D.A.B. Miller, "Logic Self-Electrooptic-Effect Devices: Optoelectronic Multiport Logic Gates, Multiplexers, Demultiplexers and Shift Registers," *IEEE J. Quantum Electron.,* Vol. 28, 1992, pp. 1539–1553.

[138] D'Asara, L. A., L.M.F. Chirovsky, E. J. Laskowski, S. S. Pei, T. K. Woodward, A. L. Lentine, R. E. Leibhenguth, M. W. Focht, J. M. Freund, G. G. Guth, and L. E. Smith, "Batch Fabrication and Operation of GaAs-$As_xGa_{1-x}$As Field-Effect Transistor-Self-Electrooptic Effect Device (FET-SEED, Smart Pixel Arrays," *IEEE J. Quantum Electron.,* Vol. 29, 1993, pp. 670–677.

[139] Kasahara, K., "VSTEP-Based Smart Pixels," *IEEE J. Quantum Electron.,* Vol. 29, 1993, pp. 757–768.

[140] Lang, R., K. Kasahara, and M. Sakaguchi, "VSTEP and Its Applications to Optical Computing," *Conf. Record of 1990 Int. Topical Meeting on Optical Computing,* Kobe, Japan, 1990, pp. 365–368.

[141] Wilmsen, C. W., F. R. Beyette, Jr., X. An, and S. A. Feld, "Smart Pixels Using the Light Amplifying Optical Switch (LAOS)," *IEEE J. Quantum Electron.,* Vol. 29, 1993, pp. 769–774.

[142] Heremans, P., M. Kuijk, R. A. Vounckx, C. V. Radehaus, J. I. Pankove, and G. Borghs, "Image

Transcription Between Arrays of n-p-n-p Optoelectronic Switches,'' *IEEE Trans. on Elec. Dev.,* Vol. 39, 1992, pp. 2248–2253.

[143] Maserjian, J., P. O. Andersson, B. R. Hancock, J. M. Iannelli, S. T. Eng, F. J. Grunthaner, K. K. Law, P. O. Holtz, R. J. Simes, L. A. Coldren, A. C. Gossard, and J. L. Merz, ''Optically Addressed Spatial Light Modulators by MBE-Grown nipi MQW Structures,'' *Applied Optics,* Vol.28, 1989, p. 4801.

[144] Chevrier, J. B., P. Cambon, R. C. Chittick, and B. Equer, ''Use of N-I-P-I-N a-SI-H Structure for Bistable Optically Addressed Spatial Light Modulator,'' 13th Int. Congress on Amorphous Semiconductor, 1991.

[145] Florence, J. M., Optical Characteristics of Deformable Mirror Spatial Light Modulators,'' in *Spatial Light Modulators and Applications,* OSA 1990 Technical Digest, Vol. 14, pp. 166–169.

[146] Boysel, R. M., ''A 128 x 128 Frame-Addressed Deformable Mirror Spatial Light Modulator,'' *Opt. Eng.,* Vol. 30, 1991, pp. 1422–1427.

[147] Boysel, R. M., T. G. McDonald, and J. B. Sampsell, ''A Linear Torsion Hinged Deformable Mirror Device for Optical Switching,'' *OSA Technical Digest: Photonics Switching,* Vol. 183, 1989.

[148] Wu, R., R. O. Gale, L. J. Hornbeck, and J. B. Sampsell, ''Electro-Optical Performance of an Improved Deformable Mirror Device,'' *Proc. SPIE,* Vol. 825, 1987, p. 24.

[149] Giles, M. K., J. Z. Smith, and J. M. Florence, ''Fourier Plane Modeling of the TI Deformable Mirror Device,'' *Proc. SPIE,* Vol. 936, 1988, p. 38.

[150] Florence, J. M., and R. O. Gale, ''Coherent Optical Correlation Using a Deformable Mirror Device Spatial Light Modulator in the Fourier Plane,'' *Applied Optics,* Vol. 27, 1988, p. 2091.

[151] Cohn, R. W., and J. B. Sampsell, ''Deformable Mirror Device Uses in Frequency Excision and Optical Switching,'' *Applied Optics,* Vol. 27, 1988, p. 937.

[152] Florence, J. M., M. K. Giles, and J. Z. Smith ''Operation of a Deformable Mirror Device as a Fourier Plane Phase Modulating Filter,'' *Proc. SPIE,* Vol. 938, 1988, p. 2.

[153] Florence, J. M., ''Joint Transform Correlator Systems Using Deformable Mirror Spatial Light Modulators,'' *Opt. Lett.,* Vol. 14, 1989, p. 341.

[154] Collins, D. R., J. B. Sampsell, L. J. Hornbeck, J. M. Florence, P. A. Penz, and M. T. Gately, ''Deformable Mirror Device Spatial Light Modulators and Their Applicability to Optical Neural Networks,'' *Applied Optics,* Vol. 28, 1989, p. 4900.

[155] Pape, D. R., and L. J. Hornbeck, ''Characteristics of the Deformable Mirror Device for Optical Information Processing,'' *Opt. Eng.,* Vol. 22, 1983, p. 675.

[156] Pape, D. R., ''Optically Addressed Membrane Spatial Light Modulator,'' *Opt. Eng.,* Vol. 24, 1985, p. 107.

[157] Somers, L. E., ''The Photoemitter Membrane Light Modulator Image Transducer,'' *Adv. Electron. Phys.,* Vol. 33A, 1972, p. 493.

[158] Fisher, A. D., L. C. Ling, J. N. Lee, and R. C. Fukuda, ''Photoemitter Membrane Light Modulator,'' *Opt. Eng.,* Vol. 25, 1986, p. 261.

[159] Roslma, P. B., J. N. Lee, T.-K. Oh, and L.-C. Ling, ''Experimental Parameters of the Photoemitter Membrane Spatial Light Modulator,'' *Applied Optics,* Vol. 28, 1989, p. 4816.

[160] Horsky, T. N., C. M. Schiller, G. J. Genetti, T. N. Tsakiris, D. M. O'Mara, R. Jurgilewicz, S.-D. Lee, and C. Warde, ''Electron-Beam-Addressed Membrane Mirror Light Modulator for Projection Display,'' *Applied Optics,* Vol. 31, 1992, pp. 3980–3990.

[161] Reizman, F., ''An Optical Spatial Phase Modulator Array Activated by Optical Signals,'' *Proc. 1969 Electro-Optical Systems Design Conf.,* New York: IEEE, 1969, pp. 225–230.

[162] Ross, W. E., D. Psaltis, and R. H. Anderson, *Opt. Eng.,* Vol. 22, 1983, p. 485. W. E. Ross, K. M. Snapp, and R. H. Anderson, ''Fundamental Characteristics of the Litton Iron Garnet Magneto-Optic Spatial Light Modulator,'' *Proc. SPIE,* Vol. 388, 1983, p. 55.

[163] Ross, W. E., D. Psaltis, and R. H. Anderson, ''2D Magneto Optic Spatial Light Modulator for Signal Processing,'' *Proc. SPIE,* Vol. 341, 1983.

[164] Psaltis, D., F. Mok, and E. G. Paek, ''On the Use of the Litton Magneto-Optic Device in Optical Processors,'' *Proc. SPIE,* Vol. 465, 1984, p.29.

[165] Pulliam, G. R., B. E. MacNeal, D. M. Warren, D. L. Cox, J. J. Fernandez de Castro, and S. A. Mills, ''Magneto-Optic Light Valve Array,'' *SPIE,* Vol. 465, 1984, pp. 1–7.

[166] Davis, J. A., J. Gamlieli, and G. W. Bach, ''Optical Transmission and Contrast Ratio Studies of the Magnetooptic Spatial Light Modulator,'' *Applied Optics,* Vol. 27, 1988, p. 5194.

# Chapter 7
# Acousto-Optic Bragg Cell Devices

**Dennis R. Pape**
*Photonic Systems Incorporated*

## 7.1 INTRODUCTION

The acousto-optic (AO) Bragg cell is a spatial light modulating device that encodes light with information contained in an RF signal. AO Bragg cells, from their historical development as the first real-time SLMs, have now matured to a level of performance unequaled among other SLM devices. Thus, the AO Bragg cell is the device of choice for most real-time optical modulation and deflection applications. Indeed, the high performance of the AO Bragg cell is directly responsible for the extensive development of optical signal processing. This chapter presents an overview of AO Bragg cells, including discussions on the AO effect and its physical realization in crystalline materials, the characteristics and operation of AO Bragg cell deflector and modulator devices, and the optical signal processing characteristics of AO Bragg cell devices.

The AO Bragg cell, shown in Figure 7.1, consists of a piezoelectric transducer mechanically bonded to a transparent crystalline material. RF signals input to the transducer through an RF matching network are converted to sound waves which travel through the AO material. The traveling ultrasonic wave manifests itself as a density wave which, through the photoelastic effect, results in a sinusoidal index of refraction variation in the AO material. Light incident on this region of index variation, or grating, is diffracted. This phenomenon was first predicted theoretically by Brillouin [1] in 1922 and observed experimentally by Lucas and Biquard [2] and Debye and Sears [3] in 1932. Maximum light diffraction into a single diffraction order occurs when the acoustic interaction length is relatively large and the incident optical beam illuminates the cell at the Bragg angle, labeled $\theta_B$ in the figure. AO devices with large interaction lengths are said to be operating in the Bragg regime and are referred to as Bragg cells. With a shorter interaction length,

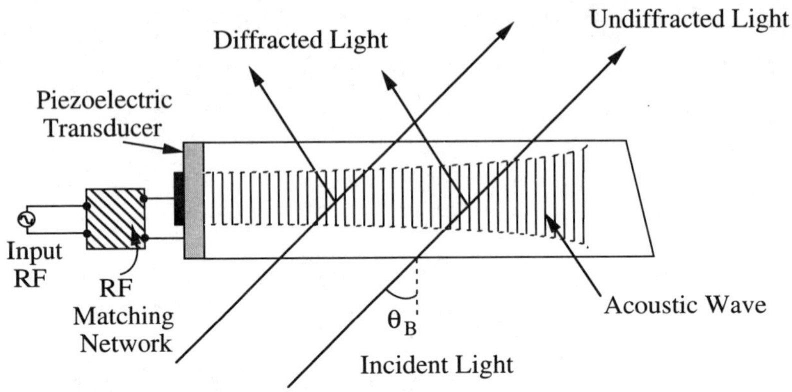


**Figure 7.1** Acousto-optic Bragg cell.

all diffraction orders are diffracted. When all diffraction orders are present, the device is said to be operating in the Raman-Nath [4] regime. Since all diffraction orders carry the same information, however, it is advantageous to suppress the upper diffraction orders and maximize the optical energy in just the first diffraction order for most optical signal processing applications. Suppression of the upper diffraction orders is accomplished by making the length of the transducer (in the direction of optical propagation) large (relative to the acoustic wavelength) [5]. In this way the acoustic waves more nearly approximate plane waves, and the optical Bragg diffraction is analogous to x-ray Bragg diffraction from crystalline lattice planes. We will confine our discussion here to AO devices operated in the Bragg regime.

Both light modulators and deflectors may be constructed using AO Bragg cell technology. AO deflectors have found wide practical application as SLMs for optical signal processing applications, where the devices can be configured to act in the time domain as optically tapped delay lines for signal correlation and, in the frequency domain, as RF frequency channelizers for spectrum analysis.

## 7.2 ACOUSTO-OPTIC BRAGG INTERACTION

The AO Bragg interaction has high efficiency only for certain allowable crystalline acoustic and optical modes. The physics of the AO interaction and the determination of the strength of the interaction are discussed next.

### 7.2.1 Acousto-Optic Effect

The propagation of an acoustic wave in a crystalline medium manifests itself as a sinusoidal perturbation in the strain field of the material. The AO effect is the change in the optical index of refraction of the material as a result of this acoustically induced strain-field perturbation.

The allowable acoustic modes in a crystalline material are found by solving the acoustic wave (or Christoffel) equation for the crystal [6]:

$$\mathbf{K}_\mathbf{a}^2 \Gamma_{ij} v_j = \rho \Omega^2 \mathbf{v}_i \tag{7.1}$$

where $\mathbf{K_a}$ is the acoustic wave vector, $v_i$ is the acoustic velocity in the $i$th direction, $\rho$ is the material density, $\Omega$ is the radian frequency of the acoustic wave, and $\Gamma_{ij}$ is the Christoffel matrix given by

$$\Gamma_{ij} = \mathbf{l}_{iK} \mathbf{c}_{KL} \mathbf{l}_{Lj} \tag{7.2}$$

where the $\mathbf{l}$ matrix is the same form as the gradient operator matrix, with each element representing a propagation direction of the acoustic wave, and where $\mathbf{c}$ is a fourth-rank tensor whose elements are the elastic stiffness constants of the material. By setting the characteristic determinant of (7.1) to 0 (using (7.2)),

$$[\mathbf{K}_\mathbf{a}^2 \Gamma_{ij}(\mathbf{l}_x, \mathbf{l}_y, \mathbf{l}_z) - \rho \Omega^2 \delta_{ij}] = 0 \tag{7.3}$$

a dispersion relation is obtained that determines the acoustic wave vector scaled by the radian frequency $\mathbf{K_a}/\Omega$ as a function of its direction $\mathbf{l}$. The surface defined by (7.3) is called the *acoustic slowness* (or *inverse velocity*) surface. It consists of three concentric surfaces, one for each of the three allowable acoustic modes—one longitudinal and two shear with mutually orthogonal polarizations. An example of a slowness surface for a common AO crystal, GaP, is shown in Figure 7.2 for acoustic propagation in the [1,-1,0] direction [7]. The inner surface corresponds to a longitudinal acoustic mode, while the outer two surfaces correspond to two different shear modes.

The strain field of a material is coupled to the index of refraction by the photoelastic effect,

$$\Delta\left(\frac{1}{n^2}\right)_{ij} = p_{ijkl} S_{kl} \qquad i,j,k,l = 1, 2, \ldots, 6 \tag{7.4}$$

where $p_{ijkl}$ are the elements of the fourth-rank photoelastic tensor coupling the mechanical strain induced by the passage of the acoustic wave $S_{kl}$ to the corresponding change in the optical index of refraction $\Delta\,(1/n^2)_{ij}$. The photoelastic tensor elements, or constants, are defined relative to the principal axes of the material. The point-group symmetry of the material determines which of the 36 tensor elements are zero. If the direction of acoustic propagation in the material does not coincide with one of the principal axes, the tensor can be transformed so that an effective photoelastic constant for the particular AO interaction can be determined.

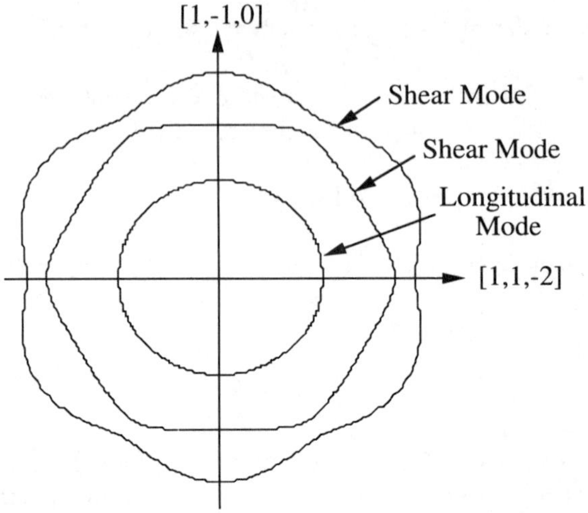


**Figure 7.2** GaP slowness surface.

The spatial variation of the index of refraction in a crystal is a 3-D ellipsoidal surface which, in the presence of a perturbed strain field due to the passage of an acoustic wave, is

$$\left(\left(\frac{1}{n^2}\right)_{ij} + p_{ijkl} S_{kl}\right) x_i x_j = 1 \tag{7.5}$$

Equation (7.5), the AO effect, couples a particular acoustic mode and direction with a particular optical mode and direction. The strength of the coupling is determined by the magnitude of the photoelastic constant linking these modes. The photoelastic constant is used to rank the AO performance of materials through the AO figure of merit.

The interaction of an acoustic wave with light couples the incident optical wave, or mode, with the diffracted optical mode. The efficiency of the AO interaction can be determined by solving the wave equation for the acousto-optically coupled optical modes [8]. (A review of this and other approaches to determining AO efficiency are given in [9].) The diffraction efficiency is

$$\frac{I_d}{I_i} = \sin^2\left(\frac{\pi L}{\sqrt{2}\,\lambda_0}\sqrt{\left[\frac{n^6 p^2}{\rho v^3}\right] P_a}\right) \tag{7.6}$$

where $L$ is the length of the AO interaction region, $\rho$ is the density of the AO material, $v$ is the acoustic velocity, and $P_a$ is the acoustic power density. A figure of merit, $M_2$[10], is defined as

$$M_2 = \frac{n^6 p_{\text{eff}}^2}{\rho v^3} \tag{7.7}$$

where $p_{\text{eff}}$ is the effective photoelastic constant of the material for the particular interaction configuration. The figure of merit combines the AO material parameters into a single constant to show the effectiveness of the material for AO applications.

The calculation of $M_2$ for a particular material thus involves first determining the allowable acoustic modes and directions using (7.3), and then determining the effective photoelastic constant for a particular incident optical mode and direction coupled to a particular diffracted optical mode and direction for the selected acoustic mode and direction using (7.5). The most commonly used AO materials are shown in Table 7.1, where the material properties along with the figures of merit for the most efficient acoustic and optical mode coupling is shown.

### 7.2.2 Interaction Geometries

Besides selecting an efficient AO material with a particular acoustic and optical mode and direction, the Bragg cell design must provide adequate bandwidth, time aperture, and dynamic range performance. Some AO materials, although providing high diffraction efficiency, cannot produce large bandwidths using the standard isotropic Bragg interaction. Other interaction geometries can be employed to enhance the bandwidth performance for a given material. These interaction geometries include the phased-array interaction, where a transducer is used to steer the acoustic beam for better efficiency, and the anisotropic interaction, where the acoustic wave couples one optical mode to an orthogonally polarized optical mode.

*Isotropic*

The Bragg interaction in an isotropic optical medium was historically the first AO interaction studied. It remains the most common interaction geometry used in AO devices. A phase (or momentum) space representation of the isotropic Bragg interaction is shown in Figure 7.3 for a single optical and acoustic plane wave. Conservation of momentum in the AO interaction requires that the momentum vector of the diffracted optical plane wave, $\mathbf{k_d}$, be equal to the vector sum of the momentum vectors of the incident optical plane wave, $\mathbf{k_i}$, and the acoustic plane wave, $\mathbf{K_a}$ (directed normal to the transducer):

$$\mathbf{k_d} = \mathbf{k_i} + \mathbf{K_a} \tag{7.8}$$

**Table 7.1**
Characteristics of Commonly Used Acousto-Optic Materials

| | *Optical Transmission (μm)* | *Refractive Index, n* | *Density, ρ ($g/cm^3$)* | *Anisotropy, b* | *Mode (Direction)* | *Acoustic Wave Velocity, v (× $10^5$ cm/s)* | *Attenuation, (dB/μs/$GHz^2$)* | *Optical Wave Polarity (Direction)* | *Figure of Merit, M2 (× 10(−18) $s^3/g$)* |
|---|---|---|---|---|---|---|---|---|---|
| Fused Silica | 0.2–4.5 | 1.46 | 2.2 | | L | 5.95 | 7.2 | Per | 1.51 |
| | | | | | S | 3.76 | | Par or Per | 0.47 |
| GaAs | 1.0–11.0 | 3.37 | 5.34 | | L[110] | 5.15 | 15.5 | Par | 104.0 |
| | | | | | S[100] | 3.32 | | Par or Per (010) | 46.3 |
| GaP | 0.6–1.0 | 3.31 | 4.13 | −0.03 | L[110] | 6.32 | 8.0 | Par | 44.6 |
| | | | | 0.487 | S[110] | 4.13 | | Par or Per (111) | 16.49 |
| Ge | 2.0–20.0 | 4 | 5.33 | | L[111] | 5.5 | 16.5 | Par | 840.0 |
| | | | | | S[100] | 3.51 | 3.2 | Par or Per | 290.0 |
| $PbMoO_4$ | 0.4–5.5 | 2.4 | 6.95 | −0.168 | L[001] | 3.75 | 5.5 | Par or Per (100) | 36.3 |
| $LiNbO_3$ | 0.4–4.5 | 2.2 | 4.64 | −0.55 | L[100] | 6.57 | 1.0 | 35yrot | 7.0 |
| | | | | −0.6 | S[100]35 | 3.6 | 1.0 | (100) | 14.0 |
| HgC12 | 0.38–28.0 | 2 | 7.18 | | L[100] | 1.62 | | Par | |
| $TeO_2$ | 0.35–5.0 | 2.26 | 6 | 0.2 | L[001] | 4.2 | 6.3 | Per (010) | 34.5 |
| | | | | −26.4 | S[110] | 0.62 | 17.9 | Cir | 793.0 |

Note: L = longitudinal; S = shear; Per = perpendicular; Par = parallel; Cir = circular.

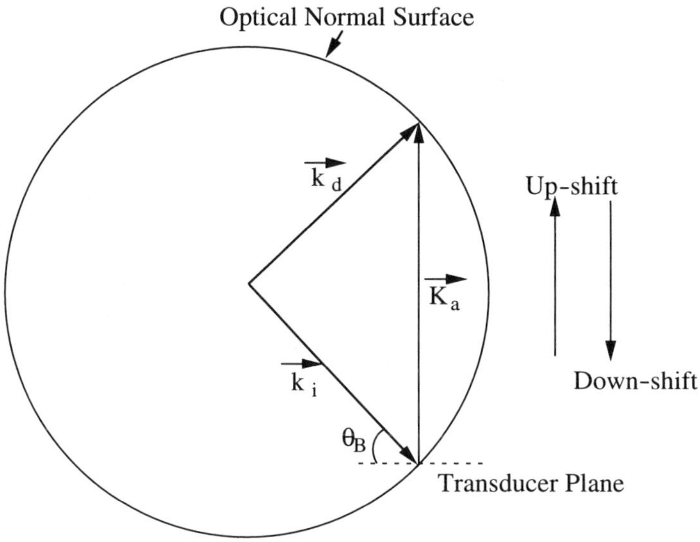


**Figure 7.3** Isotropic Bragg interaction.

The momentum vector is inversely proportional to the wavelength. Since the incident optical wavelength is typically 400 to 1,000 nm, while the acoustic wavelength is typically 1,000 to 10,000 nm, the magnitude of $\mathbf{k_d}$ is approximately that of $\mathbf{k_i}$. This is shown graphically in Figure 7.3, where $\mathbf{k_d}$ and $\mathbf{k_i}$ are drawn with constant magnitude in phase space (confined to the optical normal surface of the medium). Bragg matching is a consequence of momentum conservation, where the magnitude of the acoustic momentum vector $\mathbf{K_a}$ is given by:

$$\mathbf{K_a} = 2\mathbf{k} \sin\theta_B \tag{7.9}$$

where the angle between the incident optical beam and the acoustic beam is the Bragg angle

$$\theta_B = \sin^{-1}\frac{\lambda}{2n\Lambda} \tag{7.10}$$

Here we have used the fact that $\mathbf{k} = \mathbf{k_d} = \mathbf{k_i} = n2\pi/\lambda$, $\mathbf{K_a} = 2\pi/\Lambda$, $n$ is the index of refraction of the AO medium, $\lambda$ is the optical wavelength, and $\Lambda$ is the acoustic wavelength.

Conservation of energy in the AO interaction requires that the frequency of the diffracted optical plane wave $\omega_d$ be equal to the sum (for up-conversion) or difference (for down-conversion) of the frequencies of the incident optical plane wave $\omega_i$ and the acoustic plane wave $\Omega$:

$$\omega_d = \omega_i \pm \Omega \tag{7.11}$$

This relationship is also shown graphically in Figure 7.3, where the diffracted optical plane wave frequency is shifted up (up-shifted) or down (down-shifted) in frequency, depending on the acoustic propagation direction.

The interaction shown in Figure 7.3 describes the diffraction of a single optical plane wave by a single acoustic plane wave. If the direction or magnitude of **k** or **K** is changed such that the conservation of momentum no longer holds, no interaction takes place. A physical AO device, however, contains an angular spectrum of plane acoustic waves generated by a transducer of finite length $L$. The spatial dependence of the angular spectrum of plane acoustic waves or radiation pattern in an AO device is

$$S(\gamma) \approx \frac{\sin(\gamma L/\Lambda)}{\gamma L/\Lambda} \tag{7.12}$$

The distribution of acoustic plane waves has an angular width of $\delta\gamma = \Lambda/L$. This distribution allows Bragg matching to occur over a range of acoustic frequencies given by

$$\Delta\mathbf{K_a} = 2\mathbf{k}\partial\gamma \sin\theta_B \tag{7.13}$$

Indeed, by changing the acoustic frequency and observing the intensity of the diffracted optical beam, a bandshape is produced which is effectively a measure of the radiation pattern of the transducer, as shown in Figure 7.4. The design of a practical AO device involves the determination of the transducer length $L$ to support the required bandwidth. We will discuss this further in the section on AO deflector devices.

### *Phased Array*

The bandwidth of an AO Bragg cell is inversely proportional to the length of the transducer, as discussed in the previous section. Decreasing the transducer length decreases the width of the AO interaction region and thus decreases efficiency as well. One way to maintain a large value of $L$ and hence high diffraction efficiency, and yet still achieve Bragg matching over a wide bandwidth, is to track or steer the acoustic beam with frequency over the optical normal surface. By driving a segmented transducer array with a signal alternately phased between elements, the acoustic beam vector can be steered away from the normal to the transducer array with the changing frequency of the drive signal. The frequency-dependent change in the acoustic beam can then be made to track the Bragg angle [11].

The phased-array transducer is shown in Figure 7.5, where the transducer has $N$ elements, each of width $W$, on centers $D$. Each element is 180 deg out of phase with respect to its neighbor. For constructive interference to occur, the direction of the acoustic plane wave vector with respect to the normal to the transducer is

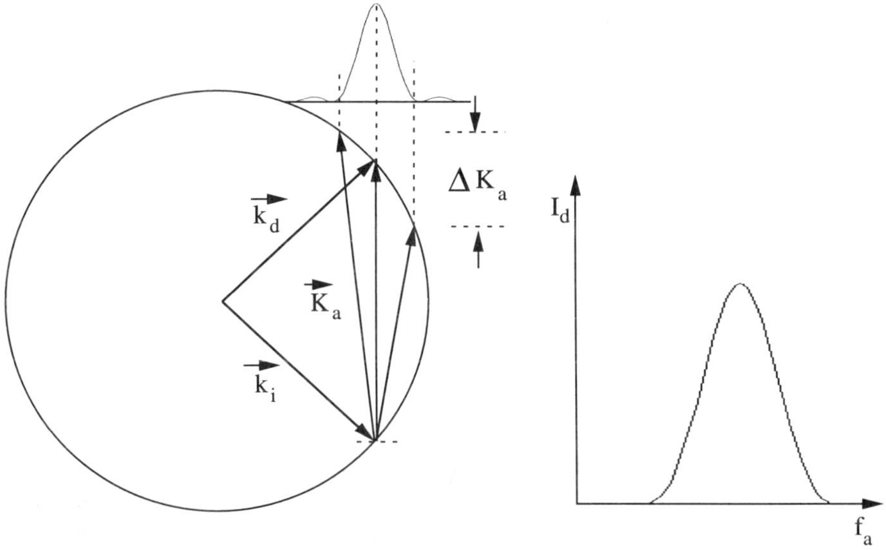


**Figure 7.4** Acousto-optic interaction: (a) plane wave distribution and (b) resulting bandshape.

$$\tan\,\theta = \frac{\Lambda/2}{D} \tag{7.14}$$

Since $\mathbf{K}_{\text{normal}} = 2\pi/\Lambda$ and $\mathbf{K}_{\mathbf{a}} = \mathbf{K}_{\text{normal}}/\cos\,\theta$, 7.14 can be rewritten as

$$\sin\,\theta = \frac{\pi/D}{\mathbf{K}_{\mathbf{a}}} \tag{7.15}$$

This behavior shown in the phase space representation of the phased-array Bragg interaction, Figure 7.6, where now, instead of the acoustic vector $\mathbf{K}_{\mathbf{a}}$ being fixed in the direction normal to the transducer, its direction tracks (as a function of frequency) along a line normal to the transducer (spaced a distance $\pi/D$ from the transducer). By arranging the tracking line to bisect the optical normal surface, perfect Bragg matching now occurs at two different frequencies (as opposed to one in the isotropic Bragg interaction), as shown in the band-shape curve in Figure 7.7. This technique allows the Bragg condition to be maintained over a wide bandwidth, even with a narrow angular spectrum of acoustic wave vectors. Hence, the interaction length $L$ can be kept at a large value for high diffraction efficiency. For a fractional bandwidth of 2/3, the gain in efficiency of a phased-array transducer over a single transducer element is approximately 3.

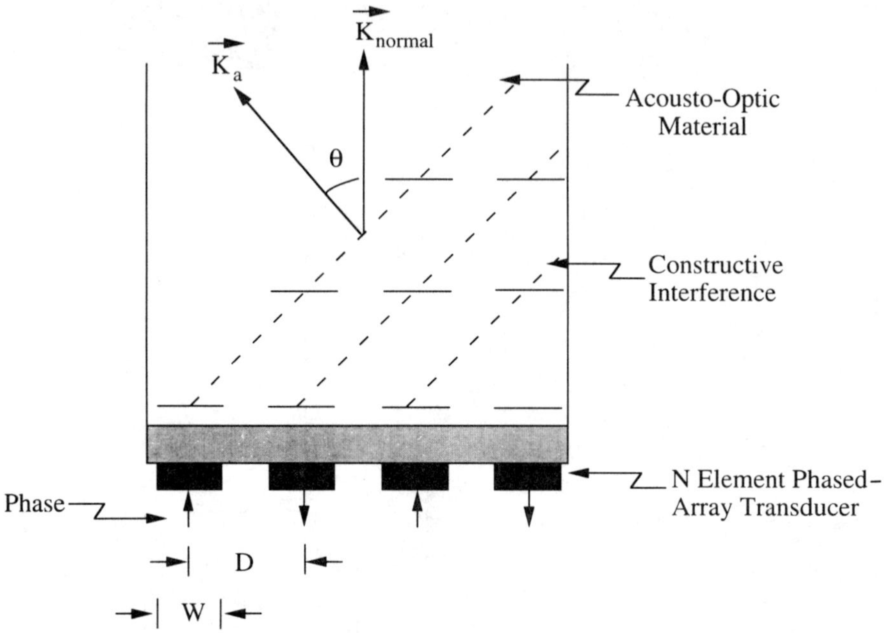


**Figure 7.5** Phased-array transducer.

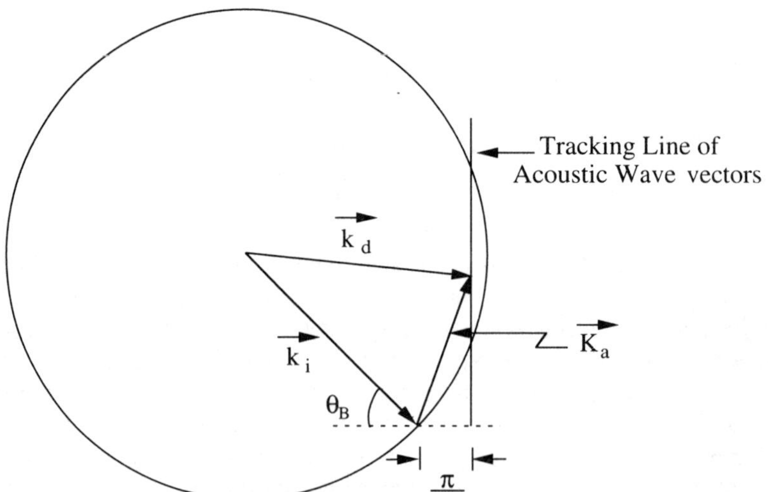


**Figure 7.6** Phased-array Bragg interaction.

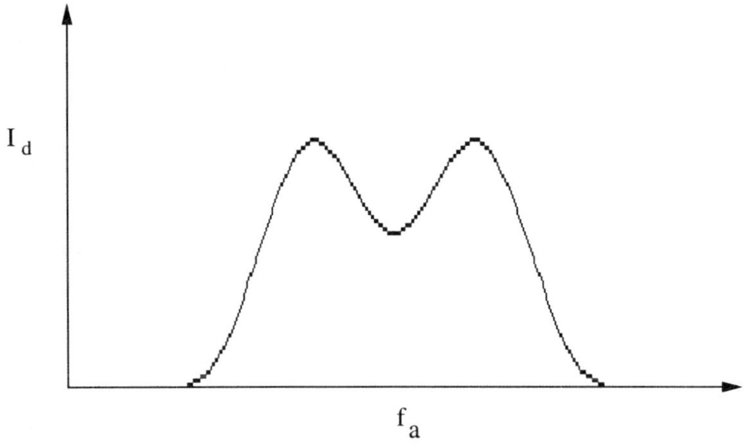


**Figure 7.7** Phased-array Bragg interaction acousto-optic band-shape curve.

*Anisotropic*

An anisotropic or birefringent material provides the opportunity to switch the polarization states by using an acoustic mode that couples an ordinary optical mode to an extraordinary optical mode [12]. The phase space representation of the anisotropic Bragg interaction is shown in Figure 7.8 for a positive uniaxial crystal. Here the index ellipsoid consists of two surfaces for an optically birefringent material; that is, two independent linearly polarized optical plane wave states are allowed. One state, the *ordinary* wave, is characterized by a directionally invariant index of refraction $n_0$. The other state, the *extraordinary*

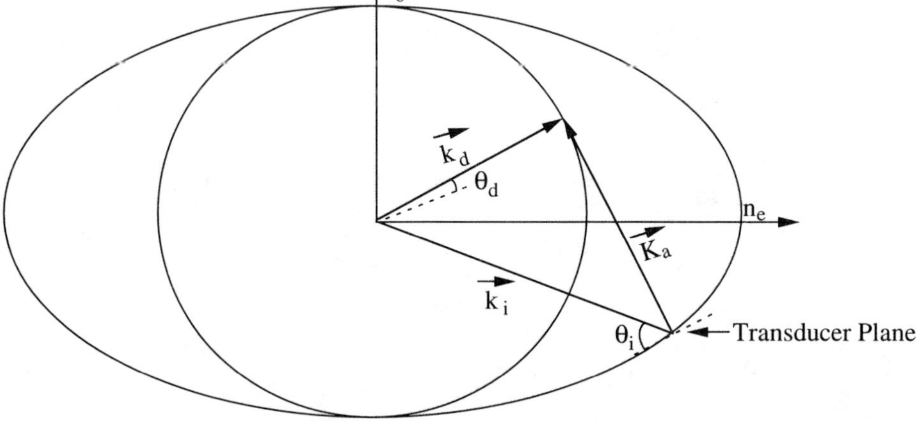


**Figure 7.8** Anisotropic Bragg interaction.

wave, is characterized by a directionally variant index $n_e$. The optical normal surface of the crystal consists of two surfaces, a sphere of radius $2\pi n_0/\lambda$ and an ellipsoid of revolution. The acoustic wave vector $\mathbf{K_a}$ is arranged to lie approximately tangential to one of the optical normal surfaces of the birefringent crystal. Hence, this interaction geometry is sometimes referred to as *tangential phase matching*. Thus, this technique also allows the Bragg condition to be maintained over a wide bandwidth, even with a narrow angular spectrum of acoustic wave vectors. From Figure 7.8, the length of the acoustic wave vector, and hence the frequency at which tangential phase matching occurs, is

$$f_t = \frac{v}{\lambda_0}\sqrt{n_i^2 - n_d^2} \tag{7.16}$$

Since the usable bandwidth limited by an octave condition and transducer conversion ability of a Bragg cell is 2/3 $f_t$, materials with high birefringence $n_e - n_0$ provide the potential for the widest bandwidth AO devices using the anisotropic Bragg interaction.

## 7.3 ACOUSTO-OPTIC BRAGG CELLS

The acoustic wave in an AO device is generated by a piezoelectric transducer bonded or deposited onto the device. As shown in Figure 7.9, the transducer consists of a piezoelectric material sandwiched between two metal layers. The bottom metal layer serves not only as an electrode but also as the bonding agent which attaches the piezoelectric material to the AO material. The top metal layer serves as the top electrode and its geometry defines the initial length and width of the sound column.

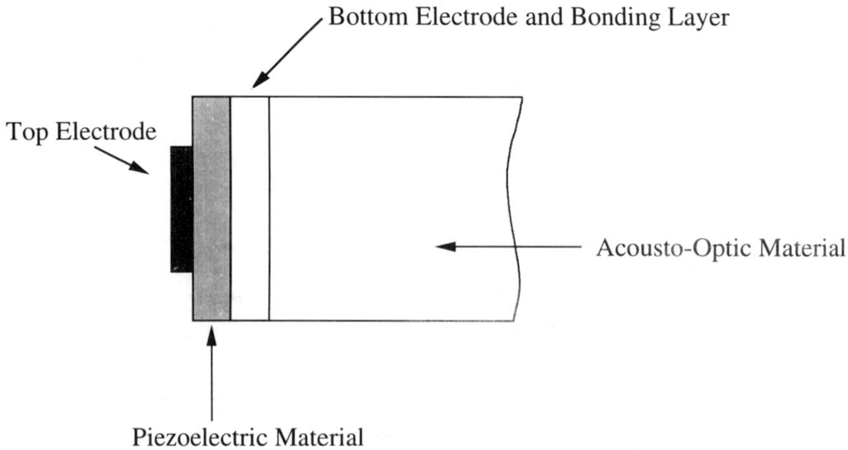


**Figure 7.9** Acousto-optic Bragg cell transducer.

Both thin-film zinc oxide and platelet lithium niobate are used as piezoelectric materials for AO Bragg cell transducers. The higher coupling coefficient and larger fractional bandwidth available from $LiNbO_3$ make it the usual choice for a transducer material.

In order to efficiently couple energy into the AO device, an electrical impedance-matching network is used to match the impedance of the AO device to the (usually $50\Omega$) source impedance of the electronic driver. Complete energy transfer from the source to the Bragg cell occurs when the impedance of the AO Bragg cell device $Z_{BC}$ is equal to the impedance of the source $Z_s$. With unequal impedances, part of the source energy is reflected back from the device, creating standing waves. A well-matched AO device will have a voltage standing wave ratio (VSWR) of less than 1.5.

A typical electrical impedance-matching network is shown in Figure 7.10. Here, series and parallel inductive, resistive and capacitive elements are used to match the source impedance to the AO device load impedance. At low frequencies (below 400 MHz), these elements are usually discrete components (*lumped* elements). At higher frequencies, parasitic impedances and inductances limit the efficiency and bandwidth of the lumped element implementation. Microstrip transmission line techniques are used at high frequencies to achieve effective impedance matching.

### 7.3.1 Device Types

Both light deflectors and modulators can be constructed using AO Bragg cell technology. In addition, because of the ability to fabricate multiple electrodes on a single transducer substrate, multichannel AO Bragg cell deflectors and modulators are possible.

*Deflectors*

An AO device arranged to deflect light is shown in Figure 7.11. Here, the optical beam is nominally collimated in the direction of acoustic propagation and focused in the trans-

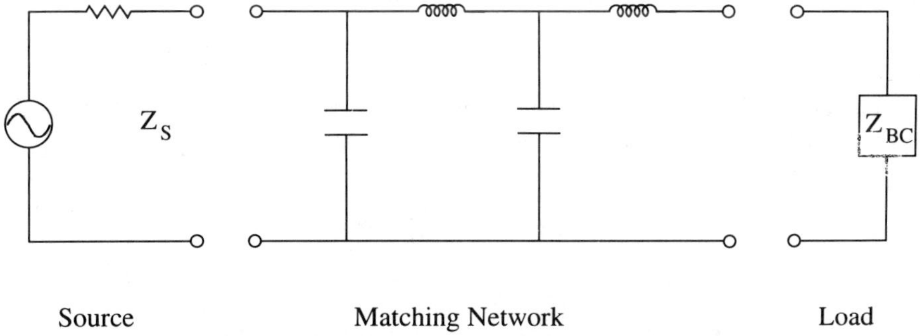


**Figure 7.10** Transducer impedance matching network.

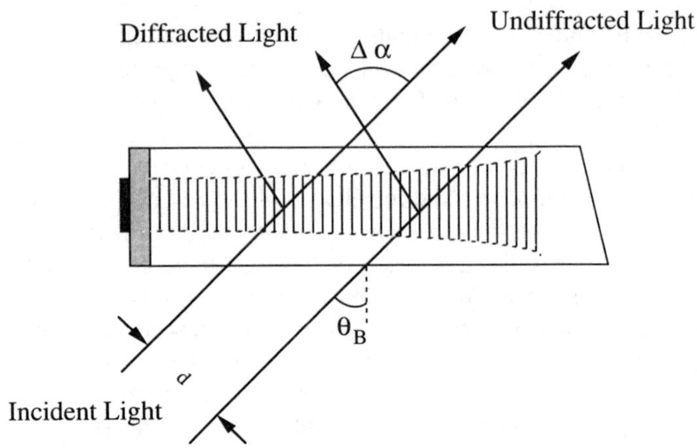


**Figure 7.11** Acousto-optic deflector.

ducer height dimension so that all of the light intersects the acoustic column. From (7.10), it can be seen that the diffraction angle of the light is linearly proportional to the frequency of the acoustic beam. Thus, beam deflection is possible by varying the input RF frequency. The maximum deflection angle $\alpha$ is determined by the maximum frequency deviation of the RF input

$$\Delta\alpha = \frac{\lambda \Delta f}{v} \tag{7.17}$$

The number of resolvable angles $N$ is just the ratio of the total deflection angle to the angular spread of the nominally collimated light. If the light has a width $d$, then

$$N = \frac{\Delta\alpha}{\lambda / d} = \frac{d}{v}\Delta f = \tau \Delta f \tag{7.18}$$

were $d/v = \tau$ is the transit time of the acoustic wave through the optical beam. The number of resolvable angles is thus equal to the time-bandwidth product of the device.

The ideal efficiency of light diffraction into the first diffraction order is, for a deflector with transducer dimensions of $LH$, from (7.6),

$$\frac{I_d}{I_i} = \sin^2\left(\frac{\pi}{\sqrt{2}\lambda_0}\sqrt{M_2 P \frac{L}{H}}\right) \tag{7.19}$$

where $H$ is the transducer height and $P = P_a LH$ is the acoustic power. By operating at low acoustic power, the efficiency will be linearly proportional to the square root of the acoustic power and thus linearly proportional to the amplitude of the input electrical signal.

The design of an AO Bragg cell deflector requires the specification of the acoustic material and mode as well as the transducer height and length. The transducer length $L$ is inversely proportional to the bandwidth of the device, as discussed in Section 7.2.2.1. However, as shown in (7.19), the efficiency of the interaction is proportional to $L$. Optimizing efficiency and bandwidth yields an optimum transducer length $L$ [13]:

$$L = \frac{8nv^2}{2\pi\lambda(\Delta f)^2} \tag{7.20}$$

The transducer height $H$ is optimized to allow for adequate spatial overlap between the optical and acoustic beams and to minimize acoustic diffraction in the device. For a rectangular transducer of height $H$, the acoustic radiation field near the transducer is approximately collimated. This collimated region, known as the *acoustic near field* or *Fresnel* region, extends a distance $D$ into the cell, given by

$$D = \frac{H^2}{\Lambda(1 - 2b)} \tag{7.21}$$

where $b$ is a measure of the acoustic anisotropy.

The diffraction efficiency is reduced from the ideal value given in (7.19) by acoustic attenuation and acoustic diffraction. The overall Bragg cell efficiency can be written as

$$\eta = \eta_0 \eta_a \eta_e \tag{7.22}$$

where $\eta_0$ is the efficiency of the AO interaction, $\eta_a$ is the reduction in efficiency due to acoustic attenuation of the signal as it traverses the Bragg cell, and $\eta_e$ is the reduction in efficiency due to transducer insertion loss.

The AO efficiency is given by (7.19). The acoustic attenuation loss is given by

$$\eta_a = \mathrm{e}^{-(\Gamma f^2 \tau/10 \log e)} \tag{7.23}$$

where $\Gamma$ is the acoustic attenuation expressed in dB/$\mu$s/GHz$^2$. The reduction in efficiency due to transducer insertion losses is usually about 50% at high frequencies.

### *Modulators*

An AO device arranged to modulate light is shown in Figure 7.12. Here, unlike the deflector arrangement, the optical beam is focused in both dimensions near the transducer.

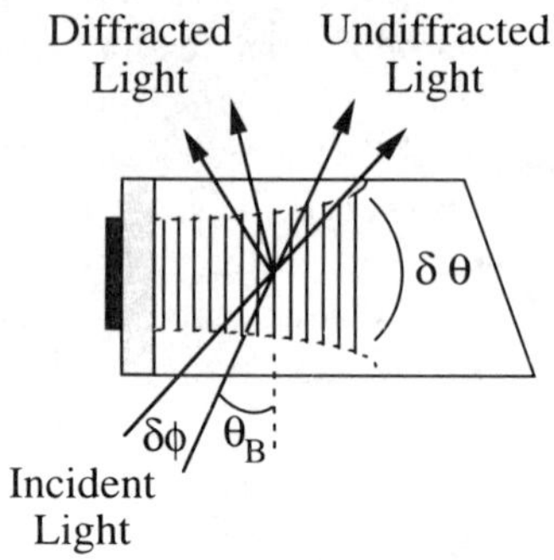


**Figure 7.12** Acousto-optic modulator.

The phase space representation of this arrangment is shown in Figure 7.13. With the optical beam focused, the angular spread of optical plane wave vectors can be made approximately equal to the angular spread of acoustic plane wave vectors. Here, unlike the deflector, the diffracted optical beam will *not* vary spatially as a function of frequency. In this way, the diffracted optical beams from a band of RF frequencies will overlap, producing modulation at a photodetector.

The ratio of the angular spread of the optical beam focused to a waist diameter $d$ ($\delta\phi = 4\lambda/\pi d$) to the acoustic beam launched from a transducer of length $L$ ($\delta\theta = \Lambda/L$) is

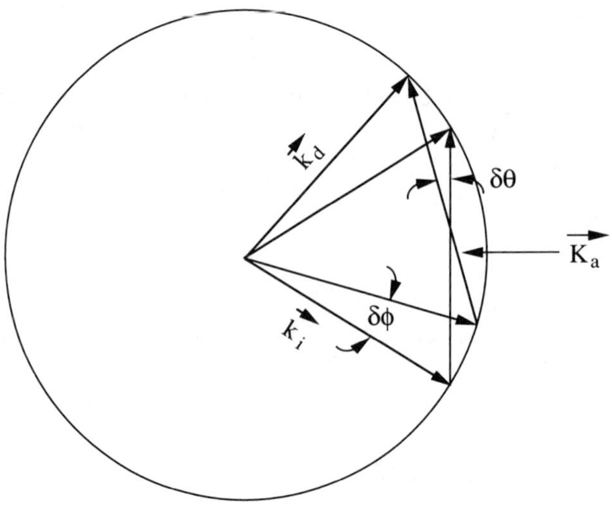


**Figure 7.13** Acousto-optic modulator interaction.

$$a = \frac{\partial \phi}{\partial \theta} = \frac{4\lambda L}{\pi d \Lambda} \tag{7.24}$$

When $a > 1$ momentum is not conserved for all of the optical plane wave vectors. The diffracted light will have an angular spread equal to the acoustic beam. The light at the edges of the incident optical beam will not be Bragg-matched, resulting in an elliptically shaped diffracted beam. On the other hand, when $a < 1$, the diffracted light outside of the angular spread of the acoustic beam will not be modulated. Thus bandwidth and efficiency are reduced. Maximizing efficiency and bandwidth yields an $a$ in the range of $2.0 < a < 1.5$ [14].

For digital modulation, with $a$ of about 1.5, modulator rise time (10% to 90%) is:

$$t_r = \frac{0.85d}{v} \tag{7.25}$$

For analog modulation, the modulation bandwidth (one-half of the modulator RF bandwidth) is [15]:

$$f_m = \frac{0.446v\sqrt{\beta}}{d} \tag{7.26}$$

where $\beta$ is the frequency response roll-off expressed in decibels

*Multichannel Devices*

Two-dimensional optical signal processing can be accomplished with multichannel Bragg cells [16]. These cells are constructed and function in the same manner as the single-channel devices described above, except that instead of a single electrode, a multielement electrode is used, as shown in Figure 7.14. The design of an individual channel of a multichannel Bragg cell is similar to that of a single-channel device. To maximize diffraction efficiency, the transducer length is maximized consistent with rise time requirements, while the transducer height is minimized consistent with diffraction spreading minimization. Multiple electrodes on a common acoustic substrate, however, impose additional design constraints not found in the design of the single-channel device. In particular, careful attention must be paid to minimizing adjacent channel crosstalk (both electrical and acoustical).

Acoustical crosstalk arises from diffraction spreading of the acoustic beam from one channel into neighboring channels. One of the most effective ways of minimizing acoustical crosstalk is through the use of acoustic anisotropy. In acoustically anisotropic materials, the magnitude of the acoustic velocity varies with the propagation direction. It can be shown that the acoustic Poynting vector **P** describing power flow in a material is

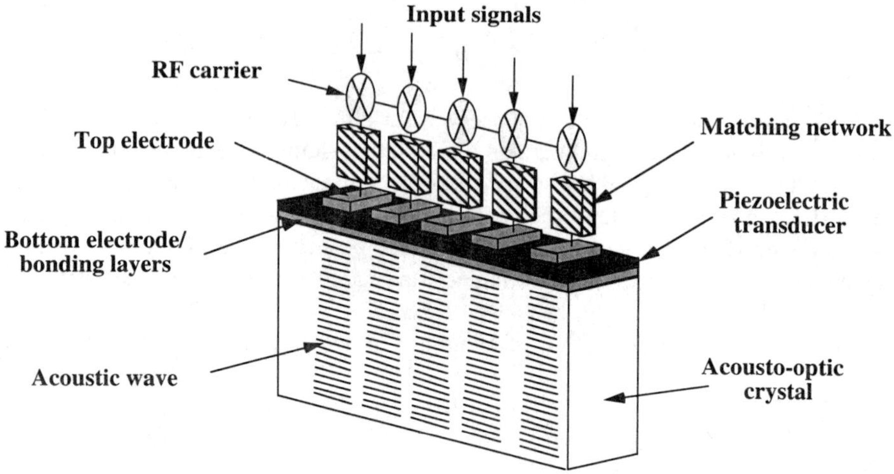


**Figure 7.14** Multichannel Bragg cell.

in a direction normal to the slowness surface (thus, in an isotropic material with a circular slowness surface, the acoustic propagation direction is the same as that of the power flow direction.) The anisotropy of material is characterized by a parameter $b$ which is zero in the isotropic case. In anisotropic materials, the power flow direction is different from the propagation direction. In some AO materials, the Poynting vector diverges from the propagation direction, and the material exhibits larger acoustic spreading than in the isotropic case. Here $b < 0$. In other materials, the Poynting vector is less divergent than the isotropic case. Here $b > 0$. Some materials have modes along directions where the slowness surface is essentially flat. In this case, the Poynting vector remains orthogonal to the transducer and the acoustic energy remains collimated. Here $b = 0.5$. These so-called *self-collimating* modes provide the means to simultaneously minimize transducer electrode spacing and adjacent channel acoustic crosstalk. Table 7.1 gives values for the most commonly used AO materials.

A representative list of commercially available AO Bragg cell devices is given in Table 7.2. This list is not meant to be exhaustive, but to show, in general, the performance of devices given the AO materials shown in Table 7.1 and the interaction geometries outlined in Section 7.2.

## 7.4 ACOUSTO-OPTIC BRAGG CELL SIGNAL PROCESSING CHARACTERISTICS

AO Bragg cells provide a means for encoding an optical beam with information contained in an RF signal in real time. The typical operation of a Bragg cell in an optical signal processing system is illustrated in Figure 7.15. The information bearing signal $s(t)$ with

**Table 7.2**
Commercially Available Acousto-Optic Devices

| | Material (Mode) | Center Frequency (MHz) | Bandwidth (Mhz) | Time Aperture ($\mu s$) | Diffraction Efficiency (%/W) | Manufacturer |
|---|---|---|---|---|---|---|
| A. Deflectors | | | | | | |
| Narrowband | $TeO_2$ (SS) | 45 | 30 | 50 | 50 | Isomet |
| | | 50 | 30 | 40–80 | 250 | CTI |
| | | 60 | 40 | 18–30 | 15–22.5 | Brimrose |
| | | 65 | 35 | 15.00 | 50 | Isomet, NEOS |
| | | 70 | 40–50 | 5–50 | 40 | IntraAction, NEOS |
| | | 75 | 50 | 2–80 | 70–150 | A et A, CTI, Inrad |
| | | 130 | 60 | 20.00 | 15 | Brimrose |
| | | 150 | 100 | 10.00 | | IntraAction |
| Midband | $PbMoO_4$ | 225 | 150 | 6.70 | 50 | A et A |
| | GaP | 250 | 100 | 0.7–2 | 50–40 | Brimrose |
| | $PbMoO_4$ | 300 | 200 | 5.00 | | Isomet |
| | $TeO_2$ (L) | 320 | 200 | 2.00 | 20 | Brimrose |
| | GaP | 350 | 200 | 2.00 | 35 | |
| | GaAs | 600 | 200 | 1.00 | 30 | |
| | GaP | 700 | 400 | 1.00 | 40 | A et A |
| Wideband | GaP | 850 | 500 | 2.00 | 25 | Brimrose |
| | $LiNbO_3$ (S) | 1,000 | 500 | 0.40 | 10 | NEOS |
| | $LiNbO_3$ (L) | 1,000 | 500 | 2.00 | 2.5 | CTI |
| | GaP L | 1,000 | 500 | 0.30 | 40 | |
| | GaP | 1,000 | 600 | 0.60 | 30 | A et A |
| | | 1,000 | 500 | 1.00 | 20 | Brimrose |
| | | 1,200 | 500 | 0.20 | 100 | Marconi |
| | | 1,500 | 1,000 | 1.00 | 10 | Brimrose |
| | | 1,500 | 560 | 0.20 | 10 | NEOS |
| | $LiNbO_3$ (L) | 2,000 | 1,000 | 1.00 | 1 | CTI |
| | GaP L | 2,000 | 1,000 | 0.15 | 30 | |
| | $LiNbO_3$ (S) | 2,100 | 500 | 1.40 | 20 | Thomson |

**Table 7.2** *(cont.)*
Commercially Available Acousto-Optic Devices

| | Material (Mode) | Center Frequency (MHz) | Bandwidth (Mhz) | Time Aperture (μs) | Diffraction Efficiency (%/W) | Manufacturer |
|---|---|---|---|---|---|---|
| | | 2,300 | 1,000 | 0.40 | 32 | CTI |
| | | 2,500 | 1,000 | 1.00 | 20 | Brimrose |
| | | 2,500 | 1,500 | 1.00 | 7 | |
| | | 2,700 | 1,500 | 0.70 | 5 | Thomson |
| | | 3,000 | 2,500 | 0.60 | 1.5 | |
| Multichannel (32 channels) | $TeO_2$ | 70 | 40 | 18.00 | 40 | Brimrose |
| | | 80 | 30 | 4.70 | 70 | |
| | GaP | 300 | 200 | 2.00 | 30 | |
| | $TeO_2$ | 300 | 200 | 2.00 | 30 | |
| | | 400 | 200 | 5.00 | 30 | CTI |
| | | 600 | 250 | 1.00 | 25 | Brimrose |
| | | 800 | 300 | 1.00 | 20 | |
| | GaP | 800 | 400 | 2.00 | 10 | |
| | | 800 | 500 | 1.00 | 30 | |
| B. Modulators | Glass | 40 | 7 | 50.00 | 85 | Anderson Labs, Inrad, IntraAction |
| | $PbMoO_4$ | 70 | 20 | 15.00 | 85 | A et A |
| | $TeO_2$ | 70 | 6 | 20.00 | 90 | |
| | Glass | 80 | 14–18 | 20–25 | 85–90 | A et A, Anderson Labs, IntraAction |
| | $PbMoO_4$ | 80 | 6–35 | 10–60 | 65–85 | A et A, CTI, Inrad |
| | $SiO_2$ | 80 | 20 | 50.00 | 70 | Brimrose |
| | $TeO_2$ | 80 | 15–17.5 | 18–20 | 65–85 | CTI, NEOS |
| | Glass | 100 | 10 | 35.00 | 85 | A et A |
| | $PbMoO_4$ | 110 | 20 | 15.00 | 85 | A et A, IntraAction |
| | $TeO_2$ | 110 | 10–28 | 18–35 | 60–70 | CTI, Inrad |
| | $TeO_2$ | 150 | 30 | 12.00 | 70–80 | Inrad |
| | $SiO_2$ | 150 | 30 | 28.00 | 75 | Brimrose |

| | | | | | | |
|---|---|---|---|---|---|---|
| | $TeO_2$ | 440 | 75 | 4.00 | 60 | |
| | $PbMoO_4$ | 200 | 35–40 | 8–10 | 90 | A et A |
| | $TeO_2$ | 200 | 40–75 | 9–25 | 50–70 | Brimrose, CTI, Inrad |
| | $TeO_2$ | 210 | 40 | 10.00 | 70 | NEOS |
| | $SiO_2$ | 210 | 40 | 20.00 | 75 | Brimrose |
| | GaP | 250 | 50–100 | 20.00 | 60 | |
| | $TeO_2$ | 260 | 40 | 10.00 | 70 | NEOS |
| | $TeO_2$ | 440 | 75 | 4.00 | 60 | |
| | $TeO_2$ | 500 | 120 | 5.00 | 50 | CTI |
| | $TeO_2$ | 600 | 200 | 10.00 | 40 | Brimrose |
| | GaP | 650 | 250 | 10.00 | 50 | |
| | GaP | 1,000 | 500 | 4.00 | 40 | |
| | $TeO_2$ | 1,000 | 300 | 6.70 | 35 | |
| | GaP | 1,500 | 1,000 | 2.67 | 10 | |
| | GaP | 1,700 | 500 | 2.35 | 40 | |
| | $TeO_2$ | 1,700 | 300 | 6.70 | 20 | |
| Multichannel: | | | | | | |
| 9 channels | Glass | 50–90 | | 200.00 | 75 | A et A |
| 32 channels | $TeO_2$ | 80 | 30 | | 70 | Brimrose |
| 9 channels | | 200–300 | | 180.00 | 75 | A et A |
| 32 channels | | 300 | 200 | | 60 | Brimrose |
| | GaP | 300 | 200 | | 50 | |
| | | 600 | 300 | | 50 | |
| | $TeO_2$ | 600 | 250 | | 45 | |
| | | 800 | 300 | | 40 | |
| | GaP | 800 | 500 | | 50 | |

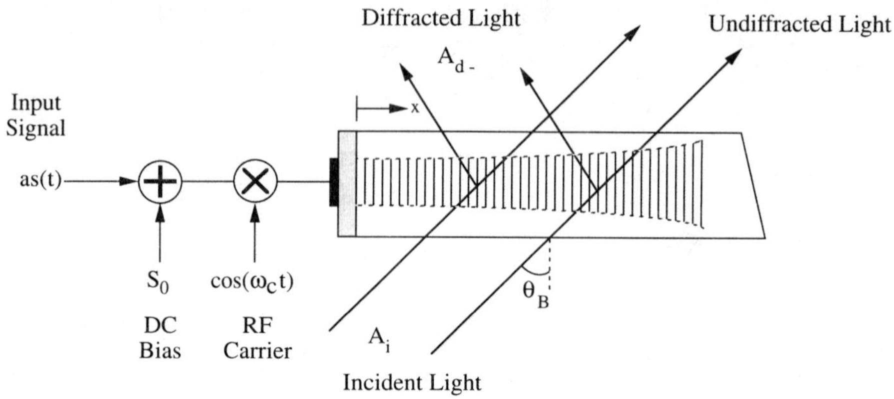


**Figure 7.15** Operation of acousto-optic Bragg cell in optical signal-processing system.

amplitude $a$ is added to a dc bias and mixed with an RF carrier of frequency $\omega_c$ and input to the Bragg cell. The level of the dc bias, as we show later, will control the linear operation of the device. The frequency band of the AO Bragg cell, as discussed in the previous section, is determined by material, AO interaction geometry, and transducer geometry considerations. AO devices have frequency bands within the frequency range of about 30 MHz to about 6 GHz. Analog signals within the frequency band of the cell can be input to the cell directly without mixing with a carrier. Analog signals outside the frequency band of the device as well as digital signals are processed by mixing the signal with an appropriate RF carrier frequency such that the resulting signal is within the frequency band of the device.

The Bragg cell transducer generates an acoustic wave, or strain-field variation, which propagates in the AO material in the $x$ direction with acoustic velocity $v$. At any given position within the cell, the strain field is modulated by the input signal delayed in time by $t = x/v$. The signal inside the cell can thus be represented as $s(t - x/v)$ and the corresponding strain-field perturbation as (where here we ignore the acoustic attenuation in the material)

$$S(x,t) = [S_0 + as(t - x/v)] \cos[\omega_c(t - x/v)] \tag{7.27}$$

where $S_0$ is the amplitude of the unmodulated RF carrier.

The strain-field perturbation induces an index of refraction variation $\Delta n$ (from (7.4)) given by

$$\Delta n = \frac{1}{2} n^3 p [S_0 + as(t - x/v)] \cos[\omega_c(t - x/v)] \tag{7.28}$$

An optical beam traversing the length $L$ of the acoustic field will undergo a phase shift given by

$$\phi = -\mathbf{k}L(n_0 + \Delta n) \tag{7.29}$$

where $\mathbf{k} = 2\pi/\lambda$ is the free-space optical wave vector. Expressing the amplitude of the incident (uniform) optical beam as

$$A_i(z = 0,\ t) = A_0 e^{i\omega_0 t} \tag{7.30}$$

where $\omega_0$ is the incident light beam frequency, the amplitude of the phase-shifted optical beam exiting the Bragg cell is

$$A_e(z = L,t) = A_0 e^{i\omega_0 t} e^{-i\mathbf{k}L(n_0+\Delta n)} \tag{7.31}$$

In the weak interaction limit, the emerging optical beam can be written as

$$A_e(x,z = L,t) = A_0 e^{i\omega_0 t} e^{-i\mathbf{k}Ln_0}\{1 + iC[S_0 + s(t - x/v)]\cos[\omega_c(t - x/v)]\} \tag{7.32}$$

where $C = 1/2kLn^3p$. Expanding the cos term into exponentials, the output is [17]:

$$A_e(x,z = L,t) = A_0 e^{i\omega_0 t} e^{-i\mathbf{k}Ln_0}\left\{1 + i\frac{C}{2}[S_0 + as(t - x/v)][e^{i[\omega_c(t-x/v)]} + e^{-i[\omega_c(t-x/v)]}]\right\} \tag{7.33}$$

The output, in general (Raman-Nath regime), consists of an undiffracted beam and two diffracted orders. Illuminating the cell at the Bragg angle preferentially selects one of these two diffracted orders.

The amplitude of the diffracted optical beam emerging from the Bragg cell is:

$$A_{d\pm}(x,t) = iC'[S_0 + as(t - x/v)][e^{\pm i[\omega_c(t-x/v)]}] \tag{7.34}$$

where $C' = (C/2)A_0 e^{i\omega_0 t} e^{-i\mathbf{k}Ln_0}$ band the sign in the exponential is determined by the Bragg illumination condition. As shown in Figure 7.16, when the Bragg angle is positive, the emerging diffracted optical beam is up-shifted in frequency and the sign in the exponential is positive. Similary, when the Bragg angle is negative, the emerging optical beam is down-shifted in frequency and the sign in the exponential is negative.

### 7.4.1 Electrical-to-Optical Spatial Transfer Function

The expression for the amplitude of the diffracted beam in (7.34) assumed that the signal propagated in the cell without attenuation and that the Bragg cell was illuminated with a

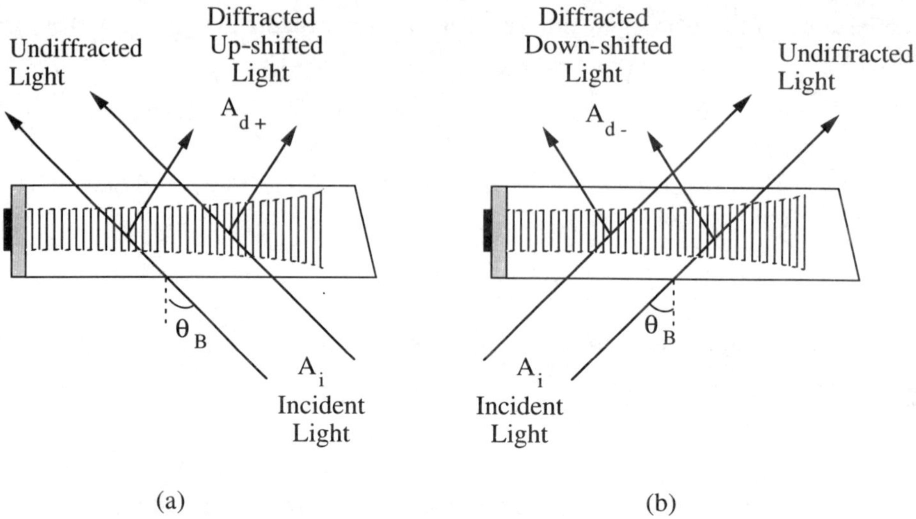


**Figure 7.16** Acousto-optic Bragg cell illumination for (a) upshift and (b) downshift operation.

uniform optical beam. However, as discussed previously, the acoustic wave, and hence the signal, is attenuated as it traverses the cell. The attenuation in spatial coordinates is (from (7.23)) given by

$$\eta_a(x) = e^{-(\alpha f^2 x/10 \log e)} \tag{7.35}$$

where $\alpha$ is the AO material acoustic attenuation expressed in dB/cm/GHz$^2$.

The optical beam illuminates a Bragg cell rectangular aperture defined by the time aperture of the cell and height of the transducer. The optical beam is also typically not uniform. The actual amplitude of the incident optical beam is then (instead of (7.30):

$$A_i(z = 0,t) = A_0 e^{i\omega_0 t} \operatorname{rect}(x/D) \operatorname{rect}(y/H) w(x,y) \tag{7.36}$$

where the rect functions define the Bragg cell aperture $D$ and transducer height $H$, and $w(x,y)$ is the intensity distribution of the optical beam [18]. The electrical-to-optical AO amplitude spatial transfer function $T(x,y,t)$ is then

$$T(x,y,t) = iC'[S_0 + as(t - x/v)][e^{\pm i[\omega_c(t-x/v)]}] \operatorname{rect}(x/D) \operatorname{rect}(y/H) w(x,y) e^{(\alpha f^2 x/10 \log e)} \tag{7.37}$$

### 7.4.2 Electrical-to-Optical Intensity Transfer Function

The AO Bragg cell can be operated so that the intensity of the diffracted optical beam is linear in either the amplitude *or* intensity of the input signal. The electrical-to-optical intensity transfer function for an AO cell is shown in Figure 7.17, where the signal amplitude $\nu$, from (7.6), is

$$\nu = \frac{\pi L}{\sqrt{2}\,\lambda_0}\sqrt{M_2 Pa} \tag{7.38}$$

where

$$\sqrt{P_a} = S_0 + as(t) \tag{7.39}$$

In the usual operating condition, $\nu$ is small (the dc bias is zero and $a$ is small) and the optical intensity is proportional to $\nu^2$ (i.e., proportional to the signal *intensity*). If $\nu$ is made large (ideally $\pi/4$), the optical intensity becomes proportional to $\nu$ (i.e., proportional to the signal *amplitude*).

### 7.4.3 Dynamic Range

The dynamic range of a Bragg cell is the ratio of the intensity of the maximum diffracted light to the intensity of spurious diffracted light. The spurious response results from optical

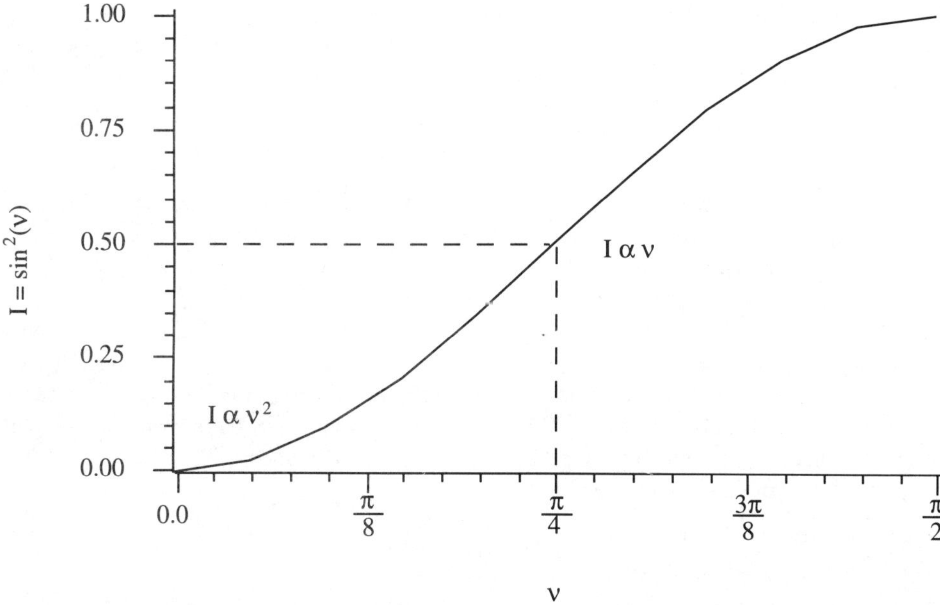


**Figure 7.17** Acousto-optic Bragg cell electrical-to-optical intensity transfer function.

noise due to scatter from the cell and unwanted light diffraction when multiple frequencies are present simultaneously within the cell. Optical noise can be minimized during Bragg cell fabrication by using AO materials of high optical quality, by maintaining high-quality optical windows, and by using an AR coating on the windows. In this way, optical noise may be kept less than 60 dB below the maximum diffracted light signal.

Spurious response from simultaneous multiple frequencies within the cell, IM products, are characterized using two input frequencies $f_1$ and $f_2$. By restricting the two frequencies to an octave band, the dominant in-band intermodulation terms are the two-tone third-order products $2f_2 - f_1$ and $2f_1 - f_2$. IM products arise from two different phenomena. First, light diffracted from one acoustic grating may be rediffracted by other acoustic gratings [19]. The magnitude of this two-tone third-order product (at $2f_1 - f_2$) is

$$I = \frac{I_1^2 I_2}{36} \tag{7.40}$$

where $I_1$ and $I_2$ are the diffraction efficiencies of the two input frequencies. A Bragg cell dynamic range of 50 dB thus requires the power in each of the two input frequency signals to result in less than 2% diffraction efficiency for each signal. The second source of IM products is the nonlinear response of the acoustic material. Harmonics of each of the input acoustic waves interact with the fundamental-frequency acoustic waves and with themselves to produce acoustic IM products [20]. These products grow as the waves propagate, and thus the effect is dependent on the illumination aperture of the Bragg cell. The effect is also strongly dependent on signal power density, frequency, and the nonlinearity of the AO material. Wide-bandwidth, high-center-frequency cells, as well as long-aperture cells, are thus particularly sensitive to these effects.

## 7.5 SUMMARY

In this brief review we have described the AO effect and shown how, in combination with isotropic, anisotropic, and phased-array transducer interaction geometries, it can be used to achieve real-time spatial light modulation. The AO Bragg cell device has other unique features, aside from its wide bandwidth, including high transmission efficiency, high diffraction efficiency, and large space-bandwidth product, making it the device of choice for many optical signal processing applications. The technology has matured, with a number of commerical vendors supplying a large range of Bragg cell devices. Research into new AO materials, improving Bragg cell performance through combinations of interaction geometries, and development of low-noise RF addressing of multichannel Bragg cells is ongoing in this field.

## REFERENCES

[1] Brillouin, L., ''Diffusion de la lumiere et des rayons X par un corps transparent homogene,'' *Ann. Phys.*, Paris, Vol. 17, 1922, pp. 88–122.

[2] Lucas, R., and P. Biquard, ''Proprietes optiques des milieux solides et liquides soumis aux vibration elastiques ultra sonores,'' *J. Phys. Rad.*, Vol. 3, 1932, pp. 464–477.

[3] Debye, P., and F. W. Sears, ''On the scattering of light by supersonic waves,'' *Proc. Nat. Acad. Sci.*, (U.S.), Vol. 18, June 1932, pp. 409–414.

[4] Raman, C. V., and N. S. Nath, ''The Diffraction of Light by High Frequency Sound Waves,'' *Proc. Ind. Acad. Sci.*, Vol. 2, 1935, pp. 406–420; Vol. 3, 1936, pp. 75–84, 119–125, pp. 459–465.

[5] Klein, W. R., and B. D. Cook, ''Unified Approach to Ultrasonic Light Diffraction,'' *IEEE Trans. Sonics and Ultrasonics* Vol. SU-14, July 1967, pp. 723–733.

[6] Auld, B. A., *Acoustic Fields and Waves in Solids*, Vol. 1, New York: John Wiley and Sons, 1973.

[7] Hecht, D. L., and G. W. Petrie, ''Acousto-Optic Diffraction From Acoustic Anisotropic Shear Modes in Gallium Phosphide,'' *Proc. IEEE Ultrasonics Symp.*, 1980, pp. 474–479.

[8] Van Cittert, P. H., ''Zur Theorie der Lichtbeugung un Ultraschallwellen,'' *Physica IV*, July 1937, pp. 590–594.

[9] Korpel, A., *Acousto-Optics*, New York: Marcel Dekker, 1988.

[10] Korpel, A., ''Acousto-Optics,'' Ch. 2 in *Applied Solid State Science*, Vol. 3, R. Wolfe, ed., New York: Academic Press, 1972, pp. 73–179.

[11] Pinnow, D. A., ''Acousto-Optic Light Deflection: Design Considerations for First Order Beam Steering Transducer,'' *IEEE Trans. Sonics and Ultrasonics*, Vol. SU-18, No. 4, 1971.

[12] Dixon, R. W., ''Acoustic Diffraction of Light in Anisotropic Media,'' *IEEE J. Quantum Electron.*, Vol. QE-3, No. 1, 1967, pp. 85–93.

[13] Gordon, E. I., ''A Review of Acoustooptical Deflection and Modulation Devices,'' *Proc. IEEE*, Vol. 54, No. 10, 1966, pp. 1391–1400.

[14] Maydan, D., ''Acoustooptical Pulse Modulators,'' *IEEE J. Quantum Electron.*, Vol. QE-6, No. 1, 1970, pp. 15–24.

[15] Young, E. H., Jr., and S.-K. Yao, ''Design Considerations for Acousto-Optic Devices,'' *Proc. IEEE*, Vol. 69, No. 1, 1981, pp. 54–64.

[16] Pape, D. R., ''Multichannel Bragg Cells: Design, Performance, and Applications,'' *Opt. Eng.*, Vol. 31, No. 10, 1992, pp. 2148–2158.

[17] Rhodes, W. T., ''Acousto-Optic Signal Processing: Convolution and Correlation,'' *Proc. IEEE*, Vol. 69, No. 1, 1981, pp. 65–79.

[18] Hecht, D. L., ''Spectrum Analysis Using Acousto-Optic Devices,'' *Opt. Eng.*, Vol. 16, No. 5, 1977, pp. 461–466.

[19] Hecht, D. L., ''Multifrequency Acoustooptic Diffraction,'' *IEEE Trans. Sonics and Ultrasonics*, Vol. SU-24, No. 1, 1977.

[20] Elston, G., and P. Kellman, ''The Effects of Acoustic Nonlinearities in Acousto-Optic Signal Processing Systems, *Proc. IEEE Ultrasonics Symp.*, 1983, pp. 449–453.

# *Chapter 8*
# *Optical Waveguide*

***Arthur Paolella***
*Army Research Laboratory*

***William D. Jemison***
*Martin Marietta*

## 8.1 INTRODUCTION

This chapter presents the fundamental properties of optical waveguides with an emphasis on their application to modern radar systems. The chapter is intended for the radar systems engineer who may have little or no previous knowledge about the potential use of optical waveguides in microwave radar systems or about the basic properties of optical waveguides themselves.

The chapter begins with a brief introduction of various radar applications where optical waveguides may be used, and outlines some of the major design considerations that must be taken into account, including both the physical and transmission characteristics of optical waveguides, which impact practical radar system implementation. Next, the fundamentals of common optical waveguides are presented, followed by a discussion of commonly used coupling techniques and waveguide materials. Finally, several examples of the current state of the art is presented, along with possible future directions for the use of optical waveguides in modern radar systems and subsystems.

## 8.2 OPTICAL WAVEGUIDES FOR RADAR SYSTEM APPLICATIONS

An optical waveguide is composed of a dielectric material whose refractive index profile is designed to confine and propagate light by the process of total internal reflection. The most common type of optical waveguide is the cylindrical optical fiber, which consists of a core and a cladding region where the index of refraction in the cladding is slightly lower than that in the core. Other types of optical waveguides include various types of rectangular guides, including channel, rib, and strip-loaded configurations, which are particularly important for integrated optics devices.

Optical waveguides may be used to distribute several types of common radar signals [1,2]. These signals include high-frequency analog microwave and millimeter-wave signals, as well as low-frequency control and monitoring signals. The specific signal characteristics such as frequency, bandwidth, modulation characteristics, power level, and manifolding requirements depend on the radar application as well as the specific radar system architecture. High-frequency analog signals are required for carrier, LO, and IF signal distribution and may range from megahertz to millimeter-wave frequencies. Analog fiber-optic links are used to distribute these signal and link gains of up to 50 dB and very low noise figure have been predicted [3]. Control and monitoring signals, on the other hand, may be either analog or digital and are required for functions such as timing, modulation, and phased-array beam steering. Optical fiber has been used to distribute analog control signals to microwave gain controllers [4], phase shifters [5], and switches [6]. In addition, high-speed interconnects have been designed to interface to high-speed digital fiber-optic links to control phased-array beam steering [7].

The use of optical waveguides for radar signal distribution presents a distinct departure from traditional radar design techniques. Therefore, it is important to understand the basic properties of optical waveguides, including both physical and transmission characteristics. Furthermore, it is important to assess the impact that optical waveguides have on the radar architecture.

Optical waveguides offer several significant advantages over traditional microwave distribution media such as waveguide and coaxial cable, which include reduced size and weight, improved flexibility (for optical fiber), and immunity from electromagnetic interference (EMI), electromagnetic pulse (EMP), and crosstalk. Typically, the optical waveguide cross section is on the order of the optical wavelength in order to provide adequate light confinement and modal characteristics. Typical fiber-optic waveguide dimensions are 9-$\mu$m core with 125-$\mu$m cladding for single-mode fiber and 50-$\mu$m core with 125-$\mu$m cladding for multimode fiber. Approximately an order of magnitude improvement is possible in weight per unit distance over coaxial waveguides due to the reduced dimensions and the dielectric composition of optical fiber. This is particularly attractive for airborne and space-based radar systems, where size and weight are critical design criteria.

In addition, fiber-optic waveguides are extremely flexible. A bend radius on the order of several millimeters may easily be achieved, which enables optical fiber to be easily routed to avoid physical interference with various radar subsystems, such as cooling, power and microwave distribution manifolds, and structural supports. The use of fiber-optic manifolds for signal distribution is particularly attractive for phased-array applications which are envisioned to employ thousands of operating elements. Signal distribution becomes increasingly difficult for these systems, particularly at high frequencies, where the interelement spacing is small.

Clearly, however, the most important distinction between optical and microwave waveguides relates to their radar signal transmission characteristics. While the physical characteristics of optical waveguides are extremely attractive from a size, weight, flexibility, and EMI point of view, their use cannot be justified if the transmission characteristics are not adequate for the radar application. Since the *carrier frequency* of a laser source is very high (350 THz for an optical wavelength of 0.855 $\mu$m), optical waveguides have an extremely large bandwidth capacity compared to traditional microwave distribution media. This allows microwave or millimeter-wave modulation of the optical carrier while using very little fractional bandwidth. Therefore, the design of the optical components in a microwave photonics system are based solely on the optical wavelength (microwave circuits, on the other hand, often employ resonant circuit techniques such as the use of quarter-wavelength lines to accommodate the bandwidth).

One of the most significant considerations in radar system design is the ability to meet specified signal and noise levels at various points in the system. The use of optical waveguides for radar signal distribution has an impact on the system design for two reasons. First, the limiting noise processes are fundamentally different for microwave and photonic systems. This is a direct result of the quantum nature of light. Second, the maximum attainable signal levels for a microwave photonic system are significantly less than can be obtained by most traditional microwave systems. While the limiting factor may often be the optical detector, the optical waveguide may also impose limits as well.

Noise may be broadly defined as spontaneous fluctuations of charge carriers (electrons and/or holes) in electronic systems. While there are several types of noise processes relevant to microwave radar system operation, the limiting noise process is thermal or Johnson noise, which is a result of the random motion of charge carriers in a conductor. The thermal noise power is given by the following equation, where k is Boltzmann's constant ($1.38 \times 10^{-23}$ W/Hz · K), $T$ is the kelvin temperature, and $B$ is the bandwidth in hertz:

$$N_{\text{th}} = \text{k}TB \tag{8.1}$$

Thus, the noise floor for microwave systems is temperature-dependent. The thermal noise floor in a microwave system at a standard temperature of 290K is −174 dBm/Hz.

Photonic systems, in contrast, are governed by the quantum nature of the photon, which has an associated energy h$\nu$ where h is Planck's constant (h = $6.626 \times 10^{-34}$ J · s) and $\nu$ is the optical frequency. The spontaneous emission of photons, common to all lightwave sources, will cause a random variation in the average light intensity. This variation results in a random generation of electron-hole pairs in the optical detector, which follows Poisson statistics and is called *shot noise.* Shot noise is the limiting noise process in photonic systems and may be expressed as

$$\overline{i_s^2} = 2qI_pB \tag{8.2}$$

where $q$ is the electronic charge in coulombs, $I_p$ is the photocurrent in amperes, and $B$ is the bandwidth in hertz. The photocurrent $I_p$ is equal to the product of the quantum efficiency (the number of electron-hole pairs generated per incident photon), the photon flux ($P_0/h\nu$), and the electronic charge of the electron:

$$I_p = \frac{\eta P_0 q}{h\nu} \tag{8.3}$$

It should be noted that the shot noise power is directly proportional to the photocurrent. Therefore, the noise in the absence of an optical signal is not defined. Typically, however, a noise floor is specified at a unity SNR. In this case, the shot noise is typically 15 to 20 dB higher than thermal noise for identical noise bandwidths and typical detector characteristics.

Optical waveguides also impact the ability to deliver microwave power. Traditional microwave sources including both tube and solid-state devices may generate average power levels from the milliwatt to megawatt range [8]. A variety of coaxial and microwave waveguide systems are available to deliver this power. While high-power laser sources exist, photonics systems are limited in their ability to deliver microwave power by either the optical detector or the optical waveguide itself. Optical detectors are subject to nonlinearities due to saturation effects, and the measurement of harmonic distortion in microwave photodetectors has been performed by Esman et al. [9]. Optical waveguides, however, are also subject to several nonlinear effects, which can lead to loss, dispersion, and possibly damage. These nonlinear effects include stimulated Raman scattering, stimulated Brillouin scattering, and self-phase modulation, which typically occur at high optical power levels in long fiber lengths [10]. Stimulated Brillouin scattering is the most likely nonlinear effect, since it has the lowest critical power threshold. This effect occurs due to scattering of light from acoustic phonons in the fiber material. The critical power for stimulated Brillouin scattering is given by [10]:

$$P_C = 4.4 \times 10^{-3} A^2 \lambda^2 \alpha \Delta\nu \text{ (watts)} \tag{8.4}$$

where $A$ is the core diameter in microns, $\lambda$ is the optical wavelength in microns, $\alpha$ is the absorption coefficient in decibels per kilometer, and $\Delta\nu$ is the smaller of the laser linewidth or the fiber Brillouin linewidth in gigahertz. For example, for a Brillouin linewidth of 100 MHz [10], wavelength of 1.3 $\mu$m, core size of 9 $\mu$m, and a fiber with 0.5-dB/km loss, the critical power is 84 mW. This power level is easily achieved with the Nd:YAG lasers commonly used for microwave modulation [11]. It should also be noted that, like all nonlinear effects, new frequency components are generated which could seriously degrade the performance of frequency-multiplexed systems.

## 8.3 TYPES OF OPTICAL WAVEGUIDE

There has been tremendous progress in the analysis, design, and fabrication of low-loss optical waveguides over the last 25 years. These waveguides fall into two general categories: optical fiber and integrated optical waveguide. Fiber-optic waveguides are typically used in long-distance digital communication systems; however, recent advances in high-speed analog fiber-optic links allow the fiber to be used to distribute analog microwave signals. In contrast, integrated optical waveguides are typically used for short-distance applications (e.g., less than several centimeters) and are often employed in active integrated circuits such as intensity modulators and optical switches, or passive devices such as optical couplers. The analysis of both types of optical waveguide is similar to that employed to analyze hollow metallic microwave waveguides. Maxwell's equations are solved subject to the appropriate boundary conditions. However, the solution for the optical waveguides is more complicated, since the fields generally exist throughout the entire waveguide structure, including both the guiding and cladding regions. This section presents the fundamental concepts of fiber-optic and integrated optical waveguides.

### 8.3.1 Fiber-Optic Waveguide

The fiber-optic waveguide is the most common type of optical waveguide and is primarily used as a transmission medium. However, fibers also find use as sensors [12], passive power splitters [13], and active devices such as fiber amplifiers [14] and fiber lasers [15]. Optical fibers are composed of a cylindrical structure of dielectric material consisting of an inner core region of refractive index $n_1$ surrounded by a cladding region of slightly lower refractive index $n_2$. The analysis of optical fibers, like all waveguiding structures, involves solving Maxwell's equations subject to appropriate boundary conditions. However, analysis using ray optics is also helpful in the understanding of basic propagation and light-gathering effects. This section will begin with a brief discussion of ray optics, and then present an overview of electromagnetic mode theory based on Maxwell's equations. Finally, the major types of optical fiber, including multimode, single-mode, and polarization-preserving fiber will be briefly discussed.

The propagation of light in an optical fiber may most easily be understood by using a ray optics approach. In general, two types of rays may propagate in the fiber. These are meridional rays, which propagate down the axis of the fiber, and skew rays, which propagate off the axis and follow a helical path. Meridional ray propagation is assumed in the following discussions. Consider an incident light ray on a dielectric interface composed of two different dielectric materials as shown in Figure 8.1. Each dielectric material has an associated index of refraction, $n_1$ and $n_2$, which is a measure of the velocity of light in a vacuum to the velocity of light in the material. The light ray at the interface will be reflected and refracted governed by the Snell law, where $\phi_1$ and $\phi_2$ are the angles of incidence and refraction, respectively:

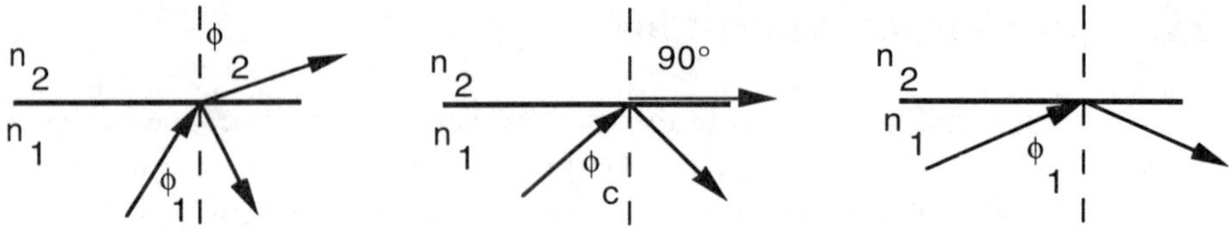


**Figure 8.1** Reflection of an optical signal at a dielectric interface.

$$n_1 \sin \phi_1 = n_2 \sin \phi_2 \tag{8.5}$$

The angle of refraction will equal 90 deg at a specific angle of incidence, called the *critical angle* or *Brewster's angle* $\phi_c$.

$$\sin \phi_c = n_2/n_1 \tag{8.6}$$

At incidence angles greater than the critical angle, the light ray experiences total internal reflection between the optical-fiber core and cladding regions, and low-loss propagation is supported.

The light-gathering capacity of the optical-fiber waveguide may also be characterized by using ray optics theory. The acceptance angle $\theta_c$ defines a cone where rays entering the end of the optical fiber in the core region are propagated by total internal reflection after entering the fiber. The numerical aperture, NA, is related to the acceptance angle and is a measure of the light-gathering capacity of the fiber based on terms of the refractive indexes of the core and cladding regions. The derivation of the numerical aperture is based on Snell's law, assuming a free-space refractive index $n_0$ outside the fiber. The numerical aperture is given by

$$\text{NA} = n_0 \sin \theta_a \cong n_1\sqrt{2\Delta} \tag{8.7}$$

when the difference between the core and cladding indices, $\Delta$ is

$$\Delta = \frac{n_1^2 - n_2^2}{2n_1^2} \cong \frac{n_1 - n_2}{n_1} \tag{8.8}$$

for $\Delta \ll 1$.

While ray optics is useful to explain basic propagation and light-gathering effects, electromagnetic theory using Maxwell's equations must be applied in order to understand the full propagation effects. The solutions to Maxwell's equations are extremely complex, since fields exist in both the core and the cladding regions. The general solution consists of both transverse electric (TE) and transverse magnetic (TM) as well as hybrid (HE and EH) modes. The hybrid modes contain both TE and TM field components, with the order of the letter designation based on the dominant transverse field component. Both TE and TM modes are a result of meridional ray propagation, while the hybrid modes are a result of skew ray propagation. Most fibers suitable for communications purposes fall under the weakly guiding approximation developed by Gloge [16], where the index of refraction between the core and cladding regions is much less than 1 (D $\ll$ 1). In this case, the hybrid modes occur in degenerate pairs which have almost the same propagation constants. The field solutions in the fiber core involve Bessel functions while those in the cladding involve modified Hankel functions. The superposition of these degenerate modes have a common propagation constant, which may be characterized by the designation of a linearly polarized ($LP_{lm}$) mode. The mode subscripts $l$ and $m$ are related to the azimuthal and radial intensity profile of a given LP mode. There are $2l$ intensity maxima in the azimuthal direction and $m$ maxima along the radial direction. The allowed regions for a given set of LP modes are given by a parameter called the *normalized frequency* or $V$ number of the fiber. The $V$ number is given by

$$V = \frac{2\pi}{\lambda} a n_1 \sqrt{2\Delta} \tag{8.9}$$

where $\Delta$ is the relative index difference, $\lambda$ is the free-space optical wavelength, and $a$ is the core radius. A plot of the allowed regions for the LP modes after Gloge [16] is shown in Figure 8.2. It should be noted that single-mode propagation may be obtained for a $V$ number of less than 2.405.

There are several important types of optical fiber, including multimode, single-mode, and polarization-preserving fiber. Multimode fiber is used for low-bandwidth signals and may have either a step or graded index of refraction profile. The graded index profile is designed to reduce the modal dispersion and thereby improve the fiber bandwidth. Irregularities and bending in the fiber can introduce mode conversion or coupling which degrades the bandwidth of the fiber, since different modes have different propagation constants. These profiles are shown in Figure 8.3, where the index profiles are given by [17]

$$n(r) = \begin{cases} n_1 & r < a \quad \text{(core)} \\ n_2 & a < r < b \quad \text{(cladding)} \end{cases} \tag{8.10}$$

for a step index profile and

$$n(r) = \begin{cases} n_1\sqrt{1 - 2\Delta(r/a)^g} & r < a \quad \text{(core)} \\ n_1\sqrt{1 - 2\Delta} & a < r < b \quad \text{(cladding)} \end{cases} \tag{8.11}$$

for a graded index profile.

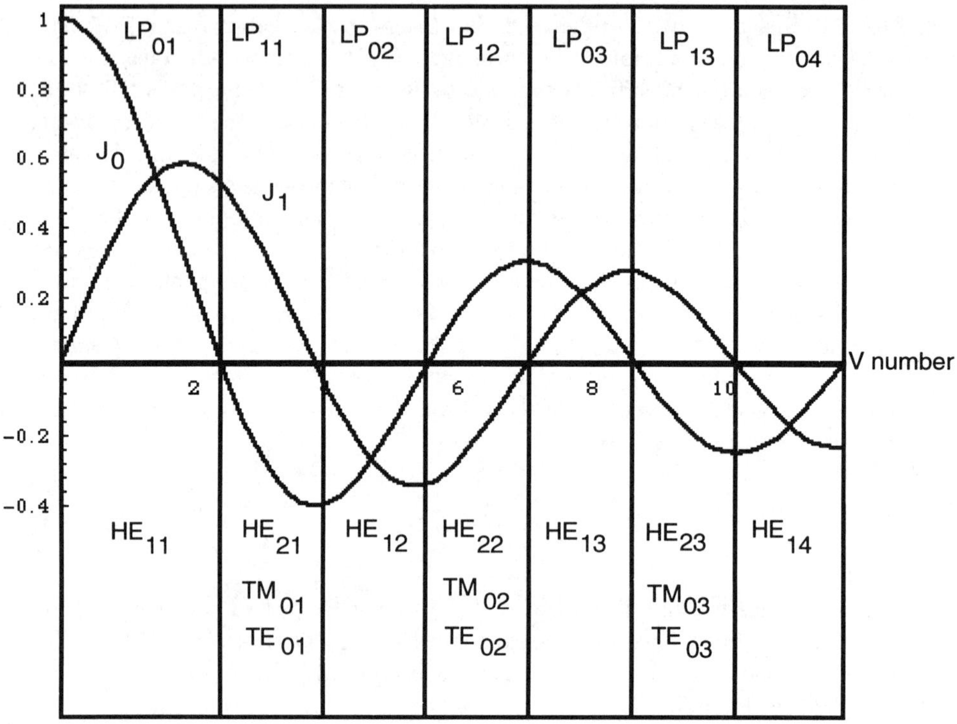


**Figure 8.2** Modes in optical fibers.

A typical multimode fiber may have a core diameter of 50 $\mu$m and a cladding diameter of 125 $\mu$m (50/125 fiber). Furthermore, a practical multimode fiber may support close to 200 guided modes [17]. Single-mode fiber is typically used for long-haul high-bandwidth applications. Single-mode fibers typically have the step index profile given above, where the core radius and index profile are designed to support only the lowest order $LP_{01}$ mode. A typical single-mode fiber may have a core diameter of 9 $\mu$m with a cladding diameter of 125 $\mu$m (9/125 fiber). It should be noted that this lowest order mode is actually a degenerate mode of orthogonal polarizations. Some applications require the use of techniques or components that are polarization-dependent. Heterodyne techniques, for example, require the linear superposition of the signal and local oscillator fields. Furthermore, many common integrated optical devices, such as electro-optic intensity modulators, are also polarization-sensitive. In these cases, polarization-maintaining fibers introducing birefringence to maintain a desired single-mode polarization are required. Birefringence means that the orthogonal polarizations do not see the same index of refraction as in an axially symmetric fiber. This effectively decouples the orthogonal degenerate modes found in the normal single-mode fiber [18]. Birefringence effects can

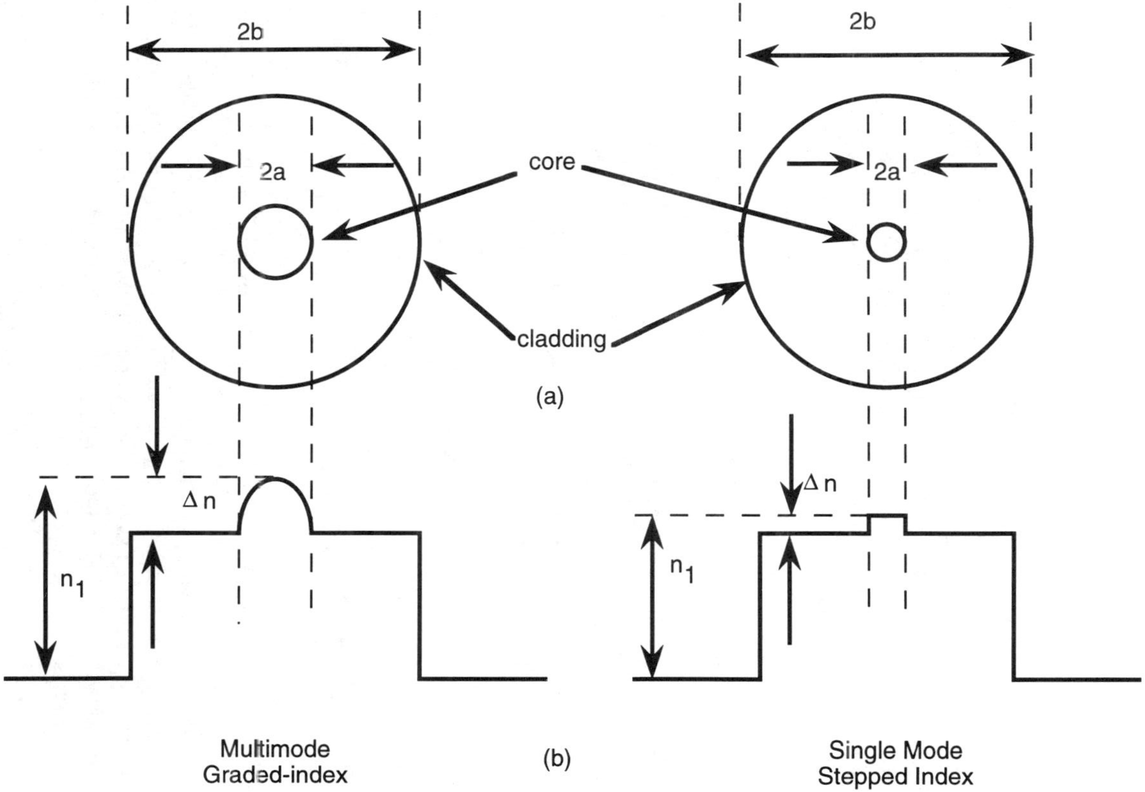


**Figure 8.3** Common optical fiber types: (a) physical construction; (b) index profile.

be induced by altering either the shape or stress of the fiber. Polarization-maintaining fibers have been fabricated using either an elliptical core, cladding, or jacket, or by introducing stress members in the fiber [20,21].

### 8.3.2 Integrated Optical Waveguide

There are several types of important integrated optical waveguides, including planar, rectangular, and semiconductor structures. The planar guide is the simplest type of guide, but it only provides optical confinement in one dimension. Rectangular dielectric waveguides include channel, strip-loaded, and rib configurations which provide light confinement in both lateral dimensions, thereby reducing loss. Integrated optical waveguides may also be fabricated using semiconductors in either silicon or various III-V compounds. The use of semiconductor integrated optical waveguides opens up the possibility of integrating both photonic and microwave circuits monolithically on the same substrate [22].

*Slab Waveguide*

The planar slab waveguide, depicted in Figure 8.4(a), is the simplest integrated optics structure for the transmission of light. The slab waveguide is composed of three regions, including the upper cladding (air), the core (channel), and the lower cladding (substrate). A uniform plane wave solution propagating in the $z$ direction is assumed where the electric field in the $y$ direction may assumed to be constant.

$$E = E(x,y)e^{j(\omega t-\beta z)} \tag{8.12}$$

This results in the following form of the Helmholtz wave equation:

$$\frac{\partial^2 E(x,y)}{\partial x^2} + (k_0^2 n_j^2 - \beta^2)E(x,y) = 0 \qquad j = 0,\ 1,\ 2 \tag{8.13}$$

where $k_0 = (2\pi/\lambda)$, and $j$ is 0 for the air region, 1 for the channel region, and 2 for the substrate region. The boundary conditions require that the electric field and its derivatives be matched at the air-channel and channel-substrate interfaces, which yield TE field solutions that are either sinusoidal or exponential. The desired traveling-wave solution in the core implies a sinusoidal transverse field dependence in the core and exponentially decaying transverse field in both the upper and lower cladding. This also gives the following wave number restrictions:

$$k_0 n_0,\ k_0 n_2 < \beta < k_0 n_1 \tag{8.14}$$

such that the required conditions for $n$ in each of the three regions for propagation are $n_1 > n_0$, $n_2$. It should be noted that while the slab waveguide is the easiest integrated

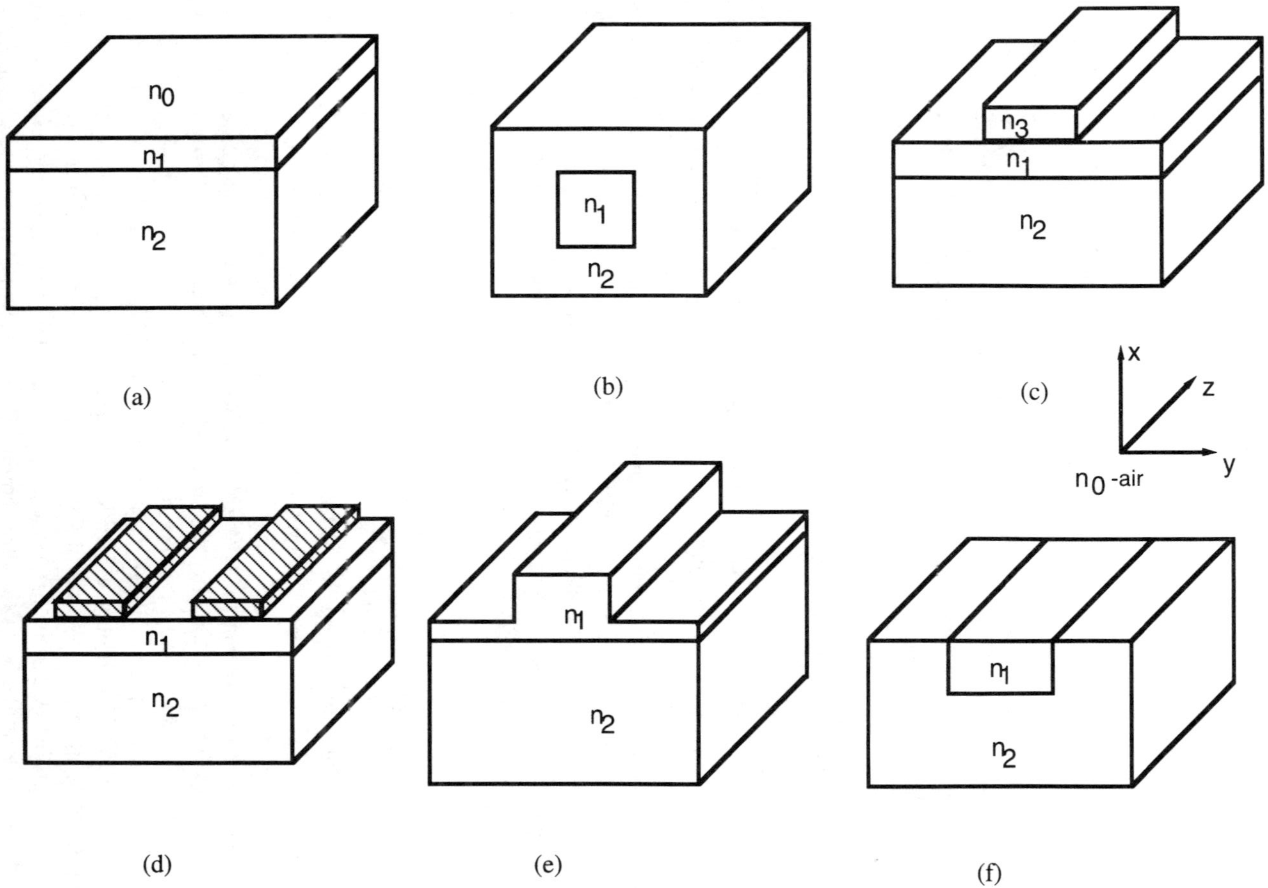


**Figure 8.4** Dielectric waveguide configurations: (a) planar; (b) channel; (c) strip loaded; (d) metal-strip loaded; (e) rib; (f) embedded strip.

optical structure to fabricate and analyze, it offers no confinement of the light in the lateral direction, which leads to high optical loss.

*Rectangular Dielectric Waveguides*

Rectangular dielectric waveguides are used to improve the light confinement and thus reduce the loss of the optical signal. The most commonly used structures are depicted in Figure 8.4(b–f). Unlike the slab waveguide, the analysis of these waveguides becomes quite complex and often requires the use of computer simulation to obtain numerical results. However, there are closed-form solutions available for the channel and strip waveguides based on the analysis by Marcatili [23]. This approximate technique is most easily understood by examining the channel waveguide.

*Channel Waveguide*

The channel waveguide, shown in Figure 8.4, has a guiding layer with an index $n_1$, surrounded on all sides by a dielectric with an index of $n_2$. In general, the electric fields exist in all regions of both the core and cladding. However, under the assumption of well-guided modes proposed by Marcatili, the fields are assumed to be vanishingly small in the cladding regions around the corners of the core, since most of the fields are in the core region. This assumption greatly simplifies the boundary conditions and leads to field solutions that have sinusoidal mode distributions in the core with exponentially decaying transverse fields in the cladding on each of the core sides. When $n_2 = n_1(1 - \Delta)$, then the modes arc essentially transverse electromagnetic (TEM) and can be grouped into $E^x_{pq}$ and $E^y_{pq}$ types [24].

*Strip-Loaded and Rib Waveguide*

Confinement of the optical signal in the lateral direction can also be obtained without the need to surround the optical waveguide with a material of lower index. This is done by fabricating a strip on top of the planar waveguide with an index $n_3$ lower than $n_1$. This structure, shown in Figure 8.4(c), is called *strip-loaded* or *optical strip line* [25]. The effect of the strip is to make the effective index $n_{eff}$ of the channel directly under the strip slightly higher than the effective index $n_{eff1}$ adjacent to it, as shown in Figure 8.5(a). Therefore, confinement is achieved in the $y$ direction as well as the $x$ direction, as described by Hunsperger [26]. The higher effective index increases the path length for a wave traveling underneath the strip, since the wave penetrates deeper at the $n_1 - n_3$ interface than at the $n_3 - n_0$ interface, since $n_3 > n_0$. This is equivalent to effectively increasing the height of the guide underneath the strip compared to the height of the guide on either side of the strip. This leads to the result

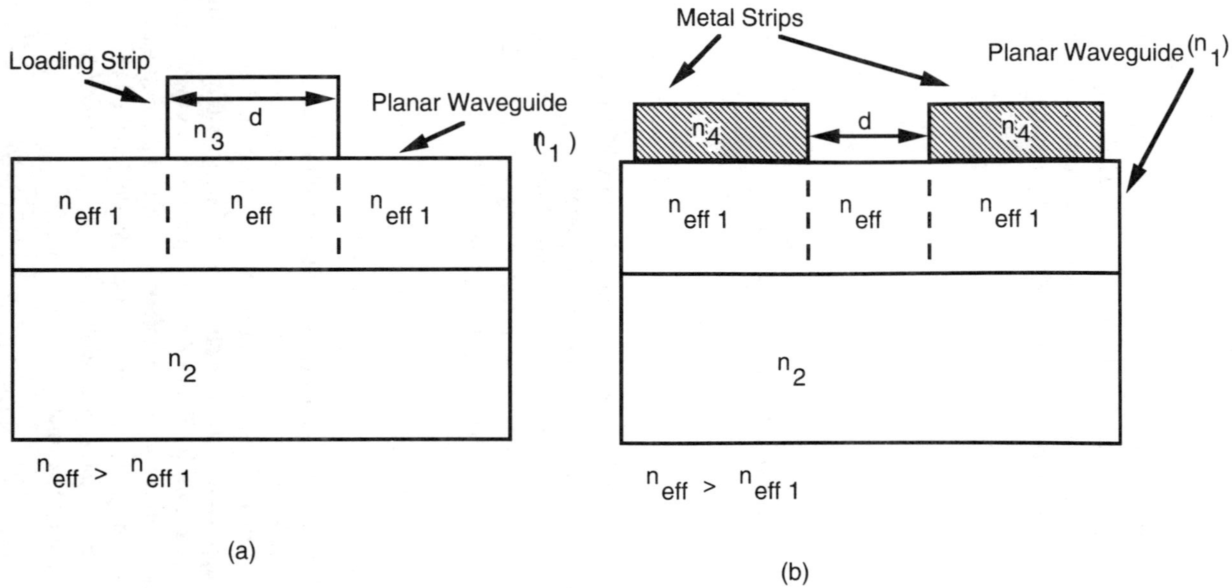


**Figure 8.5** Loading of planar waveguide with (a) dielectric and (b) electric materials.

$$n_{\text{eff1}} = \frac{\beta'}{k} > n_{\text{eff}} = \frac{\beta}{k} \tag{8.15}$$

Closed-form solutions for the strip-loaded guide may be obtained by modeling the strip-loaded guide as an effective channel guide by incorporating these effective indexes [27]. This allows the approximate method of Marceteli to be used. A strip-loaded waveguide can also be made using material with the same index as the channel. This type of waveguide called a *rib* or *ridge guide* is shown in Figure 8.4(e). The propagation is governed by the phase constant

$$\beta = kN \tag{8.16}$$

where $N$ is

$$N^2 = n_{\text{eff}}^2 + b(n_{\text{eff1}}^2 - n_{\text{eff}}^2) \tag{8.17}$$

The parameter $b$ is the normalized guide index given by

$$b = \frac{(N^2 - n_2^2)}{(n_1^2 - n_2^2)} \tag{8.18}$$

Confinement in the $y$ direction can also be accomplished with the use of metallic strips, as shown in Figure 8.4(d). The penetration of the guided waves is deeper at the $n_0 - n_1$ interface than at the $n_1 - n_3$ interface, and confinement takes place. The metal strips, as shown in Figure 8.5(b), act to reduce the effective index underneath. By reducing the effective index under the metal, the condition for confinement in the $y$ direction, $n_{\text{eff}} > n_{\text{eff1}}$, is met.

*Depletion Waveguide*

A waveguide can be formed on a semiconductor substrate by electronically depleting the free carriers. The change in carrier concentration between the substrate and the depletion region gives rise to a change in the index of the guide where the height of the guide can be controlled by the application of an electrical bias. The structure for an electronically controlled waveguide is shown in Figure 8.6. The substrate is sandwiched by two metal contacts where the top contact forms a Schottky barrier and the bottom contact forms an ohmic contact. The Schottky barrier induces a region depleted of carriers. The channel height $d$ can be controlled by the applied voltage $V_b$. The waveguide height can be solved using the abrupt junction approximation and is

$$d = \sqrt{\frac{2e_s}{qN_{\text{sub}}}\left(\phi_b - V_b - \frac{\text{k}T}{q}\right)} \tag{8.19}$$

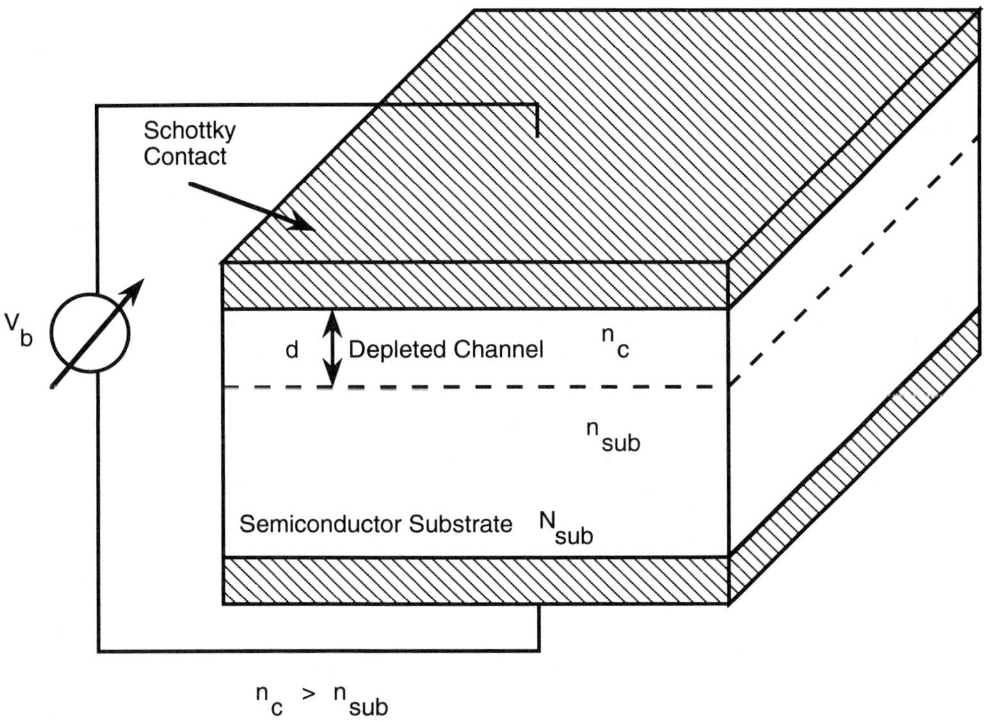


**Figure 8.6** Electronically controlled optical waveguide.

where $\phi_b$ is the barrier potential, $e_s$ is the permittivity and $N_{sub}$ is the carrier concentration of the substrate, k is the Boltzmann constant, and $T$ is the temperature.

## 8.4 OPTICAL COUPLING TECHNIQUES

In general, coupling is required in three places in a photonics system. These are coupling from the light source (laser or light-emitting diode (LED)) to the optical waveguide, coupling in and out of integrated optical devices (external modulator, power divider, spatial light modulator), and coupling from the optical waveguide to an optical detector. Unfortunately, efficient coupling is often difficult to obtain, but is particularly important for microwave photonics systems, since a 1-dB loss of optical power results in a 2-dB loss of microwave power at the photodetector output. The problem is particularly difficult for single-mode waveguides because of the small size of the guiding region. Furthermore, coupling from the integrated optical waveguide to optical fiber is also challenging because of the differing geometries and modal characteristics. Good coupling requires the reduction of reflections, as well as careful consideration of geometric and modal issues. Several standard coupling techniques are butt coupling, tapered couplers, prism couplers, and diffraction gratings. Each of these will be briefly discussed below.

### 8.4.1 Butt Coupling

The most simple optical coupling technique is the end-to-end or butt-coupled method, which involves placing two optical components in close proximity so that the optical signal of the first is incident on the second. Several factors that influence the coupling efficiency are the reflectivity of the waveguide ends, the media between the waveguides, the distance between waveguides, and the position of the waveguide centers. A typical coupling arrangement is shown in Figure 8.7 after Hammer and Neil [28]. In this figure, the two waveguides are separated by a distance $z$ with no lateral offset, and a Gaussian intensity distribution is assumed across each waveguide. The majority of the optical intensity is within the beam waist $2\omega_1$ for waveguide 1 and $2\omega_2$ for waveguide 2, where the beam waist is defined as the point where the optical intensity has been reduced to $e^{-2}$ of its maximum value. The optical coupling coefficient $\kappa$, assuming no reflectivity, is defined as [29]:

$$\kappa = 2\left(\frac{\omega_1}{\omega_2} + \frac{\omega_2}{\omega_1}\right)^{-1}\left\{1 + \left[\frac{\lambda z}{\pi(\omega_1 + \omega_2)}\right]^2\right\}^{-1/2} \tag{8.20}$$

It should be noted that a large difference in beam waist between $\omega_2$ and $\omega_1$ will cause poor optical coupling. In addition, coupling is reduced as the distance between waveguides is increased. Furthermore, the optical signal is reflected due to the discontinuity in the

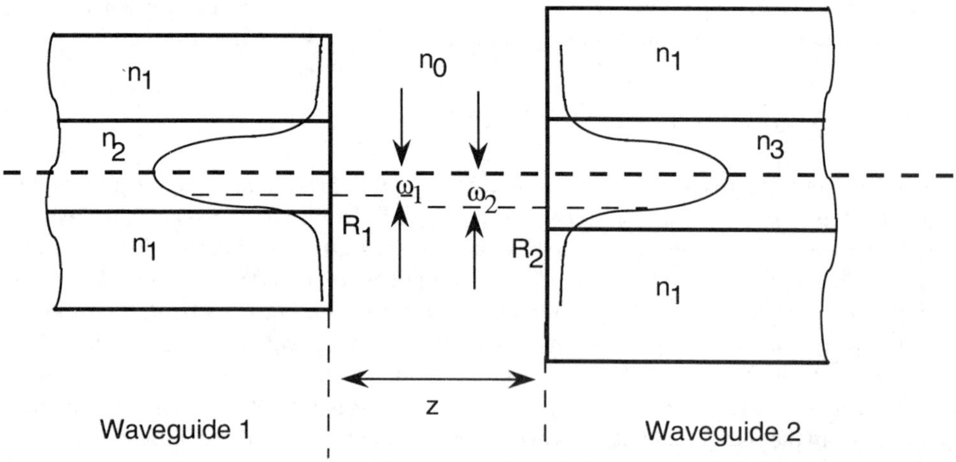


**Figure 8.7** Coupling arrangement between two optical waveguides.

index of refraction between the waveguide and the air. These discontinuities form a resonant cavity called a *Fabry-Perot interferometer.* In this case, light transmitted from one waveguide to the other will be a function of the surface reflectivities and distance between the guides for a given optical wavelength [30]. Impedance matching is therefore required to reduce the effect of mismatches in the index of refraction. This is analogous to reducing reflections in microwave systems caused by impedance discontinuities.

The calculation of the optical reflection coefficient can be derived by analyzing the reflections from a plane wave incident on a multidielectric interface. The reflection coefficient can be expressed as

$$R = \left| \frac{n_1 \Psi - n_2}{n_1 \Psi + n_2} \right|^2 \tag{8.21}$$

where $n_1$ and $n_2$ are the indexes of waveguide 1 and the matching section, respectively. The term $\Psi$ is expressed as

$$\Psi = \frac{n_2 \cos(k_2 l) + j n_3 \sin(k_2 l)}{n_3 \cos(k_2 l) + n_2 \sin(k_2 l)} \tag{8.22}$$

where $n_3$ is the index of waveguide 2, $l$ is the length of the matching section, and $k_2$ is the wave number in the second medium. A quarter-wave section of the correct index will provide a low-loss matched "circuit" for monochromatic light. A typical matching problem for photonic circuits occurs at a fiber-GaAs waveguide ($n = 3.6$) interface structure operating at 1.3 $\mu$m. Using a quarter-wavelength section of silicon nitride ($n = 2.05$) (a common passivation layer used to process microwave circuits as a matching material between the fiber and the GaAs), a transmission coefficient of 0.98 can be obtained.

### 8.4.2 Tapered Couplers

In addition to butt coupling techniques, tapered coupling techniques are also useful, since they are simple to fabricate and offer up to 70% coupling efficiency [25]. In this technique, an integrated optical waveguide is tapered down so that guided modes are eventually converted to radiation modes. This occurs once the incidence angle is less than that of the critical angle for total internal reflection. The major application of this coupler is to couple light from the thin-film waveguide to fiber, since the fiber can be placed close to the waveguide and an on-fiber lens can be used to further improve efficiency. More exotic methods include the use of tapered slab waveguides with nonlinear cladding. Numerical results show a 92% coupling efficiency between waveguides with reduced offset sensitivity [31].

Both butt and tapered coupling require strict physical alignment in order to maximize coupling efficiency and ensure good mechanical stability. The use of V-grooves has been widely accepted as a method of aligning optical fiber for coupling. In this technique, V-shaped channels are etched in a substrate. Optical fibers may then be placed in the V-grooves to obtain proper coupling alignment. An improved technique that increases the freedom of the location of the fiber is the use of U-grooves, where connection losses of 0.6 dB have been reported [32].

Butt and tapered coupling techniques are both end or edge coupling techniques, which require that the edge of the optical waveguide be accessible. However, this may not be possible in many instances. In these cases, it is desirable to couple light from the top of the waveguide. Prism couplers and diffraction gratings are the most commonly used techniques for these applications.

Efficient optical coupling between fibers and thin-film waveguides is difficult because the thin-film waveguides have a different electromagnetic field distribution compared to fibers. Coupling efficiency can be improved to over 90% by using a coupling waveguide layer between the fiber and the thin-film waveguide [33]. A coupling waveguide is made as large as the core of the optical fiber, and the index is determined so that the propagation constant is the same as the thin-film waveguide. A trade-off exists between coupling efficiency and the length of the coupling waveguide.

### 8.4.3 Prism Couplers

Coupling into a waveguide can be accomplished by using a prism with an index of refraction $n_p$ placed on top of a planar waveguide with a slightly lower index. An optical wave illuminates the prism at an angle $\theta_{\text{prism}}$ such that a standing wave is set up in the prism in a direction normal to the waveguide. This standing wave propagates in the desired $z$ direction and also has an evanescent field component which extends into the waveguide region. This evanescent field coupling can excite a traveling wave in the waveguide [25]. While the use of prism couplers is particularly useful for laboratory work, better mechanical positioning and reliability are often required for practical systems.

### 8.4.4 Diffraction Gratings

Diffraction gratings, which may be designed to periodically modulate the phase and/or the amplitude of an optical wave, may also be used to couple light through the top of an optical waveguide. These types of couplers are advantageous over prism couplers, since the grating may be constructed as an integral part of the photonics circuit, thus reducing alignment and mechanical reliability concerns. In this technique, transparent optical phase gratings are made by varying the thickness of the refractive index in a periodic fashion. The index variation may be sinusoidal, triangular, or rectangular and is constructed using either holography or photolithography techniques.

The grating perturbs the waveguide modes underneath causing each mode to have a set of spatial harmonics in the direction of propagation [34]. All of the spatial harmonics of each mode are coupled to form the complete surface wave in the grating region. A beam introduced into the grating at any one of the spatial harmonics is coupled into the fundamental mode as it travels past the grating region. Therefore, the grating can selectively transfer energy from a beam to a particular waveguide mode by choosing the proper angle of incidence.

## 8.5 WAVEGUIDE MATERIALS

Various materials are used to fabricate optical waveguides. Silica is used to fabricate fiber-optic guides, while a wide variety of materials including glass, lithium niobate, and various semiconductors have been used to fabricate integrated optical waveguides. In all cases, it is desirable to minimize optical losses through the waveguide, which arise from three primary mechanisms: absorption, scattering, and radiation.

### 8.5.1 Glass Waveguides

Most fiber-optic waveguides are made from high-purity silica ($SiO_2$). A variety of dopants can be used to produce the change in the index of refraction required between the core and cladding regions, which is on the order of 0.001 to 0.02. Typical dopants include germanium, boron, phosphorus, and titanium. Fibers with loss as low as 0.2 dB/km have been fabricated [35].

Integrated optical waveguides use materials such as silicon nitride ($Si_3N_4$) and silicon dioxide ($SiO_2$) to form core and cladding structures. Silicon dioxide waveguides with buffer, core, and cladding layers of 10, 45, and 35 $\mu$m, respectively, exhibit losses of 0.15 dB/cm at 0.6328-$\mu$m wavelength and 0.1 dB/cm at 1.3-$\mu$m wavelength [36]. An 8-$\mu$m core high-silica single-mode channel waveguide has also been fabricated with a loss of 0.5 dB [37]. Glass waveguides may also be deposited on a semiconductor substrate which offers the potential for hybrid integration. Recently, two types of waveguides on silicon have been developed by researchers at AT&T Bell Laboratories [38]. A nitride core waveguide ($n = 1.97$) is used to match to the index of a laser over optical wavelengths between 1.2 and 1.6 $\mu$m. Efficient matching requires that the waveguide have a similar mode size as the laser. This requirement is met by using a nitride core in the form of a 3-$\mu$m rib surrounded by silicon dioxide. The loss is less than 0.3 dB/cm. A *P-glass* core, made by doping silicon dioxide with phosphorus, is used to match the waveguides to a fiber. A fiber-to-fiber loss of only 0.05 dB/cm was obtained.

### 8.5.2 Lithium Niobate Waveguides

Lithium niobate ($LiNbO_3$) is a highly transparent single-crystal material commonly used in integrated optical waveguides. This material also exhibits a strong electro-optic effect which is used to build modulators and switches [39]. Lithium niobate waveguides can be formed by various processes, such as out-diffusion, in-diffusion, ion exchange, and proton exchange, to modify the refractive index of the material. Out-diffusion of $Li_2O$ from the surface produces an index gradient that increases toward the surface. Alternatively, many metals have been used for in-diffusion processes. Titanium ($Ti{:}Li_2O$), however, is the primary choice, since it increases both the ordinary and extraordinary refractive indexes ($\Delta n_o < 0.02$ and $\Delta n_e < 0.04$) [39]. Waveguides have also been produced by ion exchange with losses of less than 2 dB/cm and $\Delta n_e = 0.05$ [40]. Furthermore, theoretical analysis and experimental investigations of waveguide fabrication using proton exchange processes have recently been employed to produce low-loss waveguides. Proton exchange processes using octanoic acid and stearic acid produce associated losses of 0.4 dB/cm [41] and 1.5 dB/cm [42], respectively, at a 0.633-$\mu$m wavelength. Both processes yield an index change in a planar waveguide of 0.118. A loss of 0.5 dB at a wavelength of 0.83 $\mu$m was reported [43] using pyrophosphoric acid. The low loss was attributed to a protective mask used during processing.

### 8.5.3 Semiconductor Waveguides

Optical waveguides can be fabricated from a variety of semiconductor materials and compounds. The main advantage of using a semiconductor as a waveguiding structure is monolithic integration with lasers and detectors. The integration of these components will improve reliability and reduce cost when compared to the use of hybrid circuits. The constraint placed on the semiconductor optical waveguide is that it must be transparent to the optical signal, yet be made from material similar to that of the laser and detector, which either emits or absorbs the light. Therefore, the band gap relationship is expressed as $E_{gwaveguide} > E_{glaser} > E_{gdetector}$.

The most widely used semiconductor materials for optical waveguides are GaAs, InP, and silicon. The fabrication of semiconductor optical waveguides can range from simple ion implantation to complex heterostructures by epitaxial growth. This section reviews techniques for the growth and fabrication of semiconductor optical waveguides based on current technology.

*Silicon Waveguides*

Waveguide structures based on silicon have received much attention due to the maturity of the silicon technology and the potential for high-level integration. Silicon is transparent above 1.05 $\mu$m, which makes it suitable for use in 1.3- and 1.55-$\mu$m optical systems. Soref and Lorenzo [44] reported a lightly doped silicon waveguide grown epitaxially on a heavily doped silicon substrate. The high loss of this structure, 5 to 12 dB, was due mainly to free-carrier absorption. Improvements in loss results were achieved by placing an insulator on the substrate and then applying the silicon guiding layer. The strong difference in the index between silicon and silicon dioxide, which is 3.5 and 1.5, respectively, makes for good optical confinement waveguide structures. Losses of 8 dB/cm were obtained at 1.15 $\mu$m for a structure of silicon dioxide (0.4 $\mu$m), silicon (0.15 $\mu$m), and an epitaxially grown layer of silicon (2.0 $\mu$m) [45].

### *III-V Semiconductor Waveguides*

GaAs can be epitaxially grown on a highly doped GaAs substrate to form simple asymmetric planar waveguides. Carrier concentrations of $10^{15}$ and $10^{18}$ for the epitaxial layer and the substrate, respectively, are sufficient to produce a low-loss optical guide in the 1.0- to 1.6-$\mu$m region. Ion implantation may also be used to increase the refractive index of a highly doped substrate to form the guiding layer [46]. In either case, the difference in index of refraction can be calculated as

$$\Delta n = n_c - n_s = \frac{(N_s - N_c)e^2}{2n_c\epsilon_0 m^* \omega^2} \tag{8.23}$$

where $N_s$ and $N_c$ are the carrier concentrations of the substrate and the guiding region (channel region), respectively, $\epsilon_0$ is the dielectric constant, and $m^*$ and $e$ are the effective mass and charge of the carriers, respectively. Once the difference in carrier concentration between the guiding and substrate layers exceeds an order of magnitude, the index difference $\Delta n$ does not change significantly. For a 3-$\mu$m-thick channel, sufficient guiding can be obtained with $N_s = 2 \times 10^{18}$ cm$^{-3}$ [46].

The loss mechanisms in semiconductor waveguides are due to absorption, scattering, and leakage [47]. The most common way to reduce losses in semiconductor optical waveguide is to confine the light in the lateral direction by etching a rib into the top layer of the guide or by strip loading. Low-loss single-mode waveguides with 0.58 dB/cm at 1.3 $\mu$m have been fabricated by strip loading [48]. The improvement in the performance was due to the reduction in scattering losses because the etched walls exist in the AlGaAs strip and not in the guiding layer. It was also shown that the loss in a rib guide was a function of the rib (strip) width. Lin et al. [49] reported that the loss decreased from 3 to 2 dB/cm with a reduction of rib widths from 10 to 4 $\mu$m.

### *Semiconductor Heterostructure Waveguides*

Semiconductor heterostructures may also be used to fabricate optical waveguides. $Ga_{1-x}Al_xAs$ can be epitaxially grown on GaAs with excellent crystal quality and good lattice-matching properties. The refractive index of $Ga_{1-x}Al_xAs$ is less than that of GaAs by [50]:

$$\Delta n = -0.4x \tag{8.24}$$

Since the waveguide guiding layer requires an index higher than that of the surrounding layers, an isolation layer ($Ga_{1-x}Al_xAs$) can be epitaxially grown on top of a GaAs substrate to form a lower cladding region. The GaAs guiding layer is then grown on top to form a complete waveguide structure. The mole fraction $x$ required to provide single-mode operation in a layer of GaAs of thickness $h$ bounded by air and $Ga_{1-x}Al_xAs$ is constrained by [51]:

$$0.022(\lambda_0/h)^2 < x < 0.07(\lambda_0/h)^2 \tag{8.25}$$

where $\lambda_0$ is the wavelength of the light. For a 1-$\mu$m guide operating at 1.3 $\mu$m, the concentration of aluminum is $0.0372 < x < 0.118$. The isolation layer must be made reasonably thick to reduce losses associated with guided light that penetrates into the substrate. For a 1-$\mu$m-thick guide with $x = 0.05$, the isolation layer must be 4.5 $\mu$m to reduce losses to 4 dB/cm [52].

For the distribution of high-frequency signals in communications and radar systems, photonic components based in InP for operation at 1.3 and 1.55 $\mu$m are employed due to their higher speed operation than GaAs devices operating at 0.85 $\mu$m. For monolithic integration, GaAs waveguide, which is transparent above 1-$\mu$m wavelength, has been grown epitaxially on InP, but the resulting waveguide losses were high. Buffer layers consisting of GaAs grown over InP have improved loss performance to typically 3 dB/cm [53]. A grafting technique have also been attempted with good results [54]. Rib waveguides were fabricated on a 1.4-$\mu$m-thick GaAlAs layer by grafting on the surface of an InP substrate, producing a 7-dB/cm improvement over epitaxially grown layers.

A common technique for InP-based optical waveguides is the use of GaInAsP/InP MQWs. The GaInAsP layer has a cut-off wavelength of 1.3 $\mu$m. Waveguides were fabricated using a 2.53-$\mu$m-thick by 6-$\mu$m-wide MQW rib, which achieved 0.75-dB/cm loss [55] for operation with lasers at 1.55-$\mu$m wavelength.

## 8.6 CONCLUSIONS

The use of optical waveguides for radar signal distribution is relatively new. There are several significant advantages associated with using optical waveguides for radar signal distribution. The large bandwidth capacity of the optical waveguide allows microwave or millimeter-wave signals to be modulated onto an optical carrier providing a much smaller, lighter, and more flexible signal distribution medium than can be obtained by conventional microwave transmission techniques. Optical waveguides may also be used to distribute lower frequency control and monitoring signals to control such functions as microwave phase shifting, gain control, and switching. A wide variety of radar system approaches have already been proposed using optical waveguides as distribution and control media. These approaches include a variety of antenna remoting, phased-array manifolding, and beam steering control schemes. However, it must be emphasized that the use of photonic radar signal distribution through optical waveguides will require modifications to the system architecture due to fundamental differences between lightwave and microwave propagation characteristics. In particular, photonic systems will be limited in their ability to deliver microwave power due to either detector limitations or possibly nonlinear effects in the waveguide. However, with appropriate modifications in basic radar architecture, the use of optical waveguide for signal distribution will provide a significant contribution to the advancement of the state of the art in modern radar technology.

## REFERENCES


[1] Wallington, A. R., and J. M. Griffin, "Optical Techniques for Signal Distribution in Phased Arrays," *G.E.C. Journal of Research,* No. 2, 1984, pp. 66–75.

[2] Forrest, J., "Optical Techniques for Signal Distribution and Control in Phased Array Radars," *IEEE Int. Radar Conf.,* 1985, pp. 410–415.

[3] Lau, K. Y., "Passive Microwave Fiber-Optic Links With Gain and a Very Low Noise Figure," *IEEE Photonics Tech. Lett.,* Vol. 3, June 1991, pp. 557–559.

[4] Paolella, A., and P. R. Herczfeld, "Optical Gain Control of a GaAs MMIC Distributed Amplifier," *Microwave and Optical Tech. Lett.,* Vol. 1, No. 1, March 1988, pp. 13–16.

[5] Jemison, W. D., et al., "Optical Control of a Digital Phase Shifter," IEEE MTT-S Int. Microwave Symp. Digest, May 1990.

[6] Paolella, A., et al., "Optically Controlled GaAs MMIC Switch Using a MESFET as an Optical Detector," *IEEE MTT-S Int. Microwave Symp. Digest,* May 1990.

[7] Bahsin, K. B., et. al., "Control of a GaAs Monolithic Ka-band Phase Shifter Using a High Speed Optical Interconnect," *IEEE Trans. Microwave Theory and Techniques,* Vol. 38, May 1990, pp. 686–688.

[8] Woods, R. L., "Microwave Power Source Overview," Military Microwaves Conf., London, June 1986, pp. 349–353.

[9] Esman, R. D., and K. J. Williams, "Measurement of Harmonic Distortion in Microwave Photodetectors," *IEEE Photonics Tech. Lett.,* Vol. 2, July 1990, pp. 502–504.

[10] Stolen, R. H., "Nonlinearity in Fiber Transmission," *Proc. IEEE,* Vol. 68, Oct. 1980, pp. 1232–1236.

[11] Kane, T. J., et al., "Diode-Pumped Lasers as Sources for Fiber and Space Communications," IEEE LEOS Annual Meeting, Paper ELT3.4, 1989,

[12] Dandridge, A., "Fiber-Optic Sensors Make Waves in Acoustics, Control, and Navigation," *IEEE Circuits and Devices Magazine,* Vol. 6, Nov. 1990, pp.13–19.

[13] Stolen, R. H., et al., "Polarization-Selective Fiber Directional Coupler," *J. Lightwave Tech.,* Vol. LT-3, Oct. 1985, pp. 1125–1129.

[14] Pedersen, B., et al., "The Design of Erbium-Doped Fiber Amplifiers," *J. Lightwave Tech.,* Vol. 9, No. 9, Sept. 1991, pp. 1105–1112.

[15] Pfeiffer, T., et al., "Output Power Characteristics of Erbium-Doped Fiber Ring Lasers," *IEEE Photonics Tech. Lett.,* Vol. 4, Aug. 1992, pp. 847–849.

[16] Gloge, D., "Weakly Guiding Fibers," *Applied Optics,* Vol. 10, Oct. 1971, pp. 2252–2258.

[17] Li, T., "Structures, Parameters, and Transmission Properties of Optical Fibers," *Proc. IEEE,* Vol. 68, Oct. 1980, pp. 1175–1180.

[18] Kaminow, I. P., "Polarization in Optical Fibers," *J. Quantum Electronics,* Vol. QE-17, No. 1, Jan. 1981, pp. 15–22.

[19] Ramaswamy, V., et al., "Single Polarization Optical Fibers: Exposed Cladding Technique," *Appl. Phys. Lett.,* Vol. 33, No. 9, Nov. 1978, pp. 814–816.

[20] Stolen, R. H., et al., "High-Birefringence Optical Fibers by Preform Deformation," *J. Lightwave Tech.,* Vol. LT-2, No. 5, Oct. 1984, pp. 639–641.

[21] Shibata, N., ''Fabrication of Polarization-Maintaining and Absorbtion-Reducing Fibers,'' *J. Lightwave Tech.,* Vol. LT-1, No. 1, March 1983, pp. 38–43.

[22] Herczfeld, P. R., ''Monolithic Microwave-Photonic Integrated Circuits: A Possible Follow-up to MIMIC,'' *Microwave J.,* Vol. 35, Jan. 1992, pp. 64–78.

[23] Marcatili, E. A. J., ''Dielectric Rectangular Waveguide and Directional Coupler for Integrated Optics,'' *Bell Systems Technical J.,* Sept. 1969, pp. 2072–2088.

[24] Marcatili, E. A. J., ''Dielectric Rectangular Waveguide and Directional Coupler for Integrated Optics,'' *Bell Systems Technical J.,* Sept. 1969, pp. 2074.

[25] Hunsperger, R. G., ''*Integrated Optics: Theory and Tech.,*'' 2nd edition, Berlin: Springer-Verlag, 1985, p. 43.

[26] Hunsperger, R. G., ''*Integrated Optics: Theory and Tech.,*'' 2nd edition, Berlin: Springer-Verlag, 1985, p. 44.

[27] Furuta, H., A. Ihaya, and A. Noda, *Applied Optics,* No. 13, 1974, p. 322.

[28] Hammer, J., and C. Neil, ''Observations and Theory of High-Power Butt Coupling to $LiNbO_3$-Type Waveguides,'' *IEEE Trans. Microwave Theory and Techniques,* Vol. 30, No. 10, Oct. 1982, pp. 1739–1746.

[29] Burnes, W. K., ''Laser Diode End Fire Coupling Into Ti:$LiNbO_3$ Waveguides,'', *Applied Optics,* Vol. 19, June 1980, pp. 1847–1852.

[30] Borne, M., and E. Wolf, *Principles of Optics,* 2nd edition, New York: Macmillan, 1964, pp. 59–65.

[31] Ono, K., and H. Osawa, ''Excitation Characteristics of Fundamental Mode in Tapered Waveguides With Nonlinear Cladding,'' *Electron. Lett.,* Vol. 27, No. 8, 11 April 1991, pp. 664–666.

[32] Grand, G., H. Denis, and S. Valette, ''A New Method for Low Cost and Efficient Optical Connections Between Single-mode fibres and Silica Guides,'' *Electron. Lett.,* Vol. 27, No. 1, 3 Jan. 1991, pp.16–18.

[33] Cai, Y., T. Mizumoto, E. Ikegami, and Y. Naito, ''An Effective Method for Coupling Single-Mode Fiber to Thin-Film Waveguide,'' *J. Lightwave Tech.,* Vol. 9, No. 5, May 1991, pp. 577–583.

[34] Tien, P. K., *Applied Optics,* Vol. 10, 1971, p. 2395.

[35] Miya, T., Y. Terunuma, T. Hosaka, and T. Miyashita, ''Ulimate Low-Loss Single Mode Fibre at 1.55 mm,'' *Electron. Lett.,* Vol. 15, No. 4, 1979, pp. 106–108.

[36] Kawachi, M., M. Yasu, and Edahiro, ''Fabrication of $Sio_2$-$Tio_2$ Glass Planar Optical Waveguides by Flame Hyftolysis Depostion,'' *Electron. Lett.,* Vol. 19, No. 15, 21 July 1983, pp. 583–584.

[37] Takato, N., M. Yasu, and M. Kawachi, ''Low-Loss High-Silica Single Mode Channel Waveguides,'' *Electron. Lett.,* Vol. 22, No. 6, 13 March 1988, pp. 321,322.

[38] Henry, C. H., G. E. Blonder, and R. F. Kazarinov, ''Glass Waveguides on Silicon for Hybrid Optical Pagkaging,'' *J. Lightwave Tech.,* Vol. 7, No. 10, Oct. 1989, pp. 1530–1539.

[39] Armenise, M. N., ''Fabrication Techniques of Lithium Niobate Waveguides,'' *IEE Proc.,* Vol. 135, April 1988, pp. 85–91.

[40] Jackel, J. L., ''Ion Exchange for Optical Waveguides in Lithium Niobate and Lithium Tantalate,'' *Proc. Topical Meeting on Integrated Optics and Guided-Wave Optics,* 1980.

[41] Pun, E.Y.B., Y. O. Tse, and P. S. Chung, ''Proton-Exchanged Optical Waveguide in $LiNbO_3$ Using Octanic Acid,'' *IEEE Photonics Tech. Lett.,* Vol. 3, No. 6, June 1991, pp. 522–523.

[42] Pun, E.Y.B., S. Zhao, K. Loi, and P. S. Chung, ''Proton-Exchanged $LiNbO_3$ Optical Waveguide Using Stearic Acid,'' *IEEE Photonics Tech. Lett.,* Vol. 3, No. 11, Nov. 1991, pp. 1006–1008.

[43] Ohmori, S., Y. Kazuhisa, and T. Taniuchi, ''Low-Loss Proton-Exchanged Waveguide in MgO-Doped $LiNbO_3$ Fabricated With Pyophosphoric Acid Using $Ta_2O_5$ Protective Mask,'' *IEEE Photonics Tech. Lett.,* Vol. 3, No. 12, Dec. 1991, pp. 1099–1001.

[44] Soref, R., and J. Lorenzo, ''All Silicon Active and Passive Guided Wave Components for 1.3 and 1.6 mm,'' *IEEE J. Quantum Electronics,* Vol. QE-22, 1986, pp. 873–879.

[45] Weiss, B., G. Reed, S. Toh, R. Soref, and F. Namavar, ''Optical Waveguide in SIMOX Structures,'' *IEEE Photonics Tech. Lett.,* Vol. 3, No. 1, Jan. 1991, pp. 19–21.

[46] Gareire, E., H. Stoll, A. Yariv, and R. G. Hunsperger, ''Optical Waveguiding in Proton-Implanted GaAs,'' *Appl. Phys. Lett.,* Vol. 21, No. 3, Aug. 1972, pp. 87–88.

[47] Deri, R. J., and E. Kapon, "Low-Loss III-V Semiconductor Optical Waveguides," *IEEE J. Quantum Electron.,* Vol. 27, No. 3, March 1991, pp. 626–640.

[48] Takeuchi, H., and K. Oe, "Low-Loss Single-Mode GaAs/AlGaAs Miniature Optical Waveguides With Straight and Bending Structures,", *J. Lightwave Tech.,* Vol. 7, No. 7, July 1989, pp. 1044–1054.

[49] Lin, S., S. Wang, S. Newton, and Y. Houng, "Low Loss GaAs/GaAlAs Strip-Loaded Waveguides With High Coupling Efficiency to Single Mode Fibers," *Electron. Lett.,* Vol. 21, No. 14, 4 July 1985, pp. 597–598.

[50] Neuberger, *III-V Semiconducting Compounds,* New York: Plenum Press, 1971.

[51] Garmire, E., "Semiconductor Components for Monolithic Application," *Integrated Optics,* Vol. 7, T. Tamir, ed., Berlin: Springer-Verlag, 1975, p. 252.

[52] Garmire, E., "Semiconductor Components for Monolithic Application," *Integrated Optics,* Vol. 7, T. Tamir, ed., Berlin: Springer-Verlag, 1975, p. 254.

[53] Deri, R. J., R. Bhat, J. Harrison, M. Set, A. Yan, T. Florez, M. Koza, and Y. Lo, "Low-Loss GaAs/AlGaAs Optical Waveguides on InP Substrate," *IEEE Photonics Tech. Lett.,* Vol. 2, No. 2, Feb. 1990, pp. 116, 117.

[54] Yi-yan, A., M. Seto, T. Gmitter, D. Hwang, and L. Florez, "Grafted GaAlAs Rib Waveguide on an InP Substrate," *Electron. Lett.,* Vol. 26, No. 19, 13 Sept. 1990, pp. 1567–1569.

[55] Lynch, T., M. MacBean, R. Walling, and A. Thurlow, " Design and Fabrication of Packaged InP/InGaAsP Multiple Quantum Well Waveguide Componets With Very Low Loss," *IEE Proc.,* Oct. 1991.

# Chapter 9
# Optical Integrated Circuits

***Richard Soref***
*Rome Laboratory*

## 9.1 INTRODUCTION

The purpose of this chapter is to examine optoelectronic integrated circuits (OEICs), in which optical and electronic components are combined on the same chip or substrate. Opto-MMICs, which integrate optics with monolithic microwave ICs will be emphasized. This chapter will also touch briefly on photonic ICs (PICs), where optical components are joined on a chip, without transistors. Those optical components include electro-optic, acousto-optic, and even magneto-optic devices, as well as passive optical waveguides. This chapter will give specific examples of what has been done in optoelectronic integration and what might be done in the future for radar applications.

## 9.2 ASPECTS OF INTEGRATION

Integration can be hybrid or monolithic. The term *hybrid* means that the various components come from different materials families, such as ferroelectrics and semiconductors. The most familiar type of hybrid is the one in which discrete optical and electronic components are bonded to a ceramic substrate (Section 9.3.3). Recent hybrids offer a smoother conjunction of disparate materials. The term *monolithic* means that all components are built from the same family, such as III-V semiconductors. Components are connected continuously in monolithic circuits because the components are grown on the substrate.

Cost saving is an important goal of integration, but the main purpose is to obtain better performance (greater functionality) from the circuit than would be derived from a collection of discrete components. For example, an OEIC has lower parasitic capacitance, inductance, and resistance than a wire-bonded mixture of elements. OEICs offer compactness, weight savings, and improved reliability. Compared to bulk optic assemblies, PICs offer stable optical coupling of components and reduced coupling loss due to decreased scattering and reflections. Integration affords easier manufacturability, which is the basis of cost reduction.

### 9.2.1 Guided and Unguided Lightwaves

Optical signals or *beams* can be confined within dielectric slabs or channel-shaped optical waveguides located on the IC. Then the various light beams are routed in a spatial pattern parallel to the chip plane. Most of our discussion will deal with guided light. Alternatively, the lightwaves could propagate through the air, without dielectric confinement, and those beams would enter or exit the OEIC typically at normal incidence to the chip. In that case the light is unguided, although there might be external microlens arrays or holograms to focus the light beams. Examples of OEICs using unguided lightwaves include chips containing 2-D arrays of vertical cavity surface-emitting laser diodes (VCSELs), photodiodes, electro-optic picture elements in an SLM, magneto-optical pixels, Fabry-Perot pixels, optically controlled SLMs, infrared detector elements in an imager, and nonlinear optical devices such as bistable elements, SEEDs, or optically controlled heterojunction phototransistors (HPTs) for neural networks or optical correlators.

### 9.2.2 Applications of Integration

In addition to radar, communications and computers are OEIC application areas. For all three areas, the OEIC designs and engineering challenges are similar. In fiber-optic communication networks there is a need for low-cost optical transmitters and optical receivers coupled to glass fibers. The optical transceiver challenge is one where OEICs can produce a cost-effective solution.

A data bottleneck has been encountered in high-speed digital computers due to the speed limitations of electrical transmission lines. By using OEICs to facilitate board-to-board or even chip-to-chip optical communications within the computer, one can create fast, cost-effective optical interconnects that alleviate the bottleneck. Parallel-to-serial data conversion in the optical domain can alleviate the pin-out problem of VLSI circuits. Another computer payoff is in *optical clock distribution,* which provides synchronization of VLSI circuits with little time skew. Here, a pulsed optical clocking signal is distributed by optical waveguides to various transistor arrays. The OEICs constructed for clocking include optical guides, demodulators of light, and digital circuits.

The OEIC applications areas of optical computing, optical storage, optical signal processing, optical neural networks, and infrared imaging are tied to free-space propagation of light beams and to 2-D arrays of surface-normal photonic devices in which an addressing transistor is integrated in the corner of each pixel.

### 9.2.3. Electro-Optic Materials

Before discussing OEICs, we shall outline the electro-optic materials that play a key role in modern OEICs. For the most part, these are solid-state materials. Our discussion of microwave electronics also emphasizes solid-state devices rather than vacuum tubes. The issues for electro-optic materials are modulation speed, modulator drive power, wavelength of operation, and ease of integration with drive electronics.

Table 9.1 lists the electro-optic material families and key examples. Our view of future performance begins with GaAs and InP. These III-V semiconductors will probably maintain their strong leadership; however, silicon-based optics (such as SiGe/Si) is currently gaining momentum, and this new approach to photonic components in group IV alloys [1] should push towards the forefront by the year 2000. The organic electro-optic polymers, with modulation speeds nearing 100 GHz, have an important and competitive role to play in optical microwave ICs. Impetus for the development of II-VI optics has been given by the 1992 results on blue and blue-green laser diodes which offer submicron diffraction-limited focal spots and easy detectability. Further II-VI payoffs should result by 2000. The $LiNbO_3$ technology is mature and very useful, but the ultimate capabilities of ferroelectric materials fall short of semiconductor capabilities. In the U.S. and Japan, considerable progress has been made on crystalline organic semiconductors for optical applications, and added development can be expected in such media. Liquid crystals, especially ferroelectric smectic liquids, have a large field-induced index change and electro-optic response times of ~10 ms. These liquids have a range of "medium-speed" applications. Miniature electrical heaters on silica-on-silicon waveguides can induce the thermo-optic effect, a large change in refraction useful for millisecond optical switching. That feature reserves a place for glass waveguide devices in electro-optics.

Acousto-optic PIC materials include glass, ferroelectrics, and semiconductors. Optically controlled materials for *opto-optic* devices come from the families listed in Table 9.1 and from other categories, such as photorefractive crystals.

**Table 9.1**
Electro-Optic Materials in Order of Estimated Importance in the Year 2000

| *Material* | *Examples* |
|---|---|
| III-V semiconductors | Alloys of GaAs, InP |
| IV-IV semiconductors | SiGe, Si, SiC, SiGeC |
| Organic electro-optic polymers | Poled polyimide, PMMA |
| II-VI semiconductors | (Cd,Zn)Se, Zn(S,Se) |
| Ferroelectric crystals | $LiNbO_3$, $LiTaO_3$, $KNbO_3$, PLZT |
| Organic semiconductors | Crystalline PTCDA, NPP, MNA |
| Liquid crystals | Nematic, ferroelectric smectic |
| Glass (thermo-optic) | P-doped silica, $Si_3N_4/SiO_2$ |

### 9.2.4 Semiconductor Techniques

The hopes for monolithic optoelectronic integration reside with semiconductors. Of the materials listed, semiconductors have the most diverse and practical properties, spanning the gamut of optics, electro-optics, nonlinear optics, and electronics. Modern heterostructures promote ongoing improvement in the performance of semiconductor devices. Heteroepitaxy refers to the growth of one semiconductor alloy upon another, such as AlGaAs on GaAs, InGaAs on InAlAs, InGaAsP on InP, and SiGe on silicon grown by molecular beam epitaxy (MBE), metal-organic chemical vapor deposition, or metal-organic vapor-phase epitaxy. Heterostructures are of proven value in optical waveguides, electro-optical modulators, infrared sensors, and state-of-the-art transistors (microwave and millimeter-wave heterojunction bipolar transistors (HBTs) and high-electron mobility transistors (HEMTs)). Heterolayers may be lattice-matched (unstrained) or slightly mismatched to the substrate. If mismatched, the epilayer lattice expands or contracts in the plane of the heterointerface to accommodate the misfit without dislocations, provided that the layer thickness is less than the critical thickness.

Strained crystalline layers are known to be entirely stable for device application. Coherent strain (pseudomorphic growth) is a means for adjusting a semiconducting alloy's band gap, effective masses, and band offsets with respect to the semiconducting substrate. A quantum well (QW) is formed when a semiconductor film of 5- to 10-nm thickness is sandwiched between wider band gap 5- to 10-nm semiconductor barrier layers. The growth of multiple QWs (MQWs) is a powerful way to modify the semiconductor's properties and to create new electro-optic features. *Band gap engineering* refers to tailoring the well-and-barrier compositions and the well-and-barrier thicknesses to produce desired properties in an *artificial crystal* of strained microlayers. Quantum confinement in two or three dimensions produces quantum wires or boxes. Quantum-wire and quantum-box devices are photonic components with a large electro-optic response at specific wavelengths. Remote doping (modulation doping) in a heterolayer device can enhance the mobility of carriers in the undoped transport layer. In summary, the optical and electrical properties of semiconductor devices can be upgraded significantly with heterostructures, strained layers, remote doping, quantum confinement, and band gap engineering.

Performance payoffs have created a trend towards the use of QWs in electro-optics and electronics, although this trend has not taken hold as yet in the realm of MMICs, where a "keep it simple" philosophy has led to the choice of bulk GaAs, accompanied by indiffusion of $n^+$ or $p^+$ regions for MESFETs. The complex processing needed in QWs does indeed entail cost-benefit tradeoffs. Nonetheless, we expect that QW devices will be pervasive by the year 2000.

## 9.3 OPTOELECTRONIC INTEGRATION

A method of manufacturing PICs in the InGaAsP/InP system has been announced by AT&T. These PICs operate at the 1.3- to 1.6-$\mu$m fiber-optic wavelengths. That work was followed by British Telecom's development of a generic indium phosphide PIC-OEIC technology for 1- to 20-Gb/s digital circuits and 1- to 66-GHz analog circuits [2]. Even so, it has been difficult to identify applications that require thousands of PICs, and the lack of market demand has kept PIC manufacturing costs high. Thus, the economics of integrated optics are not currently robust, although the scientific and technical condition of PICs is very good. In Sections 9.3.1 through 9.3.7, representative PICs and OEICs are described. Methods of making OEICs will be surveyed and the integration of microwave devices with photonic components will be illustrated.

### 9.3.1 Photonic Integrated Circuits

The technical status of PICs is reviewed in the excellent textbooks of Tamir [3], Hunsperger [4], and Nishihara et al. [5]. The field of integrated optics is ripe with possibilities of combining many kinds of optical components on a chip to achieve a variety of optical processing functions. Most of those functions have not been exploited. Sophisticated PICs could be built from the following: laser diodes, light-emitting diodes, traveling-wave optical amplifiers, channel waveguides, guides with turning mirrors, directional coupler switches, wavelength-selective couplers, grating couplers, electro-optical modulators, Y-branch splitters and combiners, waveguide lenses, acousto-optic frequency shifters, acousto-optic filters, magneto-optic isolators, TE-to-TM mode converters, star couplers, channel-dropping filters, and so forth. An early example of a PIC from the 1970s is the RF spectrometer-on-a-chip, which is being "resurrected" in improved form.

One of the largest scale PICs built to date is illustrated in Figures 9.1 and 9.2. It is a waveguided 4 × 4 electro-optic matrix switch in AlGaAs/GaAs QWs [6]. Switched sections of semiconductor traveling-wave optical amplifiers are used to impart gain to an optical signal during electrical forward bias. At zero bias, the laser-like structure has the property of attenuating the optical signal at the 0.86-$\mu$m gain wavelength. Hence, the active waveguide can serve as an optical amplitude "shutter" with transparency (or gain) in the ON state and opacity in the OFF state. This amplitude control is used for optical path selection in the matrix. The advantages of the amplifying switch are lossless signal transmission due to gain compensation along each path, low optical crosstalk (30-dB extinction), and ~10-ns switching speeds. The drawbacks are amplified spontaneous emission noise, moderate power drain (drive currents >10 mA), wavelength sensitivity, polarization sensitivity, and temperature rise in the active sections. Another problem with the PICs illustrated in Figures 9.1 and 9.2 is that the connecting waveguides between shutter sections must be actively pumped to overcome their loss at zero-bias. This pumping requires extra electrical power.

In future PICs, we can see ways to create passive, unpumped optical waveguide connectors between active elements. A method suggested in [7] is to use MQW material for all waveguides and to use impurity-induced or vacancy-induced disorder of QWs along the connecting paths. QW "mixing" shifts the waveguide absorption towards the

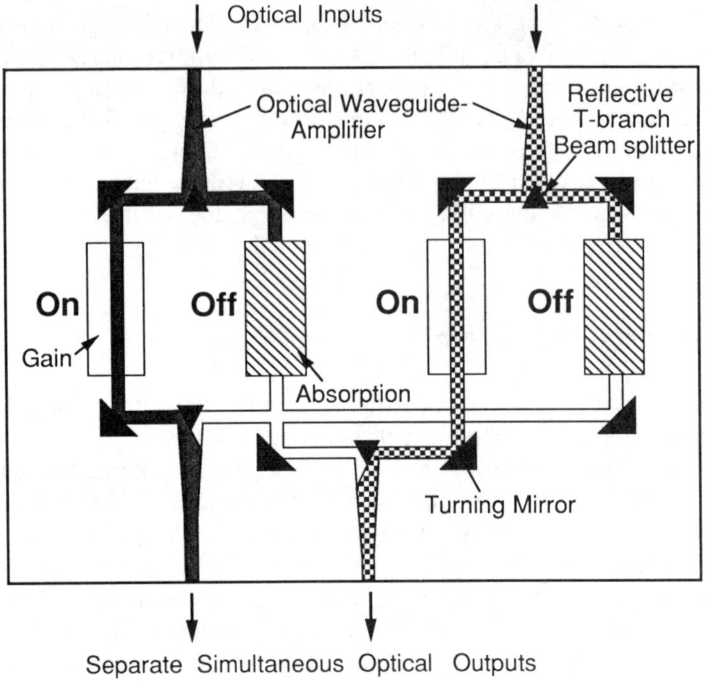


**Figure 9.1** Optical crossbar 2 × 2 switch component used in 4 × 4 matrix (courtesy of C. L. Balestra at McDonnell Douglas Electronic Systems Co.).

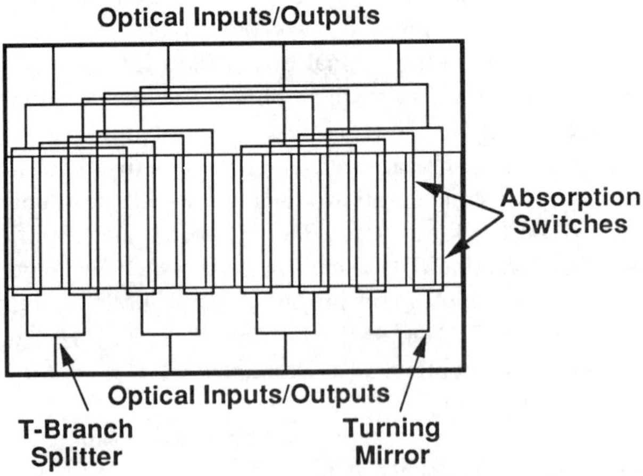


**Figure 9.2(a)** Monolithic PIC chip containing 4 × 4 matrix switch: layout of active guides.

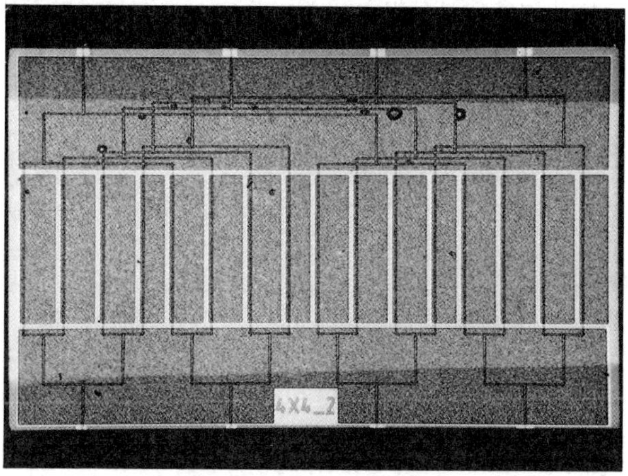


**Figure 9.2(b)** Monolithic PIC chip containing 4 × 4 matrix switch: photograph of actual device (courtesy of C. L. Balestra at McDonnell Douglas Electronic Systems Co.).

blue. Researchers at AT&T [8] made passive waveguide paths between integral laser diodes and amplifying shutters by changing the composition of the passive guides with respect to the active guide composition. The passive InGaAsP material had an optical absorption edge at 1.3 $\mu$m, while the InGaAs/InGaAsP laser diodes and amplifiers had a band gap at the 1.55-$\mu$m operating wavelength. The active cladding was etched away from the passive regions and a passive InP cladding was regrown in those areas.

### 9.3.2 Optoelectronic Integrated Circuits

Worldwide enthusiasm for OEICs is evident in the rapidly growing list of OEIC journal articles and conference presentations. The reviews of OEICs by Matsueda et al. [9], Wada [10] and Leheny [11] give details of their current status. An examination of those articles shows that the described OEICs generally fall into two categories: (1) optical transmitters, including laser diodes and drive electronics, and (2) optical receivers, including photodiodes and electronic preamplifier circuits (e.g., pin-FET). OEICs apply also to all-optical logic and switching. The DFB laser diode integrated with a fast electroabsorption modulator is a recent OEIC example. This monolithic circuit, illustrated conceptually in Figure 9.3, is based on the experiments of Aoki et al. [12]. The OEIC uses an alloy composition change in the MQW InAlAs/InGaAs alloy waveguide as the guide extends from the lasing region to the modulation region.

### 9.3.3 Hybrid Optoelectronic Integration

A significant example of the hybrid art is the 3- to 6-GHz optical transceiver created by G.E. Electronics Lab and AT&T [13]. Their hybrid OEIC was based on a gold-plated

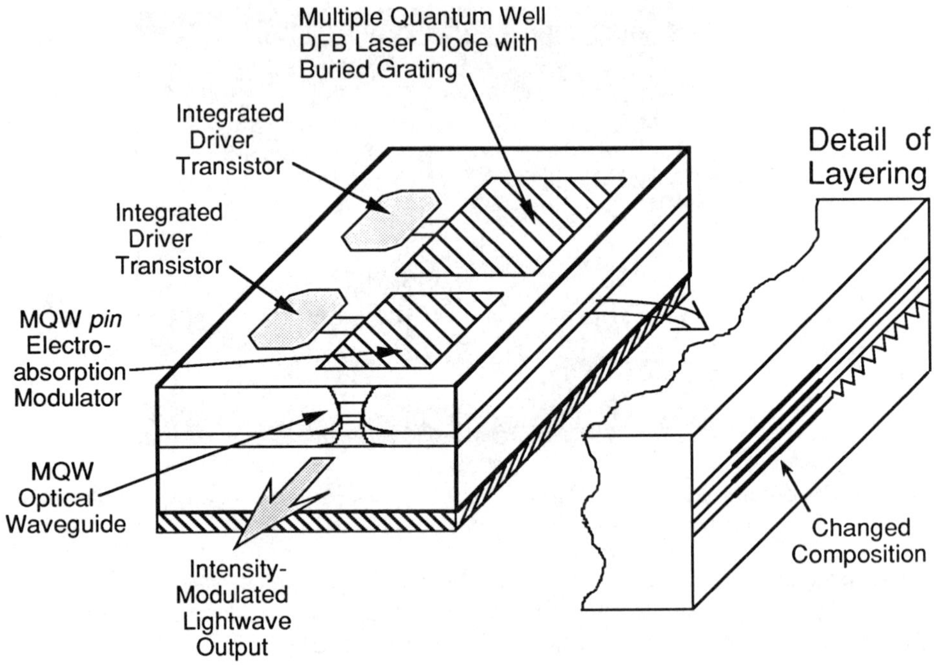


**Figure 9.3** Monolithic OEIC with MQW InGaAs/InGaAsP DFB laser diode and electroabsorption modulator (from Fig. 2, p. 581, of [12]).

alumina substrate containing microstrip lines and impedance-matching elements. A low-noise MMIC amplifier and an MMIC postamplifier were soldered to the substrate, together with an InGaAs fiber-coupled optical receiver and a fiber-coupled InGaAsP laser diode with its subassembly of glass-ball lens, YIG film optical isolator, GRIN lens, and silicon V-groove platform.

Other examples of hybrid integration include optical waveguides made from an organic electro-optic polymer or phospho-silicate glass deposited on oxidized substrates of silicon, InP, or GaAs. Those substrates often contain transistor circuits. Another hybrid is the formation of an InGaAsP/InP or AlGaAs/GaAs laser diode on a silicon electronic substrate. III-V phototransistors or photodiodes can also be created on silicon-circuit substrates, mainly by the fabrication technique of epitaxial liftoff. Epitaxial liftoff also permits semiconductor devices to be formed on $LiNbO_3$ substrates.

### 9.3.4. Grafting and Related Techniques

The leading hybrid OEIC integration techniques are heteroepitaxy, grafting, and flip-chip bonding. Heteroepitaxy, such as the growth of germanium on GaAs, InGaAsP/InP/GaP on silicon, and GaAs on silicon, is useful when the lattice mismatch of the materials is small. An example of flip-chip bonding is the attachment of laser diodes to silica optical waveguides. The flip-chip technique can use solder bumps or conductive polymer bumps.

Grafting of semiconductor thin-film devices is a powerful integration technique for joining high-quality crystalline films to substrates whose atomic structure is seemingly incompatible with the films, such as GaAs on $LiNbO_3$ or GaAs on glass. Thin semiconductor devices are selectively removed from their original host by epitaxial liftoff in which a buffer layer beneath the film is eroded, or the host substrate is removed chemically [14]. Then the thin films are reattached to the new host, held by van der Waals forces of attraction. Disparate devices can be joined intimately, then processed further. Figure 9.4, based on the discussion of Yi-Yan and Chan [14], illustrates the processing sequence for forming grafted InGaAs pin photodetectors on SiGe indiffused optical waveguides on a silicon substrate. The SiGe process is shown here, rather than the glass guides described in [14].

### 9.3.5. Monolithic Optoelectronic Integration

Monolithic integration has been very successful in the InP and GaAs optoelectronic technologies, and further refinements are coming soon. Current examples of monolithic integration include wavelength-division-multiplex receivers (multiple photodetectors) and wavelength-division-multiplex transmitters (multiple tunable QW lasers). Although silicon-based optoelectronics is in its infancy, monolithic integration of silicon and SiGe electronics with SiGe, SiC, and SiGeC photonics can be expected within 10 years [15]. Group IV monolithics are likely to be cost-competitive with Group III-V devices.

### 9.3.6 Fiber-to-Circuit Coupling

Efficient transfer of light from an optical fiber to a microwave optoelectronic circuit can be difficult. For the optomicrowave coupling task, single-mode fibers with a 6- to 8-$\mu$m core diameter are preferred. A fiber can be beveled at 45 deg to its axis, thereby reflecting fiber light out of the fiber side wall into the circuit, although the light is not well focused. End fire or butt coupling of fibers to an optical circuit is also a good optical transfer method. Here, the fibers are prealigned in precision-etched silicon V-groove chips adjacent to the circuit. Active alignment of cleaved, butt-coupled fibers is another method. In this case, the outside of the fiber is metallized and the fiber is moved with micropositioners to obtain maximum optical input to the circuit while the circuit is monitored. Then the fiber is permanently soldered into place. A third technique is end-fire coupling with lensed fibers. The most successful focusing approach has been to form an aspheric hyperboloid lens [16] on the end of a single-mode fiber. A small air gap exists between the lens and the waveguide device, as shown in Figure 9.5. The coupling efficiency is typically greater than 90%. Ultimately, a fiber-to-chip connector is needed.

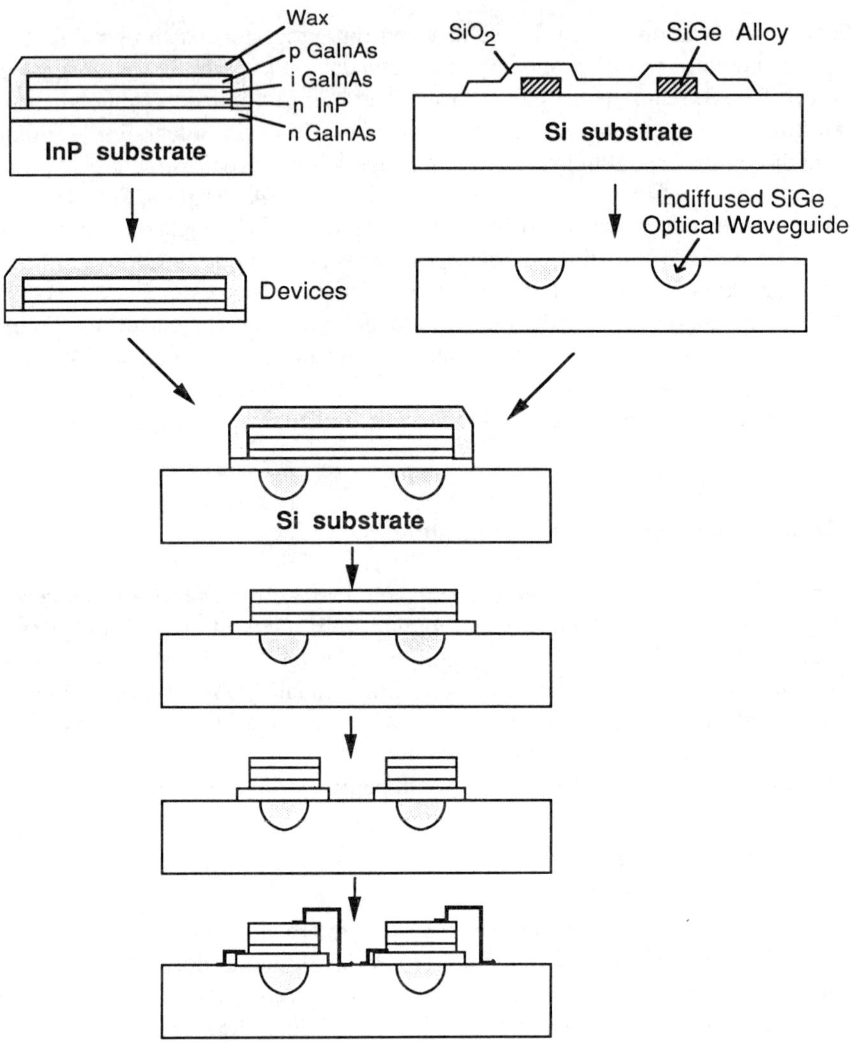


**Figure 9.4** Grafting of III-V photodiodes to a SiGe/Si waveguide substrate (from Fig. 2, p. 28, of [14]).

### 9.3.7 Optical-Microwave Circuits

An optical signal can create electron-hole pairs directly inside a microwave metal semiconductor FET (MESFET) or impact ionization avalanche transit time (IMPATT) diode in order to synchronize the device or to modify its response. Alternatively, lightwaves can impinge on a separate photodetector connected electrically to the microwave device in the manner illustrated in Figure 9.6. This figure shows conceptually the integrated optical

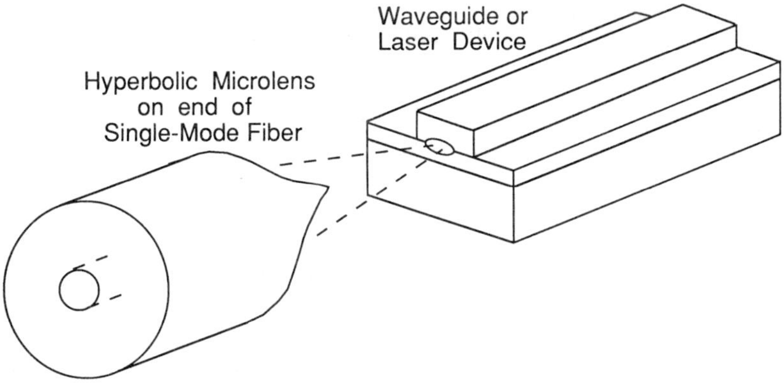


**Figure 9.5** Efficient fiber-to-circuit coupling with a microlensed fiber (from Fig. 3, p. 411, of [16]).

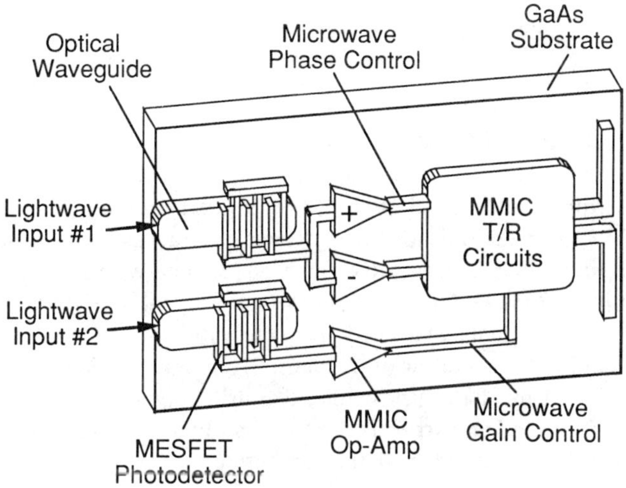


**Figure 9.6** Integrated opto-MMIC with optical waveguided control (from Fig. 5.2.1 of [17]).

control of a GaAs MMIC T/R module. Figure 9.6 is based on experimental work in Paolella's thesis [17]. The GaAs MESFET photodetectors absorb modulated light transmitted to them by integrated optical waveguides, such as silicon oxynitride channels coupled by evanescent lightwaves to the FETs. Alternatively, the light could come from optical fibers coupled in the manner detailed in the previous paragraph. The interdigital-gate electrode structure is similar in form to the low-capacitance GaAs metal-semiconductor-metal (MSM) photodiode that can be used in this application. In essence, the optical

MMIC shown in Figure 9.6 is a modular, integrated version of an optical command circuit given in Section 9.4.1.

## 9.4 PHASED-ARRAY ANTENNA OEICs

Over the past decade, Rome Laboratory has sponsored pioneering projects on OCPAR. In addition, the Advanced Research Projects Agency (ARPA) has funded the National Center for Integrated Photonic Technology, a five-university consortium that has chosen OCPAR as one of its demonstration projects. The Rome-ARPA focus has enriched the OCPAR art through the efforts of contractors like Hughes Aircraft ($1 \times N$ switched arrays of high-speed photodetectors coupled to varying-length optical-fiber delay paths) and Texas Instruments (electromechanical guided-wave optical switches on silicon for selecting optical delay paths). Both systems benefit from OEICs which replace large numbers of discrete components with small inexpensive packages. OEICs also play a key role in the new OCPAR schemes put forward by H. A. Haus, S. R. Forrest, H. R. Fetterman, and A. J. Seeds. (Further details on the application of OEICs to radar can be found in Volumes 886, 1217, 1476, and 1703 of the *SPIE Proceedings*.)

### 9.4.1 Optical Command Circuits

The 28-GHz experiment of Richard et al. [18] is a good example of optical distribution of control signals in an electronically steered phased-array antenna. A schematic of the setup is shown in Figure 9.7. A lightwave signal modulated with serialized data was sent over a fiber to a hybrid GaAs OEIC, where the light signal was demultiplexed on-chip into 16 parallel output signals that controlled the monolithic GaAs microwave phase shifters of a $K_a$-band patch-panel array antenna. By phase shifting, the antenna pattern was reconfigured in 2.3 ms and, in principle, the OEIC could have been integrated with the phase shifters. If an optical carrier is used to distribute RF, IF, and LO signals to the antenna, OEICs can help in the distribution process as well.

### 9.4.2 Optical Phase-Shift Circuits

Optical phase-shift steering refers to beam forming in the optical domain, resulting in a microwave phase shift at each antenna element. Typically, the shift is independent of microwave frequency. This works well when the instantaneous microwave bandwidth is 2% or less. The Fourier-optic beam forming scheme of Koepf [19] is a coherent optical phase-shifting technique that uses free-space optics rather than guided-wave means to form the optical interference pattern. Acousto-optic beam steering [20] also uses a free-space arrangement of components. Both of these systems do not profit much from OEICs, but other approaches are more OEIC-intensive. For example, PICs and OEICs reduce

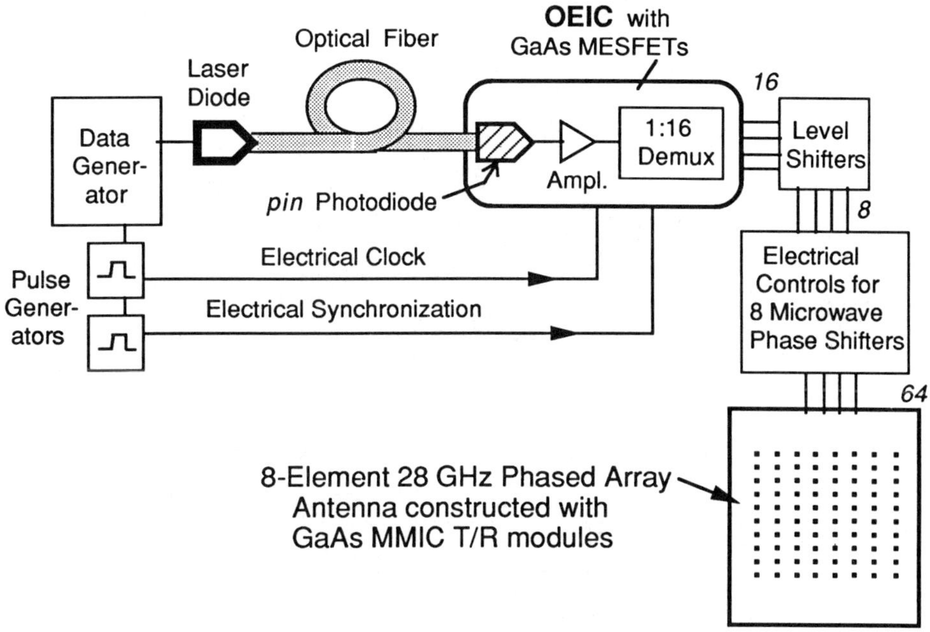


**Figure 9.7** Experimental arrangement for phased-array switching (from Fig. 2 of [18]).

size, weight, and cost in the optical heterodyne systems of Soref [21] and Matsumoto et al. [22]. We illustrate in Figure 9.8 the 0.8-GHz PIC phase shifter of [22], where the optical frequency shifter and optical phase shifter are integrated in a titanium-indiffused $LiNbO_3$ electro-optic waveguide chip. For an $N$-element antenna, researchers at Sandia Laboratory [23] are developing a more sophisticated two-stage PIC version of the Figure 9.8 device. They place two electro-optic waveguide circuits on one chip. The first contains an optical frequency translator; the second includes an $N$-fold optical phase controller with $N$ optical signal combiners (Fig. 9.3 of [23]). The approach of [23] requires only one optical frequency shifter. This is probably better than integrating N frequency shifters per chip, as in the PIC of Figure 9.5 in [21]. In lieu of single-sideband shifting of one light beam (Fig. 9.8), an internally locked two-frequency optical signal source has been proposed by G. Simonis of the U.S. Army Harry Diamond Laboratory and by W. J. Stewart and M. J. Wale of GEC-Marconi (U.K.).

### 9.4.3 Optical Time-Delay Circuits

The purpose of true time-delay steering is to allow broadband microwave operation and/or microwave frequency hopping while avoiding beam squinting. *Time shift* means that the microwave phase shift at each antenna element is proportional to the microwave

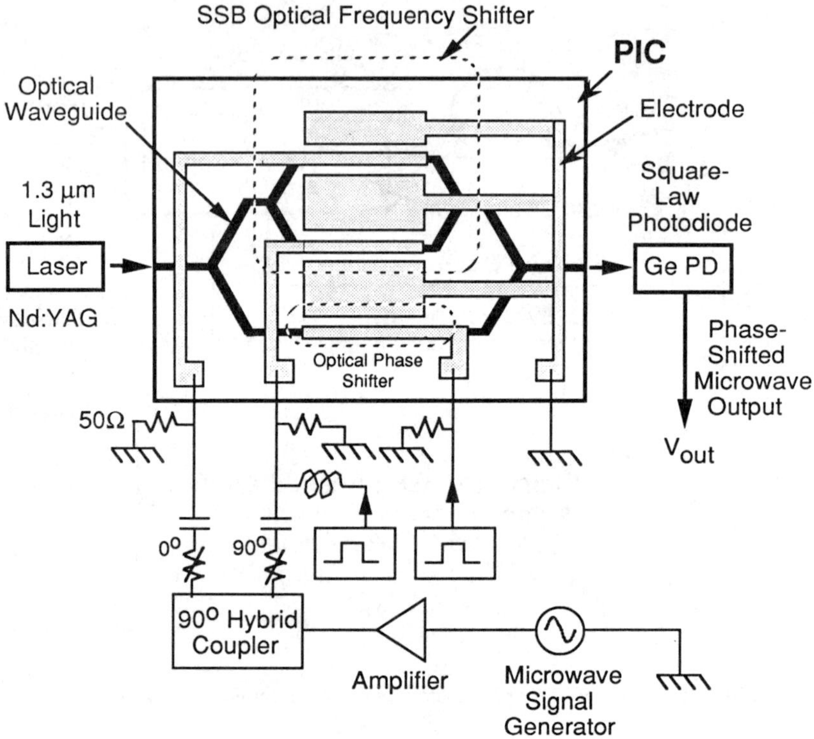


**Figure 9.8** Programmable phase shift of 800-MHz signal by optical heterodyne technique in electro-optic $LiNbO_3$ PIC (from Fig. 5, p. 1526, of [22]).

frequency $f_m$. We can show that the maximum time shift $\Delta\tau_{max}$ needed for azimuth or elevation steering of an $N \times N$ antenna array is $(Nd/c)$ sin $\Theta$, where $d$ is the uniform element spacing and $\Theta$ is the beam pointing angle. Assuming $\Theta \sim 90$ deg and $d = \lambda_m/2$, where $\lambda m$ is the microwave wavelength in free space, we find that $\Delta\tau_{max} \sim N/2f_m$.

Sullivan et al. [24] reported a PIC optical time-shift network comprised of 2 × 2 waveguide switches and cascaded binary-delay stages with on-chip delays in GaAs. The two-mode-interference electro-optic switches were 10 mm long and had a loss of 1 dB. I would like to present a modified version of Sullivan's network, in which the switches are eliminated and waveguided amplifiers are introduced for gain-or-loss switching. Y-branch waveguide splitters and combiners are added, and pairs of etched-facet 90-deg turning mirrors give serpentine folding of the various optical paths on the chip, packing a large number of delay stages within a small area, as illustrated in Figure 9.9. The active PIC electro-optic delay network is expected to have an optical insertion loss of essentially zero. If we assume that $\Delta\tau_{max} < 500$ ps, then a 1 × 2-cm PIC could house a seven-bit time shifter. A recent patent [25] shows how to get reflective time delays on a semiconductor

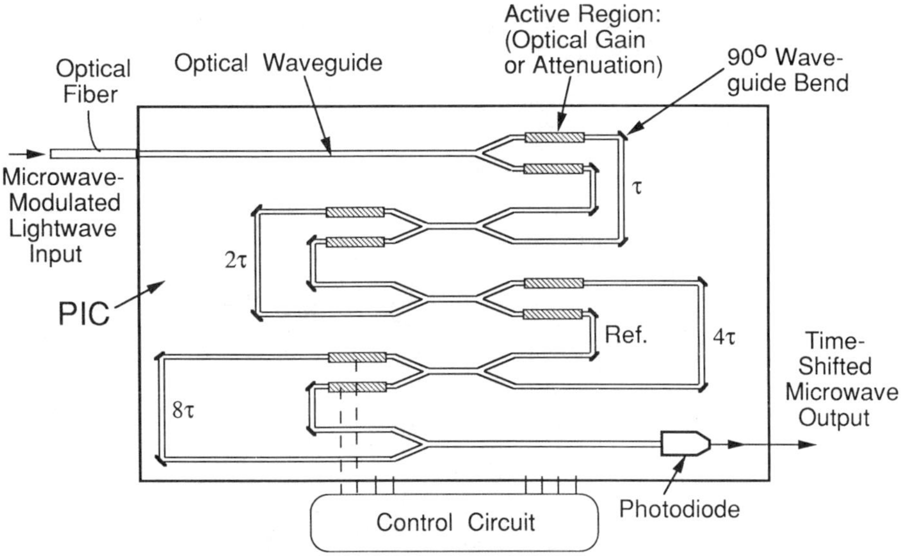


**Figure 9.9** Low-loss programmable four-bit time-shift network (16 delays) in an active semiconductor PIC.

PIC chip. Optical signals are reflected back to the input guide by one of a series of voltage-variable Fabry-Perot resonators integrated in the waveguide.

Elsewhere in this book, optical time-delay schemes are explored. Many schemes use different physical lengths of optical fiber to obtain the delays. I would like to suggest an OEIC-based alternative that uses a fixed length of fiber; that is, all fibers have the same length in the time-shift network. The idea is to use the chromatic dispersion of single-mode fibers together with an array of color-tunable laser diode sources. Because widely tunable laser diodes are available ($\Delta\lambda \sim 57$ nm) and because highly dispersive fibers have been made, this approach appears practical. As Figure 9.10 illustrates, the experimental segmented-core fiber from Corning [26] has a dispersion ($D$) of –65 ps/nm/km and a loss of 0.48 dB/km. A numerical analysis [27] shows that 1 to 2 km of fiber is usually needed; for example, at $f_m = 10$ GHz and $N = 50$, the $\Delta\tau_{\max}$ requirement of 2.5 ns is satisfied by 1 km of Corning fiber [26] with laser diodes tuned over 39 nm.

Figure 9.11 presents a schematic diagram of the proposed time-shift OCPAR. A set of PICs at the remote control station include the tunable laser diodes (TLD) electro-optical modulators (EOM), and photodiodes (PD), while an opto-MMIC at the antenna plane contains the T/R circuits as well as the aforementioned photonics. More details are given in [27] and [28]. The variable-delay line dispersion technique can be extended to new radar components such as the one shown Figure 9.12: an agile, adaptive, microwave-optical transversal filter. In this $N = 4$ example, the current sources $i1, \ldots, i4$ control the emission wavelengths $\lambda 1, \ldots, \lambda 4$ of the TLDs, typically $\lambda_2 = \lambda_1 + \Delta\lambda_g$, $\lambda_3 = \lambda_1 + 2\Delta\lambda$, and $\lambda_4 = \lambda_1 + 3\Delta\lambda$. This produces time delays of $\tau$, $\tau + \Delta\tau$, $\tau + 2\Delta\tau$, and $\tau + 3\Delta\tau$,

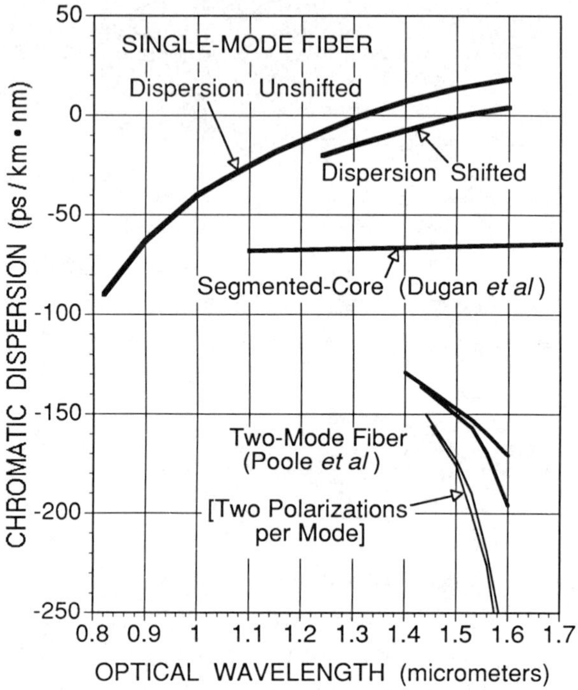


**Figure 9.10** Chromatic dispersion of commercial and special single-mode fibers.

where $\tau = (n/c)L$ and $\Delta\tau = D(\Delta\lambda)L$. For the Figure 9.12 device, the frequency passband between adjacent nulls in the frequency plane ($\Delta f = 1/4\Delta\tau$) is widely tunable. Also, the position of the frequency notches is tunable, giving variable rejection. Fiberless PIC filters are shown in Figures 7 and 8 of [7].

### 9.4.5 Millimeter-Wave OEICs

Optical generation of millimeter waves is an intriguing new development from H. R. Fetterman and colleagues at UCLA. Wafer-scale optoelectronic integration is another goal of that project. As illustrated in Figure 9.13, millimeter waves are generated by nonlinear optical mixing in HBTs and HEMTs [29]. First, a train of 2-ps, 830-nm optical pulses at a 65-GHz repetition rate was created by a mode-locked, two-section AlGaAs MQW laser diode. Then the short optical pulses were focused into an $8 \times 8$-$\mu$m window in the emitter of the AlInAs/GaInAs bipolar phototransistor. The 65-GHz radiation emerged from the twin-dipole printed circuit integrated with the HPT, making the HPT-IC a narrow-linewidth, fixed-frequency, millimeter-wave oscillator. Note that the laser diode could in principle be integrated with the HPT. In fact, Figure 9.13 is the precursor of the wafer-scale circuit illustrated in Figure 9.14. This figure represents the architecture [30] of a complete

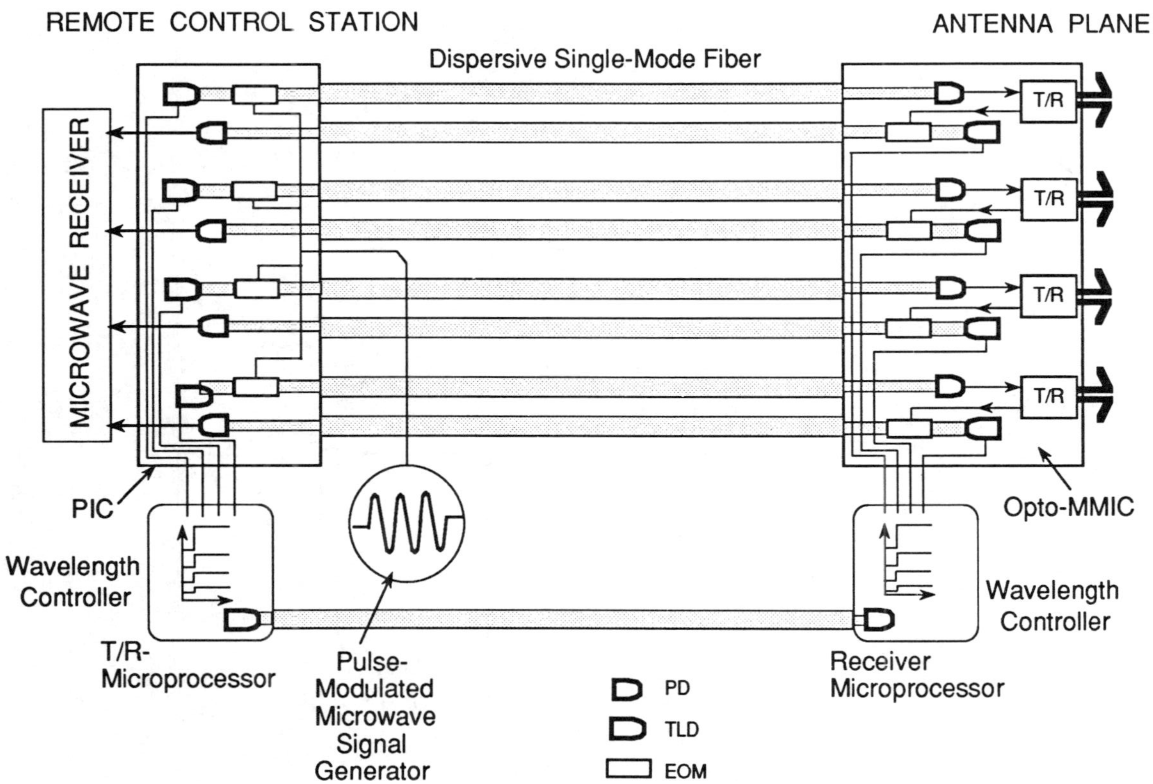


**Figure 9.11** System diagram of dispersive OCPAR beam steerer.

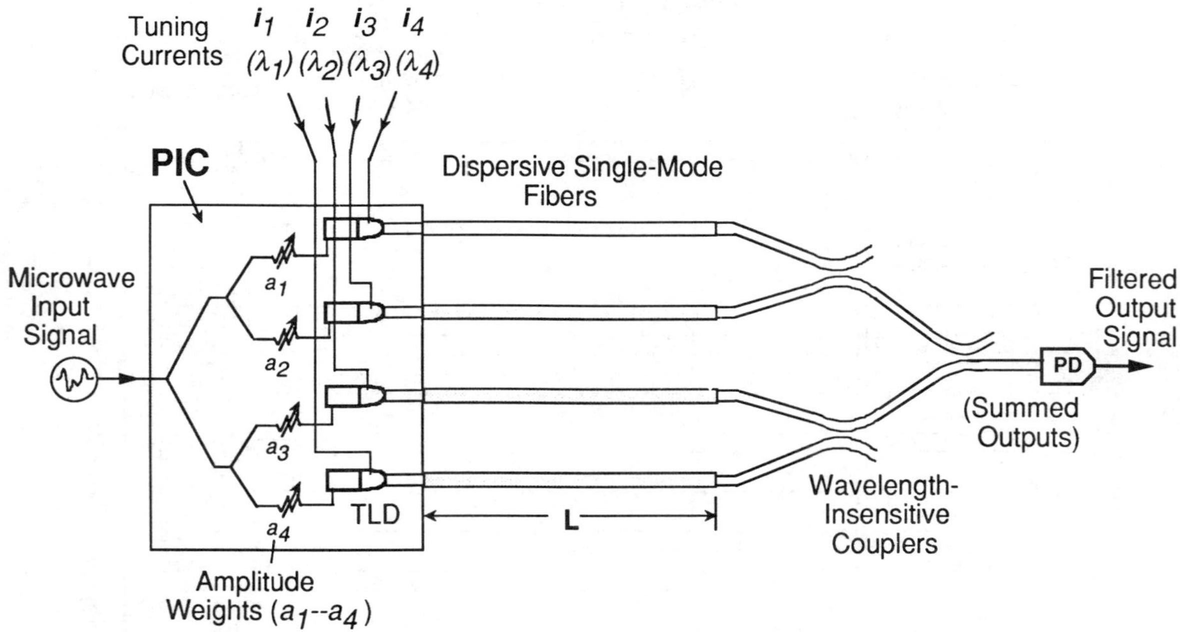


**Figure 9.12** Tunable microwave-optical transversal filter with adaptive weighting.

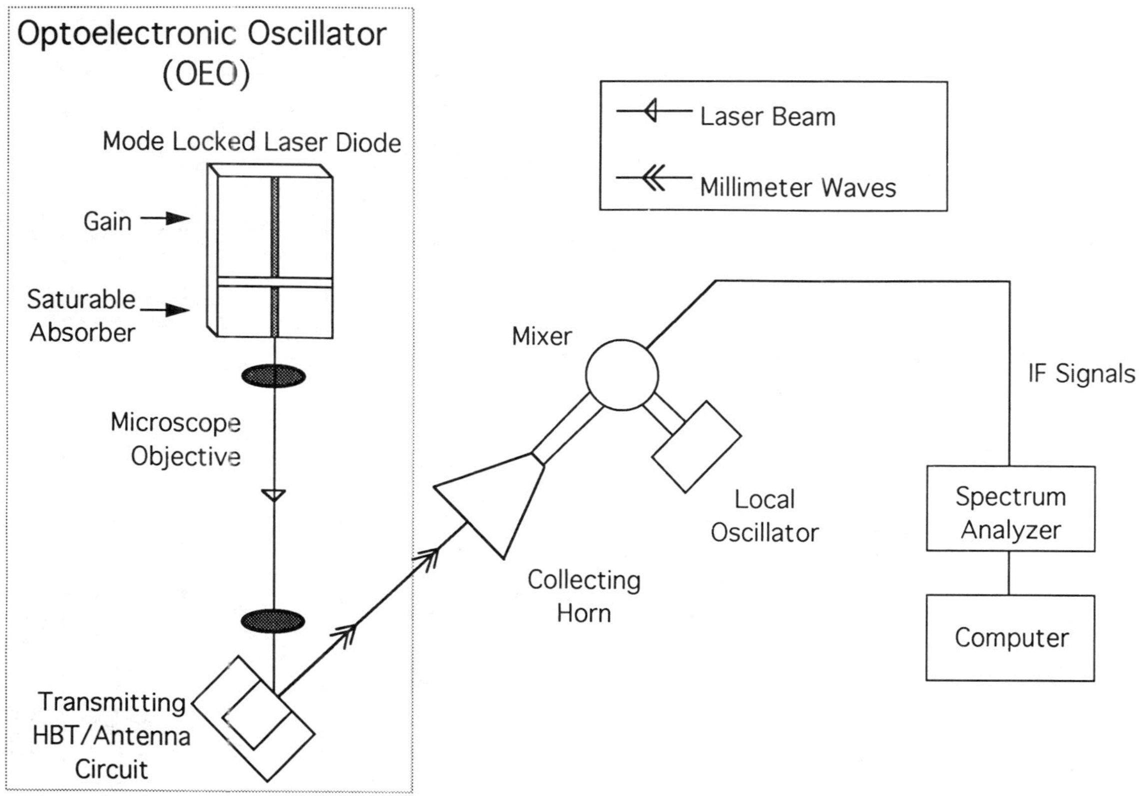


**Figure 9.13** Experimental setup of millimeter-wave optoelectronic oscillator (courtesy of D. C. Scott, D. V. Plant, and H. R. Fetterman at University of California, Los Angeles).

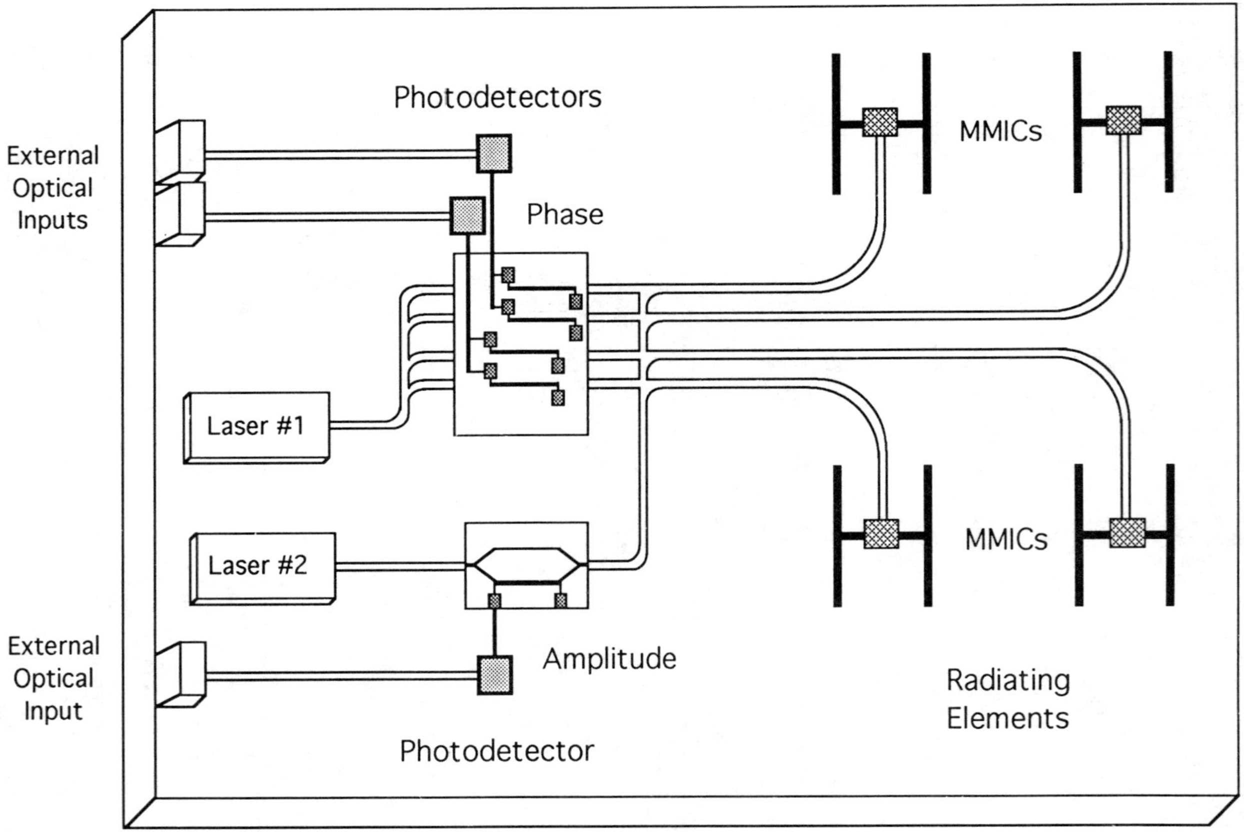


**Figure 9.14** Integrated millimeter-wave (MMW) antenna system, including optical transistors, waveguides, modulators, phase shifters, and phased MMW radiators (courtesy of H. R. Fetterman at Univesity of California, Los Angeles).

millimeter-wave phased-array antenna system on a wafer. Starting with on-chip laser diodes, this OEIC provides waveguided phase control and amplitude control in the optical domain for an integrated phased array of phototransistor radiators, each radiator being an amplifying opto-MMIC for the 60- to 100-GHz range. As before, the radiator circuits use optomilliwave conversion.

## 9.5 CONCLUSION

We have shown that there are many ways of integrating optics and electronics. New ways emerge each day. But the promise of OEICs should be tempered by caveats and cautionary notes [31]. Whether OEICs are adopted in real systems will hinge upon issues such as chip yield, wafer yield, processing costs, packaging, assembly, testing, and the compatibility of optics fabrication with electronics fabrication. The yield issue is critical. For example, in 1993, manufacturers of VLSI computer circuits have opposed the use of advanced SiGe/Si heterojunction bipolar transistors in multimegabit RAMs because chemical vapor deposition (CVD) and molecular beam epitaxy (MBE) processing for heterocircuits is currently not capable of producing less than one defect per square centimeter, unlike the processing for bipolar CMOS (BiCMOS). On the other hand, an opto-MMIC has fewer components and lower packing density than a digital VLSI chip; hence, the opto-MMIC is less likely to contain a defect. The higher threshold for defects per square centimeter suggests that opto-MMICs are feasible within the current state of the art.

## REFERENCES


[1] Soref, R. A., "Silicon-Based Optoelectronics," *Proc. DoD Fiber Optics Conf. '92,* McLean, VA, 24–27 March 1992, pp. 333–336.

[2] Steventon, A. G., et al., "Selection and Optimization of Technologies for a Generic InP OEIC Capability," *IEEE/LEOS '92 Summer Topical Meeting Digest: Integrated Optoelectronics,* Santa Barbara, CA, 5–7 Aug. 1992, p. 17.

[3] Tamir, T., ed., *Guided-Wave Optoelectronics,* Berlin: Springer-Verlag, 1988. See also T. Tamir, *Integrated Optics,* Berlin, Springer-Verlag, 1979.

[4] Hunsperger, R. G., *Integrated Optics: Theory and Techniques,* 2nd edition, New York: Springer-Verlag, 1985.

[5] Nishihara, H., M. Haruna, and T. Suhara, *Optical Integrated Circuits,* New York: McGraw-Hill, 1989.

[6] Probst, D. K., L. G. Perrymore, B. C. Johnson, R. J. Blackwell, J. A. Priest, and C. L. Balestra, "Demonstration of an Integrated, Active 4 × 4 Photonic Crossbar," *IEEE Photonic Tech. Lett.,* Vol. 4, Oct. 1992, pp. 1139–1141.

[7] Soref, R. A., "Active Semiconductor Optical Switches With Blue-Shifted Channel Waveguide," *Fiber and Integrated Optics,* Vol. 9, No. 4, 1990, pp. 313–321.

[8] Young, M. G., U. Koren, B. I. Miller, M. Chien, M. A. Newkirk, and J. M. Verdeill, "A Compact 2 × 2 Amplifier Switch With Integrated DBR Lasers Operating at 1.55 $\mu$m," *Proc. OSA Integrated Photonics Research Conf.,* New Orleans, LA, 13–15 April 1992, pp. 312–313.

[9] Matsueda, H., T. Tanaka, and M. Nakamura, "Optoelectronic Integrated Circuits," *Semiconductors and Semimetals,* Vol. 30, R. Willardson and A. Beer, eds., New York: Academic Press, 1990, pp. 231–284.

[10] Wada, O., ''Advances in Optoelectronic Integration,'' *Int. J. High Speed Electronics,* Vol. 1, 1990, pp. 47–71.

[11] Leheny, R. L., ''InP-Based Materials and Devices for Opto-Electronic Applications,'' *Proc. DoD Fiber Optics Conf. '92,* McLean, VA, 24–27 March 1992, pp. 337–340.

[12] Aoki, M., M. Takahashi, M. Suzuki, H. Sano, K. Uomi, T. Kawano, and A. Takai, ''High-Extinction-Ratio MQW Electroabsorption-Modulator Integrated DFB Laser Fabricated by in-Plane Bandgap Energy Control Technique,'' *IEEE Photonics Tech. Lett.,* Vol. 4, June 1992, pp. 580–582.

[13] Ackerman, E., S. Wanuga, K. Candela, R. Scotti, W. MacDonald, and J. Gates, ''A 3 to 6 GHz Microwave/Photonic Transceiver for Phased-Array Interconnects,'' *Microwave J.,* Vol. 35, April 1992, pp. 60–71.

[14] Yi-Yan, A., and W. K. Chan, ''Semiconductor-Film Grafting: A New Approach to OEICs,'' *IEEE Circuits and Devices Magazine,* Vol. 8, May 1992, pp. 26–30.

[15] Soref, R. A., ''Silicon-Based Optical-Microwave Integrated Circuits,'' *Microwave J.,* Vol. 35, No. 5, May 1992, pp. 230–236.

[16] Presby, H. M., ''Near 100% Efficient Fiber Microlens,'' Optical Fiber Communications Conf. '92, San Jose, CA, 2–7 Feb. 1992, *Post-Deadline Paper Digest,* pp. 408–411. (See also *IEEE Photonic Tech. Lett.,* Vol. 4, Aug. 1992, pp. 897–899).

[17] Paolella, A., ''The Photoresponse of the GaAs Metal-Semiconductor Field Effect Transistor,'' Ph.D. thesis, Drexel University, June 1992.

[18] Richard, M. A., P. C. Claspy, K. B. Bhasin, and M. Bendett, ''Optical Control of an 8-Element Ka-Band Phased Array Antenna Using a High Speed Optoelectronic Interconnect,'' *Proc. IEEE Antennas and Propagation Symp.,* Dallas, TX, 7 May 1990 (NASA Tech. Memo. 102550).

[19] Koepf, G. A., ''Optical Processor for Phased-Array Antenna Beamforming,'' *Proc. SPIE,* Vol. 477, May 1984, pp. 75–81.

[20] Riza, N. A., ''High Speed 2-D Phased Array Antenna Scanning Using Acousto-Optics'' *Proc. SPIE,* Vol. 1703, Orlando, FL, 20–23 Apr 1992, pp. 460–468.

[21] Soref, R. A., ''Electro-Optical Beamforming Network for Phased Array Antennas'' U.S. Patent 4,739,334, 19 April 1988.

[22] Matsumoto, K., M. Izutsu, and T. Sueta, ''Microwave Phase Shifter Using Optical Waveguide Structure,'' *IEEE J. Lightwave Tech.,* Vol. 9, Nov. 1991, pp. 1523–1527.

[23] Hietala, V. M., G. A. Vawter, W. J. Meyer, and S. H. Kravitz, ''Phased-Array Antenna Control by a Monolithic Photonic Integrated Circuit,'' *Proc. SPIE,* Vol. 1476, 1991, pp. 170–175.

[24] Sullivan, C. T., S. D. Mukherjee, M. K. Hibbs-Brenner, A. Gopinath, E. Kalweit, T. Marta, W. Goldberg, and R. Walterson, ''Switched Time Delay Elements based on AlGaAs/GaAs Optical Waveguide Technology at 1.32 $\mu$m for Optically Controlled Phased Array Antennas,'' *Proc. SPIE,* Vol. 1703, April 1992.

[25] Soref, R. A., and H. F. Taylor, ''Semiconductive Guided-Wave Programmable Optical Delay Lines Using Electro-optic Fabry-Perot Elements,'' U.S. Patent 5,140,651, 18 Aug. 1992.

[26] Dugan, J. M., A. J. Price, M. Ramadan, D. L. Wolf, E. F. Murphy, A. J. Antos, D. K. Smith, and D. W. Hall, ''All-Optical Fiber Based 1550 nm Dispersion Compensation 10 Gbit/s, 150 km Transmission Experiment Over 1310 nm Optimized Fiber,'' Optical Fiber Communications Conf. '92, San Jose, CA, 2–7 Feb. 1992, *Post-Deadline Paper Digest,* pp. 367–370.

[27] Soref, R. A., ''Optical Dispersion Technique for Time-Delay Beamsteering'' *Applied Optics,* Vol. 31, 10 Dec. 1992, pp. 7395–7397.

[28] Soref, R. A., ''Wavelength-Dependent Tunable Optical Time-Delay System for Electrical Signals, U.S. Patent 4,671,604, 9 June 1987.

[29] Scott, D. C., D. V. Plant, and H. R. Fetterman, ''60 GHz Sources Using Optically Driven Heterojunction Bipolar Transistors,'' *Applied Physics Lett.,* Vol. 61, 6 July 1992, pp. 1–3.

[30] Fetterman, H. R., ''High Frequency Optoelectronic Integrated Systems for AFOSR,'' 24 June 1992, unpublished.

[31] Fischetti, M. A., ''Integrated Photonics Comes Out of the Laboratory,'' *Photonics Spectra Magazine,* Vol. 26, May 1992, pp. 98–108.

# *Chapter 10*

# *Control Requirements for Dynamic Beam Steering Systems*

***Robert Mailloux***
*Rome Laboratory*

## 10.1 INTRODUCTION

Phased-array antenna systems combine the signals from as many as thousands of antenna elements to point a directive beam at some angle in space. The characteristics and angle of the beam are selected electronically using analog or digital control of the amplitude and phase of excitations across the array.

The control functions could be performed by photonic systems. The most simple task, the transmission of digital signals to control phase and amplitude weights, has already been implemented for array systems. An intermediate task for photonics is the distribution of phased microwave signals to array elements. A very advanced capability that could be accomplished by photonic systems is to provide true time-delay transmission paths for the microwave signals that are distributed to array elements. Some of the systems that perform these functions are described in other chapters.

The implications of various array requirements will be discussed in this chapter. Array technology has advanced to a point at which it is not adequate to simply demonstrate optical control of phase shifters, amplitude weights, or approximate time delay. Present analog systems provide amplitude and phase control commensurate with average sidelobe levels below −50 dB, and many applications require this precision. For these applications, the photonic system must provide ultraprecise control. The specifics of this required precision are outlined later in this chapter.

This chapter introduces the subject of phased-array antenna beam steering and describes the primary measures of system performance. It presents an analytical background to explain array scanning characteristics, to compute antenna beamwidth and directivity, and to demonstrate that phase-steered arrays have inherent bandwidth limitations. The primary contribution of the chapter is to describe the peak and average sidelobes that result from array control errors.

## 10.2 ARRAY PATTERN CHARACTERISTICS AND SYSTEM PARAMETERS

### 10.2.1 Patterns of Planar Arrays

Throughout this chapter the discussion is limited to planar, uniformly spaced arrays. More general arrays with unequal element spacing or conformal to nonplanar surfaces are treated in other texts [1,2].

The electric field at a point $(x,y,z)$ radiated from any $i$th element of an array with elements in the plane $z = 0$ (Figure 10.1) at the location

$$\overline{\mathbf{r}}_{\mathrm{i}} = \hat{\mathbf{x}}x_i + \hat{\mathbf{y}}y_i \tag{10.1}$$

is given by

$$\mathbf{E}_{\mathrm{i}}(\theta,\phi) = \overline{\mathbf{f}}_{\mathrm{i}}(\theta,\phi)\frac{e^{-jkR_i}}{R_i} \tag{10.2}$$

where

$$R_i = [(x - x_i)^2 + (y - y_i)^2 + z^2]^{1/2} \tag{10.3}$$

In this expression, $\overline{\mathbf{f}}_{\mathrm{i}}$ is the element (field) pattern of the excited element in the presence of the array, so that all interelement coupling is included in this pattern. The element pattern is, in general, a vector.

In most cases, one assumes that the antenna radiation is sampled in the far field, a distance roughly bounded by

$$R > 2(D^2/\lambda) \tag{10.4}$$

where $D$ is the largest dimension of the aperture. This allows the exponential term in the field pattern to be approximated by a simple expression.

If, in addition, the array is large and one can neglect edge effects, then another common assumption is that all of the element patterns $\overline{\mathbf{f}}_{\mathrm{i}}(\theta,\phi)$ are the same, and in this case the far-field radiation pattern is written as the product of the element pattern $\overline{\mathbf{f}}(\theta,\phi)$ and the (scalar) array factor $F(\theta,\phi)$.

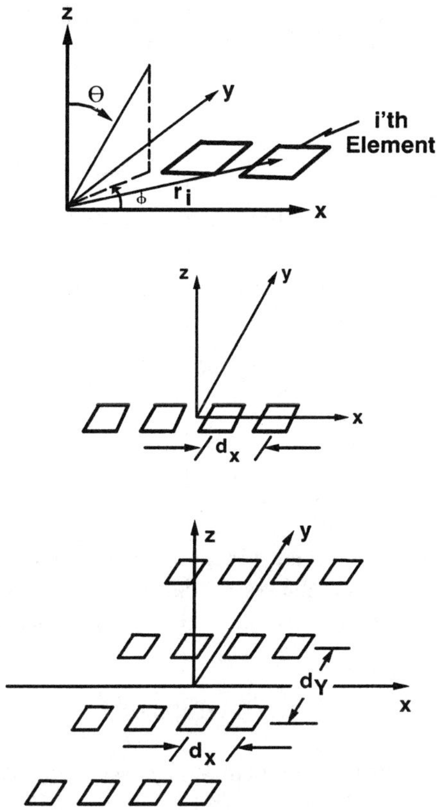


**Figure 10.1** Array geometries: (a) coordinate system; (b) linear array; (c) planar 2-D array.

$$\overline{\mathbf{E}}_{\mathrm{i}}(\theta,\phi) = \frac{\mathrm{e}^{-jkR}}{R}\overline{\mathbf{f}}(\theta,\phi)F(\theta,\phi) \tag{10.5}$$

where the array factor is written

$$F(\theta,\phi) = \sum_i a_i \mathrm{e}^{\,jk\overline{\mathbf{r}}_i\,\mathbf{r}} \tag{10.6}$$

$$\hat{\mathbf{r}} = \hat{\mathbf{x}}u + \hat{\mathbf{y}}v + \hat{\mathbf{z}}\cos\theta \tag{10.7}$$

and

$$\overline{\mathbf{r}}_{\mathrm{i}} = \hat{\mathbf{x}}x_i + \hat{\mathbf{y}}y_i \tag{10.8}$$

for

$$R = [x^2 + y^2 + z^2]^{1/2} \; k = 2\pi/\lambda \tag{10.9}$$

and

$$u = \sin\theta\cos\phi \qquad v = \sin\theta\sin\theta \tag{10.10}$$

The paramters $u$ and $v$ are the direction cosines of the $(\theta,\phi)$ location and are introduced for convenience.

To form a beam with its peak at $(\theta_0,\phi_0)$ at all frequencies, one needs to choose the excitation

$$\mathbf{a}_i = |\mathbf{a}_i| \exp(-jk\overline{\mathbf{r}}_i \cdot \hat{\mathbf{r}}_0) \tag{10.11}$$

where

$$\hat{\mathbf{r}}_0 = \hat{\mathbf{x}}u_0 + \hat{\mathbf{y}}v_0 + \hat{\mathbf{z}}\cos\theta_0 \tag{10.12}$$

The array beamwidth in either principal plane $(\theta,0)$ or $(\theta,\pi)$ is dependent upon the array dimensions. For a rectangular array with dimensions $L_x$ in one plane and uniform illumination, the half-power beamwidth $\theta$ in the plane that includes the x-axis is given by

$$\theta_3 = 0.886\lambda/L_x \tag{10.13}$$

An array with nonuiform illumination, *tapered* so that the elements near the edges have less signal, as required for sidelobe reduction, will have a broader beam. It is customary to define a beam broadening factor $B_b$, such that the beamwidth of a tapered array is normalized to the uniform array beamwidth as

$$\theta_3 = 0.886(\lambda/L_x)B_b \tag{10.14}$$

Figure 10.2 [3] shows three array patterns for arrays with Chebyshev illuminations designed to produce −20-dB, −30-dB, and −40-dB sidelobes. The Chebyshev illumination is one of a number of established techniques for synthesizing low-sidelobe antenna patterns [3]. Note the beam broadening that results from lowering the sidelobes.

With the choice of weights $a_i$ given above, the beam is stationary at $(u_0,v_0)$ for all frequencies. The required exponential dependence has a linear phase relationship with frequency that corresponds to inserting time delays or lengths of transmission line. These are chosen so that the signals radiated from all elements arrive together at some set of angles $(\theta_0,\phi_0)$ in the far field.

Although the excitation above is ideal, present time-delay devices are bulky, lossy, and expensive, so it is more common to use a steering signal that can be produced by phase shifters. The resulting excitation is

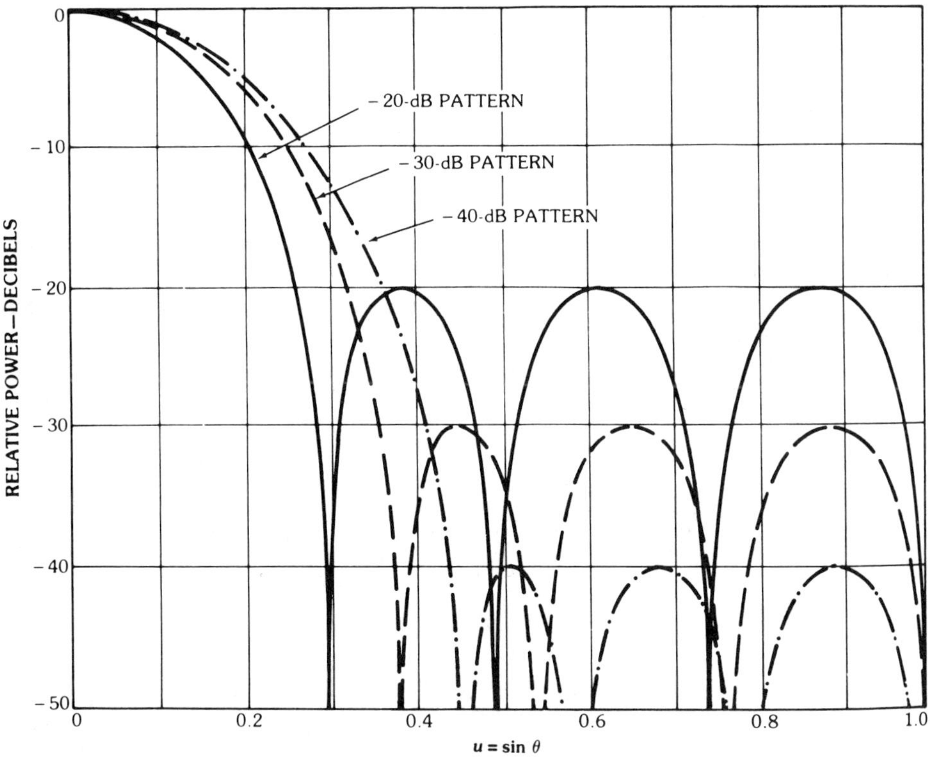


**Figure 10.2** Chebyshev patterns for eight-element arrays with varying sidelobe levels (after Mailloux [3]).

$$\mathbf{a}_i = |\mathbf{a}_i| \exp(-jk_0\overline{\mathbf{r}_i} \cdot \hat{\mathbf{r}}_0) \tag{10.15}$$

with $k_0$ defined as $k$, but with a specific fixed wavelength $\lambda_0$.

*Characteristics of One-Dimensional (Linear) Arrays*

These expressions take on a simpler form for arrays with periodic element spacings. In 1-D (Fig. 1(b)), the array has elements spaced $d_x$ apart, and the array factor for an array scanned to the direction cosine $u_0$ at frequency $f_0$ is given by

$$F(\theta,\phi) = \sum_n |\mathbf{a}_n| e^{(jnd_x 2\pi(u/\lambda - u_0/\lambda_0))} \tag{10.16}$$

This expression indicates the pattern dependence only upon $u/\lambda - u_0/\lambda_0$, so that the shape of the beam is unchanged with scan.

This expression also indicates that the pattern has at least one maximum, where the $F(\theta,\phi) = \Sigma|a_n|$, but other maxima with this same value could exist at any angle $u_p$ such that the argument of the exponent is equal to $j2\pi p$ (for integer $p$). For a pattern with principal maximum at $u_0$, the resulting angle $u_p$ is thus given as

$$u_p = u_0 + (p\lambda/d_x) \tag{10.17}$$

and this beam peak at $u_p$ is called a *grating lobe* of the array. Since $u = \sin\theta$, any existing grating lobes must lie within the region $|u_p| \le 1$, called *real space*. This leads to an expression for maximum element spacing $d_x$ in order that a beam be scanned to the angle $u_0$ with no grating lobes in real space.

$$\frac{d_x}{\lambda} \le \frac{1}{1 + \sin\theta_0} \tag{10.18}$$

A second phenomenon that is evident from (10.16) is that at frequencies other than $f_0$ the beam peak of a phase-steered array "squints" off the desired angle $\theta$ to a new angle given by

$$u = (f_0/f)u_0 \tag{10.19}$$

This result means that at frequencies other than $f_0$, the beam peak does not point directly at the desired $u_0$ direction, and since the array beamwidth is limited, it could mean that the intended target can enter the beam far from its peak. If the bandwidth is defined by the frequency limits at which the gain is reduced to half power, the resulting fractional bandwidth as limited by the beam squint is given as

$$\frac{\delta f}{f_0} = \frac{\delta u}{u_0} = \frac{\theta_3}{\sin\theta_0} \tag{10.20}$$

For a uniformly illuminated array this is related to the array size as

$$\frac{\delta f}{f_0} = 0.886B_b\frac{\lambda_0}{L\sin(\theta_0)} \tag{10.21}$$

*Characteristics of Two-Dimensional Arrays*

The 2-D array of Figure 10.1(c) has elements disposed on a rectangular grid such that their center location $r_{mn}$ is

$$\overline{r_{m,n}} = \hat{x}_m d_x + \hat{y}_n d_y \tag{10.22}$$

and the array factor at frequency $f$ for an array with time-delay steering is

$$F(\theta,\phi) = \sum_{m,n} |a_{m,n}| e^{jk[md_x(u-u_0)+nd_y(v-v_0)]} \tag{10.23}$$

Often, for a rectangular aperture, a separable amplitude distribution is chosen so that

$$a_{m,n} = b_m c_n \tag{10.24}$$

and then the array factor can be written as the product of 1-D array factors of a form similar to (10.16) in terms of $u$ and $v$.

Two-dimensional arrays also exhibit a grating lobe phenomenon. Depending on the array cell dimensions, the array factor with main beam at $u_0,v_0$ can have other beams with the same strength as the main beam at angles given by

$$u_p = u_0 + p\lambda/d_x \tag{10.25}$$

$$v_q = v_0 + q\lambda/d_y \tag{10.26}$$

These beams contribute to the radiated pattern if they are located within the unit circle:

$$u_p^2 + v_q^2 \le 1 \tag{10.27}$$

### 10.2.2 Array Directivity and Sidelobe Levels

The directivity $D_0$ of an antenna is the maximum of the power density $S(\theta,\phi)$ (in watts per square meter) at the beam peak, divided by the total radiated power.

$$D_0 = \frac{4\pi S_{\max}}{\int S(\theta,\phi) \sin\theta d\theta d\phi} \tag{10.28}$$

The directivity is seen to depend only on the antenna pattern and does not include losses or any of the circuit behavior.

For any aperture antenna with dimensions many wavelengths on a side, the maximum achievable directivity is related to the area of the aperture using the following equation.

$$D_{\text{MAX}} = 4\pi A/\lambda^2 \tag{10.29}$$

It is convenient to define an aperture efficiency $\epsilon_A$ to relate the actual directivity of an antenna to the maximum directivity.

$$D_0 = \epsilon_A D_{\text{MAX}} \tag{10.30}$$

For a large array, in which edge effects can be neglected and all array element patterns can be assumed to be the same and have directivity $d_e$, the directivity at broadside of an array of $N$ elements is given in terms of the excitation amplitudes $|a_n|$ as

$$D_0 = \frac{d_e\left[\sum |a_n|\right]^2}{\sum |a_n|^2} \tag{10.31}$$

$$D_0 = d_e \epsilon_T N \tag{10.32}$$

where here $\epsilon_T$ is the array taper loss, a term that is the same as the aperture efficiency for the large-array case. The element pattern directivity $d_e$ for a 2-D matched array is related primarily to the area $A_e$ of one cell of the rectangular grid, and is $4\pi A_e/\lambda^2$.

The array gain $G_0$ is the directivity reduced by any losses in the feed lines, circuits, and antennas themselves. The gain is given by

$$G_0 = D_0 \epsilon_L (1 - |\Gamma|^2) \tag{10.33}$$

where $\epsilon_L$ is the loss efficiency and accounts for losses in the feeds, and $\Gamma$ is the net reflection coefficient at the antenna input terminals.

As the array is scanned, the mismatch changes and the pattern broadens. Although one generally needs to use the basic equation for directivity, one can use the following approximation for an antenna that is large in both dimensions, forms a narrow beam in two orthogonal plans (a pencil beam), and is scanned to $(\theta_0, \phi_0)$.

$$D(\theta_0, \phi_0) = D_0 \cos \theta_0 \tag{10.34}$$

Similarly, in the same limit, the beamwidth in the scan plane broadens in proportion to $(\cos \theta_0)^{-1}$

The directivity is reduced as the array sidelobes are lowered. Figure 10.3 shows a relationship between the sidelobe level and taper efficiency for a 32-element linear array of elements with isotropic patterns ($f^i = 1$) using a Chebyshev taper.

### 10.2.3 The Effects of Uncorrelated Errors in Phase and Amplitude

Phased-array antenna technology has advanced to meet extremely precise performance standards. In this text, which emphasizes photonics applications to array control, it is appropriate to list the performance characteristics that set the standards for array antennas.

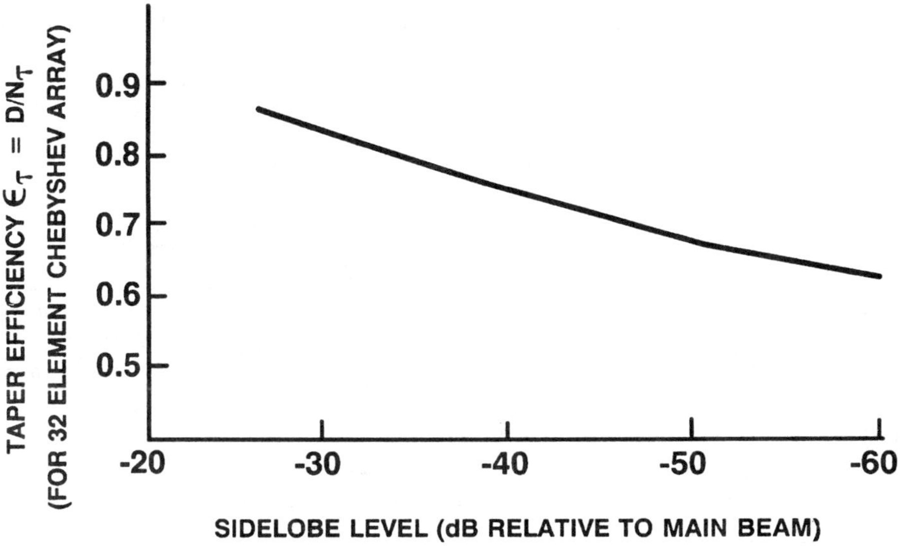


**Figure 10.3** Taper efficiency versus sidelobe level (based on 32-element Chebyshev array).

Amplitude and phase errors need to be carefully controlled if array sidelobes are a consideration. The statistical description of array error effects is well established, and only the principal results are summarized below.

*Average Sidelobe Levels*

For an array of $N$ elements with phase error variance $\overline{\Phi^2}$ and amplitude variance $\overline{\delta^2}$, the resulting ensemble average pattern is given below, normalized to the peak of the beam [4]:

$$\overline{|F_N(\theta,\phi)|^2} = |F_{0N}(\theta,\phi)|^2 + |f(\theta,\phi)|^2[\overline{\Phi^2} + \overline{\delta^2}]/g_A$$

$$g_A = \frac{\left(\sum A_n\right)^2}{\sum A_n^2} = N\epsilon_T \tag{10.35}$$

where $\epsilon_T$ is defined in (10.32).

Note that the element patterns have been assumed scalar in the above for the sake of simplicity.

Here, the ratio $\epsilon_T$ is the taper efficiency of the array factor, and $g_A$ is the array factor directivity.

The expression above includes the ideal (no error) pattern $F_{0N}$ and a constant pattern multiplied by the element pattern. The constant pattern is the residual (or average) sidelobe level, and its value, relative to the main beam peak is given by

$$\overline{\sigma^2} = \frac{\overline{\epsilon^2}}{g_A} = \frac{\overline{\Phi^2} + \overline{\delta^2}}{g_A} \tag{10.36}$$

Figure 10.4(a) [4] shows this average sidelobe level as a function of phase and amplitude error and normalized to the isotropic radiation level, which is a factor $D_0$ greater than the above. For an array with $\lambda/2$ spacing in both planes, $d_e$ is $\pi$ and so relative to the isotropic radiation level. The average sidelobe level is $\pi g_A$ times the above.

The expressions above assume that $N$ is the total number of degrees of freedom in the array, so if all the elements in a 2-D array have errors that are statistically independent, then this is the sidelobe level in the hemisphere averaged over many arrays. If a square array has $N$ columns and no errors in any columns, but independent errors in the column feed lines, the average sidelobes in the scan plane (relative to the beam peak) are

$$\overline{\sigma^2} = \frac{\pi^{1/2}\overline{\epsilon^2}}{D_A^{1/2}}$$

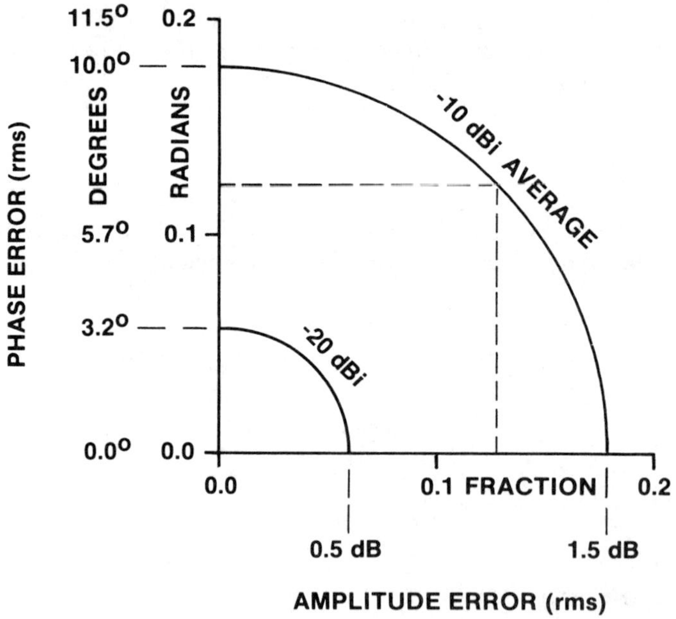


**Figure 10.4(a)** Average sidelobe levels due to amplitude and phase errors: sidelobe levels relative to isotropic radiation (element pattern gain $\pi$ assumed) (after Ruze [4]).

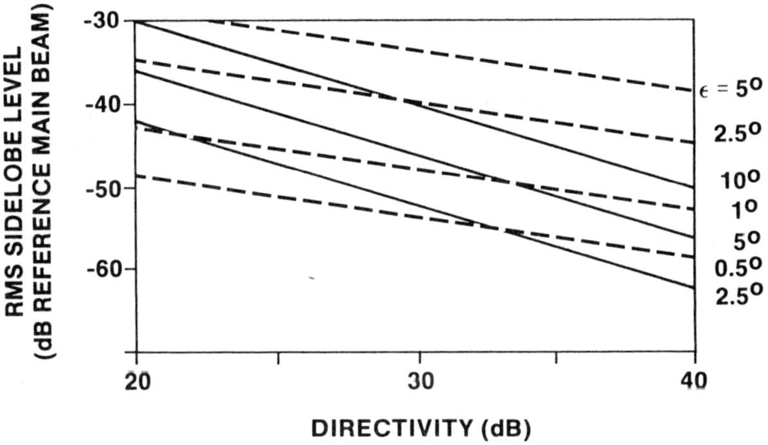


**Figure 10.4(b)** Average sidelobe levels due to amplitude and phase errors: sidelobe levels relative to main beam versus directivity of square array for phase error variance Φ. Solid lines for 2-D array. Dashed lines for array of linear arrays (after Mailloux [3]).

Figure 10.4(b) [3] shows the sidelobe level due to phase errors alone and compares 1-D and 2-D arrays. The figure indicates that extreme precision is required to maintain the same error sidelobe level for 1-D arrays (dashed), because the number of degrees of freedom is reduced.

*Peak Sidelobe Levels*

Often it is important to know the peak sidelobes associated with errored phase and amplitude. Detailed considerations of the peak sidelobe behavior are based on the statistics of the error sidelobes. At any point in space the amplitude $F(\theta, \phi)$ of the far-field pattern of an ensemble of arrays with the same statistics is given according to the following expression [5]:

$$\mathrm{p}(F) = (2F/\overline{\sigma^2})I_0[2FF_0/\overline{\sigma^2}]\exp[-(F^2 + F_0^2)/\overline{\sigma^2}] \tag{10.38}$$

where $\overline{\sigma^2}$ is the variance of an ensemble of array sidelobes, $F$ is the value of the ensemble pattern including the designed (ideal) pattern and the average pattern, and $F_0$ is the design pattern at some given angle.

In this expression, p($F$) is the probability of the field intensity at some angle lying between $F$ and $F + \Delta F$. $I_0$ is the modified Bessel function. To obtain the cumulative probability $P$ at any sidelobe being less than some specified number, one needs to integrate the probability density function over $F$ from zero to the specified number.

Figure 10.5 shows the peak sidelobe level for an array with amplitude and phase errors. This result, due to Hsiao [6], presents the cumulative probability of peak sidelobes being less than any given ''desired'' sidelobe value $S_d$ for a given $F_0$ ($S_d$) sidelobe levels as a function of the residual sidelobe level. The curves emphasize the fact that by over-designing the array (choosing extra low design sidelobes), one can achieve the desired sidelobe with a higher residual (average) sidelobe level. This has the advantage of allowing the use of lower tolerance components to achieve the same desired sidelobe level.

## 10.3 ARRAY FEEDS AND CONTROL

### 10.3.1 Analog Versus Digital Beam Forming

Analog beam forming systems use signal-combining circuits and phase control to combine signals from all array elements and to point the array main beam at some chosen direction with a specified sidelobe level. Analog systems typically have one or several ports that correspond to various sum and difference beams as required for monopulse operation, but apart from that they provide no access to the specific signals received by the array elements. Digital beam forming systems, on the other hand, do not present a combined signal, but have direct access to the signal at every element. Used primarily for receive systems, digital beam formers have amplifiers, mixers, and analog-digital converters at each element. The resulting signals are brought to a computer for processing. Digital beam formers therefore allow the added flexibility of processing these signals separately

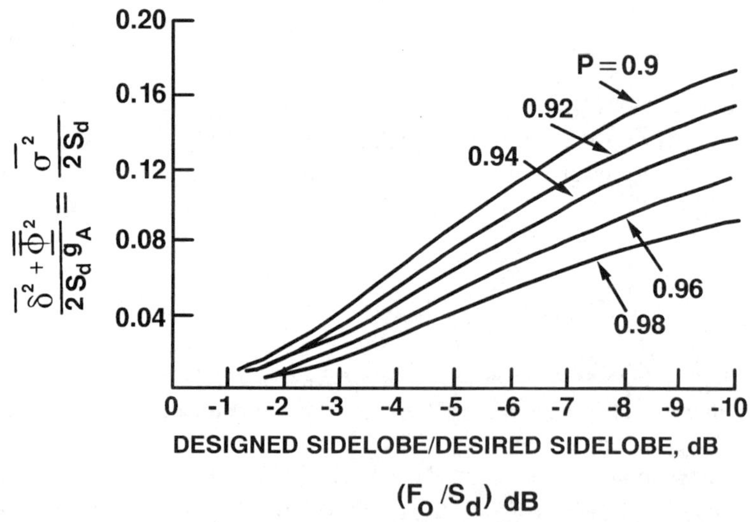


**Figure 10.5** Peak-design-to-peak-desired sidelobe ratio versus residual sidelobe level (after Hsiao [6]).

to optimize pattern functions, SNR, interference rejection, or any other performance parameter. Because digital beam forming systems can incorporate so much processing, it is possible to have more precise beam forming and pattern control than with present day analog systems. The disadvantage of digital beam forming systems is the cost of digital receivers at each element and the required processing of huge amounts of data. Optical devices and processing may someday play a role in future digital beam forming systems.

## 10.3.2 Analog Beam Forming Feeds

### *10.3.2.1 Corporate and Space Feeds With Phase or Time-Delay Steering*

Figure 10.6(a–c) depicts equal line length and series feeds that are constrained by transmission lines and power dividers, while Figure 10.6(d,e) shows two kinds of space feeds. The equal line length feeds are useful for broadband systems using phase or time-delayed control for scanning, but the series feed (Fig. 10.6(c)) causes the beam to scan with frequency and is necessarily a narrowband feed. The power distribution in space feeds like the lens feed (Fig. 10.6(d)) and the reflectarray feed (Fig. 10.6(e)) is not constrained by transmission line circuits and so can be relatively low-loss if spillover and reflection losses are minimized. Space feeds require significant volume and so are not useful whenever volume is a constraint.

Equal line length corporate or space-fed antennas form a beam at broadside for all frequencies, but are scanned with phase shifters or time-delay units. A number of different phase shifters have been built using various technologies, but basically they are intended to provide accurate phase shifts that are independent of frequency and constrained modulo $2\pi$. For a linear array with spacing $d_x$, the phase shift of the $n$th element is

$$\psi(n) = \text{MOD } 2\pi\left[\frac{2\pi u_0}{\lambda_0} n d_x\right] \tag{10.39}$$

Phase shifters often do not provide continuously variable phase, but have a number of discrete phase states. An $N$-bit phase shifter has $2^N$ phase states, with the $p$th phase state given as

$$\psi_p = p\psi_0 \text{ for } p = 0, 1, 2, \ldots, (2^N - 1) \tag{10.40}$$

related to the least significant bit

$$\psi_0 = \frac{2\pi}{2^N} \tag{10.41}$$

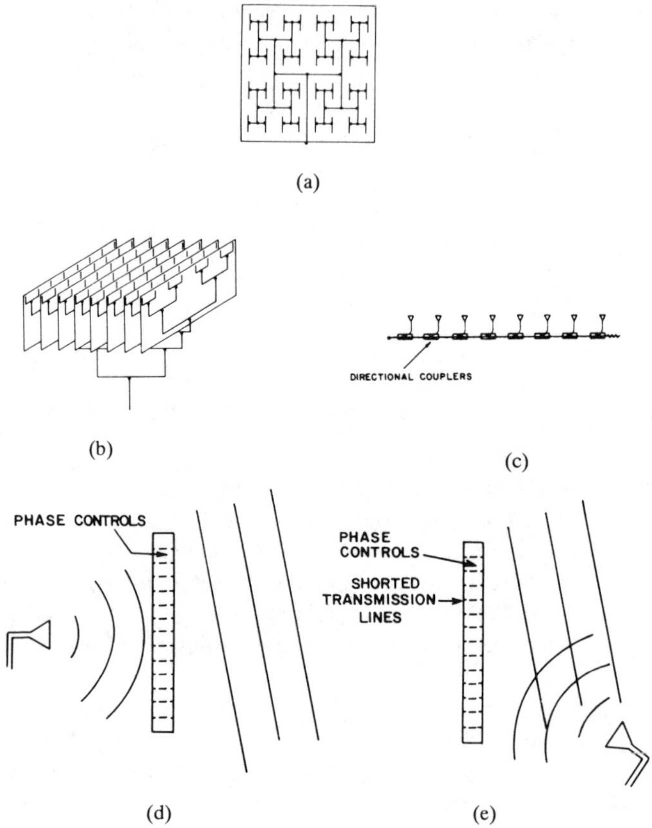


**Figure 10.6** Array feed geometries: (a) planar corporate feed; (b) volume corporate feed; (c) series-fed line array; (d) space-fed array; (e) reflectarray.

The consequence of such a stepwise approximation to a constant slope phase progression is an increase in average and peak sidelobes, as discussed in a later section. Phase shifter bit accuracies are usually kept on the order of one-half of the least significant bit, because improved tolerance would not result in reduced sidelobes. Maintaining the ultraprecise phase distribution required for some array systems requires an added analog phase trimmer bit.

Since phase shifters present less than $2\pi$ phase shift, there is no advantage in attempting to tailor the frequency dependence to anything other than constant.

To have a stable beam over wide bandwidth, however, requires that the phase shift have an inverse wavelength dependence (or vary directly with frequency) as it would have after passing through a length of line. The required phase shift is

$$\psi(n) = \frac{-2\pi u_0 n d_x}{\lambda} \tag{10.42}$$

for all values of $n$ within the array. The phase shift of a time-delayed unit is not modulo $2\pi$, but must correspond to electrical lengths on the order of the array length. Most often this is accomplished by switching transmission line lengths, and the resulting devices are bulky, expensive, lossy and may not provide true time-delay dispersion because of mismatches in the switching network.

In principle, an array with accurate time-delay units could have low sidelobes and form a beam at a location independent of frequency. However, to reduce costs, these devices are also built with discrete states and used in conjunction with phase shifters to provide some, but not all, of the benefits of true time-delay steering. This is discussed in a later section.

### *10.3.2.2 Grating Lobes Due to Periodic Phase and Time-Delay Quantization*

Phased arrays are often constructed using phase shifters with discrete phase bits or with time-delay units that switch fixed lengths of line into the circuit to provide quantized time delays. In both cases the use of discrete devices results in periodic errors in the array aperture illumination. Unless the periodicity is broken up, the array illumination will result in high sidelobes that have some of the characteristics of grating lobes. Figure 10.7 [7] illustrates these two types of periodic errors and the resulting aperture illumination and radiated patterns for several typical cases.

In the case of phase control with $N$ bits, the phase shifter has states separated by the least significant bit

$$\psi_0 = \frac{2\pi}{2^N} \tag{10.43}$$

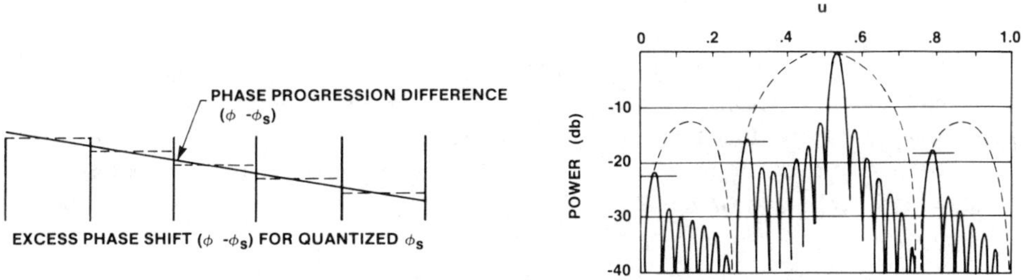


**Figure 10.7(a)** Array sidelobes due to discrete phase and time-delay states (after Mailloux [7]): phase progression and pattern distortion due to quantized phase (dashed line shows the locus of sidelobe pattern peak amplitude).

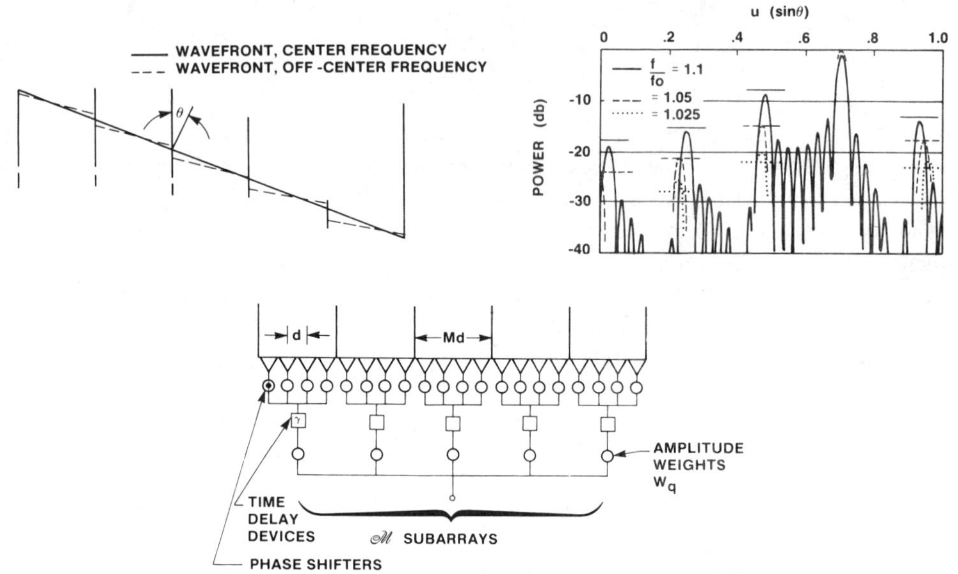


**Figure 10.7(b,c)** Array sidelobes due to discrete phase and time-delay states (after Mailloux [7]): phase progression and pattern distortion due to quantized time delay; subarray organization for time-delayed steering.

Since this discretization allows only a staircase approximation to the correct phase, the resulting radiated pattern has characteristic undesired lobes as shown in Figure 10.7(a). The average sidelobe level due to the quantization is given by

$$\overline{\sigma^2} = \frac{1}{3g_A}\frac{\pi^2}{2^{2N}} \tag{10.44}$$

Figure 10.8 due to Miller [8] shows this average sidelobe level due to phase quantization and its dependence on the number of array elements. In deriving these results, Miller used the above equation, with $g_A = N$, but reduced the array factor directivity by 2 dB to account for scan losses. Figure 10.8 therefore shows Miller's data with sidelobes approximately 2 dB higher than given by (10.44) above.

Miller also evaluated the peak level of the grating lobe nearest to the main beam and showed it to be approximately given by

$$P_{GL}\,(\text{dB}) = -6N \tag{10.45}$$

for an $N$-bit phase shifter. This formula is a good estimate and applies in the limit of a continuous array. The actual grating level may be several decibels higher [7] than given

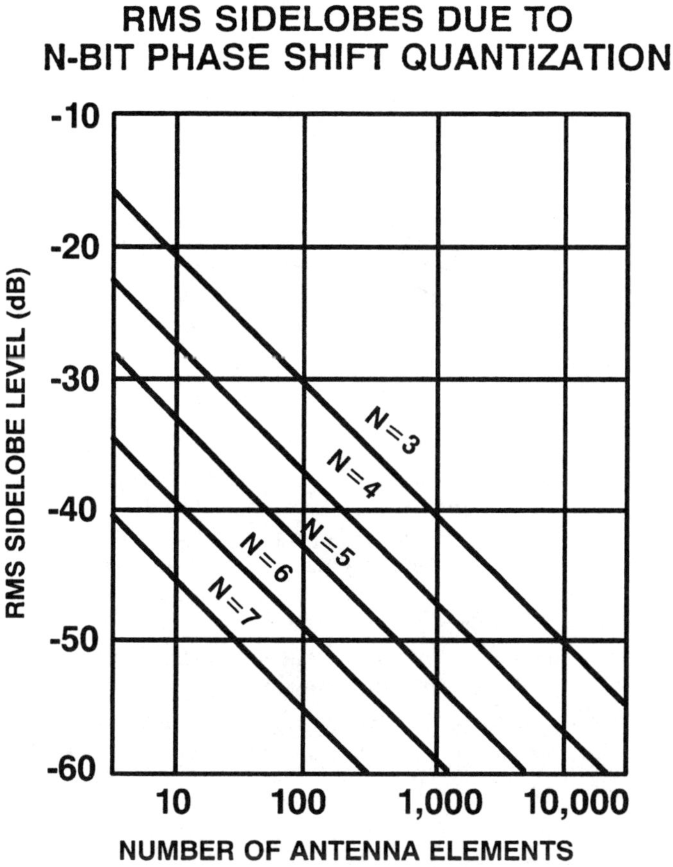


**Figure 10.8** RMS sidelobes due to phase quantization (after Miller [8]).

by this relation, as indicated in the pattern of Figure 10.7(a), obtained for a three-bit phase shifter in a 64-element array and eight quantized phase states.

*Time Delay at the Subarray Level*

Among the various possibilities for incorporating time delay into the beam steering unit of an array, the most obvious is to use time delay at the subarray level, with phase shifters at each element of array. As indicated in Figure 10.7(c) for subarrays of $M$ elements spaced $d$ apart, this leads to a periodic phase error that behaves in a manner similar to phase shift quantization in that it causes grating-lobe-type beams in the radiated pattern. Figure 10.9 [7] shows the magnitude of the $p$th grating lobe level versus the variable subarray sizes $M$. Since the smallest lobes occur for $p/M = 0.5$, only ratios of

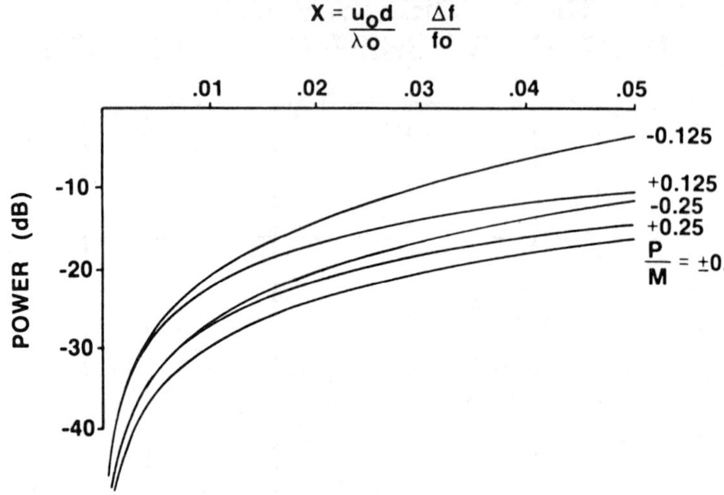


**Figure 10.9** Bandwidth parameters of time-delayed systems using contiguous subarrays (after Mailloux [7]).

*p*/*M* equal or less 0.5 this are shown. This figure indicates that significant lobes are excited for wideband signals, and essentially shows that this manner of integrating time delay into an array is inherently a high-sidelobe approach. Figure 10.7(b) shows the pattern of a 64-element array with time delay entered at eight subarrays. The array is scanned to 45 deg and results are shown for 5%, 10%, and 20% bandwidth. Data evaluated using Figure 10.9 is indicated by horizontal lines.

### *Wideband Array Feeds*

Figure 10.10 shows the several ways of incorporating time delay into an array to provide wideband control without the practical disadvantages indicated above. Figure 10.10(a) is the contiguous subarraying geometry of Figure 10.9(a). Figure 10.10(b) shows an *overlapped subarraying* system [9,10], which uses a microwave optical multiple beam feed to form subarrays. Unlike the contiguous subarraying system, the microwave optically formed subarrays overlap in the aperture, and each subarray radiates a broad flat-topped pattern. The width of this pattern is chosen to suppress the grating lobes that would have been caused by the widely spaced subarrays. Use of the overlapped subarray system allows substantially increased bandwidth and low sidelobes. With such a system, the bandwidth can be increased to approximately *M* times the phase-steered array, where *M* is the number of subarrays formed. *M* is also the minimum number of elements in the feed array.

Another special architecture for incorporating time-delay devices into an array combines a true time-delay network and a complete set of phase shifters, as shown in

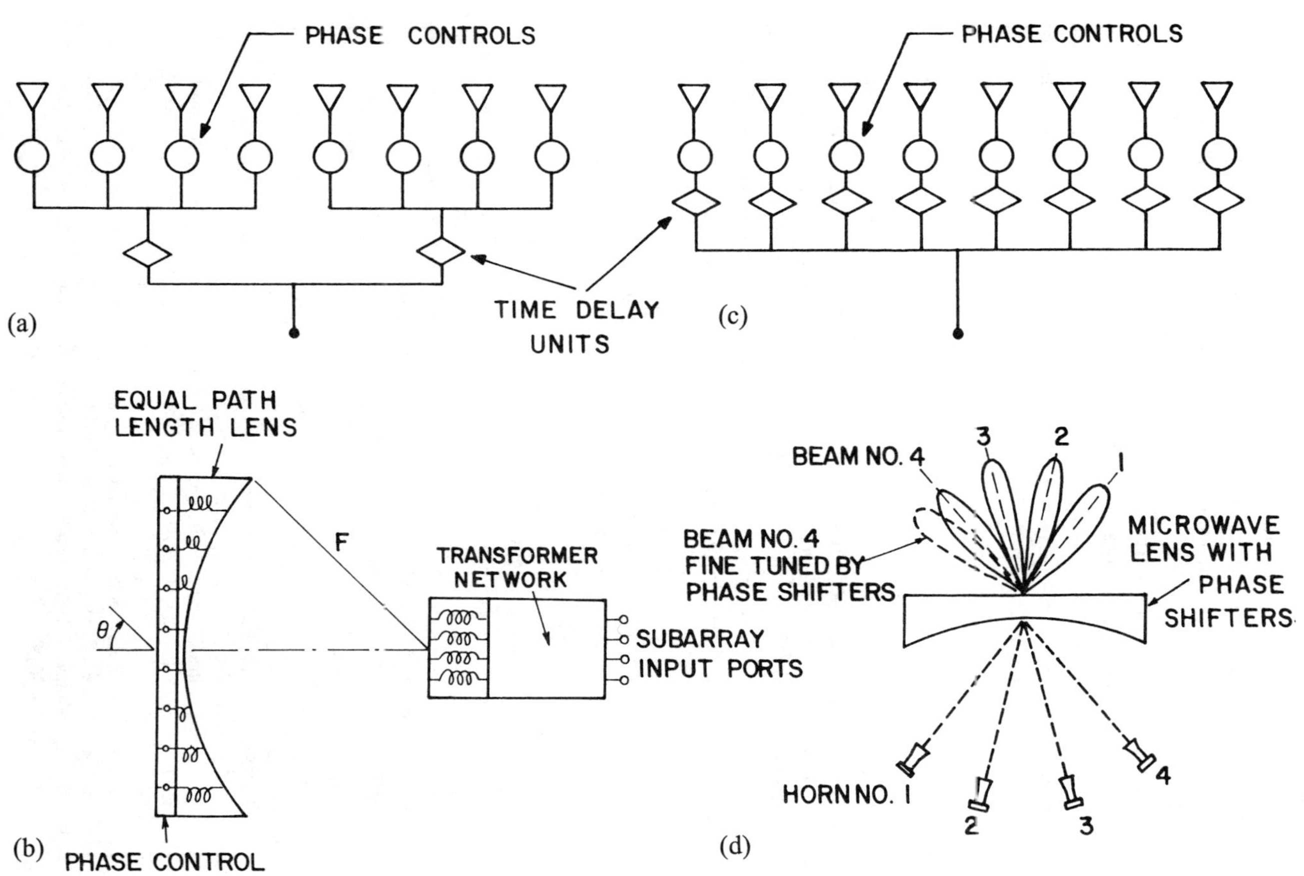


**Figure 10.10** Broadband array feeds: (a) contiguous time-delayed subarrays; (b) overlapped subarray system; (c) System with fixed time-delayed beams, array configuration; (d) system with fixed time-delayed beams, multi-beam lens configuration.

Figure 10.10(c,d). The network provides true time delay at only a small number of beam positions, as few as two to four. The time-delay network of Figure 10.10(c) is shown as a number of time-delay units; the true time delays could also be implemented using a multiple-beam network, as shown in Figure 10.10(d). With this geometry, the various separate feeds form fixed time-delayed beams in space. The phase shifters scan the beam over these small angle increments between these true time-delayed positions. In effect, the system bandwidth is multiplied by the number of true time-delayed positions. Moreover, the sidelobes behavior is not degraded by the addition of periodic errors. One disadvantage of the system is that all delay networks are different and large time-delay units must be accommodated at elements away form the array center.

## 10.4 CONCLUSION

This chapter has described the basic operation of phased-array antennas and the phenomena that characterize their performance. Fundamental array control elements, phase shifters, and time-delay units are presently designed to provide highly precise control. Their replacement by photonic technology within the constraints of system cost, packaging, complexity, and weight poses a significant challenge to photonic component development. This chapter has defined the relevent tolerances and performance criteria required of these components.

## REFERENCES


[1] Borgiotti, G. V., "Conformal Arrays," Ch. 11 in *The Handbook of Antenna Design*, Vol. 2 Rudge, Milner, Oliver, Knight, eds., Peter Peregrinus, 1987.

[2] Hansen, R. C., ed., *Conformal Antenna Array Design Handbook*, Dept. of the Navy, Air Systems Command, Sept. 1981, AD A110091.

[3] Mailloux, R. J., "Periodic Arrays", Ch. 13 in *Antenna Handbook: Theory, Applications and Design*, Y. T. Lo and S. W. Lee, ed., New York: Van Nostrand Reinhold, 1988.

[4] Ruze, J., *Pattern Distortion in Space Fed Phased Arrays*, Lincoln Laboratories Report SBR-1, Dec. 1974.

[5] Skolnik, M. I., "Nonuniform Arrays", Ch. 6 in *Antenna Theory*, Collin and Zucker, Eds., New York: McGraw-Hill, 1969, pp. 227–234.

[6] Hsiao, J. K., "Design of Error Tolerance of a Phased Array," *Electron. Lett.*, 12 Sept. 1985, Vol. 21, No. 19, pp. 834–836.

[7] Mailloux, R. J., "Array Grating Lobes Due to Periodic Phase, Amplitude and Time Delay Quantization," *IEEE Trans.*, Vol. AP-32, No. 12, Dec. 1984, pp. 1364–1368.

[8] Miller, C. J., "Minimizing the Effects of Phase Quantization Errors in an Electronically Scanned Array," *Proc. Symp. Electronically Scanned Phased Arrays and Applications*, RADC-TDR-64-225, RADC Griffiss AFB, p. 17–38.

[9] Tang, R., "Survey of Time Delay Steering Techniques," in *Phased Array Antennas: Proc. 1970 Phased Array Antenna Symp.*, Norwood, MA: Artech House, 1972, pp. 254–260.

[10] Fante, R. L., "Systems Study of Overlapped Subarrayed Scanning Antennas," *IEEE Trans.*, Vol. AP-28, No. 5, Sept. 1980, pp. 668–679.

# Chapter 11
# Spatial Optical Processing

***Denzil Stilwell***
*Naval Research Laboratory*

## 11.1 INTRODUCTION

Array antennas with independent control of each of the elemental radiators (elements) have long been recognized as the most versatile of antenna types. The resulting large number of degrees of freedom permits beam forming of the most general nature. With electronically controlled phase and amplitude of the elemental signals (perhaps even during the processes of transmitting or receiving a signal), almost arbitrary patterns may be synthesized, rapid changes in look direction are possible, interfering sources can be suppressed, and multiple functions can be integrated into a single antenna. Replacement of high-power microwave tubes by an active T/R module at each element allows a high-power aperture product to be achieved using only "driver" power amplifiers in the T/R modules. This also permits considerable waveform flexibility and graceful degradation as components fail or age over time.

Rapid improvements in MMIC capability make such an active aperture architecture very attractive for combat radar systems in a high-density target environment, for communication systems serving multiple terminals, or for space-based sensors requiring high directivity with a large power aperture product. An increase of control complexity is an attractive alternative to the use of extremely-high-power transmitters, large space-fed geometries, multiple high-order microwave power dividing/combining manifolds, or large rotating structures.

Historically, this active aperture approach suffers from extremely high implementation costs for large arrays. Detailed analyses have yielded costs of about \$3,000 to \$5,000 on a per element basis, whether the study was done in the 1960s or in the present. Adding in the rest of a system's costs (physical structure, cooling, power, signal distribution, processing, computer control, etc.) yields a bottom line which can only be afforded in the most crucial of applications, such as when the defense of highly valued assets requires the performance enhancement.

The use of photonic hardware may ameliorate some of the prohibitive impact of cost in the implementation of future systems by fully exploiting the parallelism available from active aperture designs. Conventional active apertures utilize purely electronic components (manifolds, digital phase shifters, etc.) and require a considerable amount of hardware in the backplane of the array to provide precise control of the separate modules. Photonic device alternatives exist for many of these electronic and microwave components which offer a number of advantages. Optical techniques may obviate the need for some of this hardware, such as phase shifters, vector modulators, and their associated drivers. Fiber-optic bundles may substitute for transmission lines for both digital data and for microwave signals. Lasers, modulators, and a collimating lens illuminating the ends of an array of fibers may replace the $N$-way signal division manifold required to drive an array. The use of low-loss fibers permits the remote location (up to kilometers away) of the array control system. Acousto-optic and/or electro-optic devices, and the intrinsic 2-D nature of bulk optics may allow signal processing and conditioning of array information with full parallelism. Integrated-optic devices spliced into low-loss fibers may permit individual trimming and calibration of element responses without the necessity of being located at the antenna aperture itself. Additionally, a multiplicity of optical and photonic techniques exist with which to incorporate time delays into the beam forming system to increase intrinsic array bandwidth. Optically based, or photonic, radar beam control promises significant performance improvement and enhanced capability in terms of large, agile bandwidth, increased instantaneous bandwidth, simultaneous multiple beams, multiple independent beams on independent frequencies, and beam scanning during either transmit or receive operation. Along with the characteristically small sizes associated with photonic circuits, their EMI immunity, and their radiation hardness, photonic alternatives may represent the enabling technologies required to implement radars in the 21st century.

Figure 11.1 provides a high-level systems diagram distinguishing conventional and photonic methods of implementing a beam former for an active array. Linear arrays are shown for simplicity, but in general 2-D arrays would be driven by these circuits. The fundamental functions for the optical subsystem are to impress the microwave signal on a light carrier and then steer or tilt the light beam to impose a lateral, spatial phase gradient on the light wavefront impinging on a fiber-optic array. The main distinctions in the two diagrams relate to the partitioning between backplane and control room components, where the receiver is split off the transmit chain, and where the control signals are routed. For example, control signals for conventional microwave realizations are routed to the T/R modules and must either carry all the detailed phase setting information, or there must be a lookup table at each element to interpret global beam direction commands. In the photonics case, information on the state of the system (transmit or receive) is all that needs to be communicated to the backplane. The phase gradients (or sum of phasors with gradient arguments in the case of multiple beams) are established in the optical subsystem

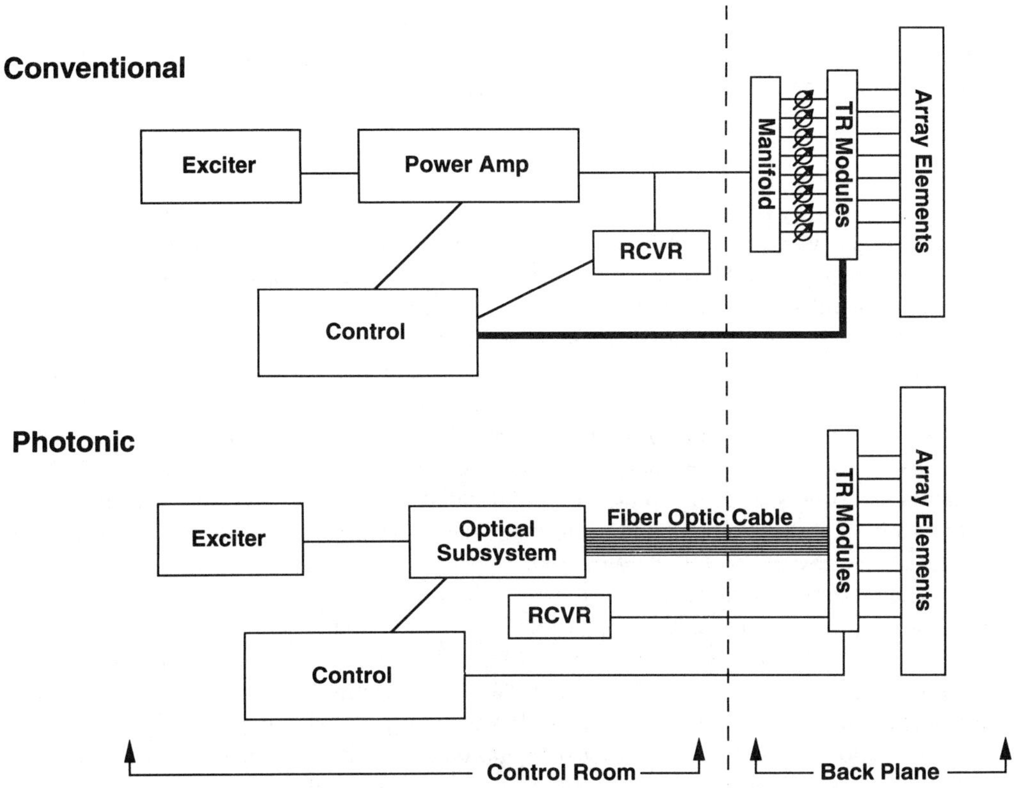


**Figure 11.1** Conventional and photonic implementations of an active phased-array system. The main distinction with the photonic alternative is the replacement of the microwave manifold and phase shifters by an array of optical fibers sampling multiple wavefronts.

and appear on the microwave signal regenerated by detection at the input to each module. In the receive mode, the optical system shifts the microwave frequency to the LO frequency and impresses a phase gradient across the set of signals such that a receive beam is generated for a particular direction. This procedure minimizes the concentration of hardware in the backplane. The antenna system should therefore be lighter, be of smaller size, require less power, and be less subject to interference. However, since optical fibers are not necessarily bidirectional transmission lines, receive beam forming may need to be accomplished by backplane circuitry and independently routed to the receiver, probably at IF.

### 11.1.1 Background—Beam Forming

The basic purposes of an array antenna are to detect or reconstruct a radiation field in the half-space normal to its surface. Its discrete set of elements sample the radiation pattern of the antenna aperture. A simple extension of Nyquist's sampling concepts to involve both space and time can be used, along with Green's theorem, to characterize the field in the half-space. Outgoing waves are characterized by the phasor

$$A(\omega)e^{-j(\mathbf{k}\cdot\mathbf{r}-\omega t)} = \sum_{i}^{N} a_i(\omega)e^{-j(\mathbf{k}\cdot(\mathbf{r}-\boldsymbol{\rho}_i)-\omega t+\phi_i(\omega))} \tag{11.1}$$

where $\mathbf{k}$ is the wave number of a sinusoidal component of the radiated field, $\omega$ is the angular frequency, $t$ is the time, and $\mathbf{r}$ is a far-field vector to the observation point. This equation is valid for omnidirectional elements, where the position of the $i$th element in the array is $\rho_i$, and $\phi_i$ is the possibly frequency-dependent phase of that element. $A(\omega)$ is the result of a sum of elemental excitation phasors and is itself a phasor. $\mathbf{k}$ is arbitrary in direction but of magnitude $2\pi/\lambda$. The peak in the antenna pattern, where the elemental phasors are most closely in phase, is governed by the $\rho_i$ and $\phi_i$. Beam forming then becomes a problem of independent control of $\phi$ and $\omega$. The angular properties of the radiation are largely determined by the spatial variation of phase over the finite aperture of the array (i.e., the projection of $\mathbf{k}$ into the plane of the array). The temporal properties are determined by the spectrum of the radiation.

The properties of an array can be defined in terms of Figure 11.2. This figure shows a uniformly spaced linear array, a phase shifter in each line, a power divider/combiner, and the elemental radiators. All the diagrams herein will be of line arrays for simplicity. A phase front is shown in the transmission line region and in the antenna half-space. The difference in the direction of the two phase fronts indicates a refraction from the transmission line region, where the velocity of propagation is less than that of light in free space, into free space. Although the free space phase front is shown as linear, it would not be so at the scale of the diagram. It would become a well-defined phase front only in the far-field region, defined at that distance beyond $2W^2/\lambda$, where $W$ is the width of the array. In the far field in a direction perpendicular to the coherent phase front, the voltages from each of the elements add, and therefore the power density is proportional to the square of the number $N$ of elements. This makes the antenna appear to have a power gain, or directivity $D$, which is many times the power actually radiated.

The beamwidth and the bandwidth of the array are defined in terms of the projected lengths of the array aperture $W$ onto the free-space phase front and orthogonal to it, respectively. Specifically, the beamwidth $\delta\Theta$ is

$$\delta\Theta = 2\pi\frac{\lambda}{W \cos \Theta} \tag{11.2}$$

and the bandwidth $\delta f$ is

$$\delta f = \frac{c}{W \sin \Theta} \tag{11.3}$$

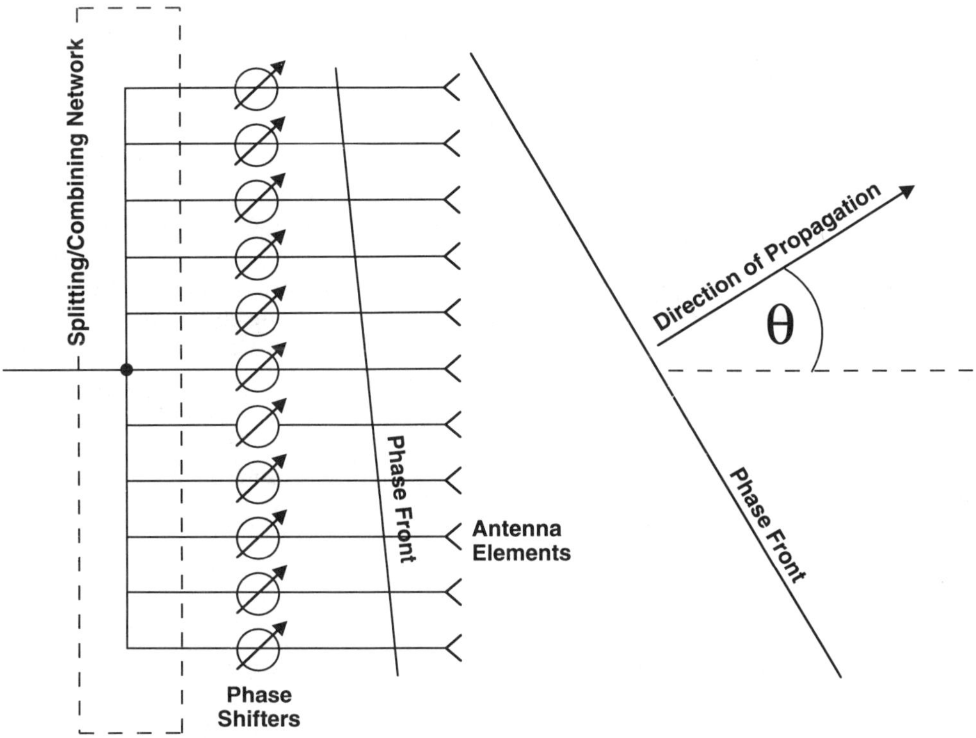


**Figure 11.2** Array of phased elements showing phase fronts after the phasing units and in free space.

The latter is valid for modulo $2\pi$ steering. For time-delay steering, the bandwidth is infinite. These definitions are for uniformly weighted arrays and are in "natural" units. Actual engineering definitions may differ by up to 20% from these. Both factors will increase if an amplitude taper is included in the aperture distribution to lower sidelobe levels. The directivity $D$ of an array is $2\pi/\delta\Theta$. The gain of the array is $LD$, where $L$ is the loss in signal passing through the array, a numeric ranging from 0 to 1. The direction $\Theta$ is related to the phase gradient (phase difference between adjacent elements in a linear array) $\Delta\phi$ by

$$\Theta = \sin^{-1}\left(\frac{\Delta\phi}{s}\frac{\lambda}{2\pi}\right) \tag{11.4}$$

where $s$ is the interelement spacing. Note that the angle at which the beam will form depends on $\lambda$ and will therefore change as the carrier frequency changes. For a spacing greater than $\lambda/2$, the array undersamples the field, and a steep phase gradient will give rise to more than one far-field phase front. Nyquist sampling in space is required to avoid the multilobed grating pattern.

The far field is proportional to the Fourier transform of the field over the aperture. Sampling theory allows one to approximate the antenna response except when one greatly oversamples. Moderate oversampling causes little problem, although the phase of the samples cannot be independent, but must reflect the effects of mutual coupling and aperture matching, features which are not a part of sampling theory. Great oversampling introduces the special case of superdirectivity and will not be pursued further. Undersampling, which causes grating lobes, is easily explained using sampling theory. Since the sampling concept does not include the effects of mutual coupling between array elements, the actual patterns will differ somewhat from that given by sampling theory alone. That difference can be expected to be small insofar as the mutual coupling effects are identical for all the elements.

The wavefront construction of (11.1) can be readily used to provide the connection of Nyquist sampling in space to the correspondence of the Fourier transform of the aperture distribution with the beam direction. The argument of the phasors on the right side of that equation can be written as

$$\mathbf{k} \cdot \mathbf{r} - \omega t + (\phi(\omega) - \mathbf{k} \cdot \boldsymbol{\rho}_i) \tag{11.5}$$

If the term in parentheses is a constant, all the elemental phasors will add coherently. Writing that term in discrete or sampled form yields

$$\phi_0 + \nabla\phi \cdot \mathbf{y}_i - \mathbf{k} \cdot (\boldsymbol{\rho}_0 + is\hat{y}) = \text{const} + (\nabla\phi - \mathbf{k}) \cdot \hat{y}is \tag{11.6}$$

Ignoring a constant phase term in the argument, this transforms the equation into a sampling form with index $i$, the position in the array. The term in parentheses on the right of (11.6) is just another expression of (11.4), but yields the direction of the pattern peak. It is the gradient of the phase in the aperture of the array that controls the beam direction in (11.4), with the correspondence

$$\frac{\partial\phi}{\partial y} = \frac{\Delta\phi}{s} \tag{11.7}$$

It is the sine of the phase in this aperture that is the spatial frequency term in the Fourier transform; thus, each sinusoidal component corresponds to a different beam direction. The wave number vector of (11.1) can be evaluated in any direction, but only in this direction will all the elemental phasors add constructively.

If the gradient in (11.7) is large, it may not be distinguished at the sampling points from different gradients counted modulo $2\pi$. This creates ambiguities in the sampling and grating lobes in the pattern. These lobes have a more serious effect than just creating an uncertainty in the arrival or transmit direction. The effect, attributed to mutual coupling between the elements and described as the *active* impedance variation of an array, causes signal energy to be reflected from the aperture. On transmission, this can reflect energy back into the output stage of the module, which may cause poor behavior of the output power amplifier. Some isolation may be needed for stable amplifier operation (e.g., the addition of a ferrite isolator between the amplifier and the radiating element). This does not help the reflection on reception; the energy is just reflected from the array back into space.

The process of beam forming for an array connected in a fully parallel manner may require a massive rate of control data flow, especially for large 2-D arrays, where the state of each module must be updated thousands of times each second. Beam forming is here meant to include the functions of electronically scanning the antenna pattern, shaping that pattern, and creating nulls in prescribed directions. Thus, beam forming is used to characterize the processes of steering, beam or pattern synthesis, and antijam adaptivity. The concept of array beam steering is quite simple: a progressive phase shift between elements (generally different in the two lateral directions) is introduced into the array excitation such that, by Huygens' principle, a wavefront is reconstructed in the desired direction. The realization of this process is far from simple. At each element, there will appear a random error in the delivery of the signal to that point which may need to be compensated. If a digital control signal is used, there is a quantization of the phase shift values available. Low-sidelobe performance may require a coordination of the strategy of selecting which adjacent level to choose. Array elements may show time-varying phase and position errors due to flexing or spatially varying line lengths resulting from thermal gradients. Multiple phase gradients of differing amplitudes must be used to synthesize a peak in the direction desired with a null in some other direction (for antijam or reduced crosstalk). The phase shifts for large scan angles may be much greater than $2\pi$. The phase shift modulo $2\pi$ may be used for narrowband operation, but the full phase shift, a time delay, must be available for broadband operation. Although all of these problems can be solved by either conventional (microwave) or photonic circuits, it is obvious that an optimal implementation of a particular radar may well involve a mix of conventional and photonic technologies.

The cost of implementing an active aperture array probably mandates that it perform in a multifunction mode. Among these modes would be surveillance, tracking, target ID, target designation, missile control, and communications. As a result, the dynamic range requirement for the receiver may be quite high, especially for the Doppler radar mode. In this mode, distant targets must be illuminated by high-PRF waveforms to unambiguously estimate a velocity (Nyquist sample the frequency difference between transmit and receive waveforms) and to allow returns to be discriminated from clutter. Unfortunately, the high PRF results in multiple range ambiguities (which may be resolved by variable PRF bursts) and folds near-range clutter in with the far-range target signal. It is not possible to use sensitivity time control (STC) to dynamically vary the receiver sensitivity proportional to $R^4$ to make do with a moderate instantaneous dynamic range. The dynamic range must exceed the range loss in the radar equation. This total dynamic range consists of two parts: the actual dynamic range of the receiver front end and that arising from Doppler processing. A dynamic range of 70 to 90 dB may be required for the front-end hardware. This means that all devices must display high linearity (e.g., a high third-order intermodulation intercept point). Likewise, sidelobes must be reduced to the extent possible so that near-range clutter integrated over all azimuth angles does not become commensurate with the unavoidable main-beam clutter. As a result, there are fairly stringent requirements on the phase and amplitude errors which can be tolerated in the individual paths to the array elements [1]. Again, these sidelobes are determined by three factors: the phase and amplitude variance over the set of T/R modules and the number of elements in the array.

### 11.1.2 Background--Photonics

With a linear array of detectors placed across the recombined wavefront of Figure 11.3 and with a tilt in one of the light beams within the Mach-Zehnder interferometer, the phase of the RF signals out of different elements of the detector array will be different. With a tilt, there will be a phase gradient across the array of detectors. This configuration, modified by using an array of fibers to sample the light field and putting detectors on the other end of the fibers, generates a set of microwave signals with a constant phase gradient. This signal set is appropriate for driving an array antenna to launch a beam in some oblique direction. A little study of such a circuit will convince one that spatial perturbations in either leg of the Mach-Zehnder interferometer on the order of a light wavelength will introduce $2\pi$ phase shifts in the output microwave signal. Perturbations introduced in the recombined beam will, however, affect both light beams in a very similar manner, and $2\pi$ phase shifts require a path length difference only on the order of the wavelength of the microwave difference frequency.

Present-day technology provides a wide selection of photonic components with which to implement the optical subsystem block of Figure 11.1. This includes diode lasers, diode-pumped Nd:YAG (yttrium aluminum garnet) components and assemblies, spatial light modulators, and many other items. The following section will describe at a functional level a few of these devices and concepts, which might be used to implement an optical beam steering subsystem. An underlying discriminator being applied is that the device be able to operate in a mode which would display both a function and a performance commensurate with a radar system's data rate, frequency, dynamic range, and bandwidth.

Fiber-optic cables are a mainstay of the optical alternatives and can be used for the transmission of digital control information, timing signals, and microwave signals. As a result of the low loss, flexibility, and small size of fibers, it is possible to locate the array at a substantial distance from the command and control center using the radar. The substitution of a set of fibers for microwave cables allows this remote siting without incurring significant RF losses and, additionally, permits a simplification of the array backplane: the elimination of a microwave manifold.

A very primitive necessity is the light transmission circuitry: lenses, mirrors, beam splitters, and the like. A particularly useful concept is that of the Fourier transform property

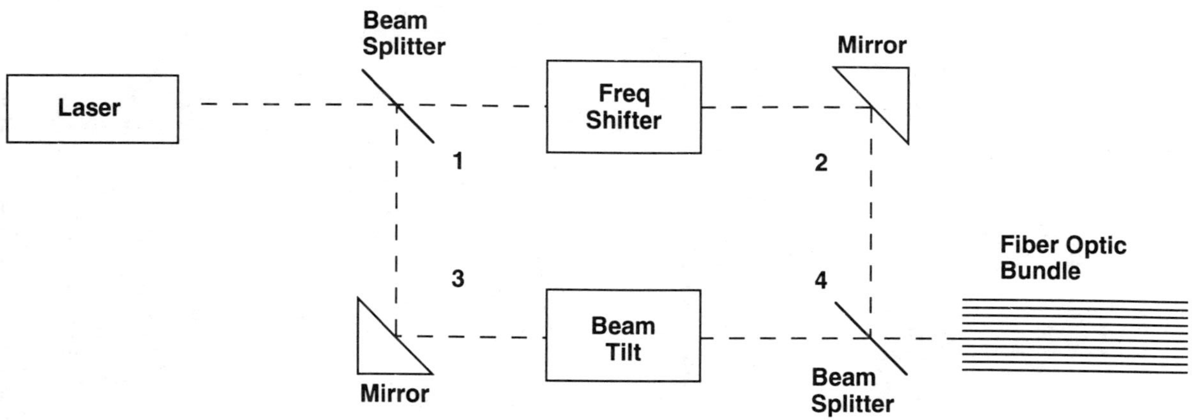


**Figure 11.3** Fundamental Mach-Zehnder means of simultaneously imposing an arbitrary frequency shift and arbitrary beam tilt.

of lenses in a *canonical cascade.* Figure 11.4 illustrates the principle. A point source of light is collimated by a lens between planes P1 and P2. This collimation process can be viewed as a simple manifestation of the Fourier transform relationship between the front and back focal planes of a lens; for example, a point source in one plane transforms to a broad distribution in the other. The next lens transforms again, continuing down the confocally positioned set of lenses. The transform relationship exists between the spatial light amplitude distributions one focal length in front of a lens to one focal length behind a lens. It converts from object spatial features to spatial frequencies, to image, and continues repetitively. Introduction of a mask or any other kind of spatial, phase, or amplitude modulation in any of the focal planes will alter the information or light distribution in all subsequent planes. Having access to the 2-D image and spatial frequency planes permits a set of very powerful signal processing operations to be executed.

It is unlikely that an optical systems designer would ever implement an optical system in this canonical manner. The mathematical and conceptual nicety is more than offset in practice by excessive path lengths, unnecessary lenses, and demanding component specifications. For example, the two-focal-length separation in the collimation regions (e.g., between lenses 1 and 2) can be decreased or increased considerably to accommodate adjustments of interferometer path length when temporal coherence is limited. Further, Figure 11.4 shows that a single lens will relay an image from P2 to P4, whereas two lenses would be required by the canonical system. The cost of these modifications is

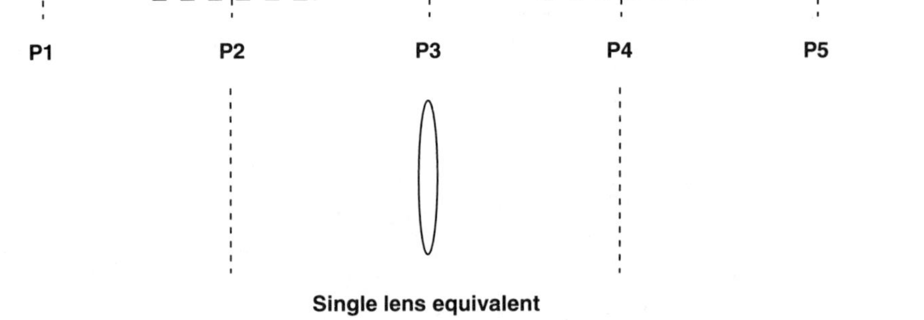


**Figure 11.4** Canonical lens cascade in which the back focal plane of one lens coincides with the front focal plane of the next. Consecutive focal planes are related by a Fourier transform relationship. Lower diagram indicates the single-lens equivalent of relay plane P2 to P4 without intermediary frequency plane.

that there is no intervening Fourier transform plane available for additional processing operations.

The beam splitters and mirrors placed in planes P2 and P4 of Figure 11.5 are seen to create and recombine two paths for the light to travel. The separation into two paths, each composed of a 2-D area on a distinct optical carrier allows parallel as opposed to sequential operations to be performed on both spatial and temporal quantities. This permits, for instance, the use of two beams for heterodyne regeneration of the microwave carrier instead of direct or external amplitude modulation of the light. Distinct operations can then be performed in the two legs of this Mach-Zehnder interferometer. Operations of interest in array signal set generation include frequency offset, beam tilting, and beam combining. The combining method can be extended to a relatively high order by the introduction of additional beam splitters. This allows the superposition of many phase fronts, which will give rise to many independent beams from the array. This provides the basis for multibeam capability when using an optical beam former. The power level decreases with the order because of the 3-dB combining loss of beam splitters.

Of particular interest is the introduction of a frequency offset between phase fronts from separate legs of a Mach-Zehnder interferometer. The combined output beam will consist of two light components which will drift in and out of phase at a rate given by the frequency offset. A detector placed anywhere in the beam will then heterodyne the two components and generate the difference frequency on the output of the detector. The power which would flow from the output of a matched detector is given by

$$p = P_1 P_2 G^2 R^2 (r/2) \tag{11.8}$$

where $P_1$ is the light power in one of the beams, $P_2$ is the light power in the other, $G$ is the gain of the detector (1 for nonavalanche types), $R$ is the responsivity of the detector in units of amps/watt, and $r$ is the characteristic impedance (normally 50 Ω) of the detector output line. The noise appearing on the output is a combination of k$TB$ noise from the matched line and the temporal noise (RIN) of the laser used. Diode lasers may generate three to five orders greater noise in the microwave band than do gas or diode-pumped YAG lasers. For the latter, RIN may be ignored.

One of the best bulk optical devices that can impress wide dynamic range microwave modulations and/or beam tilts onto optical wavefronts is the acousto-optic cell, often called a *Bragg cell*. The Bragg cell designation describes a method of excitation of an acousto-optic cell which is very efficient. Devices are available with greater than 90% at 100 MHz and in excess of 50% even at 1 GHz. The Bragg interaction is diagrammed in Figure 11.6 [2]. Acoustic phonons propagating laterally across an optical system interact with optical photons incident at an oblique angle with respect to the optical axis of the system. This photon-phonon interaction is quantized and the interaction geometry is set up to simultaneously satisfy conservation of energy ($\Omega_{\text{out}} = \Omega_{\text{in}} \pm \omega$) and conservation of momentum ($\mathbf{K}_{\text{out}} = \mathbf{K}_{\text{in}} \pm \mathbf{k}$). Since microwave frequencies are five orders of magnitude less than optical frequencies, the only way to simultaneously satisfy these two requirements

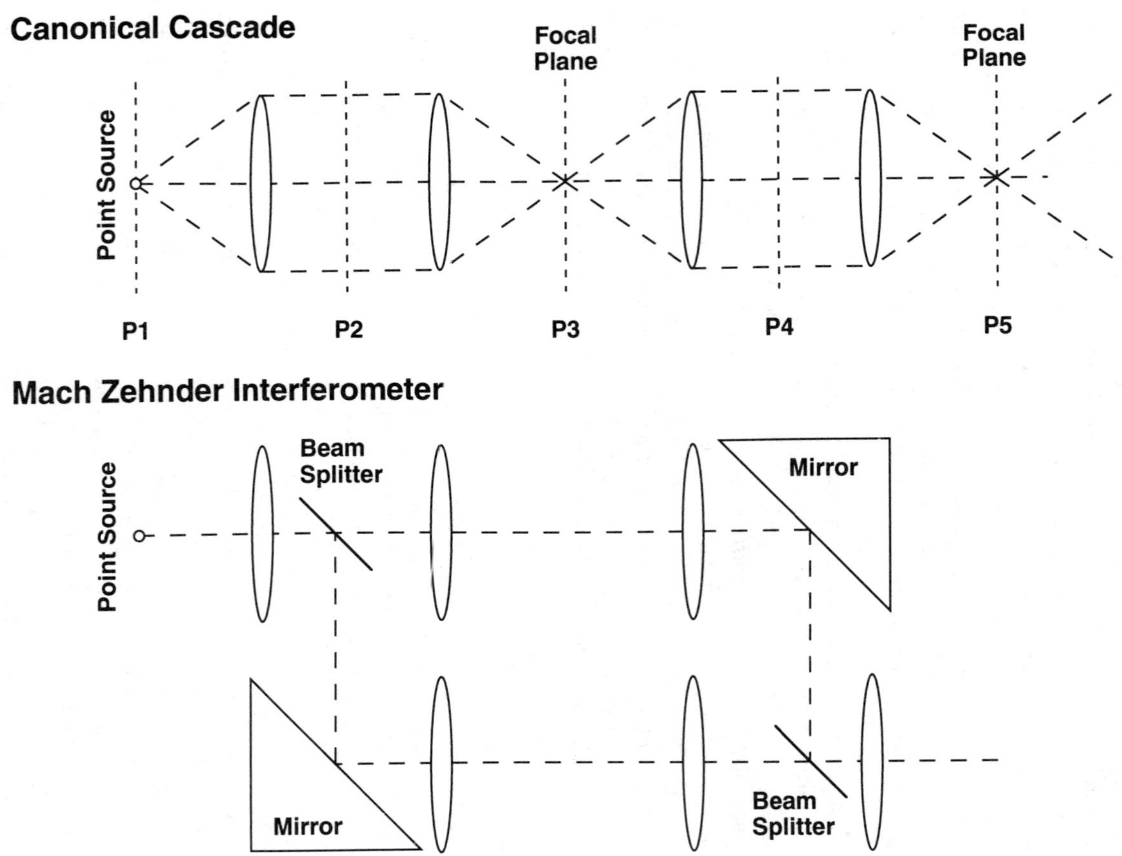


**Figure 11.5** Derivation of a Mach-Zehnder interferometer from the canonical lens configuration. Beam splitting occurs in a collimated region, which will only introduce quadratic phase errors in the Fourier transform relationships.

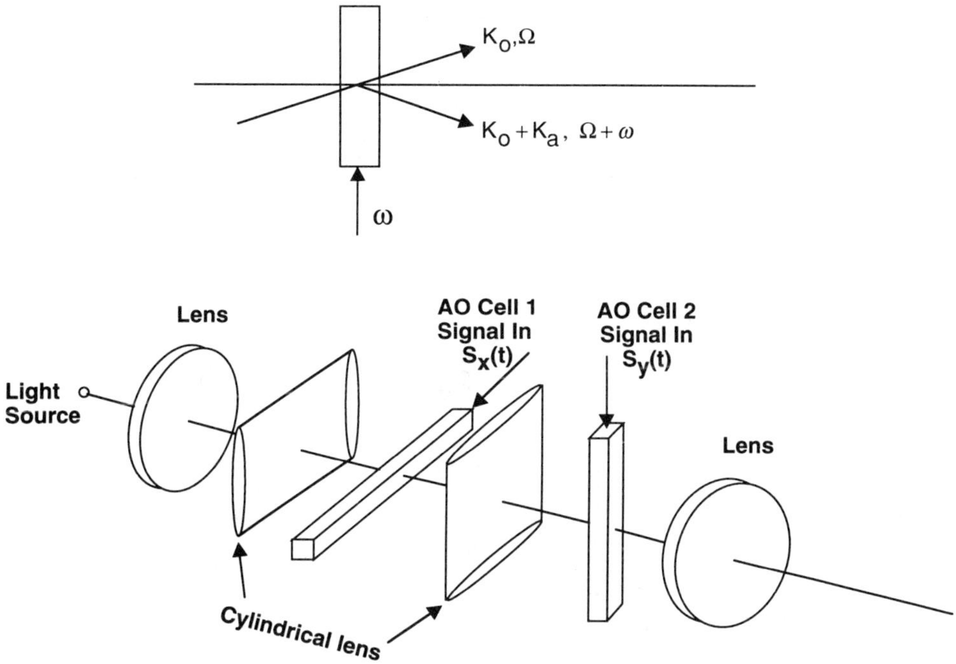


**Figure 11.6** Ray diagram of Bragg interaction (top) between an optical wavefront and an acoustic wave. Anamorphic lens configuration (bottom) to relay point source through orthogonal acousto-optic cells to create a 2-D light distribution related to the two input electronic signals.

and have an observable change in the output wave number is for $\mathbf{K}_{in}$ to undergo a rotation. Thus, the vector diagram is very nearly an equilateral triangle. The acousto-optic cell can be an extremely linear device in the performance of the above mixing operation.

Since an acousto-optic cell is intrinsically a 1-D device, it is necessary to operate two such devices, crossed and in tandem, to perform 2-D functions (e.g., 2-D beam tilts). Figure 11.6 shows the principle. It can be seen that anamorphic lenses (lenses with differing powers in different lateral directions) are used to efficiently couple to the intrinsic 1-D nature of these devices. Using such techniques, one can synthesize optical functions with a separable kernel (e.g., a spherical lens).

Tandem acousto-optic cells are also used to counteract the simultaneous modification of both the direction of light propagation $\mathbf{K}$ and angular frequency $\Omega$ in a single cell. By choosing tandem cells, the first cell can alter the light direction, and increase the light frequency. The second cell can augment the light deflection while decreasing the light frequency (Fig. 11.7). Using the same frequency to drive both cells, a tandem arrangement can yield a purely deflected light beam and act like a 1-D mirror. Using four cells, two tandem pairs which are crossed, one can obtain arbitrary 2-D light beam deflections. For

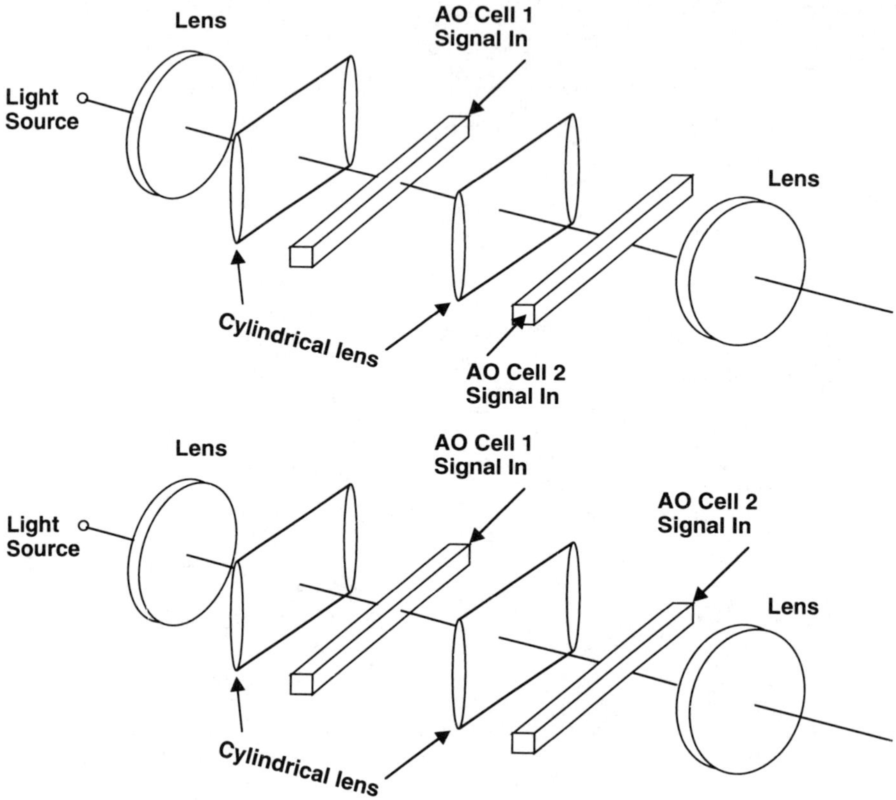


**Figure 11.7** Driving signal configuration for tandem, parallel acousto-optic cells to either double the deflection angle (top) or double the frequency shift (bottom).

light deflection, the frequency of operation of the cells is of secondary importance in that it primarily influences the cell fill time. Thus, the most efficient cells ($TeO_2$) are preferred. Independent deflections can only be obtained as fast as the cell information can be refreshed. If the time to change the deflection direction is important, then high-frequency cells may be employed. Higher frequency cells tend to be disproportionately lossy and show a reduction in the time-bandwidth product (*BT*) available. The *BT* of the cell governs the precision with which, for instance, beam directions can be resolved.

The use of identical control frequencies in tandem cells results in a single output beam at the input light frequency which has been deflected by both cells. An alternative driving configuration can be used with tandem cells such that the frequency offset augments while the beam deflections cancel (Fig. 11.7). This then results in a single output beam at an optical frequency plus twice the control frequency. Driving the cells with nonidentical control frequencies results in a partially deflected, partially frequency-shifted output beam.

Driving the second of the tandem cells with a control signal consisting of two frequencies will result in two output light beams differing in frequency by the difference in the two applied frequencies. This latter case is useful for generating a cluster of closely spaced microwave beams, as in monopulse beam forming.

## 11.2 HARDWARE REALIZATIONS

Optical subsystems that perform the spatially dependent microwave frequency signal conditioning to accomplish beam forming with an array can be implemented on a phase basis, where maximum time delay between any of the radiating element signals is on the order of a wavelength, or on a time-delay basis, where maximum time delay is on the order of the width of the array. The major distinction in terms of the beam forming is that the first, modulo $2\pi$ phasing displays an instantaneous bandwidth inversely related to the steering angle from broadside (see (11.4)), whereas time delays permit infinite bandwidth waveforms in any direction. A spread-spectrum signal might, for instance, hop over an octave band at a high rate but exhibit only a small instantaneous bandwidth. Instantaneous bandwidth is that bandpass which, at a particular instant, will pass all the information in the signal. For the hopper signal, that bandpass center would have to move at the hopping rate to avoid degradation of the signal over time. Thus, the agile bandwidth of a radar carrier might be orders of magnitude greater than the instantaneous bandwidth of the envelope signal. For a phase beam steering system, the frequency changes in a hopping signal would necessitate a change in the phase settings for the array elements between hops. This might be a reasonable thing to do on transmit, but would require considerable synchronization to accomplish in a receive mode. Generally, the hardware used for modulo $2\pi$ phase shifters is considerably simpler and cheaper than that used to implement time delays. The following section will cover a number of configurations that can be used for narrowband (i.e., modulo $2\pi$) beam steering systems and the next will present time-delay concepts.

### 11.2.1 Differential Phase Systems

There are several distinct optical methods of synthesizing the set of signals to apply to the elements in a 2-D array. These are distinguished primarily in the devices used to effect the phase distribution desired. They have in common the use of heterodyne techniques to demodulate the microwave signal from lightwave carriers, the use of an array of fibers to divide the light among the elements, and a division of the laser light into two paths. Direct modulation of a laser is rarely used, since only limited dynamic range is possible before intermodulation products adversely affect the waveform. A Mach-Zehnder interferometer structure is usually employed, although inline beams can be used (e.g., using polarization to distinguish the two beams rather than an interferometric division into completely distinct paths).

Figure 11.3 shows the basic Mach-Zehnder structure. A variation of the canonical lens configuration is assumed, in which variable spacing is allowed in collimated regions to accommodate any path length differences needed. Beam tilts can be achieved by a rotation of any of the beam splitters or mirrors. Electronic beam tilting could use acousto-optic cells (tandem or not) [3], electro-optic modulators [5], or electrically addressed SLMs [6]. In the acousto-optic case, without frequency-compensating tandem cells, different beam directions would correspond to different frequencies. This effect would make pattern synthesis quite difficult, since, for instance, weak beams directed appropriately to null out a troublesome sidelobe would not have the same frequency, and hence would not cancel as intended. With an SLM in the beam tilt leg and with a single pixel passing light (all other pixels being in the nontransmissive state), beam direction can be controlled without alteration of carrier frequency (see Fig. 11.8) [6]. Figure 11.8 also shows a modification using a reflective SLM. These masking techniques are quite wasteful of light power, but would form beams. They suffer from the problem of requiring high contrast ratios between the ON-OFF states of the pixels, since even small light contributions from all the other pixels in the control plane might contribute light energy commensurate with that from the control pixel. This would result in much energy being deposited in the far-out sidelobes and thereby degrade the gain of the beam in the desired direction. All these techniques effectively decouple the **K** and $\Omega$ terms in the light phasor (i.e., the spatial and the temporal modulation). It does so in such a manner that even though the magnitude of **K** remains $\Omega/c$, the direction may be controlled at will.

It is possible to use an inline interferometer in a canonical-like lens configuration to provide the signal conditioning to drive an array [8]. At frequency planes, the diffracted and undiffracted light beams are usually well separated. Frequency plane operation can then be used to change the direction, frequency, or amplitude distribution of the two beams. This technique, although more robust with mechanical perturbations of the interferometer, requires the entire microwave offset to be accomplished with wideband acousto-optic cells. This may cause problems at frequencies much above 4 GHz.

Instead of dividing the output of a single laser in a Mach-Zehnder circuit, two diode-pumped lasers with thermal tuning of the frequency difference may be combined to obtain a very-high-purity microwave signal. Use of two lasers is very advantageous for very-large-frequency differences and has been demonstrated at greater than 30 GHz. At these frequency offsets, direct modulation of laser diodes is quite inefficient. Figure 11.9 modifies the interferometer design to do this. This circuit decouples to a large degree the process of providing the microwave carrier and that of envelope modulation of the microwave signal. It shows a three-beam beam former using three diode-pumped Nd:YAG lasers, which can generate two completely independent (or dependent) microwave beams from the array. Many more beams are feasible by coupling in additional lasers through additional beam splitters. Two of the lasers are connected to the reference laser by frequency-locking circuitry [9]. Each frequency-locking circuit consists of a reference CW microwave signal generator, a discriminator, and a laser cavity tuning mechanism. These phase-locked-loop circuits control the dimensions of the laser cavity to maintain lock with a microwave

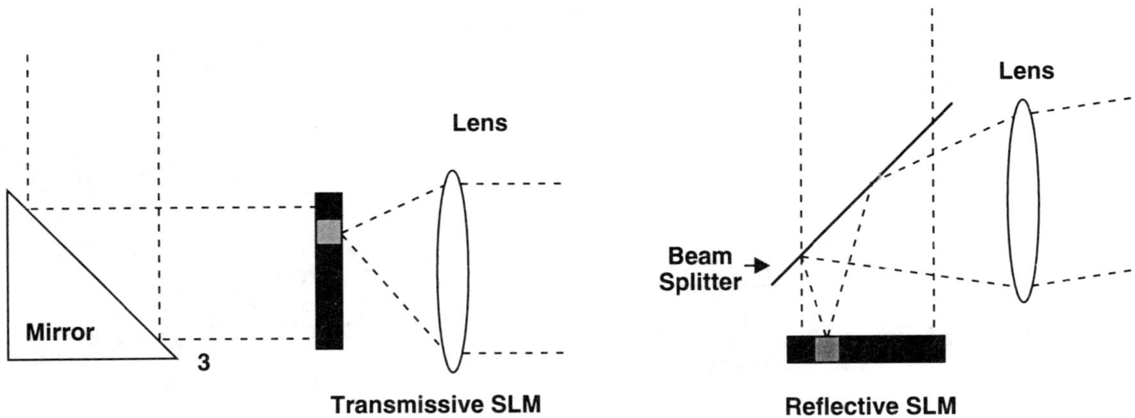


**Figure 11.8** The use of SLMs to steer optical phase fronts.

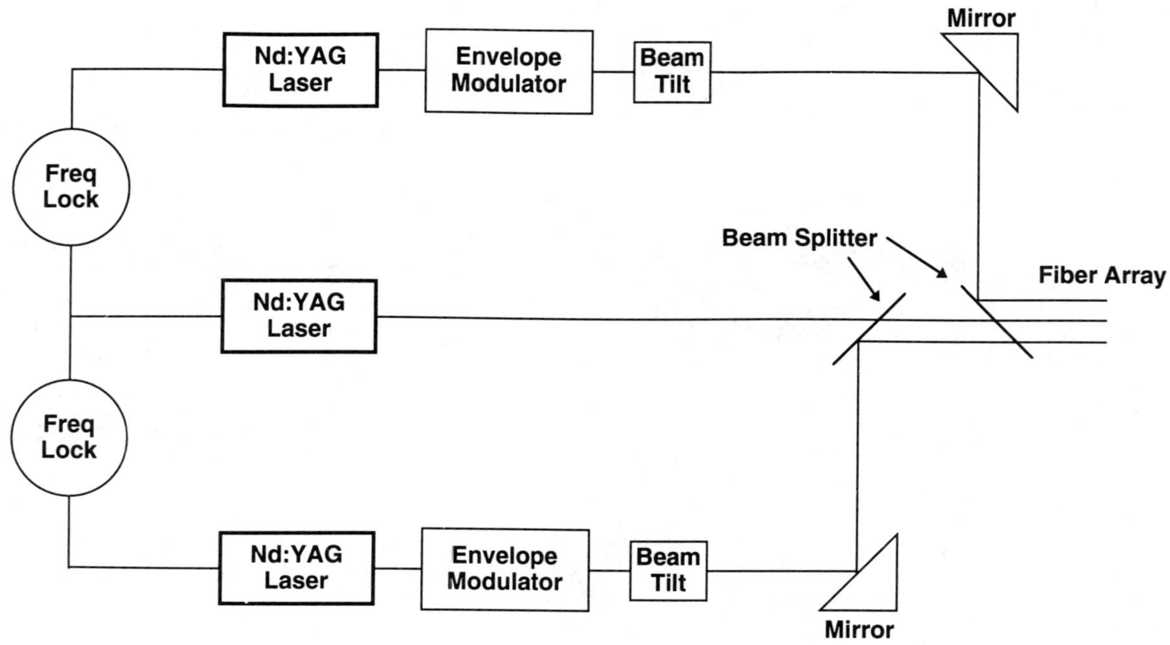


**Figure 11.9** Modified and extended Mach-Zehnder method of superimposing multiple optical wavefronts to allow multiple beams from an active array with arbitrary frequencies and beam directions.

offset frequency. It results in an extremely stable output microwave signal with low FM sidebands on the order of –125 dBc/Hz, even without the use of a stalo. The microwave carrier difference frequency is maintained by the phase-locked-loop circuitry while the envelope modulation is set by an optical frequency modulator. This permits optical devices to work at much lower modulation frequencies for the envelope, yet allows a carrier frequency from very low frequencies to millimeter waves. The solid-state lasers also produce very significant power levels, allowing the system to display a considerable dynamic range by reducing the normally expected conversion loss associated with impressing a microwave signal onto an optical carrier. Ultimately, the dynamic range may be limited by the maximum intensity of light that can be transmitted through fibers without driving the fiber into nonlinearity. The use of multimode fibers raises that limit. Usually, multimode fibers are permissible in the control of arrays with separation from the control area by only a few hundred meters.

The receive operation for arrays may be carried out in a number of distinct ways. Retaining multiple-beam capability with high dynamic range restricts the choices, however. The elemental signal at each array element must be phase-shifted and summed coherently to achieve the array gain. One method of doing this is to insert a mixer into the receive section of the T/R module so that a "steered" LO can transfer its relative phase to each of the elemental signals. The output of all the elements will be in phase from a given direction if the negative phase gradient applied for transmit is used to drive the LO. This mixer should be preceded by a preamplifier so that the conversion loss of the mixer will not degrade the SNR of the system. The mixer should be at least a doubly balanced type so that the third-order intermodulation intercept point can be as high as possible. Similarly, the preamplifier should have relatively low gain and the highest intercept possible. Since the intermodulation products will tend to be incoherent from module to module, the system dynamic range would be expected to approach the dynamic range of the T/R module plus the gain given by combining the signal from all the modules.

Figure 11.10 shows conceptual sketches of a conventional active aperture T/R module and of an optical T/R module which would permit high dynamic range and multiple-beam reception. Not shown are the manifold which divides the input signal among the many elements, the digital control lines which carry information on the state of the system, cooling structures, power supplies, and the like. The main distinctions between the two are in the replacement of the digitally controlled phase shifter with a high-dynamic-range mixer and, perhaps, the inclusion of an additional stage of amplification in the optical version. The output of the first stages of amplification is used as the local oscillator for the mixer. Thus, only the T/R state of the system is needed to convert the optical module from transmit to receive, as opposed to the phase shifter and calibration information needed by a conventional digitally controlled device. Multiple beams can be channelized onto the IF (intermediate frequency output of a mixer) by the proper choice of local oscillator carrier frequencies. The spectrum of the multiple-beam LO would show a series of "rails" or delta functions, each of which might have a distinct phase gradient from element to element. To convert a conventional module into a multibeam module

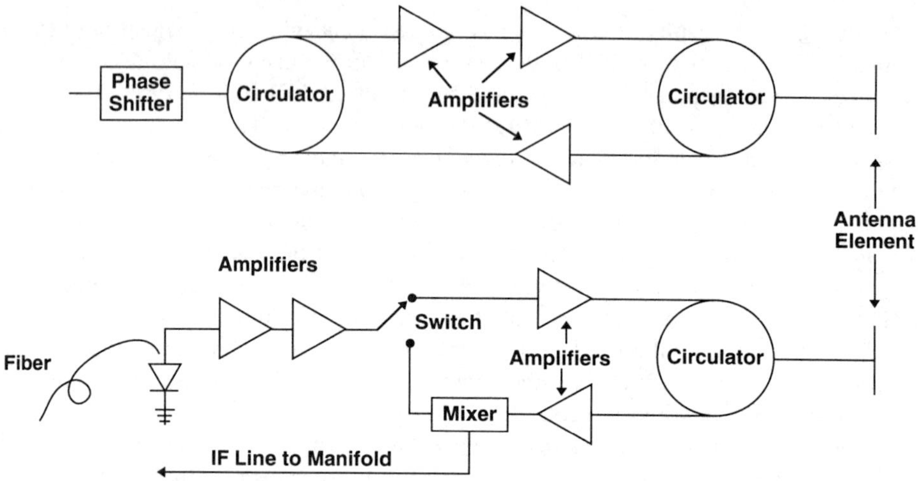


**Figure 11.10** Comparison of basic T/R module components for an all-microwave module (top) and an optical alternative (bottom).

would entail the use of a separate phase shifter in each module for each beam and a distinct manifold for each beam.

### 11.2.2 Time-Delay Systems

Total phase or time-delay beam steering systems for arrays are generally more difficult to build in terms of cost and complexity than are the narrowband phase systems discussed in the previous section. A useful technique that avoids building a time-delay unit (TDU) for each element involves the process of partitioning. One can partition in either of two ways: (1) the frequency band being time-delayed can be subdivided into subbands; (2) the array can be subdivided into subarrays.

A TDU driving each subarray or an independent control signal for each subband will allow broadband array beam forming. In either case, partitioning into 10 subarrays/subbands would increase the instantaneous bandwidth tenfold for a 1-D array, whereas only subbanding will effect a tenfold increase with 2-D arrays.

Many conventional time-delay beam forming techniques are of a line-length, switched nature. The intrinsic bandwidth of the array is usually sufficiently large so that required bandwidth is only a few times larger. With microwave-only techniques, feeding an array with a space-fed structure also provides time-delay beam forming in the primary direction of the space-fed beam. For instance, switching between feed ports of a Rotman lens will provide total phase beams at other than broadside. There are, however, several optical implementation schemes that do not require switching between $N$ separate ports, yet achieve time-delay beam control. For example, it is possible to use only a few beams from those available to transmit or receive a signal of sufficient bandwidth to provide target size identification capability for a search radar.

Acousto-optic cells were used in early computers as digital signal delay lines. Electrical transducers were attached to both ends of a cell and the acoustic transit time constituted the delay. Figure 11.11 shows a method of achieving variable delay using an optical variation on that approach. The object plane acousto-optic cell is imaged to the output image plane. A fiber (probably a single-mode fiber to avoid spatial averaging) samples the signal there and reproduces the signal waveform upon detection. By moving the fiber sampling probe to a different point in the image plane, a different delay will be obtained. Alternatively, a mirror may be inserted and rotated to reposition the image, thereby modifying the temporal delay.

There is a significant advantage to relocating the above fiber-optic pickup from an image plane to a frequency plane. In this case, a larger area of the illuminated plane may be used. All of the light contributes to the regeneration of the signal at a given instant rather than being distributed over the whole image. The time delay is determined by the relative phasing of the frequency components of the signal. This procedure can improve the SNR of the reconstructed signal by orders of magnitude. A rotation of the mirror will create a delay, not by a shift in the image position, but by the introduction of a phase gradient in the spectral information. The maximum delay is obviously still limited by the total delay in the Bragg cell.

Mirror rotation has been exploited to obtain a multiple set of time delays for a given input signal using a structure conceptually similar to Figure 11.8 [7]. A segmented mirror device (SMD) with independent controls for each submirror is used to tilt the dc term from the above acousto-optic cell. Since each submirror of the mirror is completely independent of its neighbors and the mirror is imaged onto the output fiber array, adjacent fiber-optic signals can be completely independent in terms of time delay. An ordered

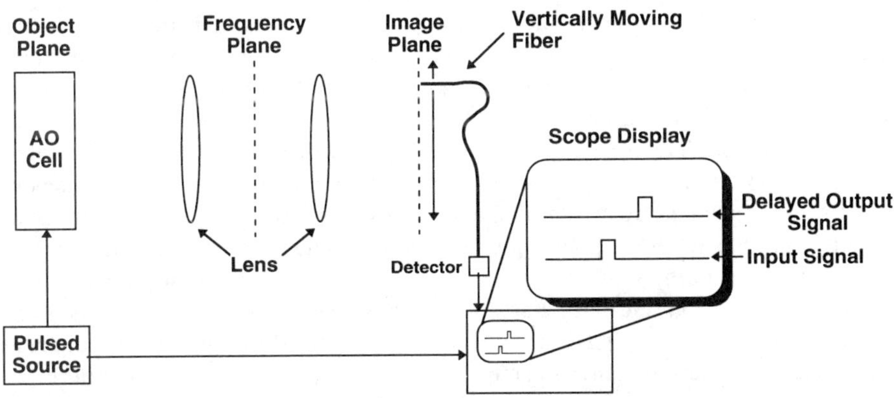


**Figure 11.11** Variable time-delay optical system. Introduction of a beam tilting mechanism between the object and image plane will allow variation of signal delay.

arrangement from the SMD would be useful in beam steering, while the arbitrary time delays would have application to many signal processing operations.

One of the classic methods of steering a microwave beam in one dimension is to change the frequency driving a tapped delay line, the taps feeding elemental antennas. A change in frequency alters the phase gradient at the taps and therefore changes the direction in which the radiated beam constructively reinforces. This technique can be performed optically by using dispersive delay lines (special fibers) such that a change of the light frequency feeding the fiber alters the delay of a microwave modulation impressed on the light carrier [11]. Each fiber is composed of a length of nondispersive fiber and a length of dispersive fiber, the ratio varying from element position to element position. Present dispersions and lasers permit delays of the order of a fraction of a nanosecond for each 100m of fiber length.

The use of separate beams to transmit contiguous fractions of the wideband signal is a straightforward method of using multiple-beam beam formers to achieve time delay. A microwave filter can be used to divide the signal into subbands on the order of the intrinsic bandwidth of the array. The subbands will not undergo severe angular spreading when radiated from an array. Each subband, when treated as a separate beam and given the proper phase gradient, will superimpose in the far field at the desired angle from array broadside. In this way, the entire frequency content is radiated in the desired direction. An added benefit of this superposition process is that the aperture dispersion disperses the sidelobes of different frequency beams and thereby tends to diminish sidelobe response.

## 11.3 NEW OR ENHANCED CAPABILITIES

The previous sections have outlined a number of photonic circuits which can be used to perform beam forming with microwave array antennas. These techniques operate on the element signals in parallel and allow the merging of the processes of beam forming and signal processing. They offer a potential for implementations that are remote, smaller, lighter, and more flexible than conventional means. In addition, they offer capabilities which can only with difficulty be implemented with purely microwave components. These novel features are discussed below.

A photonic implementation permits a multiplicity of independent signals with arbitrary relative directions to be transmitted (received). To the degree that the independent frequencies are separated by greater than the intrinsic bandwidth of the array at the angle considered, the beams corresponding to different frequencies can be orthogonal (i.e., display small cross-correlation). This is a mode of operation not easily configured in a conventional realization. Of particular importance is the case where the frequencies are contiguous and separated by the intrinsic bandwidth. This allows multiple frequencies to be broadcast in a given direction, thereby synthesizing a wideband beam. Alternatively, if the signals are at the same microwave carrier frequency, but the steering phases are distinct, the antenna pattern will have a multilobed appearance or have an adaptive null formed at some angle.

Another mode of operation involves a time-dependent variation of the steering angle during transmit. This can be done because there are no phase shifters or switches changing state discontinuously while power is flowing. A wideband chirp waveform can be steered in a given direction with all the frequency components transmitted in that direction. This can be accomplished dynamically during the transmit mode as an alternative to allocating additional beam degrees of freedom to accommodate a wideband waveform.

Most of the photonic techniques discussed here employ a fiber-optic array that resembles a scaled version of the microwave array being driven. This feature may be extended to provide global control to arrays which are quite distinct from the filled, uniform lattice arrays normally envisioned. In fact, aperiodic, sparse, conformal, and disjoint array configurations may be steered, although some degradation of beam form may occur because the optical signal conditioning does not include mutual coupling effects. For an element embedded deeply in a uniform section of the array, the effects of mutual coupling are of little or no import. Near an edge, however, the variation of mutual coupling or active impedance may significantly impact the required phase and amplitude for optimum beam forming. A priori or open-loop compensation for such effects could be considered. This is an area that will deserve attention upon the construction of a photonically driven microwave antenna of significant size and quality.

Photonic methods are largely analog in nature, both in terms of beam pointing and the signals provided on reception. The beam pointing feature is very useful, allowing infinite resolution in the beam direction and for beam synthesis. Handling analog signals is normally the bane of optical replacements for microwave equipment. In these applications, however, which are largely of the nature of synthesizing spectrally pure microwave signals using optical techniques, the optical procedures demonstrate quite useful dynamic ranges. It is more than likely that the preamplifier or the mixer in the T/R modules limits the dynamic range rather than the signal synthesis by optical heterodyne techniques.

The multiple simultaneous beam capability has an impact on the capabilities of photonic implementations in monopulse measurements and in low-sidelobe operation. Each beam of the multiple degrees of freedom can consist of a cluster of beams for monopulse operation. Instead of having the monopulse implemented by sums and differences in the manifold structure, overlapping beams can be used to convert SNR into an improvement in the estimate of a target position within a beam. This may require a frequency channelization of the different beams into the IF manifold, which connects the array to the receiver.

Transmit signal conditioning must be of an open-loop nature, but the receive mode could be closed-loop and adaptive. The low-sidelobe behavior can be further improved by using adaptive beam forming to squelch a few of the peak sidelobes which might fall on nearby clutter of significant magnitude. Another utility for multiple, independent, simultaneous beams is to mask sidelobes of a beam with a covering beam operating at a different frequency or modulation. This technique allows one to manage a communications link to avoid intercept or to deny a jammer the information about the present operating frequency. Low-sidelobe performance can be imposed on both transmit and receive.

Partitioning the band can yield additional beam forming properties. It is not necessary that the radiated frequencies of the subbands be either identical or even proportional to that given by the partitioning microwave filter. They could be given arbitrary carrier frequencies within a given octave. The relative phases of the spectral subbands, as well as their radiated frequencies, could be altered to modify the waveform reconstructed in the far field. This transformation could be undone on reception. Thus, by a scrambling of the waveform spectra, the waveform could be obscured. On receive, the separately transmitted components could be frequency-channelized to either contiguous or noncontiguous subbands in the IF band. A spectral signature of the target could be obtained from the individual channels. The IF channels could be subsequently converted to the original waveform bandwidth at a second IF. Another alternative would be to superimpose all the components in the same IF subchannel. This would degrade resolution but would coherently add in the narrowband channel for improved detection capability. The octave bandwidth constraint avoids intermodulation products out of the heterodyne detectors. With the IF band at a frequency greater than the transmitted center frequency, frequencies spanning in excess of an octave could be used. In this manner, a simple pulsed waveform could appear at the target to be a continuous carrier waveform and the beam former would function additionally as a pulse compression network or as a pulse stretching network.

## 11.4 SUMMARY

The bottom line on the acceptability of any proposed implementation of an array antenna system will be the aggregate effect on system performance, acquisition cost, life cycle cost, maintainability, weight, size, immunity to microwave and ionizing radiations, lightning or EMP, robustness, and the multiple-beam capability intrinsic to the architecture. This chapter has illustrated how photonic alternatives to conventional array beam forming techniques give rise to significant additional capability for radar or communications systems. These are in addition to that capability normally associated with electronically steered beams. Among the most important of these photonic capabilities is the ability to generate multiple simultaneous, independent/dependent beams with nearly arbitrary frequency relationships. The ramifications of multiple beams are manifold. It becomes possible to implement very powerful techniques in beam synthesis; including broadband nulls, multiple channels for data links, open and/or closed loop nulling, spoofing. The net effect is that the additional degrees of freedom allow the incorporation of signal processing operations with the beam forming, thereby opening up a plethora of potential new possibilities in remote sensing, target identification, electronic counter measures (ECM) and electronic counter-counter measures (ECCM). At the same time, the optical alternatives allow a simplification of the active T/R modules, which should reflect itself in reduced cost, improved reliability, and reduced power consumption at the backplane of the array. The different transmission line, fiber-optic versus microwave cable or waveguide, alters the susceptibility to electromagnetic fields and interference. This is all achievable with a reduction in size, weight, and with the ability to remotely locate the controlling system. Thus, photonic implementations should do well in future active aperture systems.

## REFERENCES

[1] Cheston, T. C., and J. Frank, "Phased Array Radar Antennas," Ch. 7 in *Radar Handbook,* M. I. Skolnik, ed., New York: McGraw-Hill, 1990.

[2] Yariv, A., *Optical Electronics,* 3rd edition, New York: Holt, Rinehart and Winston, 1985.

[3] Stilwell, D., M. Parent, and L. Goldberg, "Fiber Optic Feed," *NRL Memorandum Report 6741,* 6 Nov. 1990.

[4] Tamburrini, M., M. Parent, L. Goldberg, and D. Stilwell, "Optical Feed for a Phased Array Microwave Antenna," *Electron. Lett.,* Vol. 23, No. 13 June 1987, pp. 680, 681.

[5] Koepf, G. A., "Optical Processor for Phased-Array Antenna Beam Formation," *Proc. SPIE,* Vol. 477, Optical Technology for Microwave Applications, Arlington, VA, 1984.

[6] Anderson, L. P., F. Boldissar, and R. Kunath, "Antenna Beamforming Using Optical Processor," *Proc. Antennas and Propagation Symp.,* Vol. 1, Blacksburg, VA, June 1987.

[7] Zmuda, H., and E. N. Toughlian, "Adaptive Microwave Signal Processing: a Photonic Solution," *Microwave J.,* Vol. 35, No. 2, Feb. 1992, pp. 58–71.

[8] Riza, N. A., and D. Psaltis, "Acousto-Optic Signal Processors for Transmission and Reception of Phased Array Antenna Signals," *IEEE Photonics Tech. Lett.,* Vol. 4, No. 2, Feb. 1992, pp. 177–179.

[9] Esman, R. D., M. J. Monsma, J. L. Dexter, and D. G. Cooper, "Microwave True Time-Delay Modulator Using Fibre-Optic Dispersion," *Electron. Lett.,* Vol. 28, No. 20, 1992, pp. 1905,1906.

# Chapter 12
# RF Fiber-Optic Links

***Edward Ackerman***
*Martin Marietta*

## 12.1 INTRODUCTION

Future generations of phased-array radar systems will require feed and distribution to many hundreds and possibly thousands of solid-state MMIC radiating elements, as was mentioned in Section II of this book. In phased arrays operating at millimeter-wave frequencies, backplane interface and signal distribution methods will need to fulfill strict performance criteria. These include greater than 60-dB isolation from both electromagnetic interference and crosstalk between module or subarray feeds, analog frequencies of operation into the millimeter-wave range with bandwidths of an octave or more, dramatic reduction in size and weight relative to present fielded arrays, and minimal insertion loss.

The metallic waveguides and coaxial cables currently used as phased-array backplane interconnects will be unable to meet these stringent requirements. Waveguides are heavy and bulky, and coaxial cables are highly susceptible to crosstalk between adjacent lines. Coax is also extremely lossy–an 18-GHz signal, for example, attenuates 1000 dB/km in a cable with a 1-mm outer diameter. Furthermore, as higher frequency phased-array operation is pursued, element spacing will become increasingly tight, causing waveguide congestion and crosstalk at the array backplane to be serious design concerns.

At millimeter-wave frequencies, where array element spacing is tight and array backplane congestion is a major problem, distribution of the RF, control, and power signals using optical fiber offers unique advantages. These include:

- Small size (nominally 125-$\mu$m outer diameter), reducing congestion at the array backplane;
- High durability (tensile strength > $10^5$ lb/in$^2$), supreme flexibility (minimum bend radius < 1 in), and negligible weight;

- Immunity to the electromagnetic interference that pervades a densely packed millimeter-wave environment;
- Low loss (attenuation < 0.5 dB/km);
- Suitability for implementing true time delay and other beam steering and signal processing functions in the optical domain.

To realize all the benefits of optical-fiber signal distribution in a phased array, the single most critical development is the high-performance analog fiber-optic link. In this book, a *fiber-optic link* is considered a two-port electronic device in which a high-frequency analog (microwave or millimeter-wave) signal is relayed from input to output via an optical fiber. Both the input and output of a fiber-optic link are electronic; therefore, its performance is evaluated using the same ''figures of merit'' used for many other electronic devices, such as amplifiers. This chapter focuses on fiber-optic links for RF signal transmission in a radar phased array.

## 12.2 FIBER-OPTIC LINK DESIGN CRITERIA

Figure 12.1(a) shows a block diagram of a fiber-optic link. At the input to this network (port 1), a microwave or millimeter-wave signal modulates the intensity of the optical carrier launched into an optical fiber; at the other end of the fiber, the modulating signal is retrieved from the optical carrier and electronically coupled to the output of the link (port 2). Analog performance of the link, as for any other microwave or millimeter-wave two-port device, can refer to any of several figures of merit. Usually, among the most critical of these are the insertion gain or loss ($G$), the noise figure (NF), the compression-limited dynamic range (CDR), and the spurious-free dynamic range (SFDR).

Figure 12.1(b) illustrates the meaning of these performance parameters. The insertion gain $G$ is defined simply as the ratio of optical receiver output power $P_{\text{out,RX}}$ to optical transmitter input power $P_{\text{in,TX}}$ at some signal frequency or frequencies (given as $f_1$ and $f_2$ in the figure); thus,

$$G \equiv \frac{P_{\text{out,RX}}}{P_{\text{in,TX}}} \tag{12.1}$$

The noise figure NF is the extent to which the signal-to-noise power ratio ($S/N$) degrades between the input and output of the optical link:

$$\text{NF} \equiv \frac{(S/N)_{\text{in}}}{(S/N)_{\text{out}}} = \frac{N_{\text{out}}}{N_{\text{in}}G} \tag{12.2}$$

The noise power at the input of the link, $N_{\text{in}}$, is simply the thermal noise k$TB$ (where k is Boltzmann's constant, $T$ is the absolute kelvin temperature, and $B$ is the instantaneous

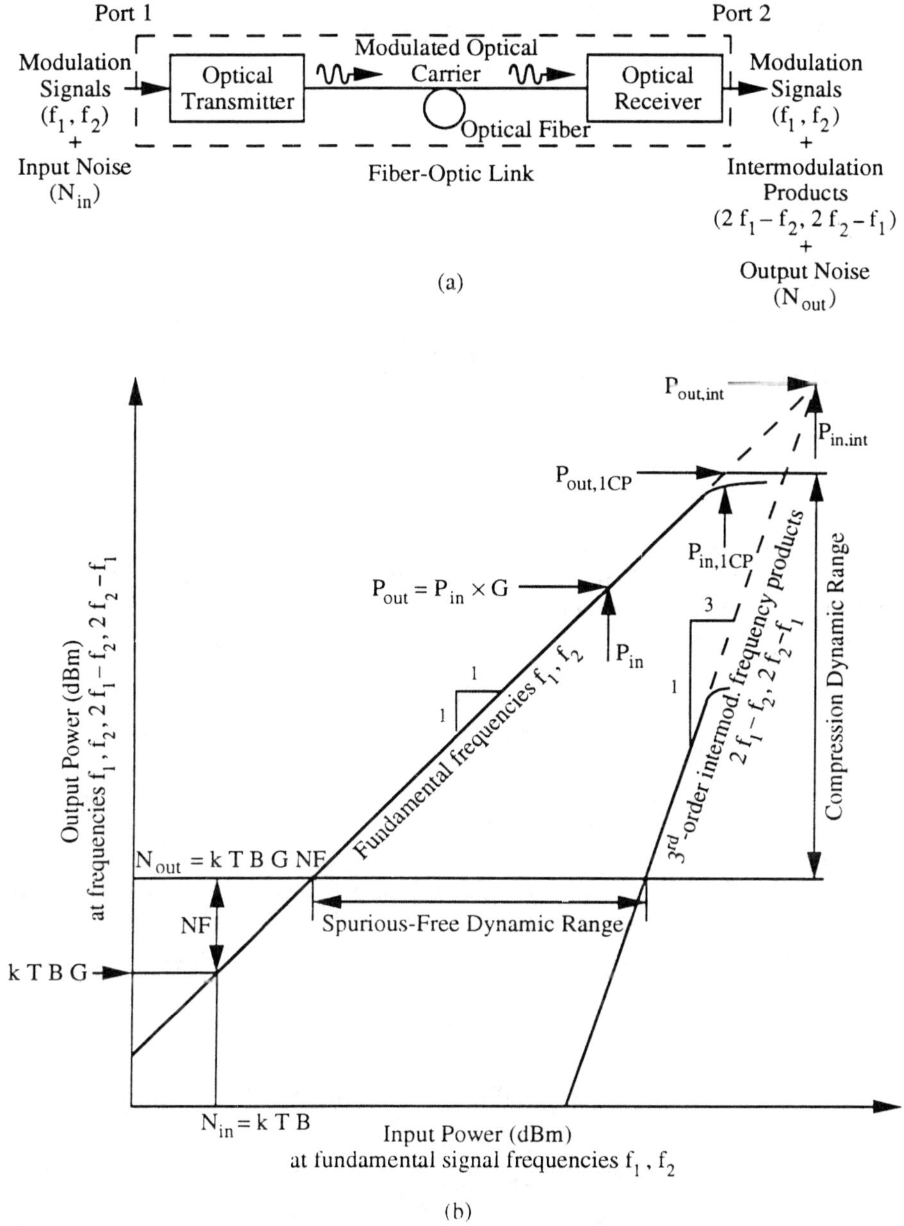


**Figure 12.1** (a) Generalized block diagram of a fiber-optic link for microwave/millimeter-wave communications applications; (b) input-output signal relationship for a fiber-optic link. Gain, noise figure, compression-limited dynamic range, and spurious-free dynamic range are all depicted graphically.

receiver bandwidth). Therefore, the insertion gain and noise figure determine $N_{out}$, the noise power at the output of the link (see Fig. 12.1(b)):

$$N_{out} = N_{in}G\text{NF} = \text{k}TBG\text{NF} \tag{12.3}$$

The output noise determines the low end of the range of usable signal power at the output of the link. The upper end of the dynamic range is limited by the nonlinearity of the active devices in the link, which, for large input power, causes both compression of the output signal amplitude and generation of unwanted signals at harmonic and intermodulation frequencies. The compression-limited dynamic range is here defined as the range of input signal power for which the output signal power is above the noise floor $N_{out}$ and is compressed by less than 1 dB relative to the small-signal response; thus,

$$\text{CDR} \equiv \frac{1.259 P_{out,1CP}}{N_{out}} = \frac{P_{in,1CP}}{\text{k}TB\text{NF}} \tag{12.4}$$

where $P_{in,1CP}$ and $P_{out,1CP}$ are the input and output signal power at the point where the output power at fundamental frequency $f_1$ is compressed by 1 dB (a factor of roughly 1.259). From a radar perspective, a more useful means of specifying the dynamic range of a system is to determine the range of input powers at some fundamental frequency for which all spurious signals at harmonic and intermodulation frequencies are below the noise power at the output of the link. The highest level and most problematic spurious signals present at the output of a link are usually the third-order intermodulation products resulting from the nonlinear mixing of two closely spaced components of the fundamental signal. That is, if the spacing of two fundamental frequencies $f_1$ and $f_2$ is smaller than the receiver resolution bandwidth B, then it is likely that one or both of the third-order intermodulation product frequencies ($2f_1 - f_2$ and $2f_2 - f_1$) will also be present at the receiver. Defining $P_{in,int}$ and $P_{out,int}$ as the input and output power at the third-order intercept (where the fundamental signal and third-order intermodulation product signal have equal output powers--see Figure 12.1(b)), the spurious-free dynamic range is given by the following expression:

$$\text{SFDR} \equiv \left(\frac{P_{out,int}}{N_{out}}\right)^{2/3} = \left(\frac{P_{in,int}}{\text{k}TB\text{NF}}\right)^{2/3} \tag{12.5}$$

## 12.3 RF-IF MIXING ARCHITECTURES

Some radar systems have stringent transmit and/or receive performance goals which may not be easily met with conventional fiber-optic links. When a phased array is operating in transmit mode, the links must deliver high-power, distortion-free RF waveforms to the subarrays to minimize the amount of amplification and processing that needs to occur at the module level. In the receive mode, the links must relay a wide dynamic range of signals from the array to the controller.

Various system architectures use mixers to up- and down-convert the signals at the input and output of links in an effort to optimize the overall system performance at RF frequencies. There are many methods of mixing baseband or IF information with a high-frequency LO to translate it into the desired RF band. Most of these methods are conceptually equivalent to either the central data mixing (CDM) or the remote data mixing (RDM) architecture.

Figure 12.2 shows block diagrams for the CDM and RDM architectures. The fundamental differences between these two architectures are the location at which RF-IF mixing is performed (as suggested by their names) and the number of links in each architecture. Note from the illustration that, whereas CDM requires two fiber-optic links to achieve bidirectional T/R interface to an array, RDM requires three.

In the CDM architecture (Fig. 12.2(a)), the single-sideband mixing of an IF or baseband signal (with frequency $f_{IF}$) with an LO (at frequency $f_{LO}$) generates the transmit RF signal ($f_{LO} + f_{IF}$). This signal is imposed on the optical carrier by either directly modulating the output of a semiconductor laser or LED, or by externally modulating an optical source using an electro-optic modulator. The same LO can be used to down-convert the received RF signal. Because mixing of the LO and IF signals occurs at the central processing unit, this architecture is also sometimes called *CPU-level data mixing.*

In the RDM architecture (Fig. 12.2(b)), the high-frequency LO and transmit IF data signals are separately imposed on two optical carriers and are mixed at the remote end of the link to construct the desired IF-modulated LO. Thus, the performance of the LO link need only be optimized at a single frequency ($f_{LO}$), and the IF data can be generated efficiently and with low noise by a semiconductor laser or modulator that is impedance-matched to yield high transducer gain across the lower frequency data bandwidth. This architecture can result in improved link performance, especially in cases where the desired RF frequency band is exceedingly high, broad, or both. Because mixing of the LO and IF signals occurs at the individual T/R modules or subarrays, this architecture is also known as *T/R-level data mixing.*

Currently, devices such as semiconductor lasers, electro-optic interferometric modulators, and pin photodetectors which exhibit high efficiency, low noise, and wide linear operating regimes at modest microwave frequencies (up to several gigahertz) are commercially available. (This was discussed in Section III of this book.) Therefore, at these frequencies the straightforward CDM architecture is likely to yield as good a system performance as the more complex (and costly) RDM architecture. For CDM, the maximum RF frequency of the system is limited to the maximum modulation bandwidth of the LED, laser, or modulator. Additionally, for a direct modulation link using a semiconductor laser as a modulated optical source, it is likely that the laser will be operating at or near its highest-noise frequencies, resulting in narrow dynamic range for the link.

For arrays operating at millimeter-wave frequencies above the rated bandwidths of commercially available electro-optic devices, it is expected that higher performance

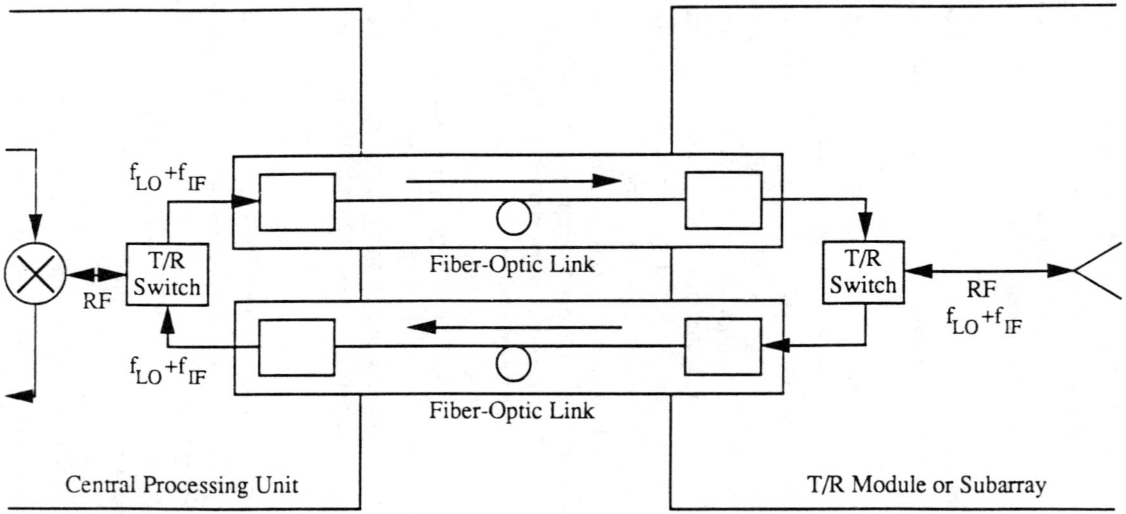


**Figure 12.2(a)** Central data mixing architecture. The single-sideband mixing of an IF or baseband signal (with frequency $f_{IF}$) with an LO (at frequency $f_{LO}$) generates and deconstructs the RF signal ($f_{LO}$ + $f_{IF}$).

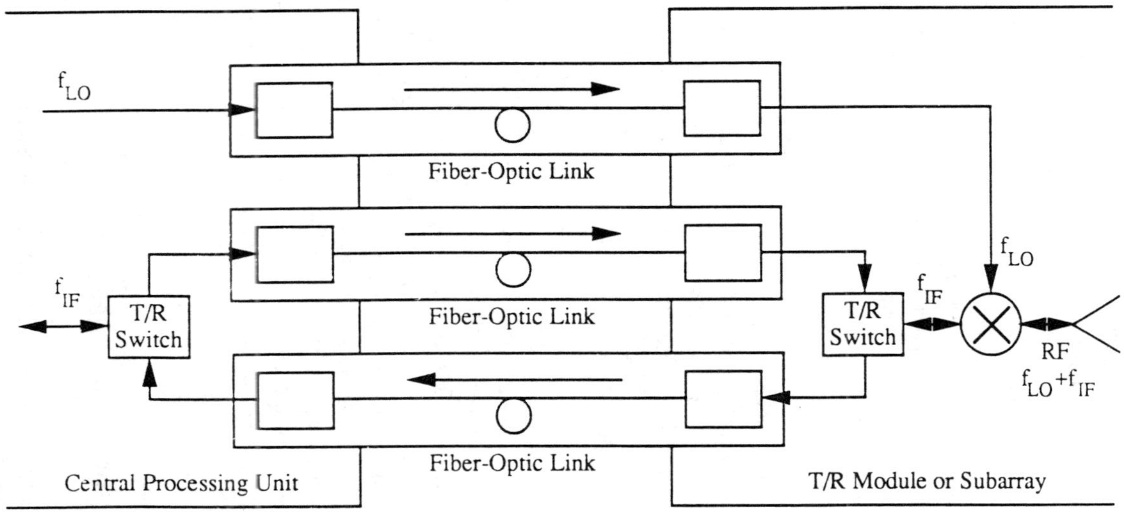


**Figure 12.2(b)** Remote data mixing architecture. The high-frequency LO and IF data signals interface at the remote end of the link separately to construct and deconstruct the desired RF signal.

interfaces will be obtained using an RDM architecture. Compared to the CDM scenario, RDM requires an additional fiber (see Fig. 12.2); also, all modules or subarrays fed by separate LO and IF links must have RF coherency established via phase locking. The payoff, however, lies in the fact that, using the nonlinearities of the laser diode, modulator, detector, or mixer, much higher modulation frequencies can be obtained. That is, at the T/R module, a millimeter-wave transmit RF signal can be constructed by mixing the optically fed IF signal with an *n*th-order harmonic of the optically fed microwave LO (using the mixer's nonlinearity to both up-convert and mix); in receive mode, a subharmonic of the RF signal from the array can be mixed with the LO to generate the IF data retrieved optically from the module. In this fashion, both the LO and IF links can be operated at modest microwave frequencies where electro-optic devices exhibit efficient, low-noise, and highly linear performance characteristics [1].

Employing the RDM scheme, Koffman et al. have demonstrated a dynamic range 22 dB greater than what was achieved using the same commercial link in a CDM architecture [1]. In further demonstrations of remote data mixing, Ogawa and Kamiya proved the viability of two RDM configurations for feeding a 10-GHz RF signal to a transmit subarray. In the first of these configurations, one of the harmonics generated in a link was filtered at the detector output and fed into a nonlinear mixer, which also served as the detector for the IF link (a MESFET or HEMT could accomplish this) to generate the RF remotely. In the second configuration, the IF data was detected with a separate photodetector and mixed in a nonlinear detector/mixer (again, a MESFET or HEMT) with one of the harmonics of the LO link frequency [2].

Several additional methods of producing carrier frequencies beyond the fundamental modulation response of laser diodes or electro-optic modulators have been proposed and investigated. These methods include mode locking of laser diodes, generating millimeter-wave beat frequencies between two wavelength-tunable lasers (heterodyning), or using the laser diode's, modulator's, or detector's nonlinearity to up-convert signals [3]. Such methods all involve feeding a signal of one frequency into the modulatable source or external modulator and retrieving a different frequency from the detector (i.e., using the link not only for the relaying of a signal, but for frequency conversion as well). The use of optical links as frequency converters requires some techniques different from those discussed in this chapter, which is concerned primarily with links having input and output signals at the same frequency.

## 12.4 RF/OPTICAL MODULATION TECHNIQUES

Selection of a technique for imposing a microwave modulation signal upon an optical carrier involves considerations of electro-optic transducer efficiency, noise, linearity, size, and cost. Modulation methods include direct modulation of semiconductor LEDs or laser diodes and external modulation of an optical source using an electro-optic or acousto-optic device.

Figure 12.3 shows a block diagram of a direct modulation fiber-optic link. This modulation method is so named because the modulating signal is impressed directly onto the dc current that forward-biases the LED or laser diode source. LEDs and laser diodes feature output optical intensities proportional to the bias current, and hence modulation of the dc bias current produces direct modulation of the optical carrier. This phenomenon is illustrated in Figure 12.4, which plots the optical power emitted from a typical semiconductor LED and laser diode as a function of the forward-bias current $I_L$. As was discussed in Section III, the laser diode behaves exactly like an LED at bias currents below the threshold $I_{\text{th}}$. This concept is also illustrated in Figure 12.4. For $I_L$ less than $I_{\text{th}}$, both the LED and laser diode have the same relationship between output optical power $P_{\text{out,op}}$ and input current:

$$P_{\text{out,op}} = \eta_{\text{LED}} I_L \tag{12.6}$$

where $\eta_{\text{LED}}$, the slope of the $P_{\text{out,op}}$-versus-$I_L$ relationship, is called the *differential quantum efficiency* of the LED. This efficiency is typically quite poor--on the order of 0.01 mA/mW. For $I_L$ greater than $I_{\text{th}}$, the LED continues to obey the relationship in (12.6), while the laser diode differential quantum efficiency $\eta_L$ increases dramatically (0.25 mW/mA or greater is typical). The optical power-to-electrical current relationship is as follows:

$$P_{\text{out,op}} = \eta_L (I_L - I_{\text{th}}) \tag{12.7}$$

In an external modulation link, an optical source (such as a semiconductor LED, semiconductor laser diode, or diode-pumped solid-state laser) generates a constant-level lightwave signal that is mode-coupled into an optical fiber. Modulation of the optical carrier is achieved via an external modulator placed on the fiber. Figure 12.5 is a block diagram of an external modulation fiber-optic link. Suitable devices to perform external modulation at microwave and millimeter-wave frequencies were discussed in Section III. Most external modulation links reported in the literature have used Mach-Zehnder interferometric electro-optic modulators fabricated in $LiNbO_3$ or GaAs [4–7]. For these external modulators, the relationship between the optical output power and the applied voltage $V_M$ is shown in Figure 12.6, where $P_{\text{in,op}}$ is the unmodulated optical power at the input of the modulator and $V_\pi$ is the voltage required for 100% modulation. $V_M$ is generalized as an RF modulation voltage of amplitude $v_m$ and frequency $f$ applied to a dc bias voltage $V_b$. It is convenient to define $V_b = 0$ as the quarter-wave bias point (where the output power is half the maximum and the $P$-versus-$V$ relationship is most linear), so that the relationship in Figure 12.6 can be expressed as follows:

$$P_{\text{out,op}} = \frac{P_{\text{in,op}}}{2}\left(1 + \sin\frac{\pi V_M}{V_\pi}\right) \tag{12.8}$$

where

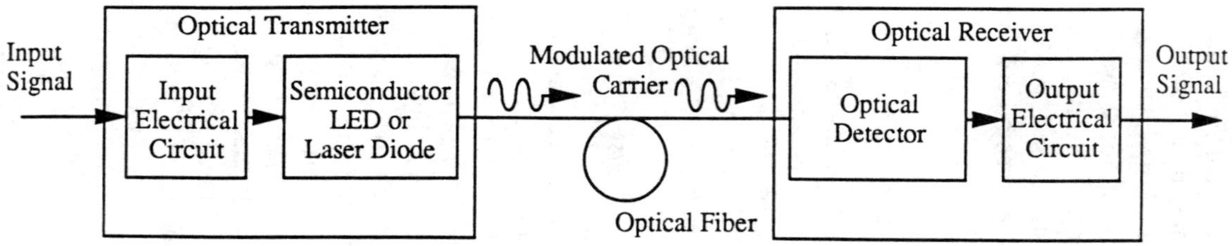


**Figure 12.3** Block diagram of a direct modulation fiber-optic link. The modulating signal is impressed directly onto the dc current that forward-biases the LED or laser diode source.

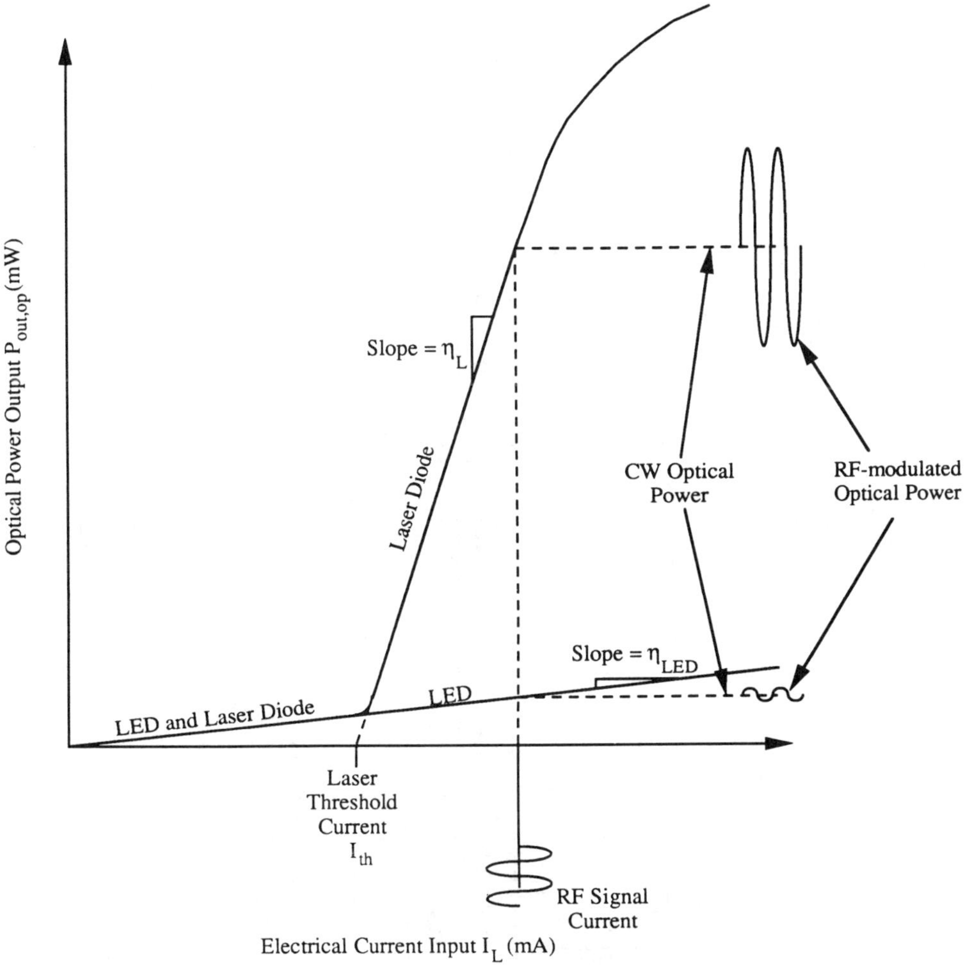


**Figure 12.4** Relationship of optical power output versus electrical current input for directly modulated semiconductor LED and semiconductor laser diode.

$$V_M = V_b + v_m \sin \omega t, \tag{12.9}$$

and $\omega = 2\pi f$.

Optical sources that can be modulated directly are generally small, with output beam sizes reasonably compatible with optical-fiber spot size. Some are reasonably inexpensive. Excellent efficiency at moderate microwave frequencies has enabled minimal insertion loss and even small levels of microwave gain to be achieved from direct modulation links [8–16]. Acceptable noise figure and high dynamic range can be achieved across substantial

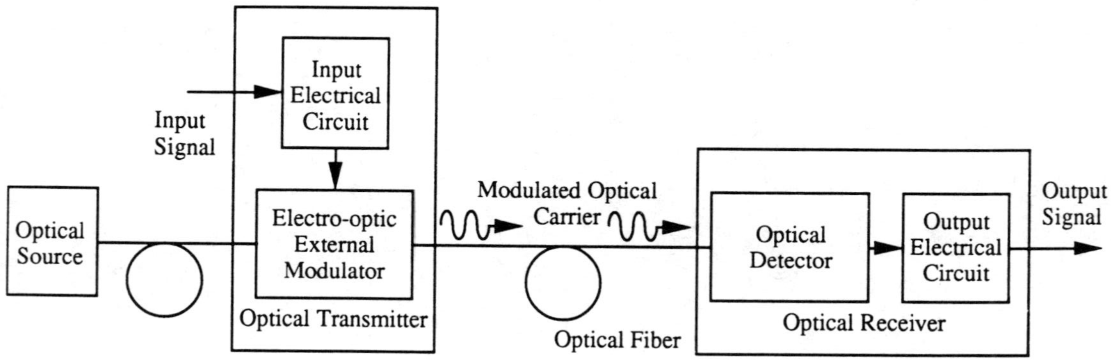


**Figure 12.5** Block diagram of an external modulation fiber-optic link. An optical source generates a constant-level lightwave signal that is mode-coupled into an optical fiber, and modulation of the optical carrier is achieved via an external modulator placed on this waveguide.

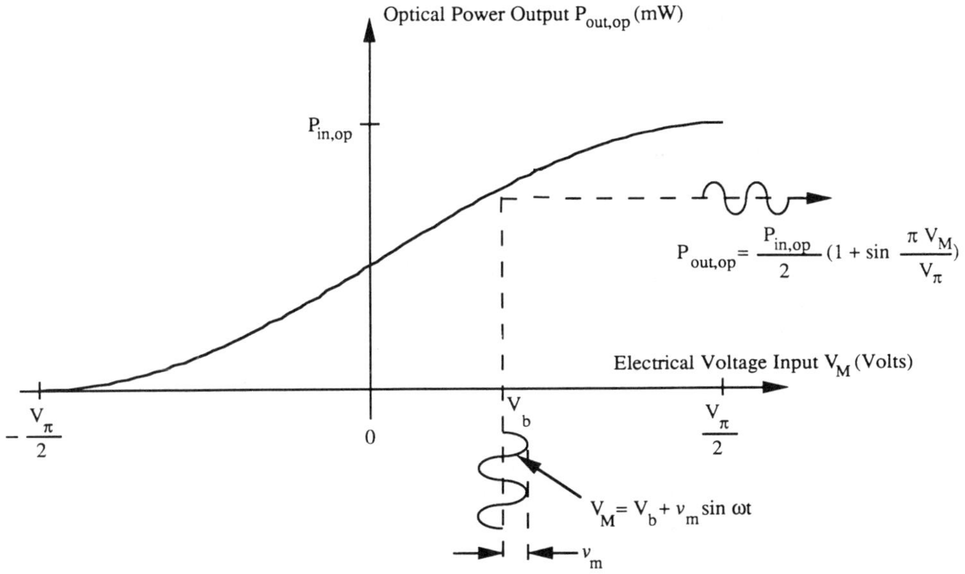


**Figure 12.6** Relationship of optical power output versus electrical voltage input for Mach-Zehnder interferometric electro-optic modulator.

ranges of frequencies when a distributed feedback (DFB) semiconductor diode laser with low RIN is used as the directly modulated optical source.

The highest frequency at which a direct modulation link can operate efficiently is limited by the relaxation oscillation frequency of the semiconductor diode—about 18 to 24 GHz for state-of-the-art MQW laser devices (see Section III and [17]). At frequencies near the laser relaxation oscillation frequency, the noisy and nonlinear behavior of the semiconductor laser—even a DFB laser—adversely affects direct modulation link performance. Semiconductor diodes also have lower optical power output and higher RIN at microwave frequencies than the solid-state lasers that can be used in external modulation links. Thus, the noise figure of a direct modulation link is generally larger than that of an external modulation link, especially at frequencies near the semiconductor laser's relaxation oscillation frequency. To reduce the noise figure of a system incorporating a direct modulation link, it is common practice to precede the laser transmitter with a low-noise amplifier with high gain.

In external modulation links, microwave modulation efficiency is dictated by the strength of the effect employed to modulate the light. When employing the electro-optic effect, as do most external modulation links [4–7], efficiency is dictated by the electro-optic coefficient of the modulator material. For both microwave efficiency and linearity goals, it is important to use a high-power source, and for noise concerns it is important that this source generate very little noise at microwave frequencies. Such concerns usually favor using a solid-state laser operating at a suitable optical wavelength, such as a diode-pumped Nd:YAG device ($\lambda$ = 1.0 or 1.3 $\mu$m), which is unfortunately much larger and more expensive than its semiconductor counterpart.

Figures 12.7 and 12.8 show the published gain and noise figure, respectively, of state-of-the-art fiber-optic links developed for operation at frequencies ranging from 60 MHz to 18 GHz [4–16].

## 12.5 OPTICAL/RF DETECTION TECHNIQUES

Detection of the RF-modulated optical carrier in fiber-optic links is usually accomplished using a semiconductor pn or pin diode. In the depletion region formed at the reverse-biased junction of the p- and n-doped layers of a semiconductor diode, an electron-hole

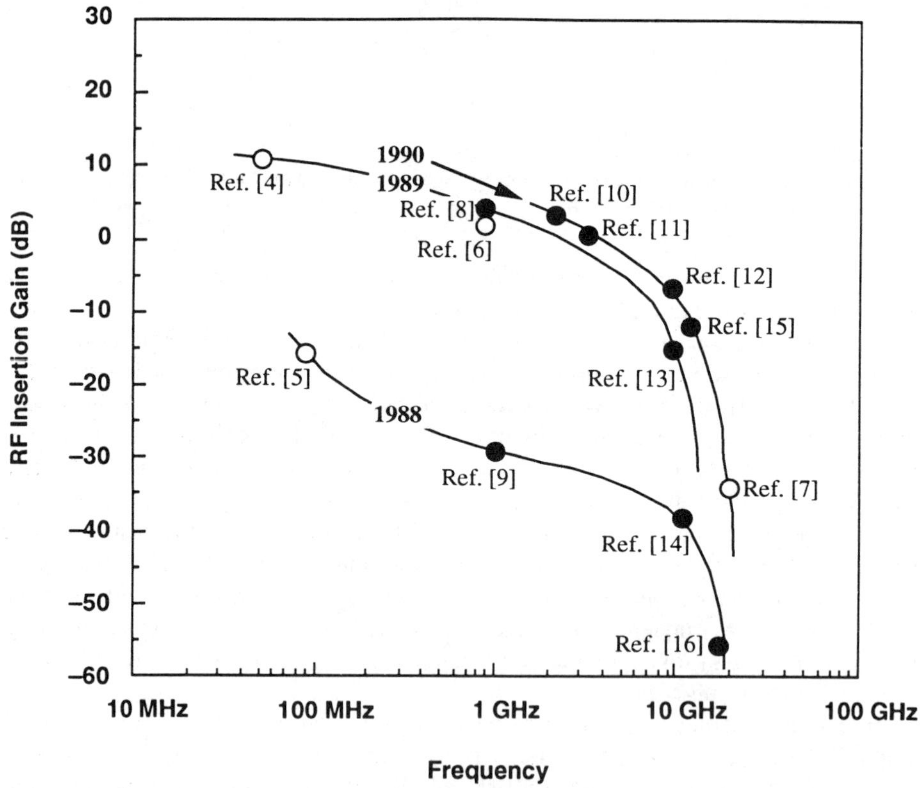


**Figure 12.7** Published insertion gain of state-of-the-art fiber-optic links at various frequencies [4–16]. Black circles indicate results for direct modulation links, and white circles denote external modulation links.

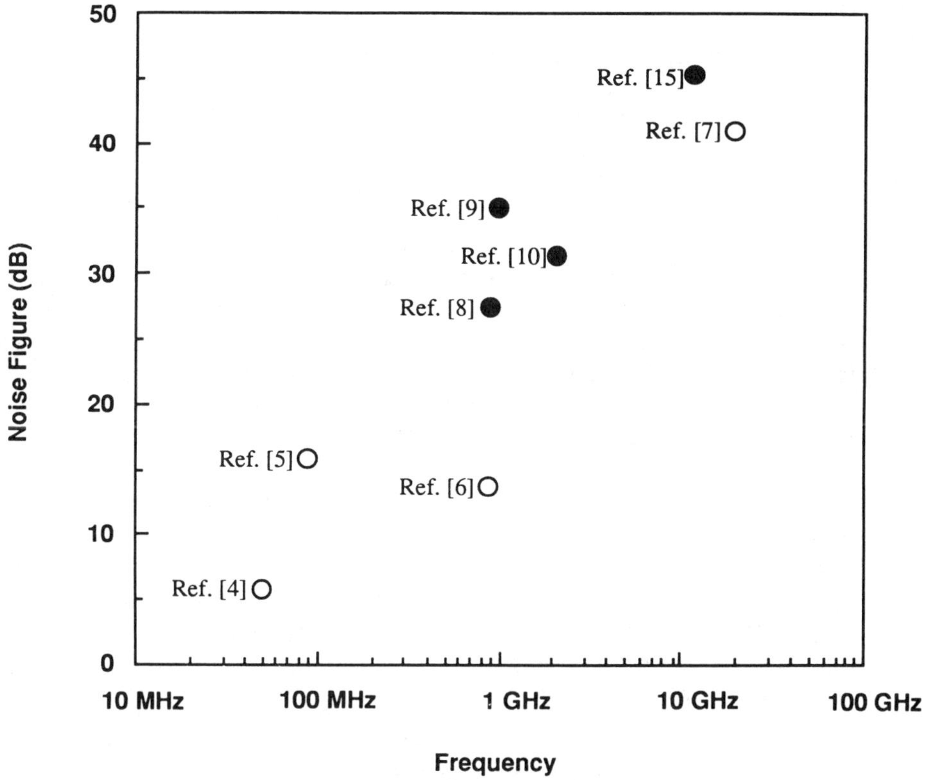


**Figure 12.8** Published noise figure of state-of-the-art fiber-optic links at various frequencies [4–10,15]. Black circles indicate results for direct modulation links, and white circles denote external modulation links.

pair is generated for each absorbed photon. This procedure is called *direct detection* because the amplitude of the photocurrent is related directly to the intensity of the incident photons. In a direct detection link, the phase of the microwave output signal is preserved; the optical phase, however, is not preserved.

If care is not taken to minimize a link's noise figure and maintain adequate optical signal amplitude throughout the fiber, it may be difficult for an optical receiver employing the direct detection technique to distinguish the optical signal from the noise floor. This is especially true when the laser RIN dominates the shot noise. Receiver sensitivity can be improved by 10 to 20 dB or more using a coherent detection configuration [18], in which both the intensity and the phase of the optical carrier are demodulated by the receiver. This technique is particularly attractive when designing long-haul fiber-optic links, in which attenuation and material dispersion of the signal as it propagates through several kilometers of optical fiber is of major concern.

In an optical receiver designed for coherent detection, an LO generates a continuous signal at or close to the carrier frequency of the received optical signal, enabling demodulation of both the amplitude and phase of the RF information. In a homodyne detection receiver, where the LO and received signal are tuned to the same frequency, the local semiconductor laser oscillator must lock onto the phase of the incoming signal, necessitating an optical phase-locked loop. A heterodyne system does not offer as high a sensitivity, since it deliberately tolerates a difference between the incoming signal and LO frequencies. By mixing the weak received field with an LO field, the detected signal can be sufficiently increased and filtered so that noise in subsequent amplification becomes negligible and the physical limit of detectability (i.e., shot noise) can be approached. The physical limit due to shot noise is not easily reached in a heterodyne detection receiver; the closer a modulation method is to the shot noise limit in the absence of phase noise, the more sensitive it is to phase noise [19]. Even more sensitive to phase noise are optical receivers employing homodyne detection. Homodyne detection is also more complex than heterodyne detection in that it requires locking of the LO phase to the received signal phase. While heterodyne detection obviates this need for a phase-locked loop, it still requires that the receiver include a highly stabilized low-noise oscillator, which is not needed for direct detection.

For the short-haul links used as backplane interconnects in radar phased arrays, fiber-optic link receiver sensitivity is rarely a primary concern, and the extra components required for homodyne or heterodyne detection techniques dictate a decreased insertion gain for a fiber-optic link employing either of these coherent detection architectures.

## 12.6 FIBER-OPTIC LINK PERFORMANCE OPTIMIZATION

This section discusses some widely used techniques for improving the microwave and millimeter-wave performance of fiber-optic links. The first of these optimization methods is RF impedance matching of the link's electro-optic transducers. It is important to match the microwave impedance of the optical source (in a direct modulation link) or modulator (in an external modulation link) and the optical detector to the system impedance $Z_0$ in order to minimize reflections of RF power at the link input and output. Such reflections reduce the link's signal-relaying efficiency and may also hamper the performance of any pre- or postlink signal-processing components (e.g., mixers, amplifiers, and phase shifters).

Obviously, to pursue this performance optimization approach, one must first know the impedance of a device at the frequencies at which matching to $Z_0$ is going to be attempted. Since it is usually not possible to connect an optical source, modulator, or detector directly to the ports of a microwave network analyzer, the transducers' microwave impedance must first be extracted from the measured one-port scattering parameters of the test fixtures in which they are housed. Measurement of a device's microwave one-port scattering parameter using a network analyzer requires the interposition of a bias tee, an RF connector, and a length of microstrip or coplanar transmission line. The two-port circuit parameters of each of these intermediary components are therefore *embedded* within the measured one-port S-parameters of the electro-optic device. The two-port circuit parameters of each intermediary element must therefore be determined and subsequently *de-embedded* from the measured scattering parameter values in order to recover the true scattering parameter of the device. Standard through-reflect line (TRL) methods [20] have been used to repeatedly determine the two-port circuit parameters of intermediary components. Figure 12.9 shows how the voltage-current transmission matrices (ABCD matrices [21]) of the intermediary elements are de-embedded from the measured reflection coefficient of an electro-optic device (laser, modulator, or detector) in its fixture ($\Gamma_{Measured}$) to obtain the device's true reflection coefficient ($\Gamma_{Device}$).

Once a device's microwave scattering parameter has been de-embedded from measured data, a physically realistic equivalent circuit model is developed. To transform the device's microwave impedance to the system impedance ($Z_0$, typically 50Ω), an impedance-matching circuit must be designed to combine the circuit elements of the device's model with external elements in such a way that the dominant resistive element in the device's

Test Fixture for Measurements:

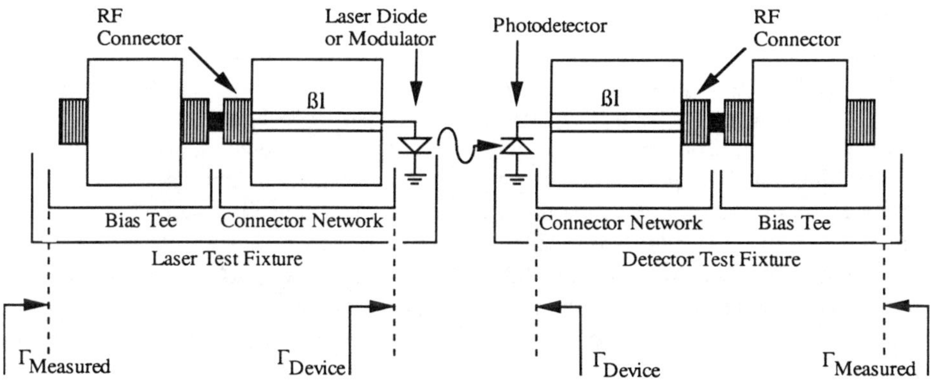


Calculate $\Gamma_{Device}$ as follows:

$$\begin{bmatrix} \dfrac{1+\Gamma_{Measured}}{1-\Gamma_{Measured}} \\ 1 \end{bmatrix} = \begin{bmatrix} A & B \\ C & D \end{bmatrix}_{bias\ tee} \times \begin{bmatrix} A & B \\ C & D \end{bmatrix}_{connector} \times \begin{bmatrix} A & B \\ C & D \end{bmatrix}_{\beta l} \times \begin{bmatrix} \dfrac{1+\Gamma_{Device}}{1-\Gamma_{Device}} \\ 1 \end{bmatrix}$$

**Figure 12.9** De-embedding to determine electro-optic device's reflection coefficient $\Gamma_{device}$ from measured reflection coefficient $\Gamma_{measured}$. Two-port voltage-current transmission matrices of the intermediary elements (bias tee, connector, and transmission line) are obtained using the TRL technique [20].

equivalent circuit is matched to $Z_0$ over the desired frequency band. Since this dominant resistive element usually has a value of only a few ohms, a common practice has been to add a resistor in series to total $Z_0$. Resistive matching accomplishes a minimization of reflections over a broad RF bandwidth, but only at the expense of efficiency as a large fraction of the RF power in the link is dissipated in the matching resistor(s) rather than being delivered to the link output.

Across the chosen range of RF frequencies, impedance-matching circuits can be designed and constructed using purely reactive elements, which maximizes the link's efficiency at that frequency. There is a distinct trade-off between the minimum acceptable return loss $|\Gamma|^2_{\min}$ to be realized and the bandwidth $\Delta f$ over which impedance matching is to be performed. This compromise is expressed by Fano [22] as follows:

$$|\Gamma|^2_{\min} \text{ (in dB)} = \frac{-27.3 f_0}{\Delta f Q_{\text{ext}}} \tag{12.10}$$

where $f_0$ is the frequency at band center and $Q_{\text{ext}}$ is the external quality factor of the device to be matched to the system impedance $Z_0$. The microwave performance characteristics of many links have been improved by replacing resistive matching circuits with reactive ones [4–13,15].

In addition to microwave coupling efficiencies, optical source-to-fiber and fiber-to-detector coupling efficiencies must be maximized in a high-efficiency fiber-optic link. Figure 12.10 illustrates several methods of device-to-fiber coupling. Shown in Figure 12.10(a) is the straightforward butt coupling technique. For semiconductor lasers, which have large maximum output divergence angles (> 30 deg, typically), it is exceedingly difficult to obtain coupling efficiencies as high as 10% using this method, even when coupling into large-core multimode fiber. To improve this efficiency, Figure 12.10(b) shows a device-to-fiber coupling method in which a microlens formed at the fiber tip magnifies the core so that more optical power can be coupled into it in a less position-sensitive manner. Optical coupling using a microlensed fiber is, however, quite angle-sensitive.

By tapering the cladding region down to just the core at the fiber tip (as shown in Fig. 12.10(c)), the effective numerical aperture of the fiber is gradually increased (due to increasing $\Delta n$), so the tapered fiber can accept light from a wider range of input angles relative to the core axis. Optical coupling to a tapered fiber end is more position-sensitive than a microlensed fiber end.

Very efficient (>50%) optical coupling—even from a semiconductor laser to a single-mode fiber with an 8-$\mu$m core diameter—can be achieved with a combination of spherical and gradient-index (GRIN) lenses, as shown in Figure 12.10(d), or with a GRIN lens having a specially curved front facet (Fig. 12.10(e)). In either case, the first convex lens surface collimates the source's divergent output beam and the cylindrical GRIN lens focuses this collimated beam down to the fiber spot size gradually enough that most light enters the fiber core at that which can propagate via total internal reflection in the fiber.

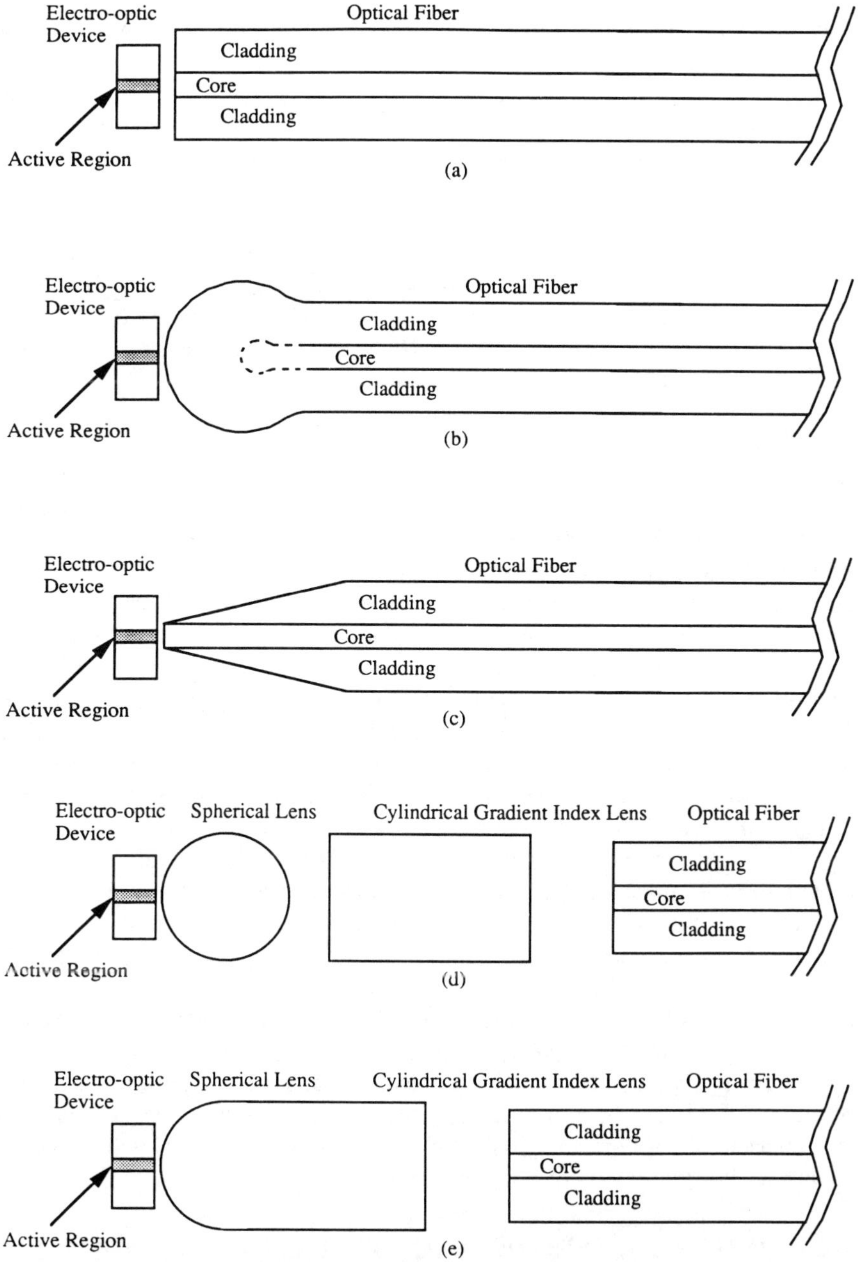


**Figure 12.10** (a) Butt coupling technique for optimizing device-to-fiber optical power coupling; (b) improved device-to-fiber coupling via a microlens formed at the fiber tip; (c) improved device-to-fiber coupling via a tapering of the fiber's cladding; (d) improved device-to-fiber coupling via a combination of spherical (collimating) and GRIN (focusing) lenses; (e) improved device-to-fiber coupling via a single plano-convex lens which approximates the function of a spherical and GRIN lens combination.

In electro-optic switches and modulators, coupling between the single-mode optical waveguides and the fiber pigtails is difficult, even when the fibers are also single-mode. This is because the refractive index profile of an optical fiber is very different from the profile of waveguides in materials such as $LiNbO_3$, GaAs, or InP. Simple butt coupling of single-mode fiber pigtails to the waveguide input and output can add 2 dB or more of loss to the overall optical insertion loss of an integrated optic device. For this reason, some of the techniques described above and illustrated in Figure 12.10 have been used to mode-match single-mode integrated optic waveguides to single-mode optical fibers.

At the fiber-detector interface, any of these same techniques can be used to improve coupling efficiency, although in the case of a single-mode fiber and a 25-, 35-, 50-$\mu$m or larger detector active area diameter, nearly 100% coupling efficiency can be obtained using the straightforward butt coupling technique.

Optical coupling efficiency throughout a fiber-optic link is also improved if all boundaries at which reflections can occur are coated with transparent materials which have dielectric constants selected to minimize reflections. Minimizing optical reflections throughout the link is vital not only from an efficiency standpoint, but also because of the effect of optical feedback on laser performance. Optical power reflected back into the laser cavity (from fiber-air interfaces at either end of the link and in connectors, as well as from lens-air and detector-air interfaces) couples with the lasing modes, thereby causing their phases to vary. This produces a modulated noise spectrum with periodicity determined by the round-trip delay of the light from the laser to the reflecting point and back again. Depending on this round-trip time, reflections can create noise peaks in the microwave or millimeter-wave frequency region, where the link is intended to have low noise [23].

There are several techniques which have been widely used to minimize the amount of optical power reflected back to the optical source in the link. One of these techniques involves depositing AR coatings on the lens and detector surfaces. Additionally, at fiber-air-fiber interfaces, index-matching gel can be used to minimize Fresnel reflections.

Minimizing the optical feedback from fiber-air interfaces involves cleaving or polishing the fiber end face at an angle sufficiently large (relative to the axis normal) to ensure that any reflected light will be deflected at too great an angle to propagate back to the optical source.

In cases where any or all of the above techniques provide inadequate protection for the optical source, a miniature optical isolator is sometimes used between the optical source and the fiber. This device consists of a linearly polarized wave plate plus a magneto-optic crystal, which rotates the polarization of the optical field by 45 deg so that a wave reflected from a surface further along in the link arrives at the wave plate with polarization perpendicular to the filter's axis of polarization.

## 12.7 FIBER-OPTIC LINK PERFORMANCE MODELS

Because the fabrication of a photonic link consumes both time and money, the ability to predict the performance of a designed link before fabrication has been sought by a number of researchers. Such a modeling technique would verify the prudence of link architectures, components, and other features selected by the designer. To accomplish this, the model must consistently and quantifiably account for all phenomena and features of the link which may contribute to its insertion gain/loss, noise figure, dynamic range, and bandwidth. Furthermore, it should apply to all devices and techniques which a designer might wish to use in the pursuit of high performance.

The fiber-optic link models presented here are those developed by Daryoush et al. [24], which express the performance parameters (gain, noise figure, dynamic range) in terms of the microwave scattering parameters of a full link equivalent circuit. The derivation of these models is too lengthy for inclusion in this chapter; the reader is referred to the literature where appropriate.

### *Direct Modulation/Direct Detection Link*

A conceptual diagram of a semiconductor-laser-based optical transmitter is given in Figure 12.11(a), which shows an impedance-matching circuit with the laser equivalent-circuit model. The semiconductor laser's microwave equivalent circuit is that of any forward-biased pn junction. The goal of the matching network, as discussed in the previous section of this chapter, is to effectively transfer current from the system input impedance (usually 50Ω) to the junction of the laser diode. That is, the matching network is designed to maximize at any given bandwidth of RF frequencies the magnitude of the current $I_L$ passing through the junction resistance $R_{JL}$. As discussed previously, the minimum transmitter input reflection coefficient attainable using reactive matching techniques is limited by the quality factor $Q$ of the laser and the percentage bandwidth over which impedance matching is performed, according to Fano's expression (12.10).

Also shown in Figure 12.11(a) is a current source with spectral density $\langle i_{th,TX,dir}^2 \rangle$ representing the thermal noise generated in the optical transmitter. This current spectral density is used in the modeling of the noise performance of the direct modulation link. The term $Y_{thL}$ in Figure 12.11(a) refers to the admittance of the transmitter circuit as seen from the transmitter output, which is used in the noise calculation that follows. Finally, the term $\Gamma_L$, which is used in the gain calculation, is the reflection coefficient of the laser junction resistance, given by the simple expression

$$\Gamma_L \equiv \frac{R_{JL} - Z_0}{R_{JL} + Z_0} \tag{12.11}$$

A conceptual diagram of an optical receiver based on a pin photodiode is given in Figure 12.12(a), showing the detector equivalent-circuit model and impedance-matching circuit. The detector's microwave equivalent circuit is that of any reverse-biased pn junction, except for the addition of an optical intensity-dependent current source across the junction. Again, the matching circuit should effectively transform the RF photocurrent to the load

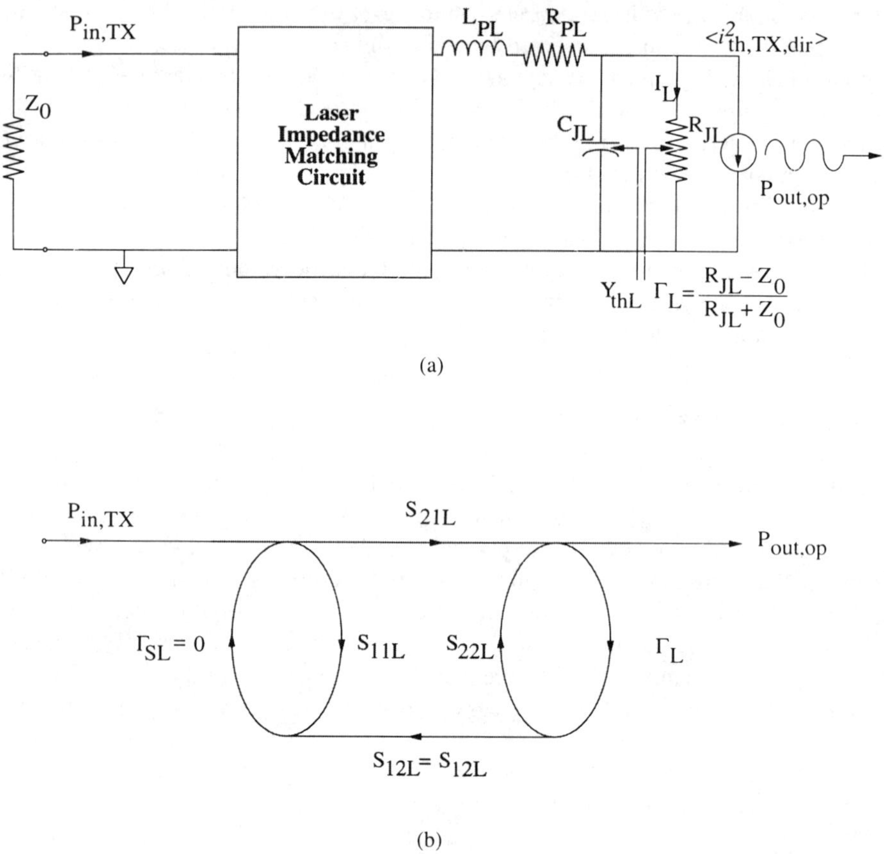


**Figure 12.11** (a) Equivalent-circuit diagram of semiconductor-laser-based optical transmitter; (b) signal flow diagram of semiconductor-laser-based optical transmitter.

impedance (typically 50Ω). Ideally, an impedance-matching network for the receiver would entail the matching of the photodiode's high reverse-biased junction resistance, $R_{JD}$ (typically on the order of 1 MΩ) to the output of the fiber-optic link. The quality factor of the photodiode is extremely high if parasitics are neglected. However, from the de-embedding and equivalent-circuit modeling of the detectors purchased from several different vendors, it has been discovered that it is not generally possible to match to the high junction resistance of the detector at many microwave frequencies, because of the device's parasitic series resistance $R_{PD}$. This series resistance most often dictates a lower $Q$ factor than the shunt $R_{JD}C_{JD}$ combination gives, and one cannot efficiently couple power to or from a high $Q$ element through an element having a lower $Q$ factor. Thus, it is the

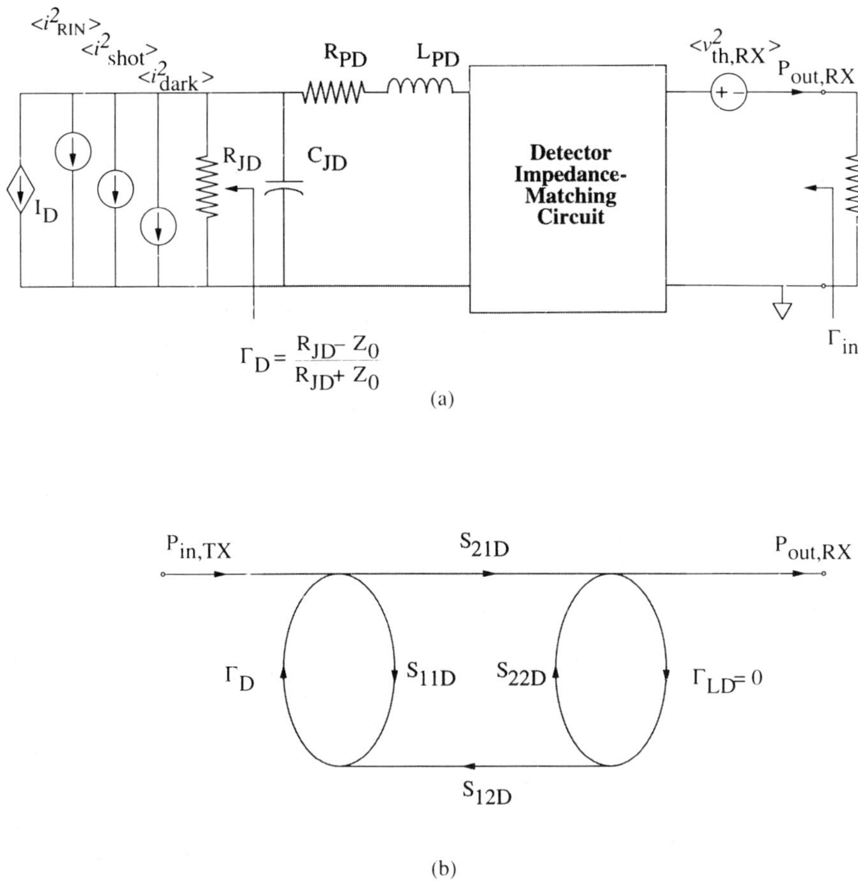


**Figure 12.12** (a) Equivalent-circuit diagram of a pin-photodiode-based optical receiver; (b) signal flow diagram of pin-photodiode-based optical receiver.

series resistance $R_{JD}$ that is transformed to $50\Omega$ in the detector's impedance-matching network.

Also shown in Figure 12.12(a) are a voltage source to indicate the presence of thermal noise in the receiver (with spectral density $<v_{th,RX}^2>$), and current sources representing three sources of optical noise in the fiber-optic link: $<i_{RIN}^2>$ is the noise current spectral density at the receiver due to the laser's relative intensity noise (RIN); $<i_{shot}^2>$ reflects the shot noise of the photodiode; and $<i_{dark}^2>$ is the contribution from the photodiode's dark current $I_{dark}$. $\Gamma_{in}$, also used to calculate noise performance, is the reflection coefficient at the output of the optical receiver. Finally, the term $\Gamma_D$ is the reflection coefficient of the photodiode junction resistance, given by the expression

$$\Gamma_D \equiv \frac{R_{JD} - Z_0}{R_{JD} + Z_0} \tag{12.12}$$

For the semiconductor laser diode, the relationship between the optical output power $P_{\text{out,op}}$ and the applied electrical current $I_L$ was given in expression (12.7). In this model it is assumed that the semiconductor laser has been chosen for the desired bandwidth, and that therefore the differential quantum efficiency at any frequency in the band of interest is $\eta_L$. The ratio of detector photocurrent $I_D$ to $P_{\text{out,op}}$ is $K_{LF}L_F K_{FD}\eta_D$, where $K_{LF}$ and $K_{FD}$ are, respectively, the laser-to-fiber and fiber-to-detector optical coupling efficiencies, $L_F$ is the total optical power loss due to attenuation in the fiber, and $\eta_D$ is the photodetector responsivity at the RF frequency. In the equations that follow, the detector is assumed to operate at a dc bias voltage large enough to ensure that it does not impose an upper limit on the RF power handling capability of the link. That is, even at 100% modulation depth in the laser, the detector remains within its linear region of operation.

The gain of a direct detection fiber-optic link can be calculated in terms of microwave scattering parameters using the signal flow diagram (SFD) technique and Mason's gain equation [25]. The SFD for the optical transmitter in a direct modulation link (an impedance-matched semiconductor laser diode) is shown in Figure 12.11(b). It is obtained by considering the forward-bias junction resistance of the laser diode to be the termination load $\Gamma_L$ at port two of a two-port network (represented by the scattering matrix $[S_L]$) consisting of the microwave impedance-matching circuit and the other device parameters of the laser. Similarly, when a reverse-biased pin photodiode is employed in the optical receiver, the SFD is obtained by considering the junction resistance of the diode to be the source termination $\Gamma_D$ of a two-port network (represented by the scattering matrix $[S_D]$) consisting of the other device parameters of the photodetector and the microwave impedance-matching circuit. Figure 12.12(b) shows the SFD for the optical receiver.

From the SFD for the semiconductor-laser-based optical transmitter and the pin-photodiode-based optical receiver—which are obtained from measurements of the device impedances—and from other measured parameters, Daryoush et al. derived the gain and noise figure as follows [24]:

$$G = (\eta_L K_{LF} L_F K_{FD} \eta_D)^2 \frac{|S_{21L}|^2 |1 - \Gamma_L|^2 |S_{21D}|^2}{|1 - S_{22L}\Gamma_L|^2 |1 - S_{11D}\Gamma_D|^2} \tag{12.13}$$

$$\begin{aligned} \text{NF} = \Bigg\{ & [I_D^2 \cdot \text{RIN} + 2q(I_D + I_{\text{dark}}) \\ & + \text{k}T(\eta_L K_{LF} L_F K_{FD} \eta_D)^2 \,\text{Re}\{Y_{\text{th}L}\}] \frac{|S_{21D}|^2 B Z_0}{|1 - S_{11D}\Gamma_D|^2} + \text{k}TB(G+1) \Bigg\} \frac{1}{\text{k}TBG} \end{aligned} \tag{12.14}$$

where $q$ is the electronic charge, RIN is the laser relative intensity noise measured at the RF frequency of interest, and all other terms have already been defined above.

The dynamic range of a direct modulation link cannot be modeled from measured parameters alone, because the laser linearity is a function of physical parameters like the laser's photon and spontaneous emission lifetimes ($\tau_p$ and $\tau_s$, respectively). The compression-limited dynamic range of a direct modulation link is given by (12.4), where

$$P_{\text{in,1CP}} = (1.00952)^2(I_L - I_{\text{th}})^2 Z_0 \frac{|1 - S_{22L}\Gamma_L|^2}{|S_{21L}|^2|1 - \Gamma_L|^2} \times \left[\left(\frac{\omega^2}{\omega_r^2} - 0.8911\right)^2 + \omega^2\tau_p^2\left(\frac{1}{\omega_r^2\tau_p\tau_s} + 1\right)^2\right] \tag{12.15}$$

and $\omega_r$ is the laser relaxation oscillation frequency $f_r$ multiplied by $2\pi$. The spurious-free dynamic range of the direct modulation link is given by (12.5), where

$$P_{\text{in,int}} = 8(I_L - I_{\text{th}})^2 Z_0 \frac{|1 - S_{22L}\Gamma_L|^2}{|S_{21L}|^2|1 - \Gamma_L|^2} \times \left[\left(\frac{\omega^2}{\omega_r^2} - 0.5631\right)^2 + \omega^2\tau_p^2\left(\frac{1}{\omega_r^2\tau_p\tau_s} + 1\right)^2\right] \tag{12.16}$$

*External Modulation/Direct Detection Link*

An equivalent-circuit model of a Mach-Zehnder electro-optic modulator–based optical transmitter is presented in Figure 12.13(a). The values of the circuit elements are based on the physical parameters of the modulator. The electrodes of the lumped-element modulator are represented by load $Z_M$ with parasitic series resistance and inductance $R_M$ and $L_M$, respectively. The impedance-matching network shown in the figure transforms the input RF current such that, for the desired bandwidth of RF frequencies, maximum voltage $V_M$ is obtained across the load $Z_M$. If the modulator electrodes can be considered as a lumped element (capacitor $C_M$)—that is, if the electrode length is much smaller than an RF wavelength—then $Z_M$ is simply $1/j\omega C_M$, the impedance of a capacitor. If, on the other hand, the electrode length is greater than a significant part of the RF wavelength, then $Z_M$ must be determined using the transmission line equation [21]:

$$Z_M = Z_C \frac{Z_t + Z_C \tanh(j\beta l + \alpha l)}{Z_C + Z_t \tanh(j\beta l + \alpha l)} \tag{12.17}$$

where $\alpha$ and $\beta$ are the real and imaginary parts, respectively, of the microwave propagation constant, $l$ is the length, $Z_C$ is the characteristic impedance, and $Z_t$ is the termination impedance of the modulator's traveling-wave electrodes. Also shown in Figure 12.13(a) is a voltage source to indicate the presence of thermal noise in the receiver (with spectral density $\langle v_{\text{th,TX,ext}}^2 \rangle$). Finally, the term $\Gamma_M$, which is used in the gain calculation, is the reflection coefficient of the modulator electrodes, given by the expression

$$\Gamma_M = \frac{Z_M - Z_0}{Z_M + Z_0} \tag{12.18}$$

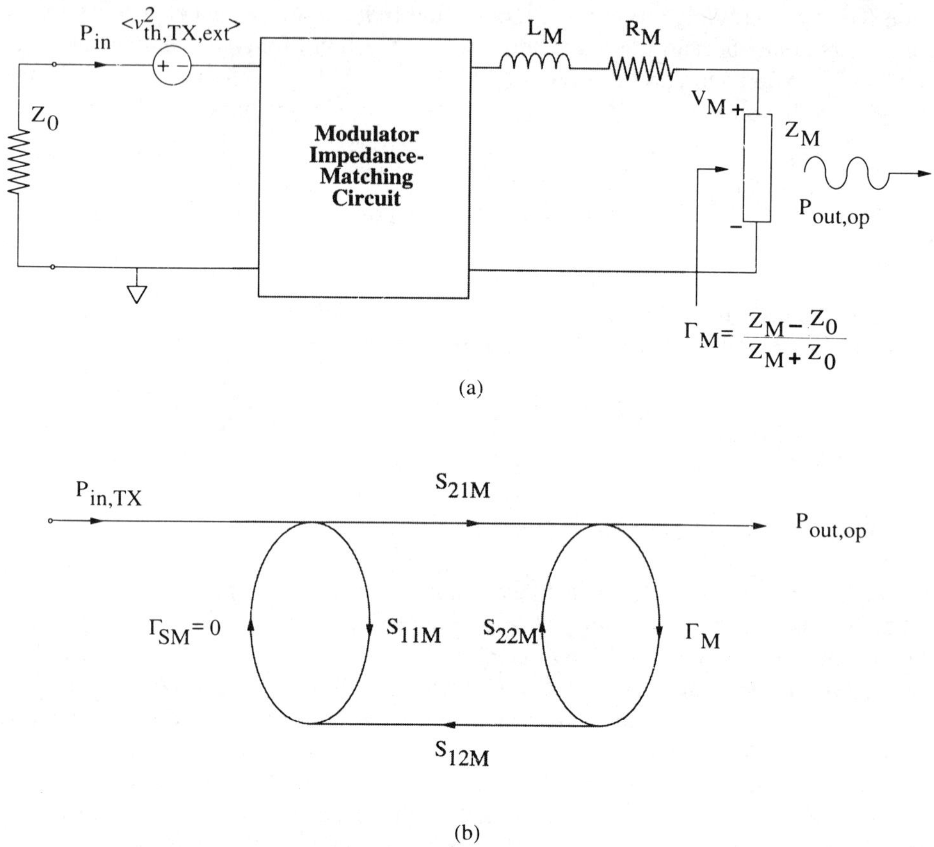


**Figure 12.13** (a) Equivalent-circuit diagram of Mach-Zehnder interferometric modulator–based optical transmitter; (b) signal flow diagram of Mach-Zehnder interferometric modulator–based optical transmitter.

The equivalent-circuit model of the direct detection optical receiver in an external modulation link is identical to that in a direct modulation link (see Fig. 12.12(a)).

For a lumped-element interferometric modulator, the relationship between the optical output power $P_{\text{out,op}}$ and the applied voltage $V_M$ was given in (12.8) and (12.9). The same expression holds true in a traveling-wave modulator, if the electronic and optical waves remain in phase (to within a few degrees) along the entire electrode length. This is likely to be the case for any traveling-wave modulator operated at or below the maximum rated frequency for the device. The ratio of detector photocurrent $I_D$ to $P_{\text{in,op}}$, the optical power at the input of the modulator, is $L_M L_F K_{FD} \eta_D$, where $L_M$ is the optical insertion loss of the modulator (the other terms have already been defined).

The RF performance of an external modulation fiber-optic link is derived using the SFD technique as was applied to direct modulation, along with the associated assumptions.

When a Mach-Zehnder interferometric modulator is employed in the optical transmitter to impress a microwave signal upon the optical carrier in a single-mode fiber, the transmitter SFD is obtained by considering the impedance $Z_M$ across the modulator terminals to be the termination load $\Gamma_M$ of a two-port network (represented by $[S_M]$), consisting of the microwave impedance-matching circuit and the other device parameters in the equivalent circuit of the modulator. Figure 12.13(b) shows the SFD for the transmitter in an external modulation fiber-optic link.

From the SFDs for the external modulator-based optical transmitter and the pin-photodiode-based optical receiver, which are obtained from measurements of the device impedances and from other measured parameters, Daryoush et al. derived the gain and noise figure as follows [24]:

$$G = \left(\frac{\pi L_M L_F K_{FD} \eta_D P_{\text{in,op}} Z_0}{2V_\pi}\right)^2 \cos^2\left(\frac{\pi V_b}{V_\pi}\right) \times \frac{|S_{21M}|^2 |1 + \Gamma_M|^2 |S_{21D}|^2}{|1 - S_{22M}\Gamma_M|^2 |1 - S_{11D}\Gamma_D|^2} \tag{12.19}$$

$$\text{NF} = \left\{\left[I_D^2 \cdot \text{RIN} + 2q(I_D + I_{\text{dark}}) + \frac{\text{k}T(\pi L_M L_F K_{FD} \eta_D P_{\text{in,op}} Z_0)^2}{4V_\pi^2 \text{Re}\{Z_M\}}\right] \frac{|S_{21D}|^2 B Z_0}{|1 - S_{11D}\Gamma_D|^2} + \text{k}TB(G + 1)\right\} \frac{1}{\text{k}TBG} \tag{12.20}$$

The nonlinearity of an interferometric external modulator can be derived (unlike a semiconductor laser) from measured parameters [24] using expressions (12.8) and (12.9). Thus, the compression-limited and spurious-free dynamic ranges of the external modulation link are given by the following expressions:

$$\text{CDR} = \frac{(0.5500)^2 V_\pi^2}{\text{k}TB\text{NF}\pi^2 Z_0} \frac{|1 - S_{22M}\Gamma_M|^2}{|S_{21M}|^2 |1 + \Gamma_M|^2} \tag{12.21}$$

$$\text{SFDR} = \left(\frac{8V_\pi^2}{\text{k}TB\text{NF}\pi^2 Z_0} \frac{|1 - S_{22M}\Gamma_M|^2}{|S_{21M}|^2 |1 + \Gamma_M|^2}\right)^{2/3} \tag{12.22}$$

## REFERENCES


[1] Koffman, I., P. R. Herczfeld, and A. S. Daryoush, "High Speed Fiber Optic Links for Short-Haul Microwave Applications," *1988 IEEE MTT-S Microwave Symp. Digest,* Vol. 2, pp. 983–986.

[2] Ogawa, H., and Y. Kamiya, "Fiber Optic Microwave Transmission Using Harmonic Modulation and Optoelectronic Mixing/Optically Pumped Mixing," *1991 IEEE MTT-S Microwave Symp. Digest,* Vol. 2, pp. 593–596.

[3] Polifko, D., and H. Ogawa, "The Merging of Photonic and Microwave Technologies," *Microwave J.,* Vol. 35, No. 3, March 1992, pp. 75–80.

[4] Betts, G. E., L. M. Johnson, C. H. Cox III, and S. D. Downey, "High-Performance Optical Analog Link Using External Modulator," *IEEE Photonics Tech. Lett.,* Vol. 1, No. 11, Nov. 1989, pp. 404–406.

[5] Betts, G. E., L. M. Johnson, and C. H. Cox, ''High-Sensitivity Bandpass RF Modulator in $LiNbO_3$,'' *Proc. SPIE,* Vol. 993: *Integrated Optical Circuit Engineering VI,* 1988, pp. 110–116.

[6] Ackerman, E., D. Kasemset, and S. Wanuga, ''External Modulation L-Band Link With 117 dB Spurious-Free Dynamic Range,'' *Microwave J.,* Vol. 34, No. 9, Sept. 1991, pp. 158–164.

[7] Betts, G. E., C. H. Cox III, and K. G. Ray, ''20-GHz Optical Analog Link Using an External Modulator,'' *IEEE Photonics Tech. Lett.,* Vol. 2, No. 12, Dec. 1990, pp. 923–925.

[8] Ackerman, E., D. Kasemset, S. Wanuga, D. Hogue, and J. Komiak, ''A High-Gain Directly Modulated L-Band Microwave Optical Link,'' *1990 MTT-S Symp. Digest,* Vol. 1, pp. 153–155.

[9] Daryoush, A. S., E. Ackerman, R. Saedi, R. Kunath, and K. Shalkhauser, ''High-Speed Fiber Optic Links for Distribution of Satellite Traffic,'' *IEEE Trans. Microwave Theory and Techniques,* Vol. 38, No. 5, May 1990, pp. 510–517.

[10] Wanuga, S., E. Ackerman, D. Kasemset, and J. Komiak, ''Integrated Microwave Optoelectronic Transceivers for Lightweight Fiber Optic Harnesses and Remote Microwave Signal Transmission,'' *1991 GOMAC Digest,* Nov. 1991, pp. 477–480.

[11] Huff, D. B., and J. P. Anthes, ''Optoelectronic Isolator for Microwave Applications,'' *IEEE Trans. Microwave Theory and Techniques,* Vol. 38, No. 5, May 1990, pp. 571–576.

[12] Blauvelt, H., D. B. Huff, G. J. Stern, and I. L. Newberg, ''Reduced Insertion Loss of X-Band RF Fiber Optic Links,'' *IEEE Trans. Microwave Theory and Techniques,* Vol. MTT-38, No. 5, May 1990, pp. 662–664.

[13] de La Chapelle, M., ''Computer-Aided Analysis and Design of Microwave Fiber Optic Links,'' *Microwave J.,* Sept. 1989, pp. 179–186.

[14] Gee, C. M., G. D. Thurmond, I. L. Newberg, H. W. Yen, D. Lafaw, and B. Spink, ''11-GHz Fiber Optic Links for Airborne Satellite Communications Applications,'' *Microwave and Optical Tech. Lett.,* Vol. 1, No. 2, April 1988, pp. 67–70.

[15] Ackerman, E., D. Kasemset, S. Wanuga, R. Boudreau, J. Schlafer, and R. Lauer, ''A Low-Loss Ku-Band Directly Modulated Fiber Optic Link,'' *IEEE Photonics Tech. Lett.,* Vol. 3, No. 2, Feb. 1991, pp. 185–187.

[16] Blauvelt, H., and K. Lau, ''High Signal to Noise Operation of Fiber Optic Links to 18 GHz,'' *1988 IEEE MTT-S Microwave Symp. Digest,* Vol. 2, pp. 979–981.

[17] Lester, L. F., W. J. Schaff, X. Song, and L. F. Eastman, ''Optical and RF Characteristics of Short-Cavity-Length Multiquantum-Well Strained-Layer Lasers,'' *IEEE Photonics Tech. Lett.,* Vol. 3, No. 12, Dec. 1991, pp. 1049–1051.

[18] Yamamoto, Y., and T. Kimura, ''Coherent Optical Fiber Transmission Systems,'' *IEEE J. Quantum Electron.,* Vol. QE-17, No. 6, June 1981, pp. 919–935.

[19] Henry, C. H., ''Phase Noise in Semiconductor Lasers,'' *IEEE J. Lightwave Tech.,* Vol. LT-4, No. 3, March 1986, pp. 298–311.

[20] Engen, G. F., and C. A. Hoer, ''Thru-Reflect-Line: An Improved Technique for Calibrating the Dual 6-Port Automatic Network Analyzer,'' *IEEE Trans. Microwave Theory and Techniques,* Vol. MTT-27, Dec. 1979, pp. 983–987.

[21] Collin, R. E., *Foundations for Microwave Engineering,* New York: McGraw-Hill, 1966, pp. 180–181.

[22] Fano, R. M., ''Theoretical Limitations on the Broadband Matching of Arbitrary Impedances,'' *J. Franklin Institute,* Vol. 249, Jan.-Feb. 1950, pp. 57–84 and 139–154.

[23] Keiser, G., *Optical Fiber Communications,* New York: McGraw-Hill, 1983.

[24] Daryoush, A. S., E. Ackerman, N. Samant, S. Wanuga, and D. Kasemset, ''Interfaces for High-Speed Fiber Optic Links: Analysis and Experiment,'' *IEEE Trans. Microwave Theory and Techniques,* Vol. 39, No. 12, Dec. 1991, pp. 2031–2044.

[25] Gonzales, G., *Microwave Transistor Amplifiers,* Englewood Cliffs, NJ: Prentice-Hall, 1984, pp. 80–87.

# Chapter 13
# Switched Fiber-Optic Delay Architectures

**Akis P. Goutzoulis and D. Kenneth Davies**
*Westinghouse Science and Technology Center*

## 13.1 INTRODUCTION

High-performance radars require the use of phased-array antennas (PAA) composed of elemental monolithic T/R modules. PAAs offer many advantages, including steering without physical movement, very accurate beam pointing, increased scan flexibility in two dimensions, precise elemental phase and amplitude control to obtain low spatial sidelobes, and reduced power consumption and weight. Such radars are required to have large scan angles (typically, ±90 deg), hundreds of megahertz of instantaneous bandwidth (BW), center frequencies anywhere from a few hundred megahertz to 20 GHz, and multiple (typically 3 to 5) simultaneous beams. The actual number of T/R modules depends on the system mission as well as the operating frequency, and is typically in the range of hundreds to several thousands for all airborne, spaceborne, shipboard, and ground-based radars. Similar requirements exist for future general-purpose, multifunction, front-end systems; moreover, such systems are generally expected to require even larger BWs because of the integration of radar, electronic counter measures (ECM), and communications functions on a common antenna.

To satisfy the wide BW requirements of future PAAs, true time-delay (TTD) steering techniques must be invoked so that efficient elemental vector summation (in the receive mode) or distribution (in the transmit mode) can be obtained that is independent of frequency. Fiber is an excellent medium for both the delay generation and signal distribution requirements of a TTD-based PAA because (1) it can store large BW analog signals (tens of gigahertz) for long times (tens of microseconds); (2) it has low attenuation (~ 0.1 dB/km), which is flat as a function of frequency up to 100 GHz; (3) it allows the remote processing of PAA signals; (4) it has excellent transmission stability by virtue of the small ratio of signal BW to optical carrier frequency; (5) it allows the utilization of optical wavelength division multiplexing (WDM) to minimize the number of lines in the PAA feed link; (6) it is a nonconducting dielectric and so does not disturb the RF field, is secure, and is immune from EMI; and (7) it is flexible and has low mass and small volume, all of which are important attributes for airborne systems.

The importance of the fiber-optic (FO) TTD has been widely recognized by both the radar and the communications community. Independent demonstrations have already verified the viability of the fiber-optic TTD concept using both laboratory equipment [1,2] and actual PAAs [3]. Several developmental fiber-optic TTD programs are currently underway worldwide, each using a different fiber-optic TTD technique.

The objective of this chapter is to categorize and describe the various classes of switched-fiber-optic TTD steering architectures, and indicate their relative advantages and disadvantages. (Information on nonswitched fiber-optic TTD techniques can be found in [4–8].) In Section 13.2, we review the basics of TTD steering, and we emphasize the large overall hardware requirements, the reduction of which is essential if fiber-optic TTD is to become practical. In Section 13.3, we develop the criteria, such as component complexity and insertion loss, to enable a quantitative evaluation and meaningful comparison of the various architectures to be made in terms of the critical component parameters. The categorization and characterization of the various fiber-optic TTD architectures are addressed in Sections 13.4 through 13.6. In Section 13.4, we discuss architectures that do not compress the hardware; in Section 13.5, we discuss architectures that compress the hardware with respect to the number of delays only; and in Section 13.6, we discuss architectures that compress the hardware with respect to both the number of delays and the number of antenna elements. For each architecture we review various versions (if applicable) and briefly describe prototype systems and their relative performance.

## 13.2 PHASED-ARRAY ANTENNA TIME-DELAY STEERING

The TTD concept is illustrated in Figure 13.1 for a $K$-element, receiving linear PAA. For this antenna, the normalized vector summation $V$ is given by

$$V = \sum_{i=1}^{K} \sin\left\{2\pi f\left[t + \frac{D_i}{v_1} + \frac{I_i}{v_2}\right]\right\} = \sum_{i=1}^{K} \sin\{2\pi f[t + (T_1)_i + (T_2)_i \sin\theta]\} \qquad (13.1)$$

where $\theta$ is the receiving angle, $I_i = d_i \sin\theta$, $d_i$ and $D_i$ are the distances of the $i$th antenna element from the wavefront, first element, and summing point, respectively, and $v_1$ and $v_2$ are the propagation velocities over $D_i$ and $I_i$, respectively. When the path delays from the wavefront to the summing point via each antenna element are equal, the sine functions in the summation of (13.1) will add in phase for all values of frequency $f$, thereby forming a beam summation in the direction $\theta$, which is independent of frequency. This condition is given by

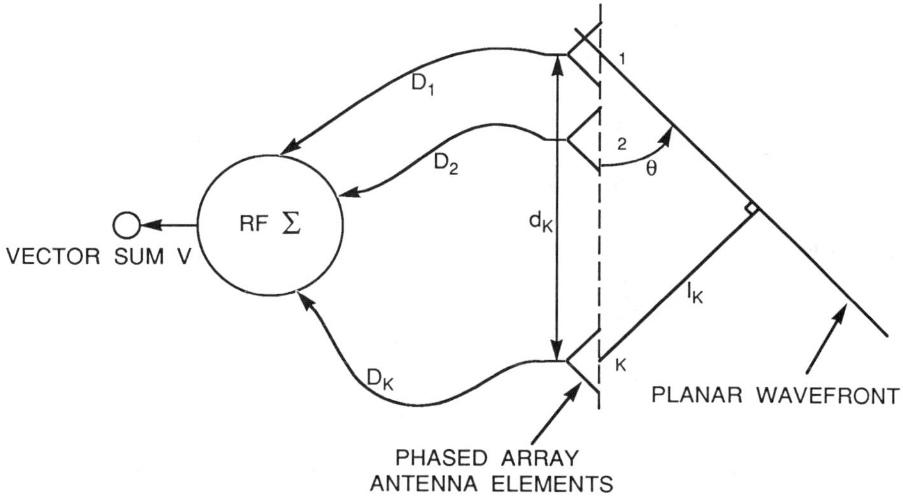


**Figure 13.1** Linear phased-array antenna geometry.

$$(T_1)_i + (T_2)_i \sin \theta = \text{constant} \tag{13.2}$$

In the transmit mode, the direction $\theta$ is varied by selecting the $D_i$ paths so that (13.2) is satisfied. Thus, for the propagation of a planar wavefront in the direction $\theta$, the delay lengths $D_i$ must satisfy the linear relationship:

$$D_{K-i} = (v_1/v_2)d_i \sin \theta \tag{13.3}$$

In a 1-D PAA with $K$ elements, the maximum time delay required by the $i$th element is given by

$$T_i = d_i \sin \theta_M/c \qquad i = 1, 2, 3, \ldots, K \tag{13.4}$$

where $\theta_M$ is the maximum scan angle and $c$ is the speed of light. For a PAA with element-to-element spacing of $d = \lambda/2$, where $\lambda$ is the wavelength of the RF radiation, the maximum possible delay $T_{i\max}$ is equal to

$$T_{i\max} = i\lambda \sin \theta_M/2c \tag{13.5}$$

Equation (13.5) indicates that the maximum delay required for a steering angle of $\theta_M$ is different for each of the $K$ elements. This is also the case for the minimum delay corresponding to the antenna angular resolution $\theta_R$:

$$T_{i\min} = i\lambda \sin \theta_R/2c \tag{13.6}$$

Equations (13.5) and (13.6) determine the total number $R$ of different delays required for steering the antenna over $\theta_M$ with resolution $\theta_R$. Since most PAAs have $10^3$ to $10^5$ elements, the number of the resulting delays C is enormous, and is approximately equal to $K \cdot R$. For example, for a 256-element 1-D PAA operating at S band (e.g., $f = 3.5$ GHz or $\lambda = 85.7$ mm) with $d = 42.9$ mm, $\theta_M = 90$ deg, and $R = 1{,}024$ (or 10 bits), a total of $C = 2.5 \times 10^5$ different delays in the 0.44-ps to 36-ns range are required. These numbers indicate not only the large delay complexity involved, but also the large delay range required, conditions which become even more severe for large 2-D PAAs. Large delay complexity results in large hardware complexity (i.e., switches and fiber segments necessary for the TTD implementation). Large hardware complexity implies not only high cost and difficulty in implementing and maintaining the TTD system, but also increased size and weight, parameters that must be minimized, especially for airborne and spaceborne platforms.

The large hardware complexity issues can be better appreciated by considering one of the early, passive (i.e., no delay-switching capability) fiber-optic TTD implementations [1], a schematic of which is shown in Figure 13.2. In this system, the RF signal from a given antenna element is used to modulate a fiber-coupled laser diode (LD), the output of which is divided into $R$ channels (or $R$ fibers) via the use of a fiber-optic one-to-$R$ channel divider (1:$R$). The length of each fiber is arranged so that it corresponds to a specific angle $\theta_i$, $i = 1, 2, \ldots, R$. This is repeated for each of the $K$ PAA elements, with the understanding that for a given angle $\theta_i$ the length of the corresponding fiber varies linearly from element to element. The outputs from all fibers that correspond to the same $\theta_i$ are combined via a $K$:1 fiber-optic combiner. The combiner output is read out by a

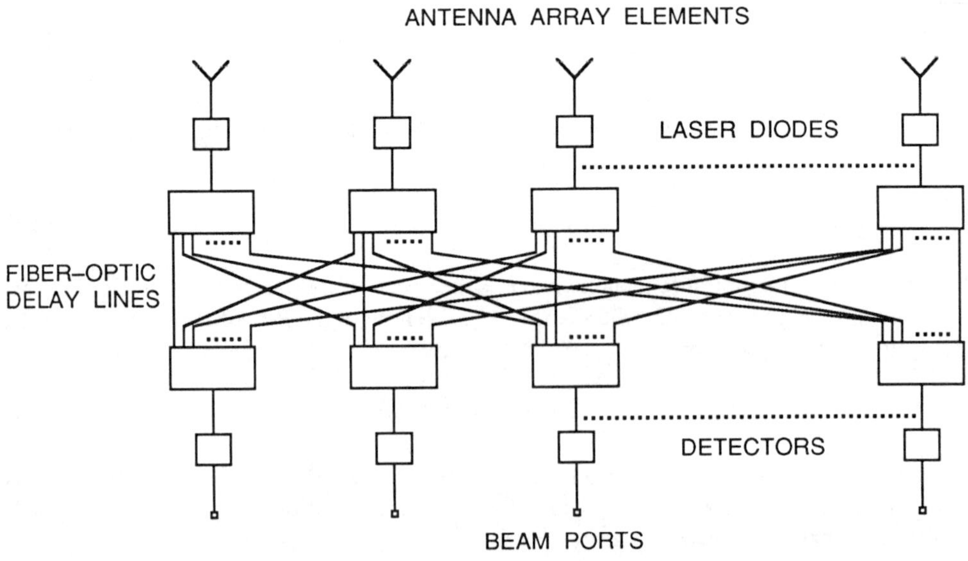


**Figure 13.2** Schematic of a passive fiber-optic TTD implementation comprising $K$ elements and $R$ beams.

detector which provides an output RF signal corresponding to that received from all PAA elements at angle $\theta_i$. Thus, each of the $R$ beam ports provides a simultaneous output of the signal received at each of the $R$ discrete directions. From this brief description and with the aid of Figure 13.2, one can easily show that this passive fiber-optic TTD scheme is hardware-intensive, since it requires a total of $K \cdot R$ fibers.

From the above discussion, we can conclude that straightforward implementations of fiber-optic TTD result in large hardware complexities and are impractical. Therefore, it is essential that ingenious, hardware-compressive, fiber-optic TTD architectures be employed for TTD steering. At the minimum, such architectures must compress the hardware with respect to the number of delays per PAA element. Ideally, however, the fiber-optic TTD architecture should be compressive with respect to both the number of delays per element and the number of elements per PAA.

## 13.3 CHARACTERIZATION PARAMETERS FOR FIBER-OPTIC TTD ARCHITECTURES

In order to compare the different fiber-optic TTD architectures meaningfully, we must develop a set of criteria that fully characterize them. Unfortunately, a meaningful comparison also requires knowledge of the actual TTD system application, given that different operating scenarios require optimization of different performance parameters. Therefore, we can only compare the various TTD architectures in a generalized fashion.

The main performance criterion is the one indicated in the previous section, namely, hardware complexity $C$. Since we are dealing with switched fiber-optic delays, $C$ can be specified with respect to the number of switches ($C_S$) and the number of different fiber segments ($C_F$) required to achieve R different delays. Depending on the architecture, $C$ will also be used with respect to the number of LDs ($C_{\text{LD}}$), detectors ($C_{\text{DET}}$), fiber-optic combiners ($C_C$), WDMs ($C_W$), and so on. Furthermore, there are cases where we will compare the number of programmable delay lines ($C_{\text{DL}}$) rather than the number of individual components.

The second criterion is the overall insertion loss (IL), which is very important, since it may limit the system SNR and dynamic range (DR). (Both SNR and DR are very important, especially in airborne and spaceborne applications.) The IL value will be given in terms of optical decibels (unless otherwise specified), and this value must be doubled in order to determine the RF decibel-based IL figure. Where necessary, IL will be derived from the IL figures of the various components. In some cases we will also specify the IL stability; that is, whether the IL varies in an extraordinary way from delay to delay.

The above performance criteria are more or less generic in the sense that they apply to most PAA scenarios. However, there are many other criteria whose degree of importance are determined almost exclusively by the system application. Examples of such criteria include amplitude and timing errors (both uncorrelated and correlated), which determine the sidelobe level of the PAA, the ease in matching the responses of the various active elements, the ease in performing active amplitude and/or timing corrections, the capability for global versus local amplification, and so on. The specification and analysis of these types of performance parameters are difficult tasks and beyond the scope of this chapter. However, where applicable, we will identify such performance characteristics.

To simplify the various descriptions and calculations, we will assume the use of lossless single-mode fiber (SMF) for all architectures, as well as (1) a 3-dB optical loss (or 6-dB RF loss) per channel for each 1:2 single-mode combiner or divider, and (2) a $10 \log_{10} R$ (in optical decibels) loss per channel for a 1:$R$ single-mode combiner or divider.

Finally, in order to provide a common basis for comparison of the actual hardware complexity involved, we will use, for reference, an example of a modest PAA with $K =$ 256 elements and $R = 1024$ (or 10 bits) delays per element.

## 13.4 NONCOMPRESSIVE FIBER-OPTIC TTD ARCHITECTURES

These TTD architectures have limited applicability for most high-performance PAA systems because of their inherent large hardware complexity. Basically, this family uses a brute-force approach, whereby a delay line comprises $R$ parallel fiber segments. A delay is selected by switching in the appropriate fiber via either an optical switch or an LD. In general, since a different delay line is required per PAA element, there is no hardware compression. However, this family of TTD architectures does not produce correlated errors, unless, of course, antenna subarray techniques are used.

### 13.4.1 Optical Switch Parallel Fiber-Optic Delay Line

This noncompressive architecture uses a set of parallel fibers to form an all-optical parallel delay line [9] (PADEL). The PADEL concept is illustrated in Figure 13.3, and requires $R$ different, fixed-length fiber segments to provide $R$ different delays. The input signal is fed through a tree-type structure of $1 \times 2$ single-pole double-throw (SPDT) optical switches to the desired delay path. The output from each delay line is combined onto a single detector via a tree structure of either 2:1 optical combiners or $2 \times 1$ SPDT optical switches. An optoelectronic PADEL variant has been described by Levine [10], where optical detectors at the output of each fiber segment are followed by an electrical summing network. A prototype optical PADEL has been demonstrated by Kondo et al. [9], who used two tree-type switch structures comprising four integrated-optic (IO) switches to form a PADEL with $R = 4$. The prototype has an IL of 22 dB and a crosstalk level of –21 dB.

For this architecture, the complexity parameters (for a $K$-element PAA) are

$$C_F = (K - 1) \cdot R \tag{13.7}$$

$$C_S = (K - 1) \cdot (R - 1) \tag{13.8}$$

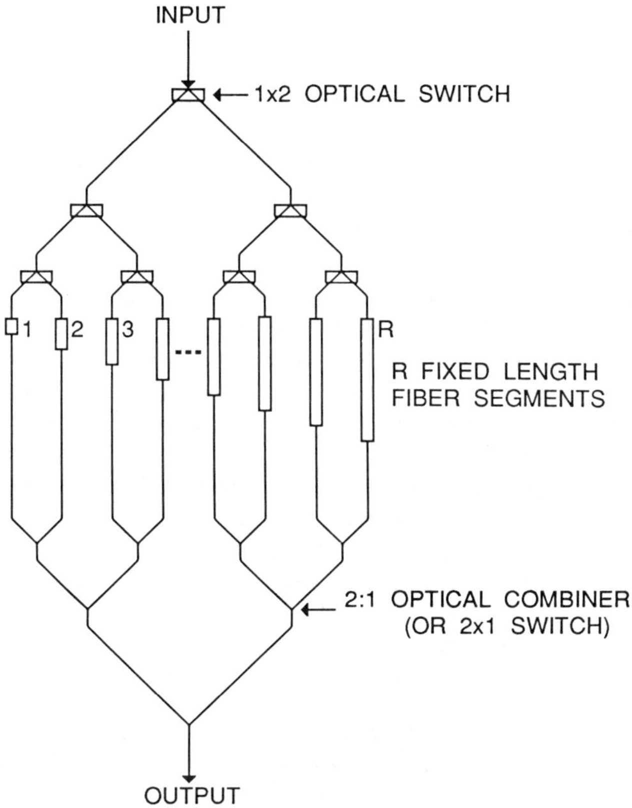


**Figure 13.3** Schematic of a parallel fiber-optic delay line (PADEL).

(or $2\ (K-1)\cdot(R-1)$ if switches are used at the output) and $C_C=(K-1)\cdot(R-1)$. In (13.7) and (13.8), the factor $(K-1)$ rather than $K$ appears because one of the $K$ antenna elements always has zero delay. The IL parameter is determined by the type of structure used at the output. Since the input signal goes through $\log_2 R$ switches to reach any fiber segment, the IL of the PADEL is

$$\mathrm{IL}=(2\log_2 R)\cdot \mathrm{IL}_S \tag{13.9}$$

if switches are used, where $\mathrm{IL}_S$ is the IL loss per switch. If optical combiners are used, then

$$\mathrm{IL}=(\log_2 R)\cdot(\mathrm{IL}_S+3) \tag{13.10}$$

where the 3-dB factor is due to the single-mode optical combiner. Since the signal always goes through the same number of switches (or combiners), the IL parameter is stable (i.e., it depends only on the switch loss variation and not on the actual delay used). For our reference PAA, $C_F = 2.6 \times 10^5$, $C_S = 2.6 \times 10^5$ (or $5.2 \times 10^5$), and $C_C = 2.6 \times 10^5$. Assuming $IL_S = 1.0$ dB, we find that IL = 20 dB if only switches are used and IL = 40 dB if switches and combiners are used. These figures show that a PADEL-based TTD system is not very practical because it requires large amounts of hardware and has large signal losses.

### 13.4.2 Laser Diode Switch Parallel Fiber-Optic Delay Line

The LD switched delay line [3,11,12] is an optoelectronic rather than an all-optical delay line. It is similar to an optical PADEL in that it uses *R* parallel fiber segments, but differs from the PADEL in that it uses *R* LDs as switches (Fig. 13.4); that is, each fiber segment has a separate LD. A specific delay is selected by switching on the bias of the LD pigtailed to the appropriate fiber segment. The LD switching is accomplished in the electrical domain via a current-controlling transistor in the bias circuit. The transistor input in one logic state allows current to flow in the LD to produce a light output, while in the other state the bias current is below threshold so that no light is emitted. The RF input is provided to the *R* LDs via an *R*-way RF power splitter. The *R* fiber segments are combined into a single output via an *R*:1 optical combiner. One of the fiber segments is used as the

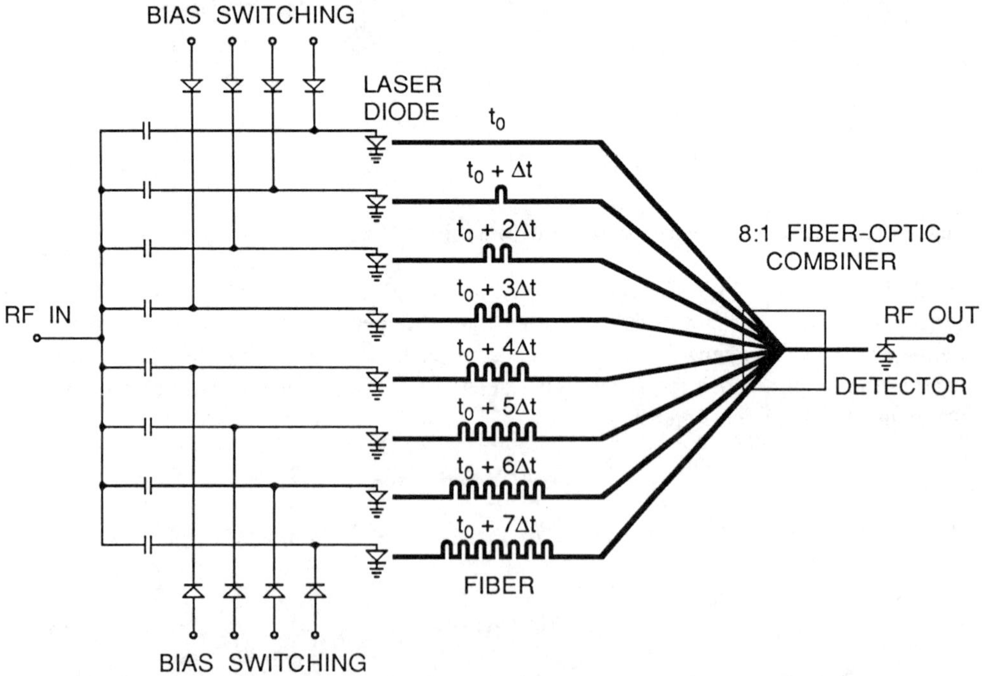


**Figure 13.4** Schematic of a switched LD parallel fiber-optic delay line.

reference delay and all other segments are used to provide delays relative to the reference path.

Since an LD and a fiber segment is required for each delay, the complexity parameters for a full fiber-optic TTD implementation are given by

$$C_{\text{LD}} = C_F = (K - 1) \cdot R \tag{13.11}$$

Clearly, the large number of LDs required is the major disadvantage of this technique. Since each fiber has its own LD, the IL is determined by the $R$:1 optical combiner and is given by

$$\text{IL} = 10 \log_{10} R \tag{13.12}$$

From (13.11) and (13.12), we find that for our reference PAA, $C_{\text{LD}} = C_F = 2.6 \times 10^5$ and IL = 30 dB.

A fiber-optic TTD network based on three-bit LD switching delay lines has been demonstrated [3] in conjunction with a dual-band (L and X) PAA consisting of four subarrays. Each L-band subarray has two monopole radiating elements, whereas each X-band subarray has eight dipole elements. Optical TTD is used only for the four subarrays, and each subarray is individually steered with phase shifters. Each of the four three-bit fiber-optic delay lines consists of eight LDs and a multimode 8:1 fiber-optic combiner. The measured IL is ~10 dB optical and the SNR at 9 GHz is ~107 dB/Hz. A photograph of the three-bit prototype is shown in Figure 13.5. The eight LD switches are shown on the left and the photodetector is shown on the top right.

## 13.5 DELAY-COMPRESSIVE FIBER-OPTIC TTD ARCHITECTURES

The second fiber-optic TTD family we describe compresses the overall hardware with respect to the number of delays. This family significantly reduces the large delay-dependent hardware complexity and it can therefore be used, in conjunction with subarray techniques, for many high-performance PAA systems. In such TTD scenarios, the subarray steering must be accomplished with all-electronic TTD delay lines, which, as we will see later, are practical for relatively small delay ranges (e.g., ~1 ns). Almost all architectures of this family are based on programmable delay lines, which use fiber segments in a serial rather than a parallel fashion. Here, too, the delay is selected by switching on the appropriate fiber via either an optical switch or an LD or a detector. As we will see, most practical PAAs with TTD can employ delay-compressive delay lines only in conjunction with subarray techniques. In this case, timing or amplitude errors in the fiber-optic part of the TTD architecture (i.e., in the optical delay lines) will result in correlated errors, and therefore high-performance fiber-optic delay lines are necessary.

Figure 13.5 Three-bit fiber-optic TTD network prototype based on the switched LD architecture. The eight LD switches are shown on the left and the photodetector is shown on the top right. (Courtesy of Hughes Aircraft Company)

### 13.5.1 Binary Fiber-Optic Delay Line

The binary fiber-optic delay line (BIFODEL) architecture, first suggested by Levine [10], is the most widely used delay-compressive delay line configuration [13–22]. In the BIFODEL architecture (Fig. 13.6(a)), the optical signal is optionally routed through $N$ fiber segments whose lengths increase successively by a power of 2. The required fiber segments are addressed using a set of $N$ $2 \times 2$ optical switches. Since each switch allows the signal to either connect or bypass a fiber segment, a delay $T$ may be inserted which can take any value, in increments of $\Delta T$, up to the maximum value $T_{max}$, given by

$$T_{max} = (2^0 + 2^1 + 2^2 + \ldots + 2^{N-1})\Delta T = (2^N - 1)\Delta T \quad (13.13)$$

For example, in order to introduce a delay of $53\Delta T$ using a six-bit BIFODEL, the signal must be routed through the first, third, fifth, and sixth fiber segments. Note that the actual delays are implemented differentially; that is, it is the difference between the delayed path and the undelayed path that sets the desired delay.

Since there exist $2^N - 1$ possible delay combinations, the number $N$ of switches and delay segments for the BIFODEL of Figure 13.6(a) is given by

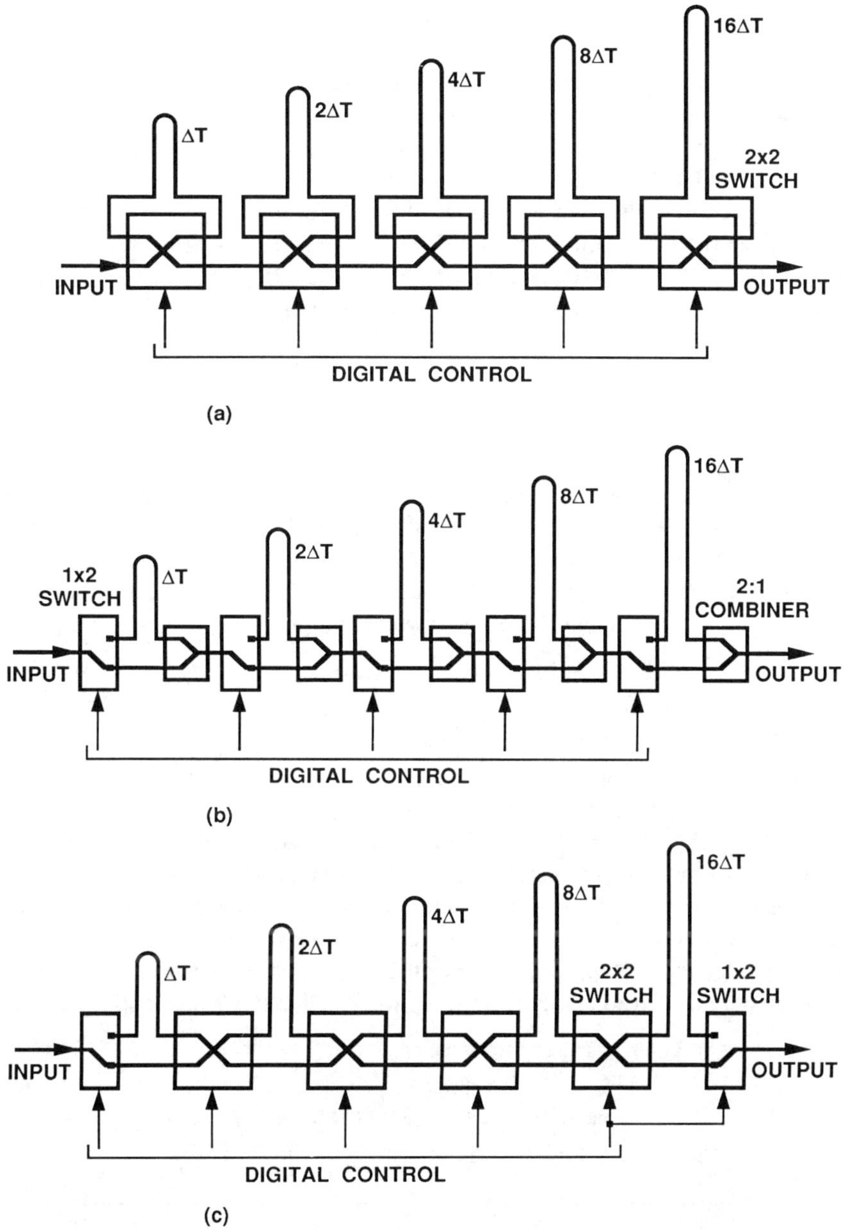


**Figure 13.6** Block diagrams for different BIFODEL architectures: (a) design 1; (b) design 2; (c) design 3.

$$N = \log_2 R \tag{13.14}$$

This results in a very significant reduction of hardware; in fact, the BIFODEL architecture is the most efficient of all delay-compressive delay lines [17]. In practice, the actual BIFODEL complexity depends on the specific design implemented [22]. There are at least three different BIFODEL architectures for which the $C$, IL, and IL stability values are different. Figure 13.6(a–c) shows three different designs for $N = 5$. Design 1 (Fig. 13.6(a)) requires the lowest component complexity of $N$ $2 \times 2$ switches. However, it has a nonstable IL that varies between $(\log_2 R) \cdot \mathrm{IL}_S$ and $2 \cdot (\log_2 R) \cdot \mathrm{IL}_S$ as the BIFODEL program changes. This is because, depending on the program, the signal might enter the same switch twice, thereby increasing the IL. Design 2 (Fig. 13.6(b)) uses $N$ $1 \times 2$ switches as well as $N$ 2:1 fiber-optic combiners. For this architecture,

$$\mathrm{IL} = (\log_2 R) \cdot (\mathrm{IL}_S + 3) \tag{13.15}$$

where the 3-dB factor is due to the insertion loss of the 2:1 fiber-optic combiner. Assuming that all switches have the same $\mathrm{IL}_S$ value and that no significant attenuation changes occur as different-length fiber segments are switched in, the loss is independent of the BIFODEL switch program; that is, the loss is stable. Design 3 (Fig. 13.6(c)) uses $(N - 1)$ $2 \times 2$ switches and two $1 \times 2$ switches, for a total switch complexity of $(\log_2 R + 1)$. This design has a stable IL given by

$$\mathrm{IL} = (\log_2 R + 1) \cdot \mathrm{IL}_S \tag{13.16}$$

Design 3 is the optimum configuration in terms of minimizing both complexity and loss. Table 13.1 shows a quantitative comparison of the performance of the three BIFODEL designs for $N = \log_2 R = 6$ and $\mathrm{IL}_S = 1$ dB. The table shows that design 3 is the best because it has minimum, stable loss and a very low complexity. The less complex design 1 can be very lossy, and, most importantly, its loss is not stable, which means that significant correction must be made (up to 12 dB in the RF domain).

Using design 3 as the optimum BIFODEL design for implementing the fiber-optic TTD K-element PAA (on a per element basis), we find that the switch and fiber complexities are

**Table 13.1**
Comparison of the Three BIFODEL Designs

| | *IL (dB)* | *Stability (dB)* | *Complexity* |
|---|---|---|---|
| Design 1 | 6–12 | ±3 | 6 |
| Design 2 | 24 | 0 | 12 |
| Design 3 | 7 | 0 | 7 |

Note: $N = 6$, $\mathrm{IL}_s = 1$ dB.

$$C_S = (K - 1) \cdot (\log_2 R + 1) \tag{13.17}$$

$$C_F = (K - 1) \quad \log_2 R \tag{13.18}$$

From (13.16) to (13.18) we find that for our reference PAA, $C_S = 2.8 \times 10^3$, $C_F = 2.6 \times 10^3$, and IL = 11 dB. These complexities are lower by about two orders of magnitude than those of the noncompressive architectures; however, they are still large for most practical PAAs. In general, BIFODELs can be practical TTD devices so long as (1) the number of PAA elements is relatively small (e.g., a few tens), or (2) they are used in conjunction with antenna subarray techniques. Since most practical PAAs have at least several hundreds of elements, the latter rather than the former scenario is more realistic.

In order to avoid ambiguities as a result of signal splitting, the minimum programmability period $T_{pr}$, defined as the time difference between the setup of two consecutive programs which occur at times $t = t_{i-1}$ and $t = t_i$, must satisfy

$$T_{pr} \geq T_s + T_{i-1} \tag{13.19}$$

where $T_s$ is the signal duration and $T_{i-1}$ is the total delay introduced at time $t = t_{i-1}$. However, there are many signal-delay scenarios for which $T_{pr}$ can be reduced. For example, consider the scenario where the signal is a pulsed waveform whose period $T_p$ is compatible with $T_{max}$ and whose pulsewidth $T_w$ is less than $2^{N-1}\Delta T$. In this case, and under the assumption that two consecutive programs require the use of the longest delay segment available, $T_{pr}$ need only satisfy

$$T_{pr} \geq T_w + T_{i-1} - 2^{N-1}\Delta T \tag{13.20}$$

BIFODELs can be implemented with a variety of delay media and/or switches including combinations of fiber and/or free space and/or IO delays. For example, a three-bit IO BIFODEL has been demonstrated by Lagerstroem et al. [14] using Ti:$LiNbO_3$, and Sullivan et al. [20] have reported a three-bit IO BIFODEL implemented on GaAs. Figure 13.7 shows a photomicrograph of the GaAs three-bit IO BIFODEL, which has delays of 250 ps, 176 ps, and 83 ps per bit. A six-bit BIFODEL based on design 3 and implemented at the Westinghouse Science and Technology Center by the authors of this chapter is shown in Figure 13.8. For this prototype, low-loss (<1.0 dB), low-crosstalk (<−70 dB) piezomechanical switches have been used, and the delays implemented are 97.8, 195.5, 391.0, 782.0, 1564.1, and 3128.2 ps. The overall accuracy in these delays is ±1.5 ps (independent of the delay value), and the measured SNR is 145 dB/Hz. Multichannel, free-space BIFODELs have been demonstrated by Dolfi et al. [23], who used 2-D SLMs to achieve parallel switching. Goutzoulis et al. [16] have reported an electronically switched seven-bit BIFODEL using fiber-optic links for implementing the delays. The switching is performed with two back-to-back 1 × 2 GaAs FET switches, and the zero delay path is implemented with a microwave strip line. The overall BW is 0.5 to 1 GHz, the maximum

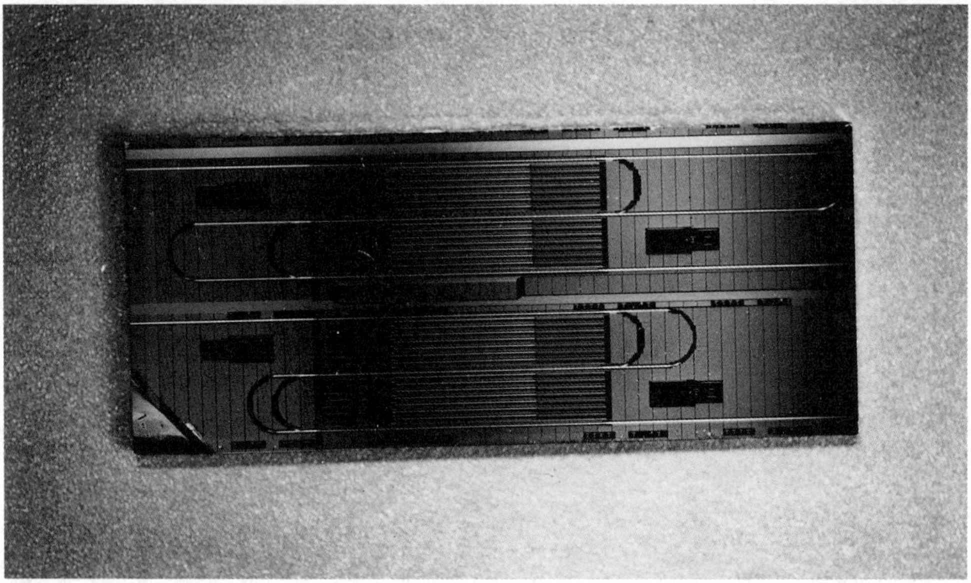

**Figure 13.7** Photomicrograph of first-generation GaAs optical time delay generators built at Honeywell Systems and Research Center [20]. This GaAs test chip is about 1-in long and is designed for three-bit delays of 250 ps, 176 ps, and 83 ps, corresponding to phase shifts of $3\pi/2$, $\pi$, and $\pi/2$ (and zero along the reference path) at 3 GHz (courtesy of Honeywell).

delay is 5 $\mu$s, the resolution is 39 ns, and the measured SNR is 123 dB/Hz. Ng et al. [24] have proposed another type of hybrid BIFODEL, where in each delay stage a different LD is used. The switching is performed in a way similar to that proposed by Levine [10]. The light from each LD is coupled into two fibers that form the reference and delayed paths, and each fiber in the pair is coupled to its own detector; the detector outputs are then combined electronically in an RF combiner. Thus, the zero or delayed path is selected by switching the voltage bias of the appropriate detector. The combiner output is then amplified and drives the LD of the next stage.

In general, all-optical BIFODELs are preferable because (1) they can be made bidirectional, and (2) they are transparent to microwave signals up to 100 GHz, and high-fidelity transmission is therefore possible [18]. Furthermore, hybrid BIFODELs require RF amplifiers per bit of delay. These amplifiers must have enough gain to recover (1) the fiber-optic splitting losses and (2) the electrical-to-optical and optical-to-electrical conversion losses. The presence of many cascaded electronic and optoelectronic components (especially LDs) may dramatically reduce the spurious-free DR of the system [16]. For delays of less than ~1 ns, IO approaches can result in smaller overall size and potentially lower cost. However, for longer delays, fiber must be used (as in the case for free-space BIFODELs [23]). IO switches are currently very lossy (about 3 dB/bit loss) and have high crosstalk (typically >−25 dB per switch).

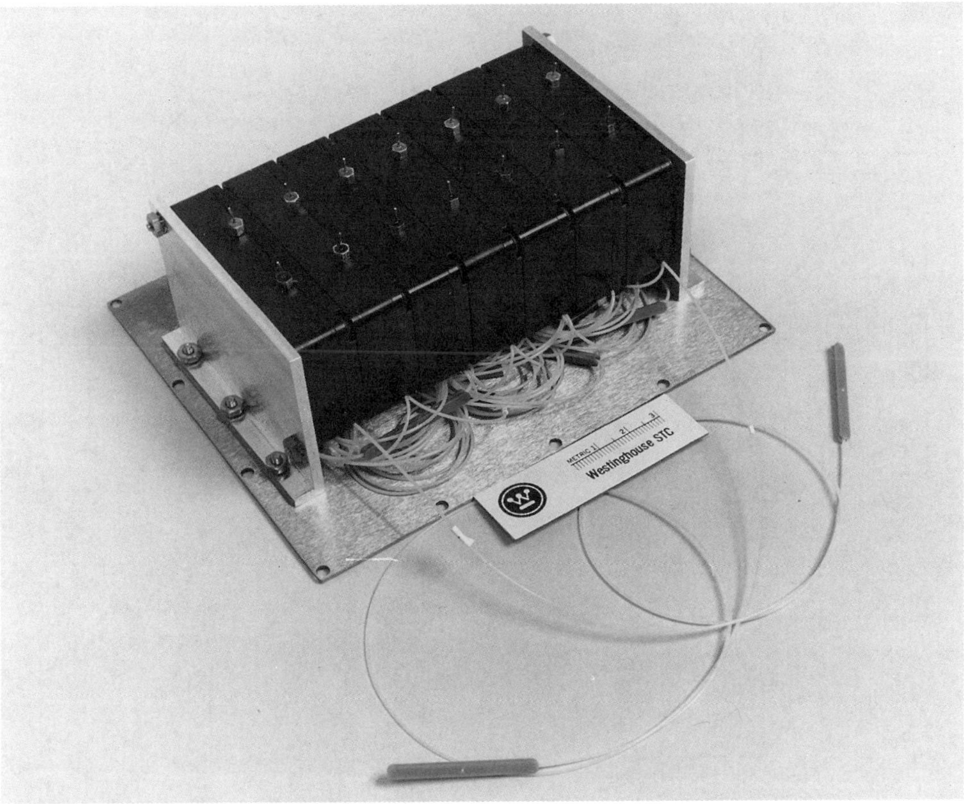


**Figure 13.8** A six-bit design 3 BIFODEL prototype using piezomechanical switches. The delays cover the range of 97.8 to 3128.2 ps with ±1.5-ps accuracy (courtesy of Westinghouse Electric Corporation).

### 13.5.2 Square-Root Fiber-Optic Delay Line

The square-root fiber-optic delay line (SRODEL) first proposed by Soref [25] and shown in Figure 13.9, is basically a dual, cascaded-segment, delay line, where the desired segment is selected from a set of $\sqrt{R}$ segments. Selection of the appropriate delay path is made using a tree-type structure of $1 \times 2$ switches at the input of each set, similar to that described for the PADEL architecture in Section 13.4.1. For the SRODEL architecture, each set of segments requires $(\sqrt{R} - 1)$ switches, and thus the total number of switches is $(2\sqrt{R} - 2)$. The number of switches traversed by the signal for each set of segments is $\log_2(\sqrt{R})$, so that the total number of switches through which the signal must pass before leaving the SRODEL is $2 \log_2(\sqrt{R})$. Thus, for this architecture, the performance parameters are

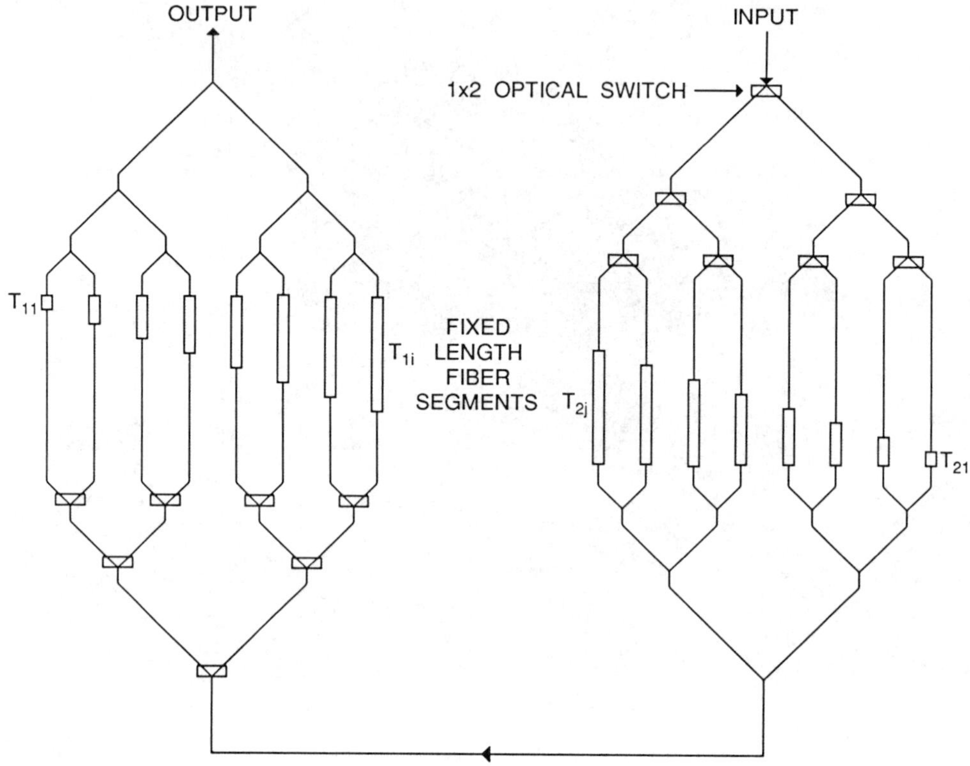


**Figure 13.9** Schematic of a SRODEL.

$$C_S = (K - 1) \cdot (2\sqrt{R} - 2) \tag{13.21}$$

$$C_F = (K - 1) \cdot 2\sqrt{R} \tag{13.22}$$

$$\text{IL} = 2\,\log_2(\sqrt{R}) \cdot \text{IL}_S \tag{13.23}$$

From (13.21) to (13.23), we find that for our reference PAA, $C_S = 1.6 \times 10^4$, $C_F = 1.6 \times 10^4$, and IL = 10 dB. Soref [25] has also proposed optimized variants of the SRODEL architecture of Figure 13.9. However, overall, the SRODEL architecture has been found to be less optimum than the BIFODEL for most practical PAA scenarios [17]. To the best of our knowledge, no SRODEL prototype has been reported in the open literature.

## 13.6 DELAY- AND ELEMENT-COMPRESSIVE FIBER-OPTIC TTD ARCHITECTURES

In this section, we discuss architectures that compress the overall hardware with respect to both the number of delays $R$ and the number of elements $K$. All architectures of this family are based on optical WDM used in conjunction with all-optical programmable delay lines, such as optical BIFODELs. Different thinking must be applied in describing the WDM fiber-optic TTD techniques for 1-D and 2-D PAAs. In general, WDM techniques can produce correlated errors, which may limit the overall system performance. Therefore, in principle, good accuracy is required in implementing the actual delays and in equalizing the losses of the various delay paths. Overall, the required accuracy is similar to that required by delay-compressive architectures used in conjunction with subarray techniques.

### 13.6.1 Wavelength-Multiplexed TTD Architectures for 1-D PAAs

Wavelength-multiplexed TTD architectures for 1-D PAAs can be achieved by partitioning the $K$ elements into $E$ sets of $M$ elements (i.e., $K = M \cdot E$) and exploiting the delay relationships among neighboring elements in the PAA. Using the indexes $j = 1, 2, \ldots, E$, and $i = 1, 2, \ldots, M$ to represent the set and element location, respectively, the equivalence between the delay required by the $k$th element $T_k$ and that required by the $ij$th element $T_{ji}$ (which is at the same location as the $k$th element) requires the following relationships:

$$j = \{(k - 1)/M\} + 1 \tag{13.24}$$

$$i = k - (j - 1)M \tag{13.25}$$

where $\{(k - 1)/M\}$ denotes the integer part of the quotient $(k - 1)/M$. Solving (13.25) with respect to $k$, we see that the delays can be expressed by

$$T_k = T_{ji} = T_i + (j - 1)T_M \tag{13.26}$$

where $T_i$ and $T_M$ are the delays of the $i$th and $M$th elements of the first set, which we will name the *reference set* (RS). Equation (13.26) shows that the delay required by the $i$th element of the $j$th set is equal to the delay of the $i$th element of the RS plus a constant (or bias) delay which is equal to $(j - 1)T_M$. We note that the bias delay depends only on $j$ and not on $i$, and thus it is common to all the elements of a given set. This situation is described graphically in Figure 13.10, which shows an $E$-way partitioned array with $M = 4$, where $T_{bj} = (j - 1)T_M$ represents the bias delay necessary for the $j$th set. Note that $E$ different bias delay lines are required, each with a total of $R$ different delays (i.e., the same number of delays as the delay lines of the RS). The bias delay resolution is equal to $(j - 1)T_{M\min}$, where $T_{M\min}$ is the delay resolution of the $M$th element of the RS.

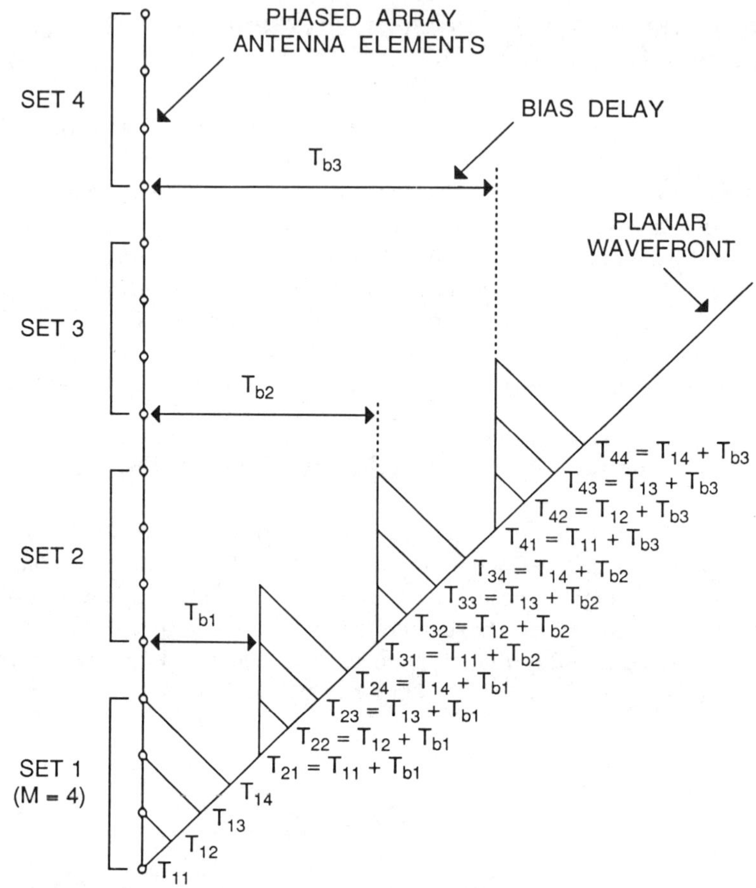


**Figure 13.10** Partitioning principles for a 1-D phased-array antenna.

In general, PAA partitioning results in very significant hardware reduction in both delay line type (i.e., delay lines with different resolution) and delay line quantity. Reduction in delay line type occurs because the total number of different types of delay lines can be as low as $M + E$ (i.e., $M$ delay lines for the RS plus $E$ bias delay lines). Reduction in the total number of delay lines occurs because only one bias delay line is required per set, and it is possible to cascade each of the $M$ delay lines of the RS to all $E$ bias delay lines and thereby address all the $M \cdot E$ elements of the PAA.

### *13.6.1.1 Fiber-Optic Wavelength-Multiplexed TTD Architecture*

The partitioned PAA concept can be uniquely implemented using optical WDM in conjunction with all-optical programmable delay lines and is reversible; that is, the same hardware can be used for both the transmit and receive PAA modes. For the remainder of this chapter, we describe and analyze the overall WDM fiber-optic TTD architectures incorporating BIFODELs.

In the transmit mode (Fig. 13.11), $M - 1$ RS BIFODELs with outputs at wavelengths $\lambda_2, \ldots, \lambda_M$ are driven in parallel by the microwave radar signal. The $M - 1$ BIFODEL outputs, along with an undelayed output at wavelength $\lambda_1$, are multiplexed via an $M$-channel multiplexer (MUX), the output of which is divided into $E$ channels via a 1:$E$-channel optical divider. All but one of the divider outputs independently drive a bias BIFODEL, each of which is followed by an optical $M$-channel demultiplexer (DEMUX). The divider output channel that does not drive a bias BIFODEL is also demultiplexed. Since the optical inputs to each bias BIFODEL contain $M$ optical wavelengths, the DEMUX output will also contain $M$ wavelengths $\lambda_1, \lambda_2, \ldots, \lambda_M$. The outputs of the nonbiased DEMUX (extreme left in Fig. 13.11) contain the $M$ progressively delayed signals required for the RS (set 1 in Fig. 13.11) which, as Figure 13.10 shows, requires no bias delay. The outputs of each of the remaining DEMUXs contain a similar set of signals (but which are further delayed via the bias BIFODELs) and correspond to a different PAA set. Similar wavelength outputs drive similar location elements in each set. All BIFODELs must have $N = \log_2 R$ cascaded segments and different time resolution $T_{i\min}$. The latter is determined by the location of the specific element, the antenna geometry, the radar characteristics, and so on. Similar comments apply to the bias BIFODELs, which have time resolutions $(j - 1)T_{M\min}$.

In the receive mode, the same architecture is used, but in reverse. Here, the output of each PAA element drives an LD of a different wavelength. Elements with similar locations in different sets drive LDs of the same wavelength. For each PAA set, the LD outputs are multiplexed and drive a bias BIFODEL. The bias BIFODELs eliminate the set-to-set bias delays (i.e., in Fig. 13.10, all $T_{bi}$ bias delays are removed), and the wavefront looks like a sawtooth having a period of a PAA set. The outputs of the bias BIFODELs are then combined via an $E$-channel optical combiner (i.e., a 1:$E$-channel optical divider operating backwards), the output of which is subsequently demultiplexed (i.e., via the $M$-channel MUX operating backwards). Each of the DEMUX outputs (which contains similar wavelengths) drives an RS BIFODEL. These RS BIFODELs eliminate the in-set delays (i.e., in Fig. 13.10, all $T_{ji}$ are now equal), and the wavefront is parallel to the antenna axis. The remaining step is to add the outputs of the reference BIFODELs via a combiner, the output of which provides the desired vector sum. The final summation can take place in either the optical or the RF domain.

We note that the partitioned PAA concept can be arranged so that it generates fully symmetric $\pm T_{ij}$ delays; that is, it can be made to operate over symmetric scan angles. The insertion of "negative" delays can be accomplished by time-biasing the antenna so that the zero delay position corresponds to the "negative" maximum scan angle. For example, consider an antenna operating over a scan angle of $\pm\theta_{sc}$ in conjunction with a TTD network that can provide a maximum delay corresponding to $\theta = 2\theta_{sc}$. We can time-bias the antenna so that when the variable TTD network provides zero delay, the antenna is steered at $\theta$

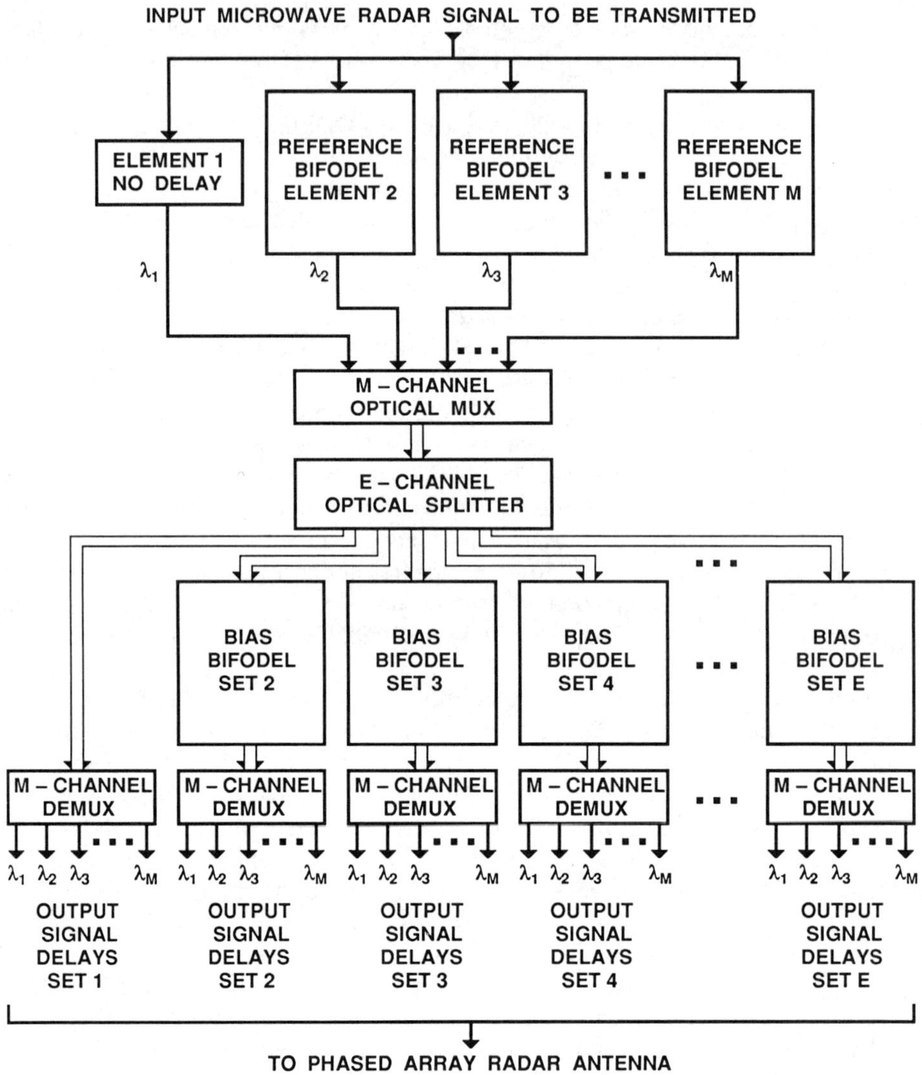


**Figure 13.11** Schematic of a BIFODEL-based delay- and element-compressive fiber-optic TTD architecture for 1-D PAAs using optical WDM.

$= -\theta_{sc}$. Thus, when the TTD network provides exactly half of the maximum delay, the antenna is steered at $\theta = 0$, and when the maximum delay is provided, the antenna is steered at $\theta = +\theta_{sc}$. This time bias can be accomplished by inserting linearly increasing but fixed time delays (e.g., fixed fiber lengths) into the paths of the various antenna elements.

From the above discussion, and for the case where $M = E = \sqrt{K}$, one can easily show that the total number of BIFODELs $C_{\text{CDL}}$ is equal to

$$C_{\text{CDL}} = 2(\sqrt{K} - 1) \tag{13.27}$$

When the partitioned PAA concept is implemented with BIFODELs, the overall switch complexity $C_{\text{CS}}$ or fiber complexity $C_{\text{CF}}$ as a function of the resolution $R$ and the number of elements $K$ (again for $M = E = \sqrt{K}$) is given by

$$C_{\text{CS}} = C_{\text{CF}} = (\log_2 R) \cdot 2(\sqrt{K} - 1) \tag{13.28}$$

The switch (or fiber) complexity for the delay- and element-compressive fiber-optic TTD architecture (i.e., (13.28)) are first compared with the switch (or fiber) complexity for a typical noncompressive fiber-optic TTD architecture for which $C_S \approx C_F \approx (K - 1) \cdot R$. In Table 13.2, we show examples of $C_{\text{CS}}$ versus $C_S$ as a function of selected values of $R$ and $K$. Table 13.2 clearly demonstrates the dramatic hardware complexity reduction (two to three orders of magnitude) that is possible using antenna partitioning in conjunction with BIFODELs.

Next, we compare the delay- and element-compressive TTD architecture complexity with that of a delay-compressive TTD. Since the optimum type of delay line for either architecture is the BIFODEL, the comparison can be simplified by using the delay line complexity (i.e., the number of BIFODELs) for each architecture. The delay line complexity for the former, $C_{\text{CDL}}$, is given by (13.27), whereas the delay line complexity for the latter, $C_{\text{NDL}}$, is equal to $K - 1$. Table 13.3 shows the values of $C_{\text{CDL}}$ and $C_{\text{NDL}}$ for selected values of $K$, as well as the percentage reduction in hardware afforded by delay and element compression. This comparison demonstrates that for typical 1-D PAAs, the delay- and element-compressive TTD architecture provides a very significant hardware reduction on the order of 80% to 90%.

In practice, the limits on the dimensionality of the WDM fiber-optic TTD architecture are dictated by the receive rather than the transmit mode, and they are determined by the

**Table 13.2**
Comparison of the Overall Switch (or Fiber) Complexity for Selected Values of $R$ and $K$

| | $C_S$ | | | $C_{\text{CS}}$ | | |
|---|---|---|---|---|---|---|
| $K$ | $R = 256$ | $R = 1{,}024$ | $R = 4{,}096$ | $R = 256$ | $R = 1{,}024$ | $R = 4{,}096$ |
| 100 | $2.5 \times 10^4$ | $1.0 \times 10^5$ | $4.1 \times 10^5$ | $1.4 \times 10^2$ | $1.8 \times 10^2$ | $2.2 \times 10^2$ |
| 256 | $6.5 \times 10^4$ | $2.6 \times 10^5$ | $1.0 \times 10^6$ | $2.4 \times 10^2$ | $3.0 \times 10^2$ | $3.6 \times 10^2$ |
| 512 | $1.3 \times 10^5$ | $5.2 \times 10^5$ | $2.1 \times 10^6$ | $3.5 \times 10^2$ | $4.4 \times 10^2$ | $5.3 \times 10^2$ |

Note: $C_{\text{CS}}$ = delay- and element-compressive architecture; $C_S$ = noncompressive architecture; $R$ = number of delays per element; $K$ = number of elements.

**Table 13.3**
Comparison of Delay Line Complexities

| $K$ | $C_{NDL}$ | $C_{CDL}$ | *Reduction (%)* |
|---|---|---|---|
| 64 | 63 | 14 | 78 |
| 128 | 127 | 21 | 83 |
| 256 | 255 | 30 | 88 |
| 512 | 511 | 44 | 91 |

Note: $C_{NDL}$ = delay-compressive true-time-delay architectures; $C_{CDL}$ = delay- and element-compressive true-time-delay architectures; $R$ = number of delays; $K$ = number of elements.

beat notes between the various optical wavelengths [22]. It can be shown that by using state-of-the-art LDs with modest temperature regulation, systems of up to 150 elements can be implemented with a full BW of over 18 GHz. Larger systems are possible at the expense of a reduction in BW.

In general, the WDM architecture has more loss than a straightforward BIFODEL because of the use of cascaded BIFODELs together with three additional optical components: MUX, 1:$E$ divider, and DEMUX. It can be shown [22] that the insertion loss can be written approximately as

$$\mathrm{IL} = 2(\log_2 R + 1)\mathrm{IL}_S + \mathrm{IL}_M + \mathrm{IL}_C + \mathrm{IL}_D \tag{13.29}$$

where $2(\log_2 R + 1)\,\mathrm{IL}_S$ is due to the cascaded BIFODELs, and $\mathrm{IL}_M$, $\mathrm{IL}_C$, and $\mathrm{IL}_D$ are the insertion losses of the MUX, 1:$E$ divider, and DEMUX, respectively. In (13.29), the factor $(\log_2 R + 1)\mathrm{IL}_S + \mathrm{IL}_M + \mathrm{IL}_C + \mathrm{IL}_D$ represents the *excess loss* of the WDM fiber-optic TTD system over a single BIFODEL. The actual value of this excess loss depends on several factors, including $R$, $K$, and the actual MUX, DEMUX, and optical divider. For practical devices, $\mathrm{IL}_M$ and $\mathrm{IL}_D$ are similar and on the order of 3.5 dB. $\mathrm{IL}_C$ depends on the number of channels ($E$) into which the optical signal must be divided (i.e., on the partitioning itself). For example, for a 16 × 16 system, a typical 1:16 optical divider will have a value of $\mathrm{IL}_C \sim 12$ dB plus ~1 dB of excess loss. However, we note that the splitting loss does not represent an energy loss, since a single optical channel drives $E$ channels. However, all the factors in (13.29) will result in lower system SNR. Most of this SNR loss can be recovered by (1) an optical amplifier positioned between the MUX and the 1:$E$ divider, or (2) transimpedance amplifiers connected at the output detectors.

#### *13.6.1.2 Hybrid Electronic Fiber-Optic Wavelength-Multiplexed TTD Architecture*

The all-optical architecture of Section 13.6.1.1 can be optimized considerably depending on the actual PAA application. For example, consider the following two PAA scenarios operating in the L and S frequency bands: (1) a 10.6m long L band ($f$ = 1.4 GHz) 1-D

PAA with $K = 100$ elements, and (2) a 12.7m-long S band ($f = 3.0$ GHz) 1-D PAA with $K = 256$ elements. Assuming that the PAAs are partitioned with $M = E = \sqrt{K}$, we find that the maximum RS delays occur for element number 10 and number 16, respectively, for the two different scenarios. For a maximum scan angle of ±45 deg in conjunction with a six-bit BIFODEL, one can show that the BIFODEL delays per bit are, for example, (1) 73, 147, 293, 586, 1173, and 2346 ps, and (2) 57, 114, 228, 456, 912, and 1824 ps.

For both the L- and S-band PAA examples, the RS delays are well within the transmission capability of microwave microstrip or strip line. They can be implemented without serious differential attenuation and/or delay dispersion effects as a function of frequency. For example, using state-of-the-art microwave board, delay lines with at least 1.5-ns delay can be fabricated with differential attenuation of ~ 0.7 dB and ±(2–4) ps delay dispersion over the 0.5- to 4-GHz band. In addition, low-cost 1 × 2 GaAs FET switches are available that operate well over the S band with very low insertion loss (<0.5 dB) and a response which is flat (to better than ±0.25 dB) over the 0.5- to 3.5-GHz band. From these data, we conclude that for typical L- and S-band PAA applications, the reference BIFODELs can be implemented using all-electronic techniques (i.e., electronic binary delay lines (DiBi) are possible). Such delay lines are often used as phase shifters in current PAA systems [26]. We emphasize that this is not necessarily the case for all PAA scenarios because, for large PAAs at higher frequencies (e.g., X band), the board and/or cable attenuation and/or dispersion may be unacceptable.

DiBis can be implemented using two back-to-back 1 × 2 FET switches in conjunction with differential delays. They are fully reversible and offer certain advantages over the equivalent BIFODELs, including (1) much lower cost (about two orders of magnitude less per delay line, the main difference being the switch cost), (2) the potential to implement the RS delays in IC form [27], and (3) much smaller size—DiBis are inherently 2-D devices, whereas BIFODELs are 3-D devices.

The use of DiBis for the RS delays and BIFODELs for the bias delays results in a hybrid electronic fiber-optic TTD WDM architecture [22]. In such a system, each DiBi is followed by an LD. The M LDs are then multiplexed and subsequently divided into E channels, each driving a different BIFODEL. This hybrid architecture has the following advantages over an all-optical approach: (1) it uses fiber-optics only where standard low-cost microwave electronic techniques cannot perform; (2) it preserves the unique features of optics: (a) WDM for implementing the element-compression architecture, and (b) use of optical fiber for implementing long delays; and (3) it reduces the optical IL by a factor of $(\log_2 R + 1)IL_S$ (because it does not use cascaded BIFODELs). We also note that it is not necessary to implement all bits of the RS delay lines in the electronic domain; one can implement as many bits as possible in the electronic domain and then revert to fiber-optic delays prior to WDM. This allows the hybrid scheme to be used for very large PAAs, for which the sole use of DiBis at the RS level may not be possible. Finally, we note that since both the DiBis and BIFODELs are reversible, the hybrid architecture is also reversible.

A prototype hybrid WDM TTD system has been designed and fabricated at the Westinghouse Science and Technology Center. The prototype, shown in Figure 13.12, has the following specifications: $K = 16$, $R = 64$, and BW = 0.6 to 1.6 GHz. The DiBis cover the delay range 8 to 1,540 ps with an accuracy of ±1.5 ps. The IL of the DiBis is 6 dB (RF), the SNR is 175 dB/Hz, and the spurious-free DR is 135 dB·Hz$^{2/3}$. The BIFODELs cover the delay range 33 to 6,159 ps, also with an accuracy of ±1.5 ps, and have an IL of ≤4.5 dB (optical). The full system SNR is 135 dB/Hz and is determined by the SNR of the fiber-optic part of the overall system.

### 13.6.2 Fiber-Optic Wavelength-Multiplexed TTD Architecture for 2-D PAAs

The WDM fiber-optic TTD concept can also be applied to 2-D PAAs, but without the necessity for antenna partitioning. The basic idea [19] is to use two sets of optical delay lines, one set with $K_x - 1$ delay lines for TTD control along the horizontal direction (called *H-control*) and another set with $K_y - 1$ delay lines for TTD control along the vertical direction (called *V-control*). The two delay line sets are interconnected such that all the delays from the H-control are combined with all delays from the V-control, thereby creating a total of $K_x \cdot K_y$ different delay combinations. These different delays are separated via the use of both WDM and spatial coding.

The concept is illustrated in the 3 × 3 BIFODEL-based transmit example shown in Figure 13.13 (the receive mode can be implemented in a similar manner). For simplicity, we will use $K_x$ and $K_y$ delay lines rather than $K_x - 1$ and $K_y - 1$ required by an actual system. In a 2-D PAA, the direction of the beam in the transmit (or receive) mode can be uniquely described by the two angular components: one along $x$ (i.e., $\theta_x$) and one along $y$ (i.e., $\theta_y$). The beam can be steered to the desired direction if independent control exists along $x$ and $y$. Control along $\theta_x$ (H-control) is established using $K_x$ different BIFODELs (three for the 3 × 3 example of Fig. 13.13). These BIFODELs are different because they address different elements along $x$, each of which requires its own delay range. All these BIFODELs have the same number of segments, $N_x = \log_2 R_x$, which is determined by the angular resolution along $x$. The RF signal to be transmitted drives $K_x$ LDs at wavelengths $\lambda_i$, $i = 1, 2, \ldots, K_x$ ($\lambda_1$, $\lambda_2$, $\lambda_3$ for the example of Fig. 13.13). Next, a MUX is used to combine the outputs of the H-control BIFODELs. The output from the MUX drives a $K_y$ channel optical divider, the outputs of which drive the $K_y$ V-control BIFODELs. (In Fig. 13.13, the MUX and 1:$K_y$ division operations are accomplished by the block entitled "3 × 3 MUX COUPLER.") Note that each of the V-control BIFODELs is different because it addresses different elements along $y$, each of which requires its own delay range. All V-control BIFODELs have the same number of segments, $N_y = \log_2 R_y$, which depends on the desired angular resolution along $y$. Unlike the 1-D PAA case, $N_x$ and $N_y$ may be different if so desired. The last step involves the separation of the $K_x$ different wavelengths at the output of each V-control BIFODEL via a 1:$K_x$ DEMUX.

Since we have a total of $K_y$ DEMUXs, each of which has $K_x$ outputs, we have the necessary $K_y \cdot K_x$ different optical signals for driving the $K_x \cdot K_y$ PAA elements. Each

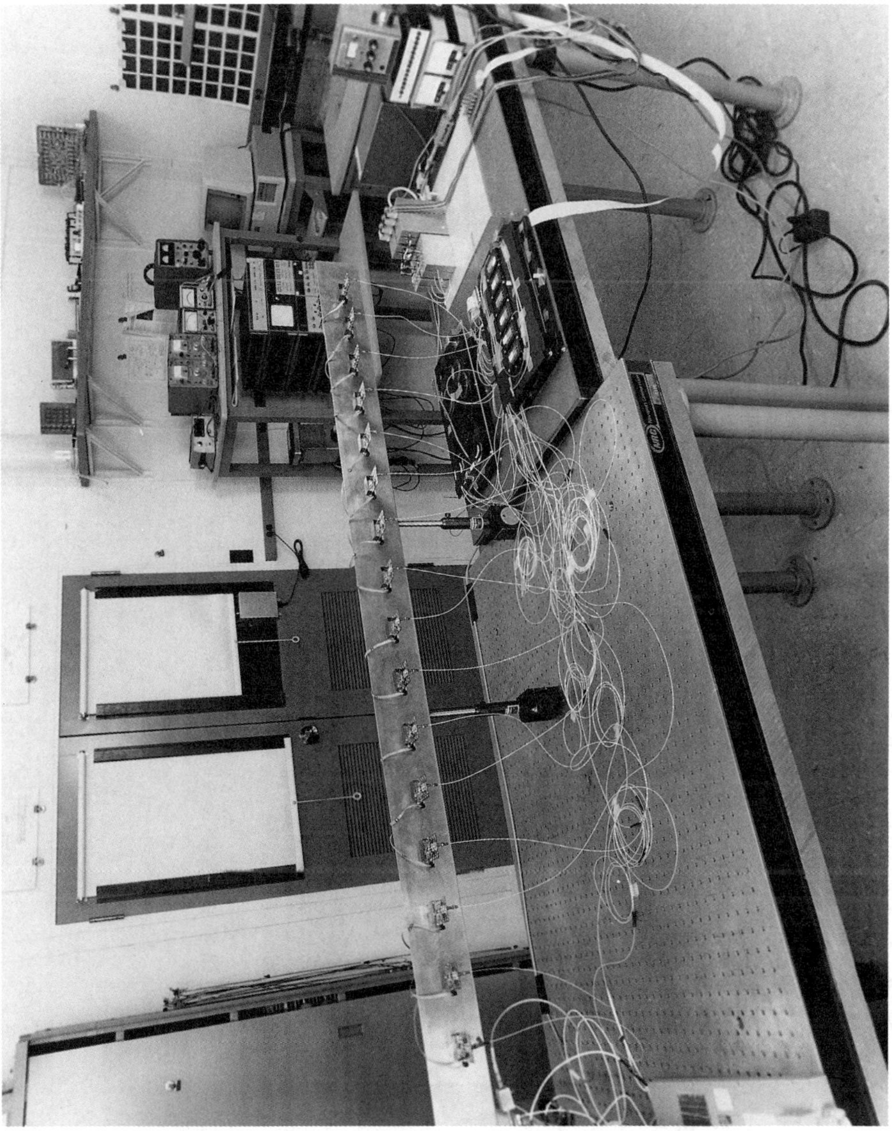

**Figure 13.12** A prototype 16-element, six-bit hybrid WDM TTD system operating over the 0.6- to 1.6-GHz band (courtesy of Westinghouse Electric Corporation).

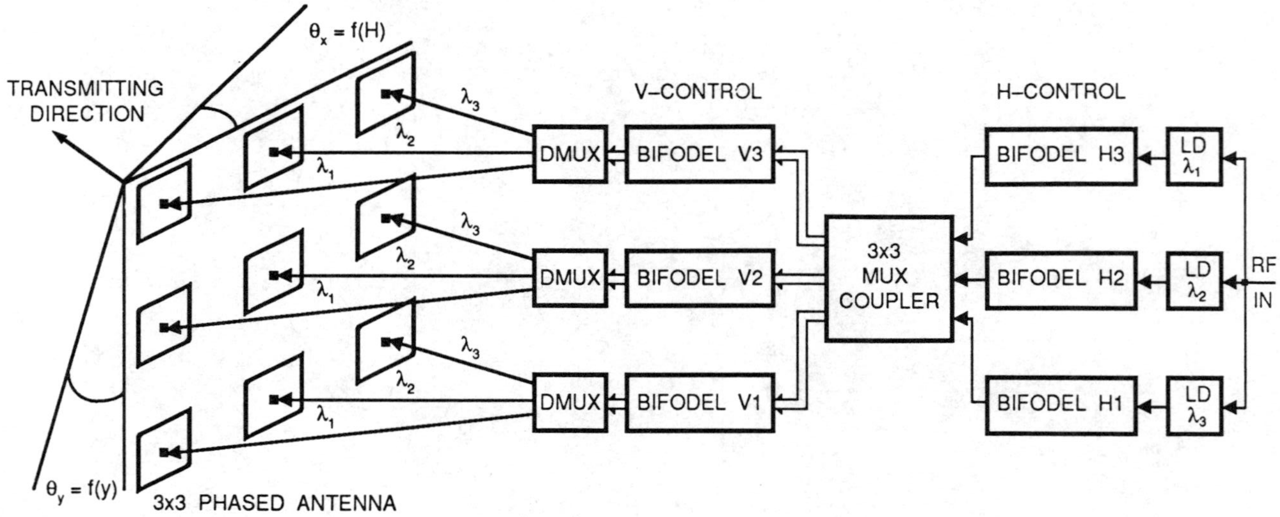


**Figure 13.13** Schematic of a BIFODEL-based delay- and element-compressive fiber-optic TTD architecture for 2-D PAAs using optical WDM.

optical signal has a unique delay $T_{ij}$, where the subscripts $i$ and $j$ depend on the location of the H- and V-control BIFODELs, respectively, and correspond to the desired $\theta_x$ and $\theta_y$ angles. Different programs are used for the H- and V-control BIFODELs in order to independently control the $\theta_x$ and $\theta_y$ angles. However, as in the case of the 1-D WDM TTD concept, the 2-D system can also operate over symmetric scan angles via time-biasing, with the understanding that the time bias now extends to two dimensions.

The 2-D WDM TTD system has the advantage (similar to the 1-D WDM TTD system) of very low BIFODEL complexity $C_{2\text{CDL}}$, given by

$$C_{2\text{CDL}} = K_x + K_y - 2 \tag{13.30}$$

Equation (13.30) should be compared with the delay line complexity $C_{2\text{NCDL}}$ required by the noncompressive 2-D architecture, which is equal to $(K_x - 1) \cdot (K_y - 1)$. The savings are similar to those shown in Table 13.3, and the IL is similar to that of the 1-D case (13.29).

It is of interest to note that the 2-D WDM TTD system allows the formation of multiple beams at different frequencies steered in the same direction. However, it only allows for one delay configuration at a time, which implies that only one direction at a time can be formed in either the transmit or the receive mode. Multiple simultaneous beams at different directions require the simultaneous use of different delay combinations, which is obviously not possible with BIFODELs that employ two-state switches. The straightforward solution is to double the hardware for two beams or quadruple it for four beams. However, less hardware is required if we partition the 2-D PAA. Figure 13.14 shows an example of a 16 × 16-element partitioned array for generating four independent beams in both frequency and direction. The idea is to use the same number of H-control BIFODELs (i.e., 16 for our example) and double the number of the V-control BIFODELs every time we need to double the number of independent beams. For the four-beam example, we use four sets of 4 H-control BIFODELs (for an H-control total of 16) in conjunction with four sets of 16 V-control BIFODELs (for a V-control total of 64). In this case, the array is divided into four 4 × 16 fully independent sections. Using this technique, we find that the BIFODEL complexity $C_m$ is given by

$$C_m = (K_x - 1) + m(K_y - 1) \tag{13.31}$$

where $m$ is the number of independent beams. Equation (13.31) shows that the beam complexity factor $m$ multiplies only $K_y - 1$ and not $K_x + K_y - 2$, which would be the case for the straightforward implementation. However, in demanding PAA applications, the straightforward implementation does provide the advantages of full TTD redundancy and multiple-beam formation without array partitioning.

In closing this section, we note that, unlike the WDM system for 1-D PAAs, hybrid techniques cannot be used efficiently in conjunction with the WDM TTD of 2-D PAAs. This is because, in the 2-D case, each BIFODEL contains both small and large delays,

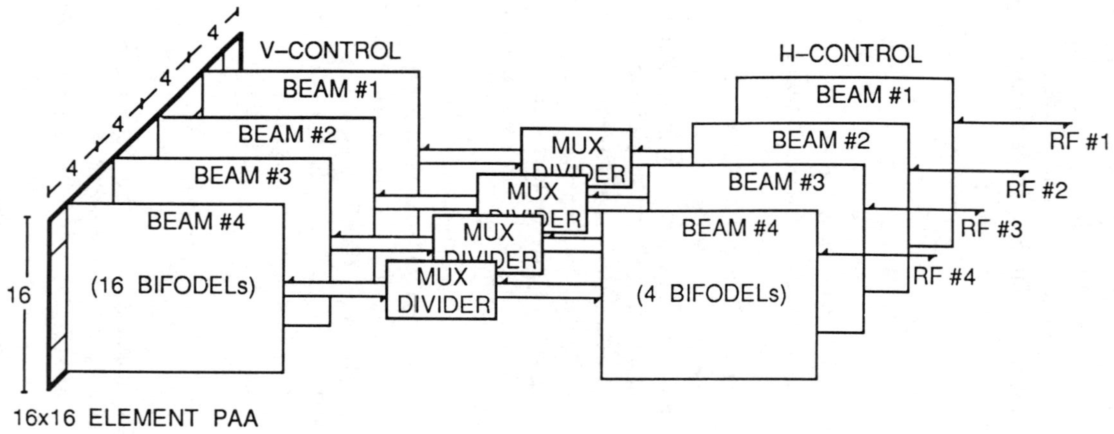


**Figure 13.14** Partitioning technique for a 2-D phased-array antenna with independent multibeam formation.

and therefore it must contain fiber to a significant degree. Thus, at best, we can implement only a few bits of some of the BIFODELs electronically.

## ACKNOWLEDGMENTS

The authors wish to thank Dr. G. L. Tangonan of Hughes Research Laboratories for providing the photograph in Figure 13.5, and Dr. C. T. Sullivan of Honeywell Systems and Research Center for providing the photograph in Figure 13.7.

## REFERENCES

[1] Cardone, L., ''Ultra-Wideband Microwave Beamforming Technique,'' *Microwave J.,* April 1985, pp. 121–131.

[2] Curtis, D., ''Frequency Domain Analysis and Performance of a True-Time Delay Fiber Optic Beamforming Network for Array Antennas,'' *Proc. SPIE,* Vol. 1217, 1990, pp. 226–234.

[3] Ng, W., A. Walston, G. Tangonan, J. Lee, I. Newberg, and N. Berstein, ''The First Demonstration of an Optically Steered Microwave Phased Array Antenna Using True-Time Delay,'' *J. Lightwave Tech.,* Vol. 9, No. 9, 1991, pp. 1124–1131.

[4] Koepf, G. A., ''Optical Processor for Phased Array Antenna Beam Formation,'' *Proc. SPIE,* Vol. 477, 1984, pp. 75–81.

[5] Herczfeld, P. R., and A. S. Daryoush, ''Fiber-Optic Feed Network for Large Aperture Phased Array Antennas,'' *Microwave J.,* August 1987, pp. 160–166.

[6] Toughlian, E. D., and H. Zmuda, ''A Photonic Variable RF Delay Line for Phased Array Antennas,'' *J. Lightwave Tech.,* Vol. 8, No. 12, 1990, pp. 1824–1828.

[7] Esman, R. D., M. J. Monsma, J. L. Dexter, and D. G. Cooper, ''Microwave True-Time Delay Modulator Using Fiberoptic Dispersion,'' *Electron. Lett.,* Vol. 28, No. 20, 1992, pp. 1905–1907.

[8] Soref, R., ''Optical Dispersion Technique for Time-Delay Beam Steering,'' *Applied Optics,* Vol. 31, No. 35, 1992, pp. 7395–7397.

[9] Kondo, M., K. Komatsu, Y. Ohta, S. Suzuki, K. Nagashima, and H. Goto, ''High-Speed Optical Time Switch With Integrated Optical 1x4 Switches and Single-Polarization Fiber Delay Lines,'' *4th Int. Conf. on Integrated Optics and Optical Fiber Communication,* Tokyo, Japan, June 1983, E. I. Conf. No. 04967.

[10] Levine, A. M., ''Fiber-Optic Phased Array Antenna System for Rf Transmission,'' U.S. Patent No. 4028702, 7 June 1977; ''Fiber Optics for Radar and Data Systems,'' *Proc. SPIE,* Vol. 150, 1978, pp. 185–192.

[11] Ng, W., A. Walston, G. Tangonan, I. Newberg, and J. J. Lee, ''Wideband Fiber-Optic Delay Network for Phased Array Antenna Steering,'' *Electron. Lett.,* Vol. 25, No. 21, 1989, pp. 1456–1457.

[12] Ng, W., G. Tangonan, A. Walston, I. Newberg, J. J. Lee, and N. Bernstein, ''True-Time Delay Steering of Dual Band Phased Array Antenna Using Laser Switched Optical Beam Forming Networks,'' *Proc. SPIE,* Vol. 1371, 1990, pp. 205–211.

[13] Taylor, H. F., ''Optical Fiber Devices for Signal Processing,'' *Proc. SPIE,* Vol. 209, 1979, pp. 159–165.

[14] Lagerstroem, B., P. Svensson, A. Djupsjoebacka, and L. Thylen, ''Integrated Optical Delay Line Signal Processor,'' *Proc. OFC/IOOC,* Paper WK2, 1987, p. 176.

[15] Thompson, R. A., ''Optimizing Photonic Variable-Integer Delay Circuits,'' OSA Topical Meeting on Photon Switching, Lake Tahoe, Paper FD4-1, March 1987.

[16] Goutzoulis, A. P., D. K. Davies, and J. M. Zomp, ''Prototype Binary Fiber Optic Delay Line,'' *Opt. Eng.,* Vol. 28, No. 11, 1989, pp. 1193–1202.

[17] Goutzoulis, A. P., and D. K. Davies, ''Hardware-Compressive 2-D Fiber Optic Delay Line Architecture for Time Steering of Phased Array Antennas,'' *Applied Optics,* Vol. 29, No. 36, 1990, pp. 5353–5359.

[18] Baldwin, D. L., and A. G. Garas, ''Architectures and Performance of Laser Links in Microwave Phased Array Antenna Systems,'' *Proc. SPIE,* Vol. 1217, 1990, pp. 235–246.

[19] Goutzoulis, A. P., and D. K. Davies, ''All-Optical Hardware-Compressive Wavelength-Multiplexed Fiber-Optic Architecture for True Time Delay Steering of 2-D Phased Array Antennas,'' *Proc. SPIE,* Vol. 1703, 1992, pp. 604–614.

[20] Sullivan, C. T., et al., ''Switched Time Delay Elements Based on AlGaAs/GaAs Optical Waveguide Technology at 1.32 $\mu$m for Optically Controlled Phased Array Antennas,'' *Proc. SPIE,* Vol. 1703, 1992, pp. 264–271.

[21] Magel, G. A., and J. L. Leonard, ''Phosphosilicate Glass Waveguides for Phased-Array Radar Time Delay,'' *Proc. SPIE,* Vol. 1703, 1992, pp. 373–378.

[22] Goutzoulis, A. P., D. K. Davies, and J. M. Zomp, ''Hybrid Electronic Fiber-Optic Wavelength-Multiplexed System for True-Time Delay Steering of Phased Array Antennas,'' *Opt. Eng.,* Vol. 31, No. 11, 1992, pp. 2312–2322.

[23] Dolfi, D., F. Michel-Gabriel, S. Bann, and J. P. Huignard, ''Two-Dimensional Optical Architecture for True-Time Delay Beam Forming in a Phased Array Antenna,'' *Opt. Lett.,* Vol. 16, No. 4, 1991, pp. 255–257.

[24] Ng, W., D. Yap, A. Narayanan, R. Hayes, and A. Walston, ''GaAs Optical Time-Shift Network for Steering a Dual Band Microwave Phased Array Antenna,'' *Proc. SPIE,* Vol. 1703, 1992, pp. 379–383.

[25] Soref, R. A., ''Programmable Time-Delay Devices,'' *Applied Optics,* Vol. 23, No. 21, 1984, pp. 3736–3737.

[26] Koul, S., and B. Bhat, ''Microwave and Millimeter Wave Phase Shifters,'' Vol. II (Semiconductor and Delay Line Phase Shifters), Norwood, MA: Artech House, 1991.

[27] Bois, J. R., S. Lazar, and P. L. Ouvrard, ''Wide Band MMIC Delay Line Phase Shifter (2–20 GHz),'' *Proc. IEEE GaAs Int. Symp.,* 1990, pp. 156–157.

# Chapter 14
# Optical Processing of Synthetic Aperture Radar Data

**Emmett N. Leith**
*The University of Michigan*

## 14.1 INTRODUCTION

The application of optical processing of radar data, especially synthetic aperture radar data, in the 1950s was a major contribution to optical technology and at the same time a major contribution to radar technology. The import of that union was great, and remains significant even today.

Synthetic aperture radar (SAR), originated by Wiley, was a major new concept in imaging radar [1,2]. However, its implementation was a major problem. The SAR process held promise of radar imagery far superior to any that had previously been obtained. Especially ambitious was the advanced form envisioned by a group led by Cutrona and Vivian [2]. However, the realization of such imagery involved a massive data processing problem. The SAR process did not immediately produce an image; instead, it generated a large volume of raw data that had to be processed. The processing was a monumental task. To achieve such imagery as envisioned in the Cutrona-Vivian system was far beyond the capabilities of the existing imaging processing techniques. Digital computers were in their infancy at that time and their capabilities fell far short of the SAR task. Other devices, such as electronic correlators, could only in a modest way aid in the problem.

In the search for a method that might enable the realization of their ambitious program, Cutrona and Vivian considered using an optical correlator. This was a gamble; the optical processing system they proposed was an untried and radically different approach. The optical processing system in its original form had to evolve a long way before it became suitable. In its original form it probably would not have worked. But this we can never know, since by the time the radar system had been built and data became available for processing, the system was already radically different from the original concept. For example, the original system was based on incoherent optics—the use of conventional, spatially, and temporally incoherent light. The final system used monochromatic, spatially coherent light [2]. After the system had been perfected, it was seen in retrospect to have been a very elegant solution indeed. The coherent optical processing system was not only a very powerful technique, but also a very natural one in that the optical system was in some crucial ways an analog of the radar imaging system. The close analog between the SAR system (a coherent microwave system) and the coherent optical processing system was the basis for the great power of the coherent optical processing system [3].

In addition to the data processing problem, there was also the one of data storage. The large amount of data produced by the SAR system had to be stored, since processing the data as it was acquired, without buffer storage, was not feasible. In the 1950s, photographic film offered an extremely high storage density, large storage capacity, and excellent dynamic range. No other storage medium came close in meeting this combination of requirements. Even if optical processing had never been considered, data storage on photographic film would still have been an attractive option; and, of course, the storage of the radar data on photographic film added considerably to the attractiveness of optical processing.

Several quite unique optical processing systems have been developed over the years. The titled plane optical processor [4] of 1962 was the most effective and most sophisticated; we describe it in detail here.

## 14.2 SYNTHETIC ANTENNA PRINCIPLE

The azimuthal resolution of a radar system is determined by the beamwidth, which in turn is determined by the antenna size through the relation $\Delta x = \lambda r/D$, where $\Delta x$ is the resolution, $\lambda$ the wavelength, $r$ the distance to the object, and $D$ the antenna size. To obtain resolution comparable to what is obtainable from an aerial reconnaissance camera, the antenna would have to be extremely long, since the wavelength of a typical radar system is thousands of times larger than the wavelength of visible light. It would be unthinkable to carry such an antenna on an aircraft; the absurdity of such an undertaking is portrayed in Figure 14.1.

The synthetic aperture technique circumvents this basic problem. The aircraft carries a small antenna, producing a beam that is rather wide in the horizontal, or azimuthal, dimension (Figure 14.2). The forward motion of the aircraft carries the small antenna along, so that it occupies in sequence all of the positions that would have been occupied, at a given instant, by the elements of the large antenna, here assumed to be an array of small dipole elements. At each position, the radar emits a pulse, then receives the signal reflected from the object field. This signal is stored coherently; that is, both the phase

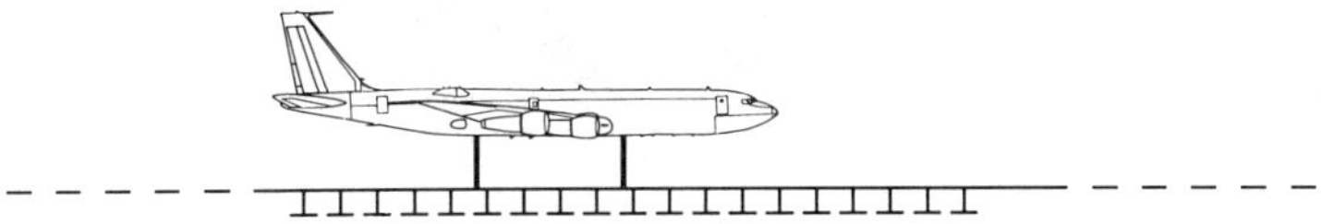

$$\text{BEAMWIDTH (IN RADIANS)} = \frac{\lambda}{D}$$

$$D = \text{ARRAY LENGTH}$$

**Figure 14.1** Aircraft carrying large antenna to get narrow beam and good resolution.

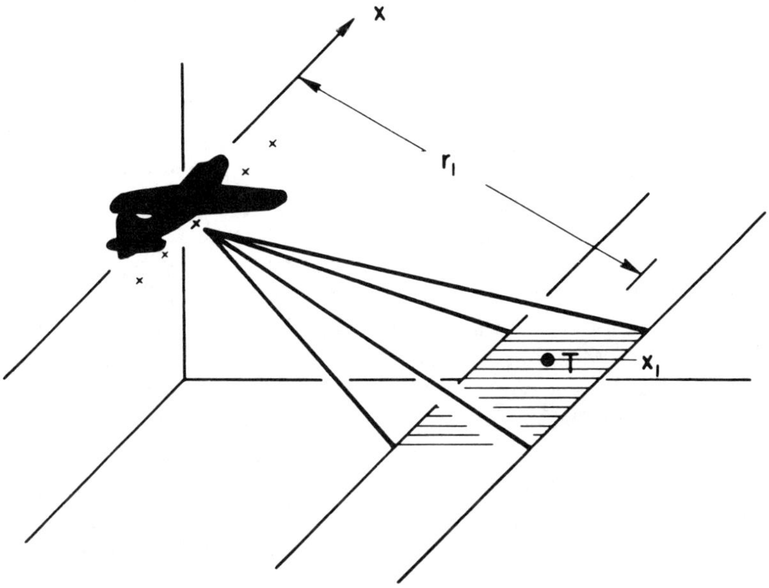


**Figure 14.2(a)** The synthetic aperture system: airplane illuminates swath of terrain. *T* is a point object.

and amplitude are recorded. The antenna is then carried to the next dipole element position, and the process is repeated. The stored data is essentially the same as the signal that would have been received by the large array. The task is then to process the data in a way analogous to the way the large array would have handled the data. The resulting image is formed as if it had been produced by the large antenna. The larger effective aperture is termed a *synthetic aperture,* and the large antenna simulated by this process is called a *synthetic antenna.*

The manner of processing the data depends on the type of antenna being simulated. In the basic, uniform array (i.e., an array focused at infinity), the signals received on the

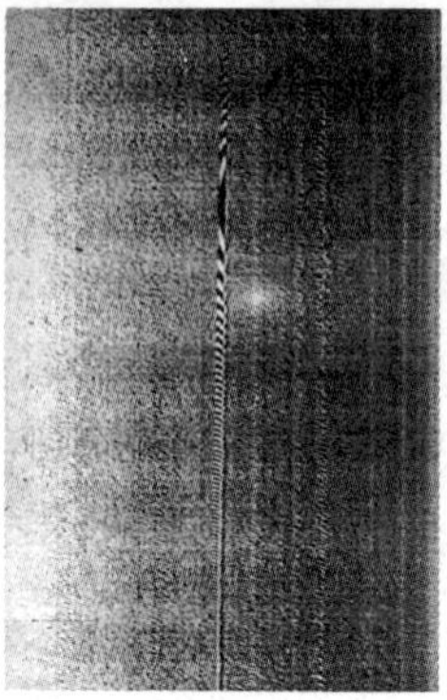

**Figure 14.2(b)** The synthetic aperture system: the signal record showing the return from object *T*.

various array elements are summed linearly. Such summation leads to a point spread function of the form (sin $cx$)$x$, with high sidelobes. A more sophisticated array would be weighted (or, in optical terminology, apodized); in this process, the signal radiated and received on the outer array elements is attenuated. This process broadens the mainlobe of the antenna pattern somewhat, but the sidelobes are considerably reduced.

The next step in the level of sophistication is the focused antenna. If an antenna is very large, it will not achieve its theoretical resolution until it is focused. The situation is quite analogous to a conventional camera. For a lens of sufficiently large f-number, the camera can be focused at infinity, and all distances, up to some rather close distance, will be approximately in focus. As the aperture is increased, the depth of focus decreases, and sharp resolution is obtained at a given distance only by focusing the camera at that distance. Similarly, the large apertures that can be generated synthetically have a limited depth of focus. Thus, the synthetic antenna must be focused in order to achieve the theoretical resolution.

Focusing the array means introducing appropriate delays on each of the various elements of the array so that the signals radiated and received on these elements are summed linearly, as before, but with appropriate phase shifts. The antenna can thus be focused at a finite range instead of at infinity.

For a radar that is to image a range interval of considerable extent, the optimum arrangement is to focus the antenna at the midrange of the interval. Other nearby ranges will be approximately in focus, although ranges at the edges of the interval may be noticeably out of focus. The situation is quite analogous to a camera with a very low f-number, say f-1. The camera can be in sharp focus only for a rather small depth interval.

The synthetic antenna has a peculiar advantage here. Since the signals are stored and processed by computer (digital electronic or optical analog), data from each range can be processed independently to produce an antenna focused at that range. Thus, the synthetic antenna can be focused at all ranges simultaneously. This is a powerful advantage for the synthetic antenna, which is not available for the conventional, real antenna [2].

## 14.3 RADAR SIGNAL

An SAR system is typically used for imaging terrain. Usually, the small SAR antenna is carried on an aircraft, although other vehicles, such as satellites, have been used. A range interval, or swath, is illuminated. The antenna beam must have a vertical pattern sufficiently broad to illuminate the entire interval to be imaged. The horizontal pattern must be wide enough to allow sufficient data to be collected to form a synthetic aperture of the desired size. The radar system may be pulsed with simple, uncoded pulses to produce resolution in range. Alternately, as is common with radar systems, a coded waveform may be radiated and pulse compression methods then used to separate data from different ranges.

Figure 14.3(a) describes the data acquisition process. Two object points, A and B, are about to enter the beam. When they first enter the beam, they have a radial velocity component, relative to the aircraft, directed toward the aircraft, and thus the radar signal reflected from them is Doppler-shifted upward in frequency. As the object points progress across the beam, the Doppler shift decreases, becoming zero when an object point is directly abeam of the aircraft. When the objects pass this point, they acquire a radial velocity component directed away from the aircraft, and the Doppler shift is downward in frequency. Finally, the object points leave the beam. This Doppler frequency as a function of time is shown in Figure 14.3(b). For a beamwidth of a few degrees, the change in frequency is linear with time. The two object point undergo the same Doppler shifts, except that the Doppler shift for *B* occurs at a later time, with the zero Doppler shift occurring at a time $t_b$ instead of $t_a$. Thus, every object entering the beam at the range $r_1$ undergoes the same linear Doppler shift, but with a time displacement corresponding to the object coordinate.

To find the form of the signal returned to the radar antenna, consider a point object located at a position $x_1$ along the ground track and at a range $r_1$ from the aircraft flight

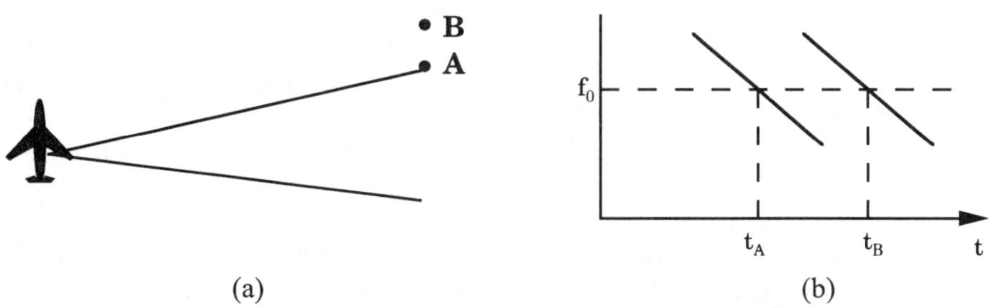


**Figure 14.3** (a) Two object points about to enter beam; (b) the Doppler history, showing Doppler frequencies as a function of time for the two object points *A* and *B*; $f_o$ is the radiated frequency.

path. Let a signal $\exp(i2\pi f_0 t)$ be transmitted. The signal returned to the aircraft has the form

$$s_1 = \sigma \exp[i2\pi f_0(t - 2r/c)] \tag{14.1}$$

where $r$ is the range from aircraft to object, $c$ is the velocity of light, and $\sigma$ is the complex amplitude of the signal as received by the radar. By using the approximation

$$r = [r_2^2 + (x - x_1)^2]^{1/2} = r_1 + (x - x_1)^2/2r_1 \tag{14.2}$$

(14.1) can be written as

$$s_1 = \sigma \exp\{i[2\pi f_0 t - 4\pi r_1/\lambda - (2\pi/\lambda)(x - x_1)^2/r_1]\} \tag{14.3}$$

The velocity ($v$) of the aircraft links the temporal behavior of the signal with the spatial coordinates of the object through the relation $x = vt$, where we have chosen the origins of both time and space coordinate systems so that $x = 0$ at $t = 0$. If the terrain at range $r_1$ is regarded as a collection of numerous point scatterers, each point scatterer gives a return signal of the form of (14.3).

We substitute the relation $x = vt$ into (14.3) and differentiate the phase term. Since the instantaneous frequency of a signal is the time derivative of its phase, differentiation of the phase term of (14.3) gives, for an object located at coordinates $x_1, r_1$, the Doppler-shifted frequency

$$f = \mathrm{d}\phi/\mathrm{d}t = f_0 - 4\pi v(vt - x_1)/\lambda r_1 \tag{14.4}$$

showing the Doppler shift to be a linear function of time, as depicted in Figure 14.3. Most of the various processing methods for converting these data into a high-resolution radar image have been based on the analysis of this Doppler shift. Such methods include Doppler filtering, cross-correlation, and matched filtering. However, as the optical processing technique developed, a more optically oriented theory for SAR emerged and eventually dominated. In this view, the SAR system, with optical processing, is a radar analog of holography.

The holographic viewpoint can be developed by returning to (14.3) and recognizing that this signal describes a spherical wave reflected from the object point ($r_1$, $x_1$) and impinging on the synthetic aperture [3, 5]. We may, using a suitable electrical-to-optical transducer such as a cathode ray tube, record the received signal on photographic film, in which case we produce a hologram. In the detection and recording process, the original high-frequency $f_o$ is shifted to a relatively low value $f_c$ suitable for recording on photographic film, and the radar space dimensions are scaled downward; that is, a distance $x$ along the flight path is reduced to a distance $x/p$ on the photographic film, where $p$ is a scaling factor, typically chosen so that a few hundred feet of travel along the flight path is reduced to a few millimeters on the photographic record. Similarly, the range coordinate is reduced to a shorter distance $r/q$, typically chosen so that a range interval of several miles is recorded across the width of a strip of 70-mm film. The photographic record for a point object thus becomes

$$s_1 = s_b + \cos[2\pi f_c x - 4\pi r_1/\lambda - 2\pi p^2(x - x_1/p)^2/(\lambda r_1)] \tag{14.5}$$

where $s_b$ is a bias term, added to the signal so that positive and negative signal values can be recorded. Figure 14.2(b) shows a section of such a photographic record, showing in particular a recorded signal from a strong scattering point in the radar field. This record can be treated like a conventional hologram. A hologram of a point object is, of course, a zone plate or diffraction lens, and like a conventional refractive lens, it has a focal length. The structure of (14.5) has focal length [2]

$$\mathrm{F} = \pm(l/p^2)(\lambda/\lambda_l)r_1 \tag{14.6}$$

where $\lambda_l$ is the wavelength of the light that illuminates the hologram. Note that the focal length is proportional to the range $r_1$ of the object point that produced the zone plate; this feature is of considerable significance in the following discussion of optical processing.

We have discussed the theory as if the radar signal were CW. However, so that resolution in the range dimension can be achieved, it is necessary to pulse the radar, either by radiating a simple, short pulse, or, equivalently, by radiating a coded pulse with a short autocorrelation function. For an object at a specific range, the pulsing does not essentially alter the theory given above. The pulsing can be regarded as a sampling of the signal collected along the flight path, and if the sampling rate (i.e., the pulse repetition frequency) is sufficiently high, the sample values are equivalent to the continuous signal. The pulsing allows the recording process thus described to be carried out simultaneously for a large number of range elements, and the data from different ranges can be separated.

At each sample position, the radar radiates a pulse and the return signal is displayed as a linear trace on a cathode ray tube (Figure 14.4). Each position along this line represents the reflected signal from a different range. This line trace is recorded as a line across a photographic film strip, which then becomes a hologram (i.e., a photographic record of both the phase and amplitude of the signal).

In the holographic viewpoint, the aircraft records the wave field reflected from the terrain being irradiated, and records it, in miniature, on a photographic record. Under coherent illumination, the wave field is regenerated at optical frequencies. When the waves are focused, a miniaturized image of the radar-illuminated terrain is formed [3, 5]. This viewpoint is illustrated in Figure 14.5, showing how, in an imaging process, the radar wave from an object point is converted into a corresponding miniaturized optical wave, and continues on its way to forming a focused spot, an optical image of the original object point in radar space.

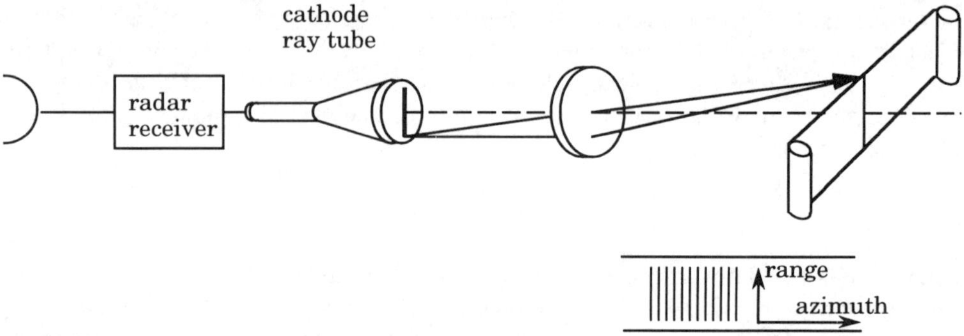


**Figure 14.4** The data recording process, from receiving antenna to the signal record.

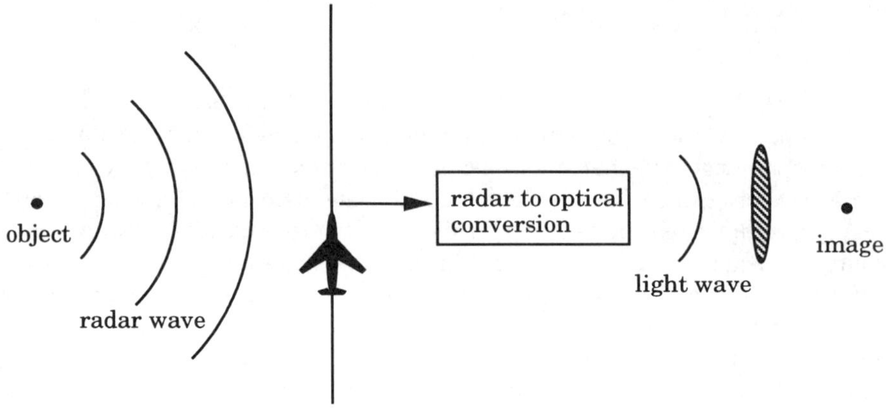


**Figure 14.5** The basic radar-to-optics conversion process.

The image from a hologram is typically formed by the illumination of the hologram with a beam of coherent light. Coherent illumination of the radar-produced hologram forms, for each object point, a corresponding, sharply focused image point. The resolution of the image depends on the length of the signal from each object point. Here, the length of the signal is the length of the synthetic apeture thus produced. The holographic process thus converts the radar synthetic aperture into a corresponding real optical aperture. The size of the synthetic aperture is thus the size of the beam that illuminates the terrain. Three interesting observations follow.

First, since the size of the beam, $L$, is inversely related to the size $a$ of the real antenna, through the well-known relation $L = \lambda r/a$, we have the interesting conclusion that the smaller the real antenna, the larger the beam width, and therefore the larger the synthetic antenna. So, contrary to conventional law, the smaller the real aperture, the better the resolution [2].

Second, since the beamwidth is proportional to range $r$, the size of the synthetic antenna grows with range. Since the resolution $\Delta x$ of an imaging system is proportional to range and inversely proportional to aperture, these factors exactly cancel, giving, in theory, resolution independent of range, in contrast to conventional, fixed-aperture systems, where the resolution becomes coarser with object distance [2]. This resolution invariance, though often touted, is in practice mostly illusory. Imaging systems tend to be limited by aberration rather than being diffraction-limited. It can be shown that when the resolution is limited by aberrations, the resolution width $\Delta x$ becomes proportional to range, even if the aperture grows with range.

Finally, there is one more point worth noting, which we illustrate in Figure 14.6. Figure 14.6(a) shows a spherical wave emanating from a point and falling on an aperture a distance $r_1$ away. The field at the aperture can be approximated as $u = \exp(i\pi x^2/\lambda z)$, and if the wave is imaged through the aperture $L$, the theoretical resolution is $\lambda r_1/L$. From (14.3), it can be seen that the wave impinging on the aircraft flight path has, in the quadratic term, an additional factor 2, which we can identify with the range $r_1$ by rewriting the quadratic phase term of (14.3) as $\pi(x - x_1)^2/[\lambda(r_1/2)]$. The impinging wavefront has a curvature that gives the appearance that the object point was only a distance $r_1/2$ away. When this wavefront of twice the curvature (half the radius of curvature) is imaged through the aperture $L$, the resolution is $\lambda r_1/2L$, which is finer by a factor 2 than would occur for a conventional aperture of the same size. This extra factor 2 of resolution improvement is an additional benefit of synthetic antennas and results from the fact that the quadratic phase term, from which the resolution is derived, is produced from a round-trip phase delay, from the real antenna to the object and back again, instead of only from object to the aperture.

## 14.4 FORMATION OF THE HOLOGRAPHIC IMAGE

The SAR holographic process has two defects that require the use of an anamorphic lens system to give a corrected image. An anamorphic lens is one with focal properties that are different in the two lateral dimensions. The simplest example is a cylindrical lens. The most common example is eyeglasses that correct for astigmatism by having slightly different focal properties in the vertical and horizontal directions.

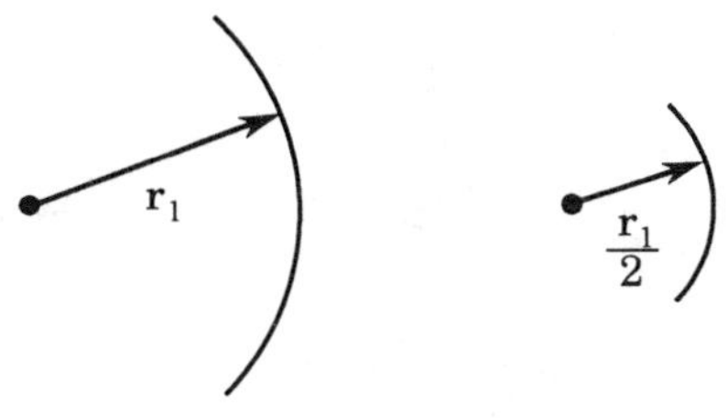


**Figure 14.6** The synthetic aperture process makes object at range $r_1$ appear to be at range $r_1/2$.

First, the holographic zone plate produced from a point object has lens-like or focusing properties only along the $x$-axis. Along this direction, the zone plate focuses to form sharp images. However, in the vertical direction, there is no focal power at all; the zone plates are thus cylindrical lenses. In the vertical, or range, direction, resolution is obtained by pulse modulation of the radar, and the required resolution in the range direction is therefore present on the film record itself.

Second, signals from different ranges produce zone plates with different focal lengths, the focal lengths being proportional to range. Thus, in the $x$ direction, objects from long ranges produce images that focus relatively far from the hologram, whereas objects from shorter ranges form images close to the hologram, as depicted in Figure 14.7. An observer viewing the holographic image from a point object would see on the hologram an image that is sharp vertically, but smeared horizontally, whereas the observer would see at the zone plate focus an image that is sharp horizontally, but unfocused vertically. The image can be said to have severe astigmatism, which must be corrected.

The locus of focal points, for collimated illumination, of all the signal histories is two planes tilted with respect to the signal-film plane. One plane is a virtual image produced by the negative focal length of the zone plate signals; the other is a real image produced by the positive focal length of the signals. These planes are called the azimuth focal planes, since they are the planes in which the signals are in sharp azimuth focus.

In the range direction, two situations commonly occur. The first, already described, arises when a narrow, uncoded pulse is radiated. In the other situation, chirp pulse compression techniques are used; the chirp pulse is recorded and then is compressed by optical means [6,7].

In the chirp method, the radiated waveform is a relatively long pulse with frequency changing linearly across the pulse (i.e., the pulse has the form $\mathrm{rect}(t/T)\exp[i(2\pi f_0 t + 1/2st^2)]$, and the instantaneous frequency is $f_0 + st$, where $T$ is the pulse duration and $s$ is the frequency modulation rate. This waveform is identical in form to the Doppler signals generated by the synthetic aperture process, and, similarly, this signal, when recorded, becomes a Fresnel zone plate structure. Thus, when a synthetic aperture system uses a chirp pulse, the recorded signal from a point object has zone plate properties in both the

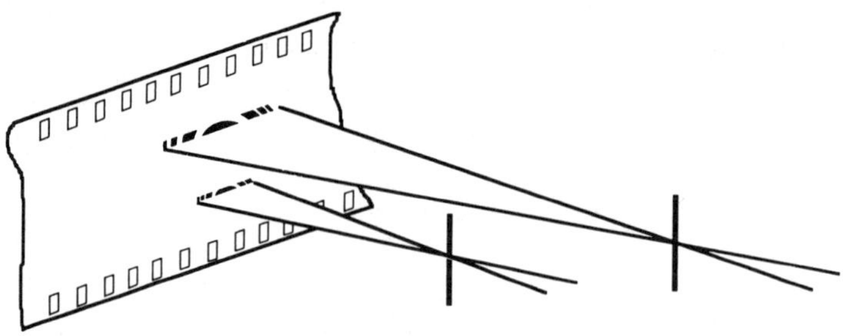

**Figure 14.7** The inherent astigmatism of the recorded zone plate signals.

range and azimuth dimensions, although the focal lengths are generally different for the two dimensions. In the range dimension, the focal length is

$$F_r = \pm(\pi c^2/q^2\lambda_1 s) \tag{14.7}$$

where $c$ is the velocity of propagation of the radar signal [7]. In the range direction, the focal lengths are not range-dependent. For collimated illumination, the locus of the focal points for the range direction is therefore two planes parallel to the signal record and distance $f_r$ from it, one plane corresponding to the positive focal length of the zone plate and the other to the negative (Figure 14.8). If optical pulse compression is not used or if the compressed pulse is recorded on the film, the position at which the signals are sharply

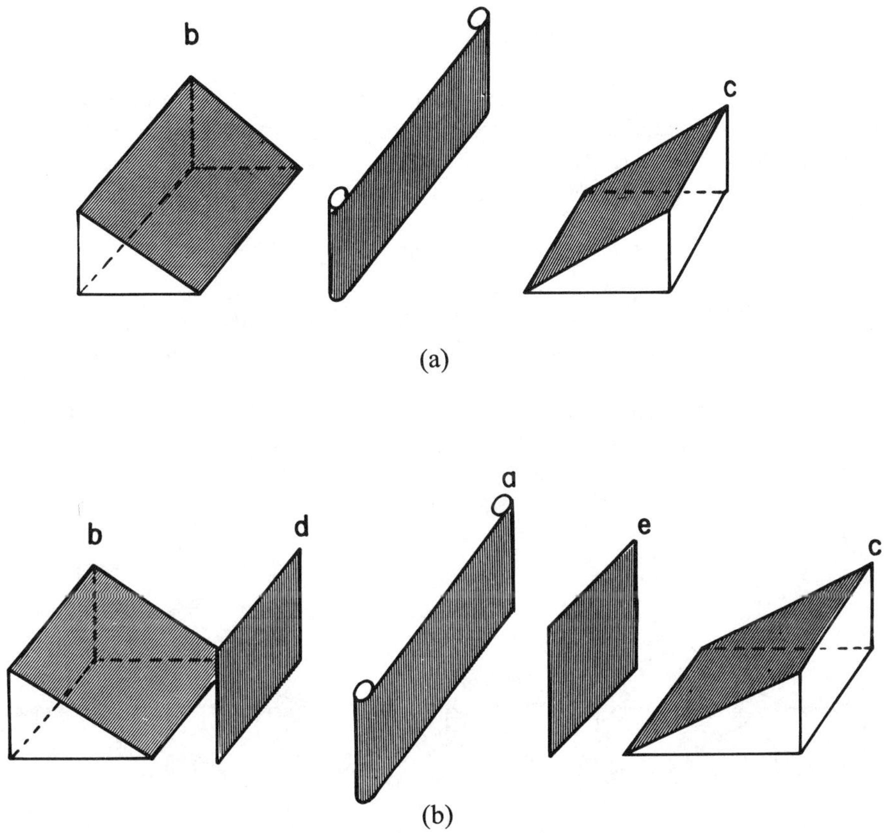


**Figure 14.8** The signal record and the planes of focus: (a) uncoded pulse (range image is the signal record itself, whereas azimuth image is the tilted planes; (b) using a chirp pulse so that range images focus away from the signal record.

focused in the range direction is the signal film itself (Figure 14.8(a)). In either case, the azimuth focal planes are tilted with respect to the signal-film plane and the range focal planes are parallel to, or coincident with, the signal-film plane. The optical processing system brings one azimuth and one range focal plane into coincidence so that sharp azimuth focus and range focus occur together. Put otherwise, the astigmatism is removed. The images contained on the other set of planes are rejected by spatial filtering.

Several systems have been devised for producing a sharply focused, undistorted, and nonaberrant image. The earliest widely used one was a simple three-lens system consisting of a spherical lens, a cylindrical lens, and a conical lens derived from an axicon [3]. This served for several years until it was replaced by a new lens system, the tilted plane processor [4]. This final one was introduced in 1962 and continues in use today.

## 14.5 TILTED-PLANE PROCESSOR

The tilted-plane optical processor concept, which is both subtle and ingenious, is based on the properties of telescope systems [3]. A simple telescopic system consists of two lenses separated by the sum of their focal lengths (Figure 14.9). For this configuration, one can readily show that image position is directly proportional to object position:

$$z_i = -(f_2/f_1)^2 z_o \tag{14.8}$$

where $z_o$ is the axial coordinate of the object measured from the front focal plane of the first lens, and $z_i$ is the axial coordinate of the image, measured from the back focal plane of the second lens.

The lateral magnification $M_x$, the magnification in a direction normal to the optical axis, is, for a two-lens telescope, just the ratio of the focal lengths, or

$$M_x = -(f_2/f_1) \tag{14.9}$$

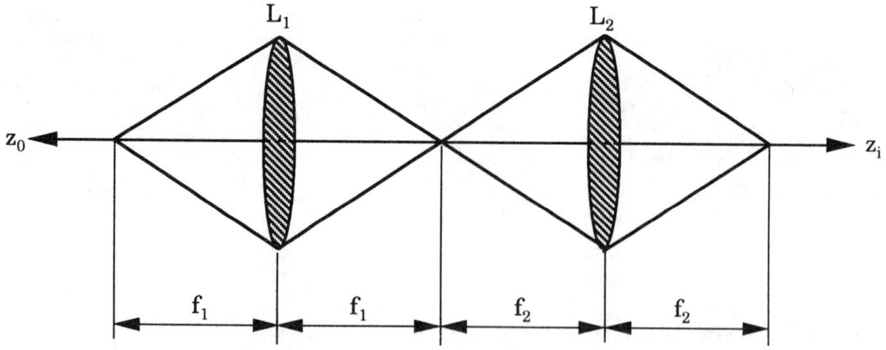


**Figure 14.9** Simple telescope with lenses $L_1$ and $L_2$ having focal lengths $f_1$ and $f_2$.

where the minus sign indicates an image inversion. The longitudinal magnification, the magnification along the optical axis, or direction of light propagation, is the square of the lateral:

$$M_z = M_x^2 = (f_2/f_1)^2 \tag{14.10}$$

Both lateral and axial magnifications are independent of object position. Consequently, a rectangular parallelopiped is always imaged into another parallelopiped, having generally a different shape (Figure 14.10). Within this parallelopiped, we inscribe a diagonal plane from the upper left edge to the lower right edge. The image of the plane will be another plane, with a different slope (or tilt), but with magnification $M_x$ that is uniform over the entire plane. If $f_1 > f_2$, there is demagnification, and since the longitudinal demagnification is greater than the lateral demagnification, the tilt of the inscribed plane is decreased.

If the imaging system is not telescopic, the magnification $M_x$ and $M_z$ will be $z$-dependent, with the result that the image will no longer be parallelopiped and the magnification will not be uniform over the inscribed tilted plane. The positional independence of magnification of a telescopic system is the basis for the tilted-plane processor concept.

In Figure 14.11, $P_o$ and $P_o'$, the previously described planes of azimuth focus, serve as the object from the telescopic lens system. The resultant image planes $P_i$ and $P_i'$ have uniform lateral magnification over the entire area.

By choosing $f_1 > f_2$, (say, $f_1f_2$ about 10) we greatly reduce the tilt of the azimuth image plane. Our objective, of course, would be to remove the tilt completely. With the

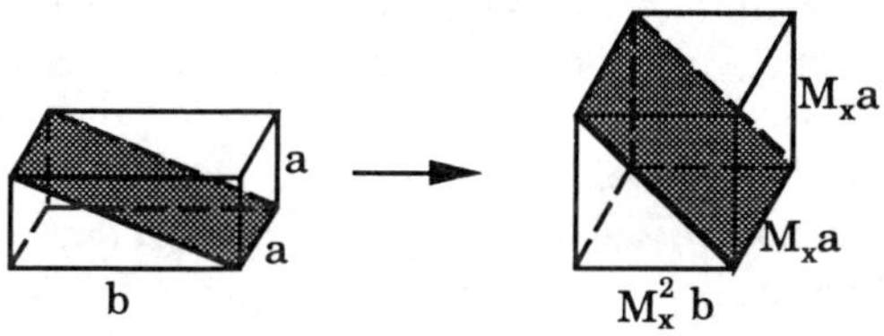


**Figure 14.10** Effect of demagnification on tilted plane.

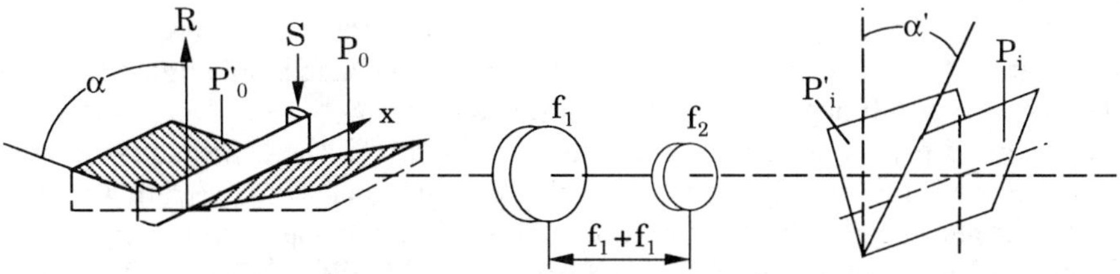


**Figure 14.11** Geometry for tilted-plane analysis, showing tilted planes $P_0$, $P_0'$, the images $P_i$ and $P_i'$, the signal record $S$, the range and azimuth coordinates $R$ and $X$, and lenses with focal length of $f_1$ and $f_2$.

simple telescope, we obviously cannot accomplish this objective entirely, but we can accomplish the greater part of it. The problem of the residual tilt can be solved by additional modification.

We have now almost removed the tilt of the azimuth focal plane. Next is the task of range focusing (i.e., of bringing the azimuth and range image planes into coincidence). Figure 14.12 shows an optical system that incorporates the telescopic system for reduction of tilt and also provides range focusing. Spherical lenses $L_1$ and $L_3$ form a simple telescope. Lenses $L_2$ and $L_4$ are cylindrical lenses, oriented to provide focusing power in the azimuth dimension. These lenses are positioned so that $L_1$ and $L_2$ form a simple telescope, as do $L_3$ and $L_4$. The optical system is consequently anamorphic; it has different focal properties in the range and azimuth dimensions.

In the range direction, lenses $L_1$ and $L_3$ image the range focal plane with lateral magnification $-f_3f_1$ (the magnification in the two-lens telescope is the ratio of the focal lengths of the two lenses) and axial magnification $(f_3/f_1)^2$. Because it is generally desirable to have unity magnification in the range dimension, we choose $f_1 = f_3$. In the azimuth dimension, all lenses are active, producing lateral and axial magnifications $f_4/f_3 \times f_2/f_1)$ and $(f_4/f_3 \times f_2/f_1)^2$, respectively.

Another problem is that at this stage the image will be elongated in the horizontal direction, because in the recording process the radar data was compressed more in the range direction than in the azimuth direction; that is, the $x$-compression factor $p$ was less than the range compression factor $q$. This situation came about because the spatial carrier $f_c$ (14.5) was applied in the $x$-direction, thus increasing the resolution requirements in this direction.

The final image should have the same scale factor in both the range and azimuth dimensions, so that the image will not be stretched in one dimension. Put otherwise, the final image should have an aspect ratio of unity. Since we have independent control over the range and azimuth dimensions, we can use the optical system to produce a unity aspect ratio. We set the ratio $f_4/f_3$ equal to the recording aspect ratio $q/p$. Thus, the azimuth dimension image is demagnified so that its scale matches that of the range dimension, giving an image with unity aspect ratio. The tilt of the azimuth focal plane is reduced by the factor $K^2$ (a typical value for $K$ is 10). In addition, the system is arranged so that the range focal plane image $R'$ best fits the now only slightly tilted azimuth focal plane $A'$ (Figure 14.13).

Next we deliberately tilt the range focal plane (i.e., tilt the data record) in order to tilt its image $R'$ into coincidence with $A'$. The resulting image, although slightly tilted typically about 1 to 10 deg, is completely corrected.

In the tilted-plane optical processor, shown in Figure 14.14, the film drive containing the signal record is thus tilted somewhat, so that the final image is free from astigmatism. Because the final image is on a tilted plane, the recording camera is also tilted.

This optical processor was an ideal solution to a quite difficult optical imaging problem. For about a decade, this most ingenious optical processor dominated the field of SAR data processing, producing some truly marvelous imagery; an example is shown

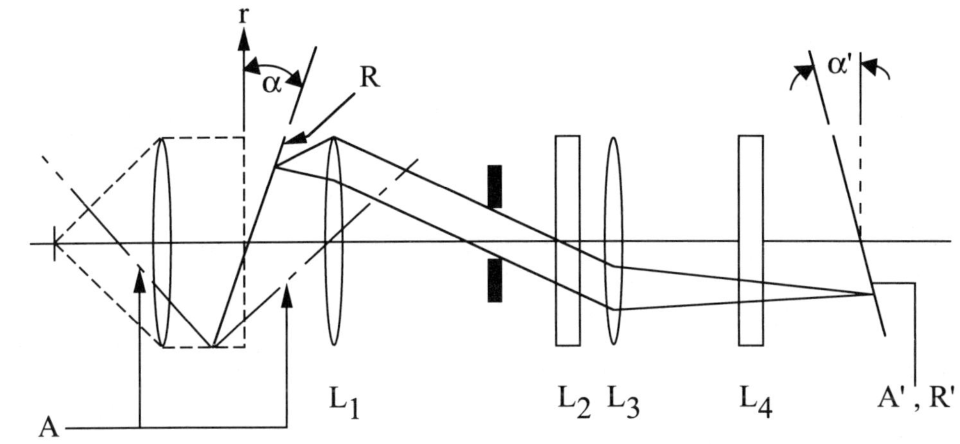


(a)

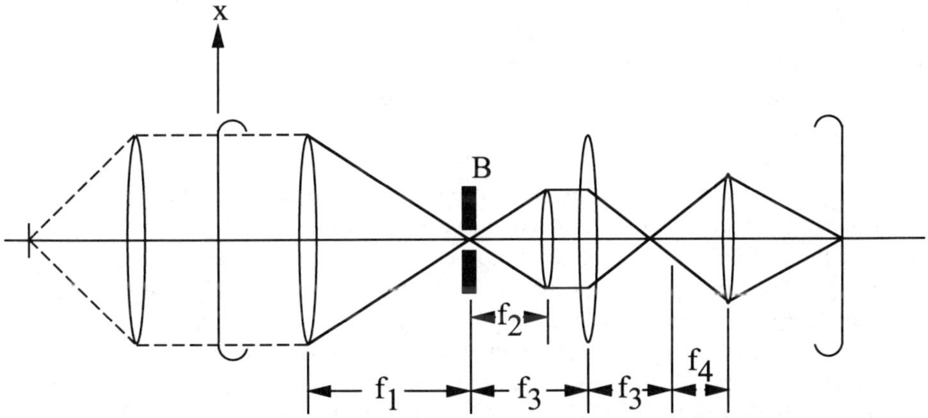


(b)

**Figure 14.12** Operation of the tilted-plane optical processor showing range focal plane $R$ tilted at angle $\alpha$ (the tilted angle of film transport) and azimuth focal plane $A$. The system reduces the azimuth plane tilt, and brings the range and azimuth image plane into coincidence: (a) range dimensions, in which lenses $L_2$ and $L_4$ are inactive; (b) azimuth dimension, in which all lenses are active.

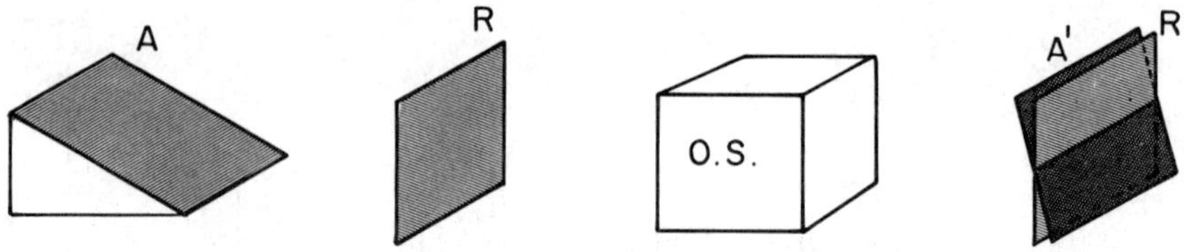


**Figure 14.13** An optical system (OS) for imaging $A$ and $R$ into approximate superposition.

in Figure 14.15. During those years, alternatives to optical processing made only a small dent in SAR data processing technology. One of the high points of the SAR optical processing combination occurred with the lunar orbiter of 1972 [8]. An SAR system was installed in the orbiter to map the surface of the moon. SAR data was recorded on photographic film and was returned to Earth for optical processing. The experiment was quite successful.

## 14.6 DIGITAL COMPUTER COMPETITION

However, as time passed, the situation slowly changed. Digital computing technology became ever more powerful, and gradually digital computing techniques were applied to SAR data processing. At first, the digital methods were secondary to optical processing and were used on a small scale and only for situations that were not overly ambitious, where the data rates were low and the resolution requirements were modest. However, the ability of digital computers to handle SAR data continued to grow. By the 1980s, digital methods had become equal to the SAR data processing problem and began displacing the optical processing methods.

The optical processing systems had limitations compared to the digital. The optical method is, as a practical matter, restricted to linear operations. The digital methods offered greater flexibility in distortion and aberration correction, image enhancement, and a host of other capabilities that are part of the standard repertory of digital image processing. It became evident that digital techniques could make superior SAR pictures.

During this time, magnetic tape continually improved, both in dynamic range and recording density, and eventually became superior to photographic film as a storage medium for SAR data. With tape, the data can be recorded instantly, whereas film requires chemical development. The tape technology is readily adapted to digital storage and as an input to the digital computer. The superior capabilities of magnetic tape compared to photographic film further tipped the balance toward digital processing.

The advantages of digital processing were considerable, so digital processing became dominant in the 1980s. Nevertheless, the optical method retained the advantages of speed and cost, so it maintained a role. The optical technique is used where large-batch processing is required. Data that are sufficiently important can then be processed digitally, leading

**Figure 14.14** The tilted-plane optical process.

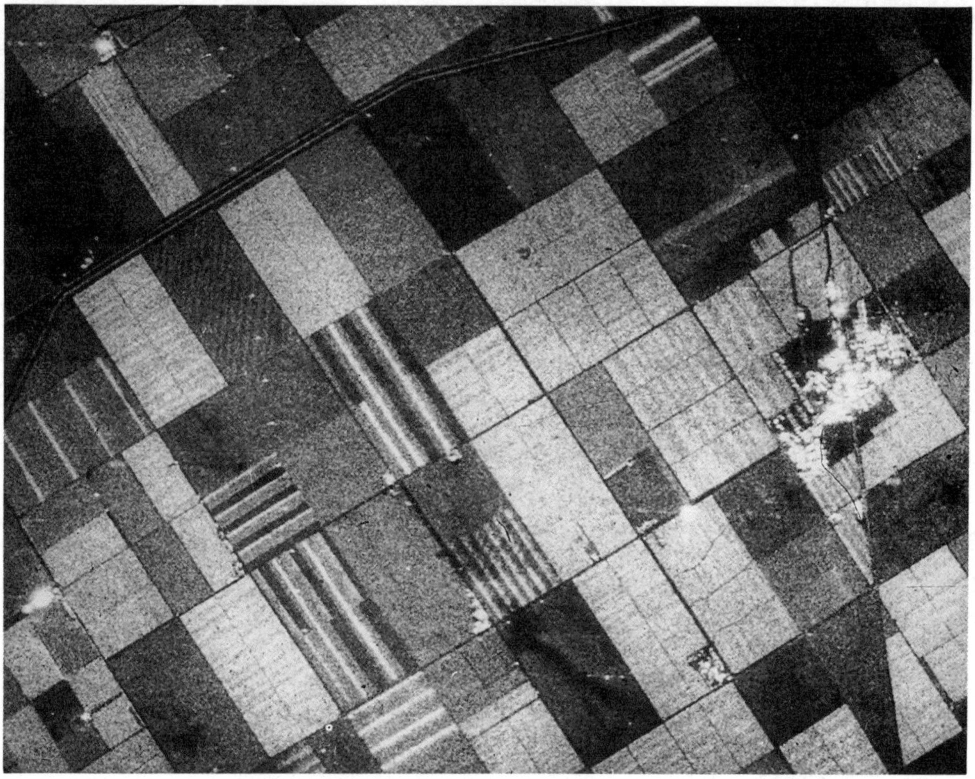

**Figure 14.15** A radar image produced with tilted plane processor, showing the Washita River and the rural area south of Ravia, Oklahoma.

to higher quality imagery. It is significant that the tilted-plane optical processors of the 1960s are in use today, having survived now for 30 years.

## 14.7 THE LEGACY

The legacy of the SAR optical processing systems remains today, both in general and in specific ways. First, the SAR was the first major application of the optical processing of data of nonoptical origin, and it spurred the growth of optical processing, which continues today. Second, the theory of holography was significantly advanced as a consequence of SAR optical data processing [9]. Finally, one of the major applications of holography, the holographic construction of spatial filters with full phase and amplitude control, and in particular spatial matched filters, was developed in SAR optical processing research [10–12]. These areas have grown to proportions far greater than SAR optical processing.

In addition, optical processing made possible the great advances of the 1950s and 1960s in SAR, in an era when no other technology could have accomplished this feat.

The interaction between holography and SAR in their earlier development days was considerable, and the exchange went both ways. The holographic viewpoint of SAR, developed in the 1955–56 period, was a new and radically different way of describing the SAR process, especially when combined with optical processing [5]. The previous theories of SAR, including Doppler filtering, cross-correlation, and matched filtering, were supplanted by the holographic viewpoint, which by 1960 had become dominant [5]. The application of holographic theory led to highly sophisticated and ingenious optical systems, including the tilted-plane optical processor [4], which could hardly have arisen except through the application of holographic theory. Without the aid of holography, the development of SAR would have been much slower. On the other side, SAR contributed much to holography. The theory of carrier frequency holography, better known as *off-axis holography,* was developed in the 1955–56 period in connection with SAR [5]. In the 1960–61 period [9], this theory was extended to optical holography by, among other innovations, introducing the reference beam at an angle relative to the object beam. As a consequence, holographic imagery of high quality was achieved for the first time, and holographic development proceeded rapidly in the next few years.

One of the most important applications of holography, the holographic complex spatial filter [10–14], was developed for the optical processing of radar data; today this technique is widely practiced and has many applications. In 1960, in the midst of the enormous success of the application of optical processing to SAR, ways were sought to expand the applications by applying new and more sophisticated techniques to SAR processing and to the optically processing of data from other forms of radar. Pulse compression technology was of especial interest. SARs at that time were beginning to use waveforms more complicated than simple pulses, such as chirp waveforms or, at an even more sophisticated level, pulse-coded waveforms. To achieve resolution in range, the pulses had to be compressed to give the effect of having radiated a short pulse. There are considerable advantages to radiating the more complicated waveforms.

The conventional way to compress a coded radar pulse is to pass it through a matched filter, which is often a very complicated filter, having both amplitude and phase components in the transfer function. The possibility of forming the required matched filter as a spatial filter and carry out pulse compression by optical processing was considered.

Spatial filtering was even at that time an old, well-established technique, some of the most important applications being dark-field and phase contrast, techniques that had long been used by microscopists. These were simple filters. Here, the aim was to make spatial filters of far greater complexity. The basic idea, due to C. Palermo, was in essence to use a Fourier transform hologram of an object as a spatial matched filter for that object [12]. It can be shown that when a signal (an object distribution) is imaged through the hologram, the conjugate term of the hologram forms an image which is the original object distribution filtered by a complex transfer function embodied in the hologram spatial filter.

In the case of the coded-pulse signal, a hologram is made of the Fourier transform of whatever pulse code is used for coding the radar pulse. When the radar signal reflected from the object field is recorded photographically and is spatially filtered by the holographic spatial filter, the conjugate term of the hologram forms an image in which the radar pulse is decoded, resulting in a sharp, well-focused image. The filter is in this case a matched filter for the coded waveform. This initial demonstration of holographic spatial matched filtering was given in 1961 by Kozma and Kelly, using the well-known shift register sequence of +1s and −1s as the coded waveform [10]. The dramatic results they achieved spurred the search for additional applications. The next application (1962) was the use of a spatial matched filter to correct for the range curvature problem [13], a second-order defect in SAR imaging, which occurs when the synthetic aperture being generated is very long, so that the range to the object changes by an amount significantly greater than the range resolution, so that a zone plate produced by an object point is recorded along an arc on the data film instead of along a straight line. This application was also highly successful.

VanderLugt, of the same laboratory, in 1962 applied the new holographic spatial filtering method to pattern recognition, again with spectacular results [14]. The holographic complex spatial filtering technique now broke away from its radar data processing origin and stood on its own as a technique of broad application.

The marriage of SAR and coherent optical processing has thus been incredibly fruitful and has left its mark on both technologies. Finally, we note that SAR development was done under military secrecy, and almost all of the original publications were classified documents. Unclassified versions emerged much later, sometimes with a time lapse of as much as 5 to 10 years. The journal references given here thus generally are not the seminal papers, having generally been written long after the work was done. Although SAR was developed at many places throughout the U.S., the optical processing technology originated and was largely developed at the University of Michigan's Willow Run Laboratories (now separate from the university and called the Environmental Research Institute of Michigan (ERIM)). I was closely involved in this work over the entire period of optical processing development, from 1954 to about 1970.

## REFERENCES

[1] Sherwin, C. W., J. P. Ruina, and R. D. Rawcliffe, "Some Early Developments in Synthetic Aperture Radar Systems," *IRE Trans. Mil. Electron.*, Vol. MIL-6, 1962, pp. 111–115.

[2] Cutrona, L. J., W. E. Vivian, E. N. Leith, and G. O. Hall, "A High-Resolution Radar Combat-Surveillance System," *IRE Trans. Mil. Electron.*, Vol. MIL-5, 1961, pp. 127–131.

[3] Cutrona, L. J., E. N. Leith, L. J. Porcello, and W. E. Vivian, "On the Application of Coherent Optical Processing Techniques to Synthetic-Aperture Radar," *Proc. IEEE,* Vol. 54, 1966, pp. 1026–1032.

[4] Kozma, A., E. N. Leith, and N. G. Massey, "Tilted-Plane Optical Processor," *Appl. Opt.*, Vol. 11, 1972, pp. 1766–1777.

[5] Leith, E. N., and A. L. Ingalls, "Synthetic Antenna Data Processing by Wavefront Reconstruction," *Appl. Opt.*, Vol. 7, 1968, 539–544.

[6] Leith, E. N., "Quasi-Holographic Techniques in the Microwave Region," *Proc. IEEE,* Vol. 59, 1971, pp. 1305–1318.

[7] Leith, E. N., "Optical Processing Techniques for Simultaneous Pulse Compression and Beam Sharpening," *IEEE Trans. Aerosp. Electron. Syst.,* Vol. AES-4, 1968, pp. 879–885.

[8] Porcello, L. J., C. E. Heerema, and N. G. Massey, "Optical Processing of Planetary Radar Data: Preliminary Results," *J. Geophys. Res.,* Vol. 74, 1969, p. 27.

[9] Leith, E., and J. Upatnieks, "Reconstructed Wavefronts and Communication Theory," *J. Opt. Soc. Am.,* Vol. 52, 1962, pp. 1123–1130.

[10] Kozma, A., and D. Kelly, "Spatial Filtering for Detection of Signals with Additive Noise," *J. Opt. Soc. Am.,* Vol. 54, 1964, p. 1395.

[11] Kozma, A., and D. Kelly, "Spatial Filtering for Detection of Signals Submerged in Noise," *Appl. Opt.,* Vol. 4, 1965, p. 385.

[12] Kozma, A., "Coherent Optical Systems for Data Processing, Spatial Filter and Wavefront Reconstruction," Ch. 8 of *Optical and Electro-Optical Information Processing,* Tippett, et al., eds., MIT Press, 1965.

[13] Leith, E. N., "Range-Azimuth-Coupling Aberrations in Pulse-Scanned Imaging Systems," *J. Opt. Soc. Am.,* Vol. 63, 1973, pp. 119–126.

[14] VanderLugt, A., "Signal Detection by Complex Spatial Filtering," *IEEE Trans. Information Theory,* Vol. IT-10, 1964, pp. 139–145.

# Chapter 15
# Correlation Systems

**William A. Penn**
*Syracuse University*

## 15.1 PURPOSE AND MATHEMATICAL DESCRIPTION OF OPTICAL CORRELATION

Interest in the idea of performing correlation with optics first developed early in this century. This early work was performed at several institutions. These first implementations used noncoherent optical imaging, but typically, such research was extended to coherent architectures [1].

Figure 15.1 depicts a simple system which can accomplish spatial correlation with noncoherent light. To develop a complete correlation function with this system, the correlation has to be repeated with different imaged offsets between the two functions, the first of which will be called the *signal function,* and the second will be called the *reference function.*

To clarify this, consider an elementary mathematical expression that describes the correlation process:

$$R(y) = \int f_1(x) \cdot f_2(y + x)\mathrm{d}x \tag{15.1}$$

The correlated result, $R(y)$, becomes a function of the offset between $f_1$ and $f_2$ only, designated by the quantity $y$. The complete correlation function is a function of $y$, and thus its construction requires the indicated execution of the integration for each of the different desired values of $y$. In optical correlators, however, it is usually possible and convenient to derive a continuous spatial or temporal presentation of this correlation function. This feature is a result of the use of the optical spatial dimensions to provide a

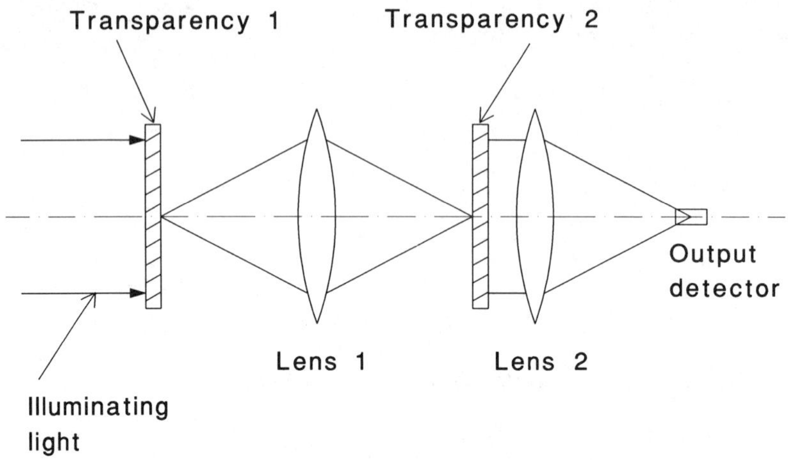


**Figure 15.1** Basic optical image correlator.

high degree of parallelism in the signal processing, which greatly augments the single temporal dimension in electronic systems. This parallelism and the continuous nature of the output presentation can represent a great advantage over correlators using other technologies.

The realization of a correlation function has many applications. Particularly common applications using optical realizations are high-resolution holographic radar image reconstruction, matched filtering of waveforms, image recognition, and, of more recent interest, correlation of waveforms to derive a weight function in adaptive systems.

## 15.2 ROLE OF OPTICAL IMAGING AND OPTICAL DIFFRACTION IN ACHIEVING OPTICAL CORRELATION

Correlation in analog optical systems is achieved primarily by the use of imaging and optical diffraction effects. In the optical correlation process, the dependent variables of the correlation can be represented as optical intensity, or, alternatively, as coherent amplitude. In the former case, the system is characterized as being optically noncoherent, and in the latter, as optically coherent.

In a noncoherent optical system, the correlation integral is performed by optical imaging. The signal function, which is positioned in one optical aperture as a spatial pattern, is imaged on a mating spatial pattern which represents the reference function. As the imaged signal passes optically through the reference (transparency) an intensity multiplication occurs between the two functions. The light is then collected as a sum of intensities to produce the correlation integral typified in (15.1). To evaluate other points in the correlation function, the signal image must be shifted spatially with respect to the reference, and the new value obtained. This feature is often a disadvantage of noncoherent techniques.

Another awkward characteristic of noncoherent systems is the representation of functions by optical intensity rather than coherent amplitude. Thus, the physical quantity being used to represent function amplitudes is unipolar, and negative values are forbidden. As a result, an optical *bias*, or average light level, is required, around which bipolar functions can be represented. This bias tends to obscure the desired coherent function.

With the advent of the laser in the 1960s, a more practical source of coherent light became available. This marked the beginning of interest in coherent optical correlators. With the use of coherent light, the dependent variables in (15.1) are represented by the amplitude of a coherent light beam instead of intensity. Bipolar values are now possible. The light components that are added to form the correlation integral add interferometrically to produce the desired coherent sums of the bipolar quantities. Moreover, the effects of optical diffraction now create a spatial Fourier plane, where the 2-D Fourier transform of the optical spatial pattern of the preceding aperture can be found. This opens up the possibility of transform correlation by multiplying spatial Fourier spectra.

## 15.3 FUNDAMENTAL TYPES OF OPTICAL CORRELATION ARCHITECTURE AND THE INTERPRETATION OF THEIR OUTPUT SPATIAL AND TIME COORDINATES

Optical analog correlators fall into several classifications. They can involve spatially fixed input data patterns or moving data; they can execute the required integration by collecting the integral in space or in time; and they can execute the correlation by direct overlay of the functions or by multiplication of spatial spectra. These various possibilities will now be considered.

### 15.3.1 Spatial Correlators (Space-Integrating)

Space-integrating correlators use at least one of the spatial dimensions of the optical aperture to achieve the required integration for correlation. Thus, the light over a spatial area is collected to produce the integral. Within this classification, the process can be carried out in the original data domain or by means of the Fourier spatial spectrum of this data.

#### *15.3.1.1 Direct Imaging*

Correlation can be achieved in the input data domain by directly imaging the first function on a transparency, which contains the second function to the same scale, as pictured in Figure 15.2. As this image passes through the transparency, the two functions are

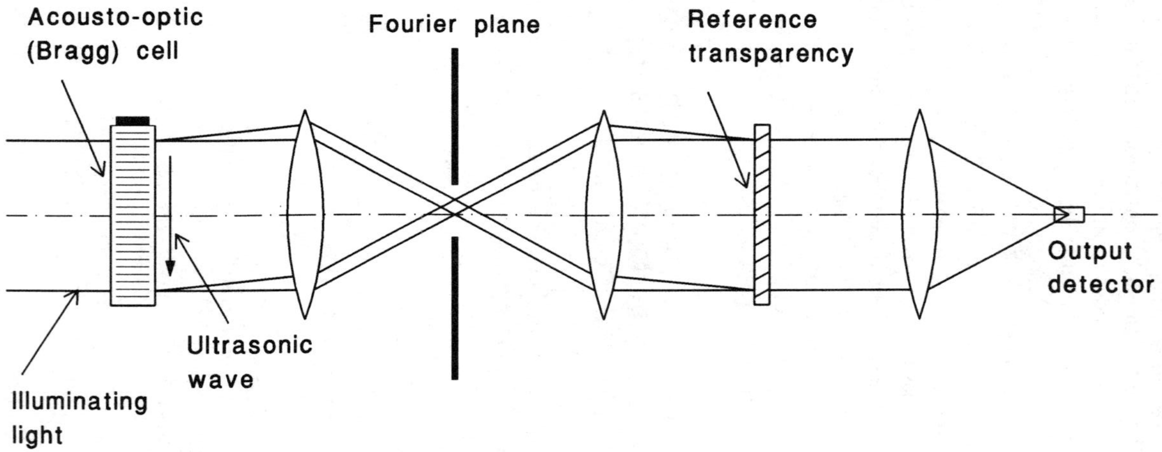


**Figure 15.2** Direct imaging coherent correlator.

multiplied. The emerging light, which represents the product function, is then integrated by a lens that collects this light and delivers the sum to an output detector. The independent variable of the resulting correlation function (i.e., the offset between the two input functions) is varied by shifting the relative positions of the first function image with respect to the second function transparency. This variable can be swept in time by imparting a uniform motion to the first function, such as is provided by an acousto-optical delay line. If the optical beam illuminating the input modulator is noncoherent, this collection of light is an addition of optical intensities. If the input illumination is coherent, the process gives the addition of optical amplitudes. Even in the optically noncoherent case, the process can represent coherent correlation by providing the input function on spatial carrier frequencies. This processor is equivalent to a *transversal filter,* in which the reference transparency plays the role of tap weights.

In the case of the optically coherent correlator, an optical Fourier plane is found between the two telecentric imaging lenses pictured in Figure 15.2. This plane can be used to stop the zero order light and retain the first order of diffraction from the spatial carrier provided in the input plane. Here, the diffracted optical beam carries the coherent phase information of the input function, since signal phase appears in the first aperture as a shift of the carrier grating pattern. This effect causes the signal phase to be imparted to the first diffracted order, as demonstrated in the following expression.

$$
\begin{aligned}
U_{out} &= U_0 + 2U_1(x)\cos[2\pi f_x x + \phi(x)] \\
&= U_0 + U_1(x) \cdot [e^{j2\pi f_x x} e^{j\phi(x)} + e^{-j2\pi f_x x} e^{j\phi(x)}]
\end{aligned}
\tag{15.2}
$$

for first order:

$$U_{1^{st}\text{order}} = e^{j2\pi f_x x} \cdot [U_1(x)e^{j\phi(x)}]$$

where: $U_1(x)$ denotes the amplitude of optical modulation (amplitude of envelope of spatial grating); $U_0$ denotes the optical zero order amplitude; $f_x$ denotes the spatial carrier (grating) frequency; $\phi(x)$ denotes the phase function of (grating) modulation (i.e., that of the acoustic signal).

### *15.3.1.2 Transform Correlation*

In Fourier transform theory, correlation of two functions in the data domain corresponds to multiplication of the spectra of these two functions in the frequency domain, where one of the spectra is presented in the form of its complex conjugate (convolution corresponds to multiplication of the spectra without the complex conjugation). This operation can be accomplished by creating and storing the spatial spectrum of one function on a transparency or modulator as suggested in Figure 15.3. This spectrum is recorded by interfering with an optical reference beam with the desired spectrum so as to produce a spatial carrier

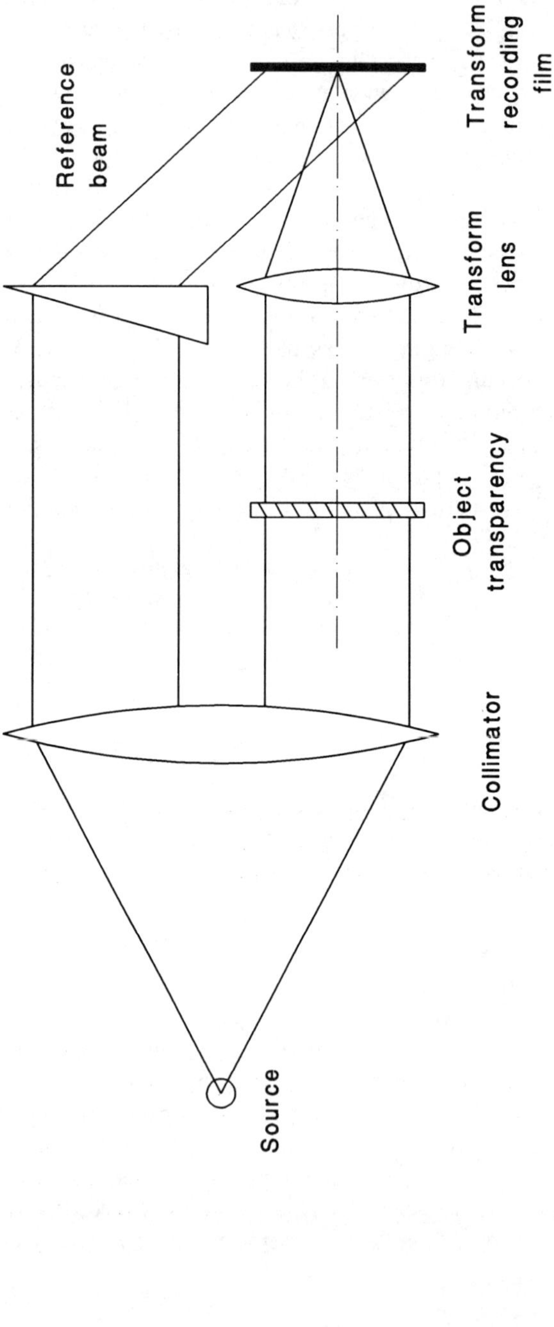


**Figure 15.3** System for creating Vander Lugt filter.

frequency grating to carry the coherent optical phase information. The grating frequency is determined by the angle between the two beams. This recording can be viewed as a *Fourier hologram* of the input function. This type of spatial filter is commonly referred to as a *Vander Lugt filter,* in honor of the original inventor of this technique. The transform thus recorded (the Vander Lugt filter) is obtained as the optical diffraction pattern projected into the back focal plane of the lens when the front focal plane is occupied by the desired first function, as pictured in Figure 15.3.

As has been shown, the first-order diffracted beam from such a grating carries the desired coherent version, including the phase information of the spectrum, in the optical beam. In the correlation operation shown in Figure 15.4, the spectrum of the second function is constructed in the back focal plane of a Fourier transform lens. This is projected on the stored spectrum of the first function, and a multiplication occurs. The product is then inverse-Fourier-transformed to form the correlation function in the back focal plane of this second lens as a spatial presentation. The correlation is obtained, as opposed to a convolution, by selecting the proper order out of the two conjugate first orders to invoke the conjugate spectrum of the stored spectrum. A classic description and thorough mathematical analysis of this system is presented by Goodman [2].

#### *15.3.1.3 Radar Applications of Space-Integrating Correlators*

It is obvious that if the two functions in either of these systems (direct imaging or transform correlation) are the same function, the result is autocorrelation of the function, and if the two functions are different, the result is cross-correlation. In the latter case, the operation can be viewed as a matched filtering of data introduced into the first aperture in Figure 15.4, where the reference function is the stored transform hologram.

In radar applications, both architectures can be used to perform matched filtering of a coded transmitted radar waveform. In the direct imaging device, the transmitted function becomes the reference transparency. In the transform correlation device, a Fourier transform hologram is required as the Vander Lugt filter. Many typical radar-coded waveforms can be handled in this way.

One very common radar waveform, the linear frequency modulated pulse, or chirp pulse, has a particularly simple optical filter. As demonstrated in (15.3), a waveform exhibiting linear frequency modulation is also characterized by a quadratic phase variation:

$$
\begin{aligned}
U &= U_m \cos\left[2\pi(f_0 + \frac{v}{2}t)t\right] \\
&= \mathrm{Re}\left[U_m e^{j2\pi(f_0 t + \frac{v}{2}t^2)}\right] \\
&= \mathrm{Re}[U_m e^{j\phi}] \\
f(t) &= \frac{1}{2\pi}\frac{\mathrm{d}\phi}{\mathrm{d}t} = f_0 + vt
\end{aligned}
\qquad (15.3)
$$

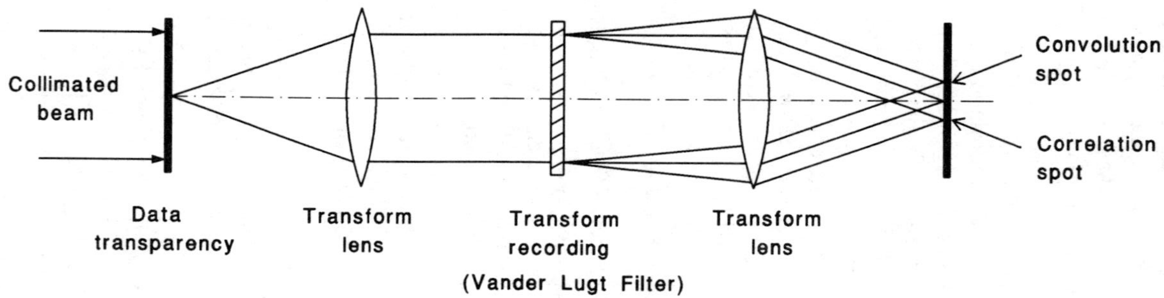


**Figure 15.4** Operation with Vander Lugt filter.

In this expression, $x$, the position in the optical aperture, is given by $x = v_a t$, where $v_a$ is the scan velocity with which the data was recorded, or, in the case of an acousto-optic delay line, it is the acoustic velocity.

The coherent transfer function of a cylindrical lens of focal length $F$ is given as

$$T(x) = e^{j\frac{\pi}{\lambda F}x^2} \tag{15.4}$$

where $F$ is the focal length of the lens and $\lambda$ is the operating wavelength.

Comparison of this to (15.3) shows that the chirp recording, in the direction of its recording axis, has a self-focusing lens effect on the first order of diffraction. The relation between the chirp frequency rate $\nu$ and the focal length of the equivalent lens $F$ is given by the following expression:

$$F = v_a^2/\lambda\nu \tag{15.5}$$

Figure 15.5 shows a conceptual diagram of such a correlator. The input recording is made to move by an acousto-optic Bragg cell. The first diffracted order is self-focused and moves with the motion of the input data. If an optical slit, matched in width to the focused spot, is provided in the focal plane of the moving grating, a compressed pulse will be obtained as the spot moves past the slit. The width of this pulse will correspond to the reciprocal bandwidth of the input waveform. It can be noted that if a formal quadratic phase reference transparency was used to compress such a chirp pulse, as in the reference

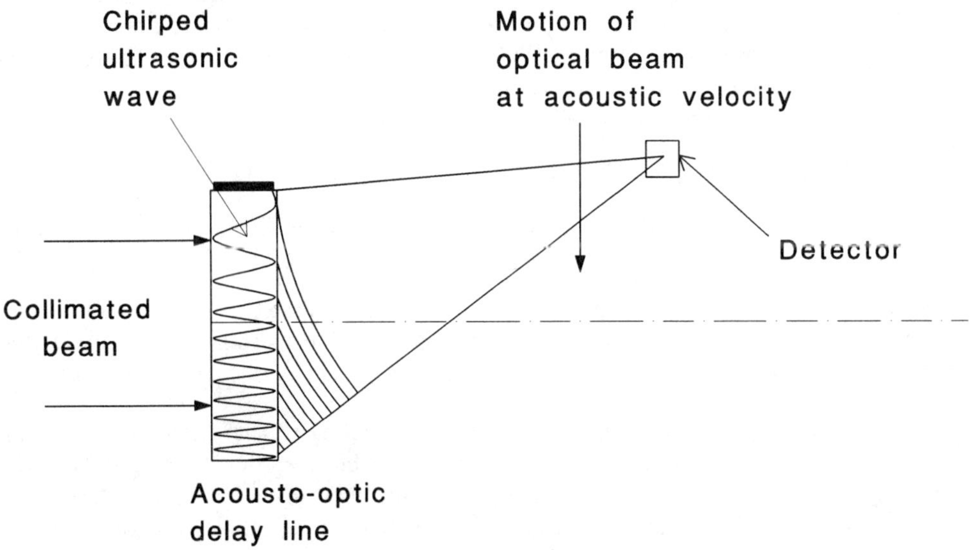


**Figure 15.5** Self-focusing correlator for chirp waveform.

position in Figure 15.2, a time mismatch between the signal and reference would result in a constant spatial frequency in the image product plane. This would cause the light to diffract at an angle. At the time that the signal image exactly matches the reference, diffraction would occur at the zero Doppler frequency angle, which would direct the light to the slit location in the output plane. Thus, even in this case, the behavior can be viewed as a moving focused spot traveling past a matching slit. This motion reflects the time-frequency ambiguity that characterizes a chirp pulse.

### 15.3.2 Time-Domain Correlators (Time-Integrating)

In the preceding systems, the integration required to perform correlation, as indicated in (15.1), is performed in the spatial domain by a lens operating on the spatial distributions of the signal and reference. Another class of optical correlators perform this integration in the time domain. This temporal integration can be supplied by the output detector of the integrator or by the electronics which follow it. Many output detectors, including detector arrays and types of retinal cameras, have an intrinsic time constant. This time constant then represents the integration time of the optical correlator.

A conceptual illustration of a time-integrating correlator is given in Figure 15.6. In this type of correlator, it is essential that the input waveform move linearly with time. Usually, the most convenient method of providing this is the use of an acousto-optic cell. The input waveform is used to modulate the light source, which illuminates the entire correlator. If the light source is a semiconductor laser, the required modulation can be achieved by direct electrical modulation of the laser. For other types of lasers, an external optical modulator is required, such as an electro-optical or acousto-optical point modulator. This time modulated light is passed through the acousto-optic cell in which the reference data are applied. The resulting product is imaged by an optical imaging stagc to thc output detector or a detector array in which the time integration occurs. This integration is expressed as

$$W(x) = \int g(\tau)h\left(\tau - \frac{x}{v_a}\right)\rho_T(t - \tau)\mathrm{d}\tau \tag{15.6}$$

where $g(t)$ denotes the light modulator function; $h(t)$ denotes the acousto-optic ultrasonic waveform; $x$ denotes the position of the output detector (image plane); $\rho_T$ denotes the time integration weighting function of output detectors(s) (usually an exponential function).

It should be appreciated that each image point, coming from a different location in the acousto-optic cell, represents a different time offset between the input signal applied to the light modulator and the reference applied to the acousto-optic cell. It follows that the product function imaged to the detector is a spatial pattern representing a continual spatial presentation of the correlation function after the time integration. Thus, the output spatial coordinate represents time offset between the two correlated functions.

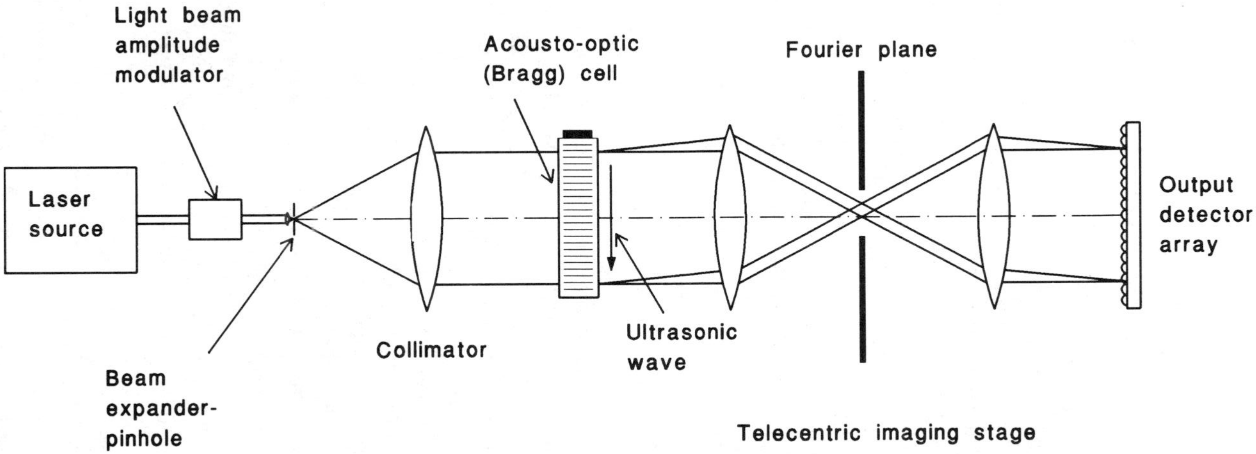


**Figure 15.6** Time-integrating correlator.

### 15.3.3 Comparison of Fixed- and Moving-Window Configurations

#### *15.3.3.1 Moving-Window Configurations*

The direct imaging correlator implies the use of a moving window for the input data, although a single correlation could be achieved without this. When the input data do move, time then represents the offset between the two correlated functions, presuming that the reference pattern is fixed (see Fig. 15.2). Thus, the moving window is the offset scan mechanism. In the particular case of an optically coherent direct imaging system, the space-integrating output lens can be viewed as a Fourier transform lens. Not only does this lens deliver the integration of the product to a detector on the optical axis, it also provides integration with a built-in frequency shift between the two correlated functions at other points in the detector array. This frequency shift is proportional to the distance off-axis according to the Fourier transform action of this lens. Thus, the output spatial dimension that is in the direction of motion is a measure of the frequency offset between the correlated signals. In radar applications, this would usually represent target Doppler frequency.

#### *15.3.3.2 Fixed-Window Configurations*

Conversely, the transform correlator configuration implies the use of a fixed window. Scanning of the offset variable is provided by interrogating the Vander Lugt–filtered image as variable points in the output plane. The data in this plane are to be regarded as a matched filtered image of the input data. With the fixed nature of the data, filtering can be performed in the two spatial dimensions of the optical apertures. In this way, 2-D correlative pattern recognition can be accomplished. Accommodation to a varying signal Doppler frequency is now provided by physical motion of the Vander Lugt filter itself. If the filter is moved linearly with time, one now finds dual behavior in this system when compared to the direct imaging technique; that is, the spatial position now corresponds to time offset, and the time of the output corresponds to the Doppler frequency. A typical use of this arrangement in a radar would involve the writing of successive range sweeps in the optical aperture side by side. Fourier transformation in a direction transverse to the sweeps would now provide multipulse Doppler analysis, and the inline direction, as described before, would be used for waveform matched filtering.

#### *15.3.3.3 Interpretation of Output Coordinates*

Table 15.1 summarizes the meaning of the time and space output coordinates for the various discussed configurations.

**Table 15.1**
Correlator Types and Output

| *Type of Correlator* | *Meaning of Output Spatial Position* | *Meaning of Output Time* |
|---|---|---|
| Moving-window direct imaging correlator (Figure 15.2) | Carrier (Doppler) frequency differential between data and reference | Time offset between data and reference |
| Fixed-window Vander Lugt filter correlator (Figure 15.4) | Spatial offset between data and reference, representing time offset if data scanned in at $v_a$ | Carrier (Doppler) frequency differential if Vander Lugt filter is moved linearly with time |
| Time-integrating correlator (Figure 15.6) | Time offset between data and reference | Integration period starting time; not significant if correlation function is time invariant |

### *15.3.3.4 Use of Acousto-Optic Devices To Provide the Moving Window*

The most effective device for providing an optical moving window in radar applications is the acousto-optic Bragg cell. As a general rule, Bragg cells are well matched to radar parameters. This is true in window delay length, where delays in the 1- to 20-$\mu$s range are typical, and in device bandwidth, where bandwidths up to the region of 1 GHz or more are possible. Provision of bandwidths of 100 MHz or less is fairly common.

In optically coherent systems, the use of a Bragg cell, rather than the thinner configuration, commonly called a *Raman-Nath* optical delay line, is indicated. As shown in the following equation, the amount of optical phase modulation provided by an acousto-optical medium is proportional to the interaction thickness of the line (i.e., the distance that the optical path intersects the acoustic beam). This phase modulation is defined as that which is experienced by an optical ray which is aligned with an acoustic wavefront throughout the interaction length, regardless of whether such alignment is actually achieved. The phase modulation, defined in this sense, is given by

$$\phi_a = kL\Delta n = \frac{2\pi L}{\lambda}\Delta n \tag{15.7}$$

where $k$ denotes the optical free-space wave number; $L$ denotes the interaction length between acoustic beam and light path; $\lambda$ denotes the optical wavelength; $\Delta n$ denotes the modulation of optical index of refraction.

In the case of the Bragg regime, this alignment is not maintained, since diffraction causes deviation of the rays immediately upon entry in the interaction region. As this process progresses, the ray drifts out of alignment with the wavefront. This is the essential issue in the definition of the boundary between Bragg and Raman-Nath operation. In the

Raman-Nath operation, the interaction layer is thin enough so that no important ray drift can occur. The boundary between the two regimes is defined as follows:

$$L_0 = \frac{n\Lambda^2}{2\lambda} \tag{15.8}$$

where $\Lambda$ denotes the acoustic wavelength; $n$ denotes the delay line material index of refraction; $L > L_0$ implies the Bragg regime; $L < L_0$ implies the Raman-Nath regime.

In (15.7), $\Delta n$ can be evaluated in terms of the acoustic power density (intensity) as

$$\Delta n = \sqrt{M_2(I_a/2)} \tag{15.9}$$

where $I_a$ denotes the acoustic intensity; $M_2$ denotes the figure of merit of acoustic delay line material. This is a commonly published parameter of delay line materials.

Given an amount of phase modulation $\Delta\phi$, the strengths of the zero, first, and second orders for the Raman-Nath operation are illustrated in Figure 15.7. It can be seen that energy is depleted from the zero order as the phase modulation becomes stronger,

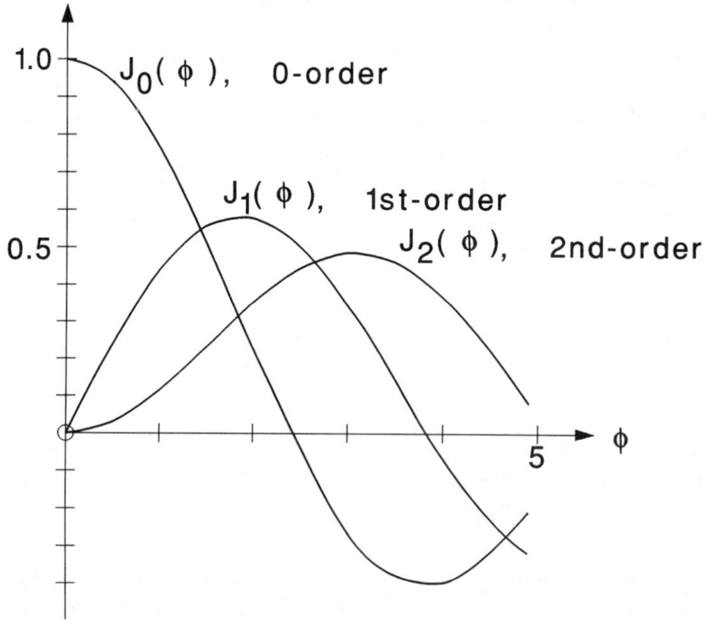


**Figure 15.7** Amplitudes of optical orders versus optical phase modulation for Raman-Nath diffraction.

and this energy is directed into the diffracted orders. The first order is of particular interest, since its behavior is linear for low amounts of phase modulation. It finally saturates at a peak amplitude value of 58% at peak phase modulation of 1.85 rad. This is the essential feature of the first order, which makes it the signal beam of choice in correlator systems.

The behavior of Bragg cells is given in a similar illustration in Figure 15.8. For optimum operation, a Bragg cell should be oriented in at the *Bragg angle,* which is a bisection of the directions of the zero and first orders. In this configuration, a Bragg cell will alternately deflect energy from the zero order to the first order and back again as an optical beam progresses through the interaction length. A first maximum of, theoretically, 100% transfer of optical energy from the zero to the first order will occur when the phase modulation reaches $\pi$ rad.

Not only is the transfer somewhat better and more linear in the case of Bragg diffraction for a given amount of phase modulation, but the given phase modulation is achieved with less acoustic power density, as given by (15.7) and (15.9), since the interaction length can be much larger in this regime. One might initially be concerned

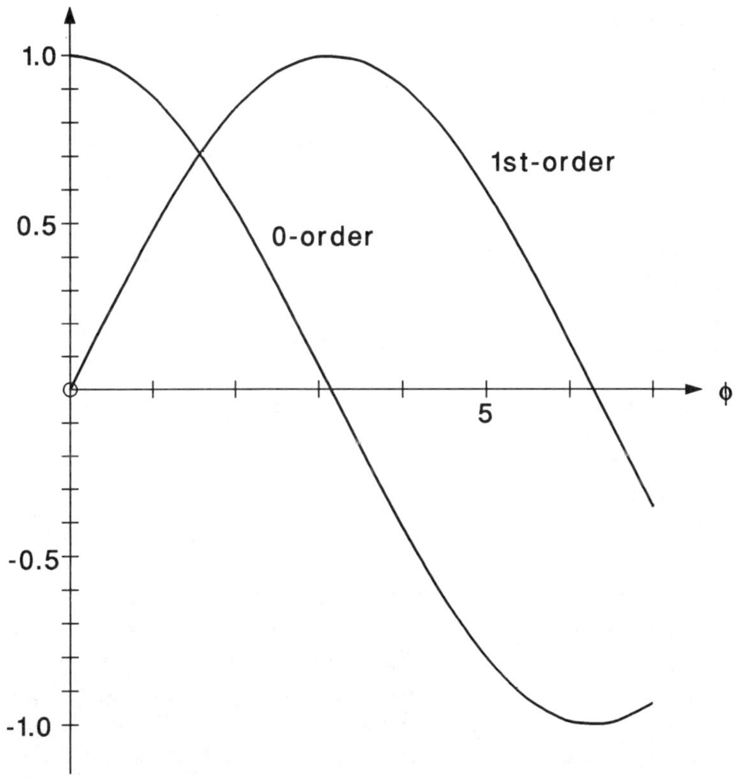


**Figure 15.8** Amplitudes of optical orders versus optical phase modulation for Bragg diffraction.

that it would be impossible to image such a thick medium into an output plane of detection or a reference transparency. However, since the entire optical beam consists only of the two orders, which typically span a very small angle, the resulting *depth of field* is adequate to span the interaction length when the fractional bandwidth of the signal is substantially lower than the acoustic carrier frequency. The limit in bandwidth due to this depth of field effect is given by

$$\Delta f \leq \frac{nv_a^2}{\lambda L f_a} \tag{15.10}$$

where: $v_a$ denotes the acoustic velocity of propagation in delay line material; $\Delta f$ denotes the acoustic waveform bandwidth; $f_a$ denotes the acoustic center frequency.

With only two strong orders in a Bragg cell, the pure phase modulation provided by the acoustic wave is imaged as a combination of equal energies of phase and amplitude modulation at a spatial frequency corresponding to the acoustic carrier frequency. To preserve pure phase modulation, the zero and both first orders would be required. With the device at the Bragg angle, the "negative" first order is effectively suppressed. The sensor of the Bragg cell image is an optical square-law detector that relies on optical intensity rather than phase modulation. The intensity modulation is thus provided automatically by the Bragg cell without any further manipulation in the spatial frequency plane of the imaging optics.

## 15.4 USE OF TIME-INTEGRATING DEVICES TO PRODUCE SPATIAL CORRELATION OUTPUT FUNCTIONS

### 15.4.1 Process of Detection and Time Integration

A suitable device to provide time integration is essential in time-integrating systems. This can be an ordinary point detector with an electronic output circuit characterized by a suitable time constant. However, if it is desired to present a continuous spatial pattern representing the correlation function, a continuously distributed medium is required with the necessary time constant as an intrinsic feature. Ordinary photographic film can serve this function.

In the time-integrating correlator, the light incident on the acousto-optic Bragg cell (Fig. 15.6) is time-modulated by the input signal, which should include an appropriate carrier frequency of the same value as the acoustic carrier used in the Bragg cell. This modulation causes a *stroboscopic* effect on the signal being carried by acoustic propagation through the Bragg cell. The resulting product produces a fixed amplitude grating when subjected to the square-law transfer function and integration of the output device.

Another point of view of this process can be understood by regarding the output of the time modulator as three coaxial optical beams representing the optical zero order and the two optical sideband frequencies, which are displaced from the zero-order optical frequency by the acoustic carrier frequency, creating both a lower and upper sideband. Depending on the direction in which the Bragg angle of the following acousto-optic cell has been adjusted, these beams will suffer either a frequency up-shift or down-shift when diffracted into the spatial first-order position of the Bragg cell. One of the sideband beams will then come back to the zero-order optical frequency, and coherently interfere with the zero-order beam. Because this diffracted beam is at an angle (twice the Bragg angle) with the zero order, a spatial beat frequency (i.e., the spatial difference frequency) of these two beams will be formed. This will be transformed into a component with that real spatial frequency by the square-law action as

$$\begin{aligned} I_D &= \gamma P_s = \mathrm{Re}[\gamma \bar{a}_s(t) r_o^*] \\ &= \mathrm{Re}\left[\bar{a}_s e^{j2\pi\left[f_o t + f_a \frac{x}{V_a}\right]} \cdot (r_o e^{j2\pi f_o t})^*\right] \\ &= \gamma \bar{a}_s R_o \cos\left[2\pi f_a \frac{x}{V_a} + \phi_a\right] \end{aligned} \tag{15.11}$$

where * denotes conjugation and where:

$I_D$ = output detector current signal;
$P_s$ = optical input signal power to detector;
$a_s$ = optical amplitude of diffracted beam (root power);
$r_o$ = optical reference beam amplitude (root power);
$f_o$ = optical wave frequency;
$\phi_a$ = phase information of acoustic beam;
$\gamma$ = detector current responsivity.

### 15.4.2 Correlator Architectures Using Other Devices, Such as Liquid Crystal Light Valves

In its simplest model, photographic film provides this square-law and integration effect. The integration time is, of course, the length of time to which the film is exposed to the light pattern before chemical development. Because of the necessary delay in chemical development, photographic film cannot provide real-time operation. Usually, an alternative is needed which will integrate without time lapse and for which the time constant fulfills the system requirements. One example of such a device is the liquid crystal integrator, or liquid crystal light valve (LCLV) [3]. This device is an electrically biased sandwich of a photoconductive input layer, married to a liquid crystal layer in intimate proximity to the photoconductor, which provides the desired spatial modulation of a readout beam. These two layers are separated by a light barrier, and the coupling between them is purely electrical. In the same way photographic film could be used to modulate a coherent read beam, the LCLV provides this capability in real time when the read beam is passed through the liquid crystal layer by reflection.

Other devices can be used for this detector-integration function, such as the image converter, which displays a time constant in its phosphor-emitting layer.

### 15.4.3 Correlator Architectures Using Photorefractive Devices

Recently, much interest has developed in the use of photorefractive devices to provide correlative integration. A photorefractive device can serve in the role of the devices previously described. This device is based on nonlinear effects in certain optical materials in which the index of refraction of the material depends on the light level. Popular materials for this purpose are bismuth silicon oxide (BSO) and barium titanate (BaTiO3).

#### *15.4.3.1 Description of the Process as Holographic Effect*

Photorefraction is the process of spatial modulation of the optical index of refraction by a redistribution of charge in the medium, caused by an incident light pattern. Thus, an interference grating is formed by two interfering light beams under a square-law response. This is analogous, for example, to the grating formed in the photoconductor in the LCLV. The most fundamental distinction of the photorefractive grating is that it is a volume effect. The index of refraction modulation can thus be viewed as a volume hologram. This volume hologram then diffracts the read beam into a first-order Bragg beam which displays the desired correlation function.

#### *15.4.3.2 Advantages of Photorefractors*

The photorefractive device offers some desirable flexibility, since the time constant of the device depends on the light levels of the incident beams. This often means that more desirable values of the time constant can be achieved to suit the radar application. The LCLV, for example, tends to have a long time constant—in the range of 25 to 100 ms. Photorefractors can provide faster response with adequate light levels using BSO as the medium.

The optical index typically responds to a fairly narrow optical spectral range, typically in the green band. If the light is turned off in this band, the medium tends to hold the integrated values of index of refraction for a relatively long time. Thus, correlation can be executed during the short time of a radar window, and the result can be held for many radar interpulse periods; indeed, for many seconds of time.

Another feature of the holographic process which characterizes the photorefractive technique follows from the usual rules of holographic reconstruction. The reconstruction of the correlation pattern can be achieved from the wavefront interference pattern obtained at points other than the image plane.

#### *15.4.3.3 Special Problems Associated With Photorefraction*

When the read beam is the same wavelength as the incident signal beam, the time constant becomes shorter, and the needs of the radar correlator may not be satisfied. This effect can be reduced by using a different color for readout outside of the optical band to which the medium is sensitive. This typically means light of a wavelength that is deeper in the red end of the spectrum. Use of the redder light also means that the correlation pattern can be held for long periods of time while it is being read. All this is at the expense of providing two colors to the system.

Current photorefractive crystals are characterized by a rather low modulation of the optical index. Thus, they tend to provide somewhat inefficient Bragg diffraction compared to acousto-optic cells of like thickness. This diffraction efficiency can be improved by making the photorefractive crystal thicker, but that tends to exacerbate the restriction on bandwidth due to Bragg angular mismatch. This mismatch, for most arrangements, is only a problem when a different color is being used for readout. Thus, under the conditions of a two-color system and a given system bandwidth, there may be a seriously low limitation in the diffraction efficiency of the device.

The physics of many photorefractive media, such as BSO, are such that the diffraction efficiency improves when the volume hologram moves at a slow rate. The optimum fringe rate might be in the range of 2 to 20 Hz. This improvement only occurs when the physical precession of the fringes is in the correct direction with respect to the electrical bias. The response of a conventional integrator can be modeled as a single-pole Lorentzian frequency response centered at zero frequency. The photorefractor can apparently be modeled in the same way, with the change of the center frequency of the response. This effect can be removed by a proper single-sideband shift between the two correlated signals. Under these conditions the system will work in the conventional manner, with the integration effect determined by the bandwidth of the response of the photorefractor to varying fringe precession rates [4].

## 15.5 ADAPTIVE SYSTEMS USING OPTICAL CORRELATORS

### 15.5.1 Characterization of Adaptive Systems

Adaptive signal processors are those in which parameters or weights are adjusted according to attributes of the input signals. In one type of architecture, two classes of signals are provided, one of which is used as a *reference*. In this case, the operation on the signal is dictated by the information contained in the reference. Prediction or cancellation of components in the signal can be accomplished by finding correlative similarities between the signal and reference. This comparison is often accomplished by a correlation process. Many of the operations required to construct an adaptive process, including the often essential correlation, are realizable in analog optical form. Thus, the problem in the radar arena of suppressing unwanted interference from external sources can be effectively addressed by an optical correlator in conjunction with several other optical or electronic stages.

### 15.5.2 Example of a Least Mean Squares Algorithm Adaptive System Using Cascaded Optical Correlators

Here we present an example of an optical adaptive signal processor which has been demonstrated by several investigators [5–9] and is under current development to cancel radar interference such as jammers. This system is an optical realization of the *Howells-Applebaum* least mean squares (LMS) adaptive algorithm [10–12].

#### *15.5.2.1 Description of Closed-Loop Correlator-Convolver System*

A diagram of this adaptive system is shown in Figure 15.9. The interference-canceled radar main antenna signal is applied on a suitable IF carrier as amplitude modulation on a laser illumination beam to a time-integrating correlator. This time-modulated light illuminates an acousto-optic cell, which carries the reference, or auxiliary signal, as the ultrasonic wave. The product function produced is imaged to a detector-integrator, which has a square-law response to its input light pattern. One component of this result represents the slowly varying part which forms an optical grating, on which carrier the adaptive correlation or weight function appears. The weight function is thus presented as a coherent and continuous function thanks to the availability of the optical spatial dimension to carry it.

The weight function is then imaged to a second acousto-optic cell, which carries an identical traveling version of the auxiliary signal as on the first acousto-optic cell. Again, a product function is derived. Now the product function is collected by an output lens, and the sum is delivered to an output detector. The choice between conjugate order beams is made so that the process can be viewed as convolution. The entire operation between the integrator and output detector is a type of space-integrating correlator which performs the function of a transversal filter.

If the diagram in Figure 15.9 is studied, one will appreciate that for each time delay resulting in a high correlation or weight component out of the time-integrating correlator, a correspondingly delayed version of the auxiliary signal will be selected by the system and delivered to the output detector. The sum of all these highly correlated time-delay signal components exiting the detector is a cancellation signal which can be subtracted from the radar main channel signal to form the processed radar output. The cancellation

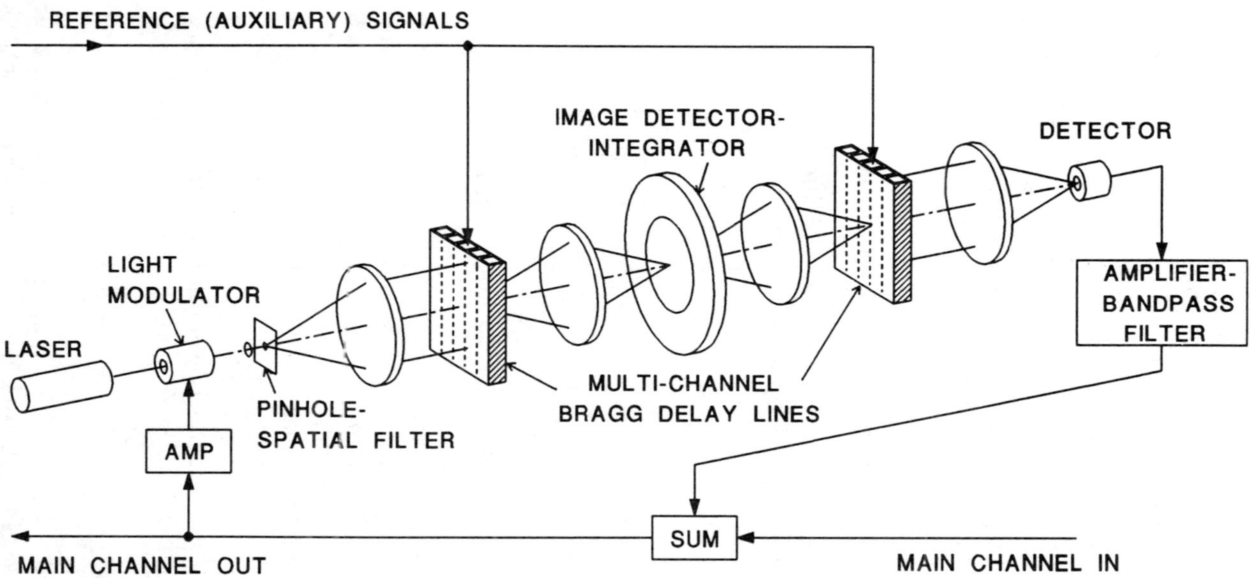


**Figure 15.9** Adaptive interference canceler using time and space integrating correlators.

is simply accomplished in an electronic subtractor. The result of this subtraction is the signal used to modulate the light modulator which was first described.

It is evident from the preceding description that the system resembles a closed-loop control system or servo-mechanism. In closed-loop operation, the correlation weights build up until cancellation is nearly complete. Like a "type 0" servo-mechanism, a slight error signal is required to support the steady-state weight that is developed. This "error" is the output radar signal. Jammer interference and other similar disturbances are canceled strongly if the auxiliary signal is obtained from a broad-beam, essentially omnidirectional antenna. The target cancellation is negligible when the main radar antenna is trained on this target, causing a strong target component in the main channel signal. The target is weak in the auxiliary outputs, so that correlation is weak between the two channels in regard to the target component. The advantage of this system is the cancellation ratio of interference components compared to target.

Because the system is like a closed-loop control system, alignment between the two acousto-optic cells is critical. If spatial misalignment of some fraction of an ultrasonic carrier cycle occurs, that fraction of 360 deg will be introduced as a loop phase shift. If the misalignment is in the neighborhood of a 90-deg shift, the system will go into self-oscillation. The cancellation ratio is primarily a result of achievable open loop gain.

The function performed by the open-loop optical section of this adaptive processor involving the two optical correlators/convolvers is expressed by

$$s_p(t) = C_o \cdot \int \left[ \int s(\tau) r^*\left(\tau - \frac{x}{v_a}\right) \rho_T(t-\tau) \mathrm{d}\tau \right] r\left(t - \frac{x}{v_a}\right) \rho_x(x) \mathrm{d}x \tag{15.12}$$

where:

$s_p(t)$ = output signal of optical section, which is the feedback cancellation or predictive signal to an input subtractor;
$s(t)$ = input main channel signal;
$r(t)$ = reference (auxiliary) signal from typical omnidirectional reference antenna;
$\rho_T(t)$ = time weighting function provided by the integration effect of the image detector-integrator (time constant);
$\rho_x(x)$ = spatial weighting function provided by the summation (output) lens and aperture in front of the detector.

The function described by (15.12) can be described in the frequency domain by performing a Fourier transform of that function. The result is expressed by

$$\hat{S}_p(f) = C_o \cdot \hat{S}(f)\hat{R}(f)\hat{R}^*(f) \tag{15.13}$$

where $\hat{S}_p(f)$ denotes the predictive (cancellation) signal; $\hat{S}(f)$ denotes the input signal; $\hat{R}(f)$ denotes the reference (auxiliary) signal. This describes the cancellation signal in the frequency domain as the input signal filtered by a frequency function which is the squared magnitude of the reference spectrum. At frequencies where the reference spectrum is large, the gain of the loop is larger, resulting in stronger cancellation of these frequency components. Thus, the interference represented by the reference signal is suppressed.

This adaptive system is thus an interesting example of the use of cascaded optical correlators, which in this case is a time-integrating correlator, followed by a space-integrating correlator, married together by the essential detector-integrator.

#### *15.5.2.2 Implementation With Conventional Integrator*

The requirement of an integrator can be fulfilled with a planar device that imitates photographic film, but that operates in real time. Thus, the liquid crystal device previously described is a good candidate for this function. In the system, the ultrasonic reference signal propagating in the first acousto-optic Bragg cell is imaged via the two orders dominant for Bragg diffraction. This results in a combination of carrier amplitude and phase modulation on the electrically biased input photoconductive surface of the LCLV. This pattern is transferred by the resulting electric field pattern across a light barrier to the liquid crystal layer. The molecular orientation in the liquid crystal responds to this with the time constant already described, creating the integration effect.

A second collimated laser beam is then phase-modulated by the liquid crystal layer, creating new orders used to create an image of the correlation weight function on the second acousto-optic cell.

One problem with the LCLV is that there exists some inflexibility of the time constant, which is determined largely by composition, temperature, and bias conditions. Another problem is that the response of the device tends to be asymmetrical for rising and falling input light levels. Additionally, the dynamic range of the device is rather limited. The conventional dynamic range is that range in which a square-law transfer essentially exists. A more appropriate definition of dynamic range applies when the device is optimally biased in input light level and the system noise floor is compared to the maximum modulation possible within the square-law range for the definition. Even with this definition, the dynamic range is probably limited to the 30- to 40-dB range.

#### *15.5.2.3 Implementation With Photorefractive Integrator*

Using a photorefractive device as an integrator for the adaptive system offers the achievement of faster time constants, with the proper selection of material, with the simultaneous possibility of ''locking up'' the weights with long storage in the device. This seemingly contradictory combination of behaviors is due to the response of the device to the bias value of writing illumination, since the time constant is strongly dependent on this level. To achieve the weight storage, then, it is likely that a different wavelength (more red) must be used for readout.

Typically, the diffraction efficiency resulting from the maximum contrast input (strong orders which are nearly matched in amplitude) is rather low, probably less than 0.05. It is thus necessary to design the remainder of the system carefully to minimize the noise floor, which will compete with the diffracted information from the photorefractive element.

### 15.5.3 Other Adaptive Architectures

Other optical architectures can be constructed to implement the function expressed by (15.12) or, equivalently, (15.13). One such architecture has been suggested by Vander Lugt [13]. In this version, two acousto-optic Bragg lines are positioned transversely to the optical axis and at right angles to each other. The second line is in a transform plane of the first and a multiplication of signals occurs in this second line. The input signal $s(t)$ is applied to the first line and the reference signal $r(t)$ is applied to the second line.

With the use of cylindrical optics, a 2-D array of cross-correlation values is constructed. This plane is an image plane of both acousto-optic lines. Correlation values of equal offset between the two acousto-optic signals appear placed in straight diagonal lines in this plane. These are integrated in the direction of equal offset and Fourier-transformed simultaneously by another Fourier transform lens. The optical spatial function thus obtained is the transform of the tap weights corresponding to the first two factors in (15.13). The spectrum of $r(t)$ is obtained from a third acousto-optic line, and this is added to the above output and is square-law-detected to get the triple product expressed in (15.13). The integration time is now determined by essentially the delay time of the acousto-optic lines, rather than the time constant of the detector-integrator pictured in Figure 15.9. Now, as before, the acousto-optic lines serve the functions of signal-reference correlation and a transversal filter to construct the cancellation signal from the correlation weights thus obtained. This operation is again most directly understood by the expression in (15.12).

## REFERENCES


[1] King, M., W. R. Bennett, L. B. Lambert, and M. Arm, "Real-Time Electro-Optical Signal Processors With Coherent Detection," *Applied Optics,* Vol. 6, No. 8, August 1967, pp. 1367–1375.

[2] Goodman, J. W., *Introduction to Fourier Optics,* New York: McGraw-Hill, 1968.

[3] Bleha, W. P., et al., "Application of the Liquid Crystal Light Valve to Real Time Optical Data Processing," *Opt. Eng.,* Vol. 17, No. 4, July-August 1978, pp. 371–384.

[4] Refregier, P., L. Solymar, and H. Rajbenbach, "Two-Beam Coupling in Photorefractive $Bi_{12}SiO_{20}$ Crystals with Moving Grating: Theory and Experiments," *J. Applied Physics,* Vol. 58, July 1985, pp. 45–57.

[5] Rhodes, J. F., "Adaptive Filter With a Time-Domain Implementation Using Correlation Cancellation Loops," *Applied Optics,* Vol. 22, No. 2, 15 Jan. 1983.

[6] Dickey, F. R., Jr., and W. A. Penn, "Adaptive Signal Processor for Interference Cancellation," Patent 4608569, 26 Aug. 1986 (filed 9 Sept. 1983).

[7] Penn, W. A., "Status Report on the Acousto-Optic Adaptive Processor," *IEEE Record of the 20th Asilomar Conf. on Signals, Systems and Computers,* 10–12 Nov. 1986.

[8] Penn, W. A., R. Wasiewicz, and R. M. Iodice, ''Optical Adaptive Multipath Canceller for Surveillance Radar,'' *SPIE Proc.,* Los Angeles, Vol. 1217, 16–17 Jan. 1990, pp. 151–160.

[9] Montgomery, R. M., and M. R. Lange, ''Photorefractive Adaptive Filter Structure With 40-dB Interference Rejection,'' *Applied Optics,* Vol. 30, No. 20, 10 July 1991.

[10] Howells, P. W., ''Intermediate Frequency Side-Lobe Canceler,'' Patent 3202990, 24 Aug. 1956 (filed 4 May 1959).

[11] Howells, P. W., ''Explorations in Fixed and Adaptive Resolution at GE and SURC,'' *IEEE Trans. Antennas and Propagation,* Vol. AP-24, No. 5, Sept. 1976, pp. 575–584.

[12] Applebaum, S. P., ''Adaptive Arrays,'' Syracuse University Research Corp., Rep. SPL-709, June 1964. Reprinted in *IEEE Trans. Antennas and Propagation,* Vol. AP-24, Sept. 1976, pp. 573–598.

[13] Vander Lugt, A., ''Optical Transversal Processor for Notch Filtering,'' *Opt. Eng.,* Vol. 23, No. 3, May-June 1984, pp. 312–318.

# Chapter 16
# Acousto-Optic Processing for Receiver Technologies

*I. C. Chang*
*Aurora Associates*

## 16.1 INTRODUCTION

Radar, communication, and other electronic warfare (EW) applications involve signal reception, analysis, identification, and monitoring of wideband electromagnetic signals. An EW receiver must provide the amplitude, frequency, direction, and time of arrival of all signals over a large range of amplitudes and frequencies with nearly 100% probability of intercept (POI). In recent years, the signal environment has become increasingly dense and complex. A significant number of these emitters are high-repetition-rate types such as pulsed Doppler radars. Conventional EW receivers are no longer capable of handling such scenarios. The inherent wide-bandwidth and parallel-processing capability of acousto-optics makes it potentially well suited to this application. In addition, the acousto-optics-based receiver provides the important advantages of small size, a requirement that is important to airborne applications.

This chapter will discuss the application of acousto-optic signal processing techniques to EW receivers. Prior to the detailed discussion of the acousto-optic approach, it is pertinent to review the conventional EW receivers based on microwave technology.

## 16.2 CONVENTIONAL RECEIVER TECHNOLOGIES

Conventional EW receivers based on microwave technology have evolved from the simple crystal video, superheterodyne, and instantaneous frequency measurement (IFM) receivers to more sophisticated architectures of compressed and channelized receivers. Each of these receivers has its advantages as well as its limitations. The choice must be based on specific system requirements. For a detailed discussion of receiver technologies, the reader is referred to [1].

### 16.2.1 Crystal Video Receiver

The crystal video receiver (CVR) is the simplest but most commonly used type of EW receiver. A block diagram for a typical CVR (four-channel in this case) can be seen in Figure 16.1. With such a receiver, the input signal is divided into bands by the multiplexer, amplified, and applied to a diode detector. The detector voltage output, proportional to the envelope of the RF input power, is amplified and output to measurement circuits. The most current approach to implementation of CVRs is the use of custom-designed monolithic chips configured in a hybrid package including detector and voltage regulators. With this technique, volumes of less than 4 in$^3$ can be realized.

A CVR is primarily used to obtain amplitude and time of arrival (TOA) information. In fact, the CVR provides no frequency information whatsoever. The primary drawback to this technique, aside from the lack of frequency measuring capability, is its performance in high signal density environments. Because it operates in the time domain, strong high-duty-cycle signal can virtually blind this type of device to other signals. The most efficient method of enhancing performance is to limit the input bandwidth and thus reduce the possibility of signal overlap.

For EW receiver applications, the CVR could be used in conjunction with a wideband frequency measurement channel. Obtained amplitude and TOA information would be used primarily for frequency measurement triggering and/or data correlation with other measuring devices.

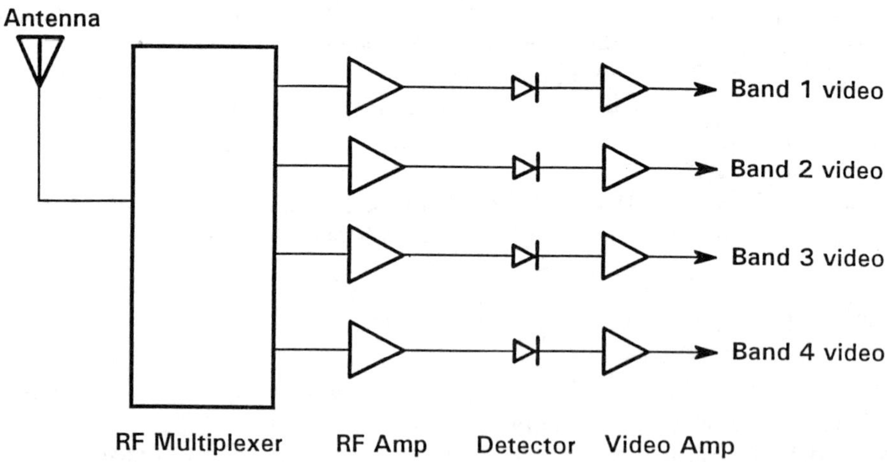


**Figure 16.1** Crystal video receiver.

### 16.2.2 Superheterodyne Receiver

The block diagram for a superheterodyne receiver (SHR) is shown in Figure 16.2. With such a receiver, the input signal is applied to a bank of tunable/selectable RF bandpass filters that provide the preselection function (one is shown in the figure). The IF band, 500 MHz, is then down-converted to some nominal IF (1 GHz). With this relatively high first IF, the tunable LO is required to cover slightly less than one octave centered in the band of receiver operation. The filter following the first mixer further limits the receiver bandwidth before the second conversion to the final IF (e.g., 160 MHz). Amplification is provided in the signal path to maintain an acceptable noise figure and set the overall receiver gain to the desired value.

Signal detection is accomplished with a logarithmic video amplifier and frequency discriminator. Aside from the known tuned frequency of a signal, outputs in the form of amplitude and frequency offset (both in video form) are obtained. An SHR such as this can be implemented in microwave integrated circuit (MIC) technology. Total volumes for the tuner and IF assemblies are less than 30 $in^3$ each.

The primary functions of an SHR are to provide highly accurate frequency and amplitude data and measure modulation characteristics (TOA, pulse width, scan rate, etc.). Sensitivity, dynamic range, and signal selectivity are typically very high. Because of its design, the SHR is not appropriate for wideband frequency measurement. POI is low and signal acquisition time is long due to the narrow bandwidth and finite LO tuning speeds. The SHR as described above could be used to perform several receiver functions. Some of these functions include CW signal measurement, low-band omnidirectional signal measurement, detailed pulse signal analysis (ambiguity resolution), CW direction finding (DF), and exotic emitter signal acquisition.

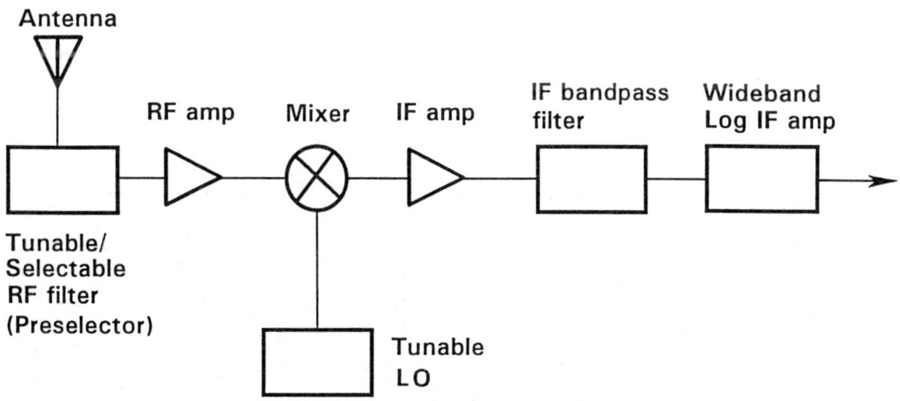


**Figure 16.2** Superheterodyne receiver.

### 16.2.3 Instantaneous Frequency Measurement Receiver

A block diagram for a typical IFM receiver can be seen in Figure 16.3. With such a receiver, the input signal is first applied to a limiting amplifier to compress the input dynamic range and capture the strongest signal, and then divided equally between two channels. One channel has a known amount of delay compared to the other. Next, the two signals pass through a phase discriminator, the outputs of which are in the form of $\sin(\omega\tau)$ and $\cos(\omega\tau)$ (where $\tau$ is the relative time delay and $\omega$ is the angular signal frequency). Based on the known delay , the $\sin(\omega\tau)$ and $\cos(\omega\tau)$ signals are decoded to yield frequency data. The output is in the form of an $N$-bit digital word.

The only output obtained from an IFM receiver is frequency data. Accuracy can be very high (2 to 20 MHz typically) and is limited primarily by the ratio of bandwidth per number of bits. Pulse and CW signals can be measured over a dynamic range in excess of 70 dB with achievable sensitivities of lower than –60 dBm. With an IFM, it is conceivable to obtain frequency data on every received pulse over a broad frequency range in less than 100 ns.

With the current technology, IFMs capable of distinguishing between pulse and CW signals are available. Additionally, circuits can be incorporated that provide information on the validity of the frequency data, thus allowing invalid measurements to be discarded. Package volumes of less than 30 $in^3$ have been achieved for some IFM receivers.

Like the CVR, an IFM receiver operates in the time domain and is, therefore, hampered by simultaneous-signal conditions. When two or more signals are present at the receiver input, there must be a 3- to 4-dB difference in relative level in order for an accurate measurement to be made. In all cases, only the strongest signal is reported.

For the EW receiver application, IFMs could be used as a primary, wideband, frequency measuring device. The use of a band-limited preselection device (e.g., SHR) is required to overcome the simultaneous-signal problem and still allow fast frequency measurement over a reasonably wide range.

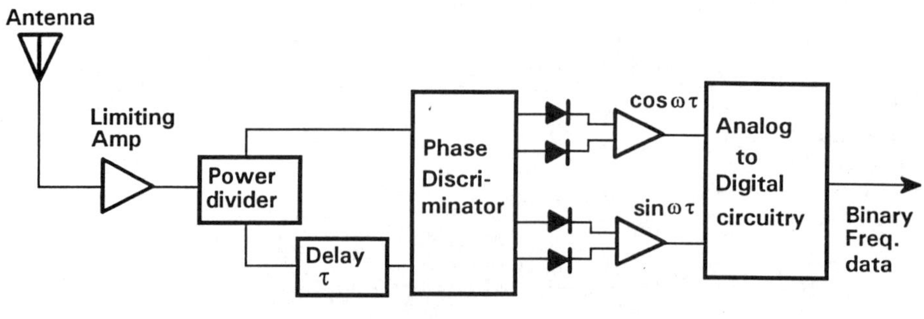


**Figure 16.3** Instantaneous frequency measurement receiver.

### 16.2.4 Compressive Receiver

A block diagram for a compressive (microscan) receiver is shown in Figure 16.4. The input is first down-converted to a selected IF. The IF signals are then mixed with a swept LO signal to generate an FM (chirp) output. This chirp signal is then applied to a dispersive delay line (DDL) with a delay-versus-frequency characteristic that is the inverse of the LO sweep characteristic. The result of this operation is that all the frequency components of the chirp arrive at the output of the DDL at approximately the same time. In fact, the response at this point is the Fourier transform of the input signal. A weighting filter is used to suppress sinc-function sidelobe responses and present a virtual impulse to the measurement circuits. Frequency information is obtained by correlating the position (with respect to frequency) of the sweeping LO at the instant of time when an output pulse occurs.

In order to maintain a high POI, the sweeping LO signal is created by applying an impulse train to a second DDL. The frequency versus delay characteristic of this device is exactly the inverse of the DDL described above. This has the effect of spreading the impulse frequency spectrum in time. The output appears as a swept frequency signal. The PRI of the impulse generator is adjusted so that the swept frequency segments at the DDL output are virtually continuous. In this way, POIs approaching 100% can be obtained using a single receiver.

The most notable attribute of the compressive receiver is its ability to frequency-resolve multiple time-coincident signals. Resolutions of better than 10 MHz are achievable over the 500-MHz bandwidth in approximately 200 ns (the *sweep* time).

The achievable bandwidth for a compressive receiver is limited by the performance characteristics of the DDL, which is implemented using a surface acoustic wave (SAW).

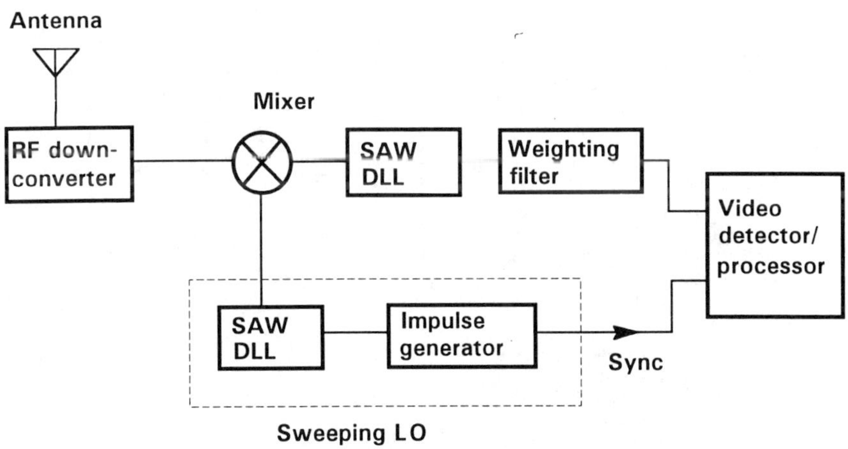


**Figure 16.4** Compressive receiver.

The two most significant problems with this receiver technique are the difficulty in deriving TOA information from the output (since output "time" corresponds to the frequency) and the method of handling the large volume of data produced.

### 16.2.5 Channelized Receiver

A block diagram for a typical channelized receiver is shown in Figure 16.5. The input signal is first applied to a multiplexer or active divider network such that the input frequency range is broken up into *N* bands, each having a bandwidth equal to that of the basic channelizer module. Each of the *N* bands is then down-converted to a common IF. The *N* IF signals are applied to separate channelizer modules (identical for each channel), which further divide the input bandwidth into *M* slots. The width of each of the *M* slots roughly determines receiver resolution. Detectors follow the slot filters, providing the basic "presence of a signal" indication and/or slot signal amplitude.

Channelized receivers can be implemented using either MIC filters or SAW filters. The MIC channelized receiver is bulky and costly to manufacture. For fine frequency channelization, the MIC approach becomes less practical, since the *Q* of the filter must be very high. A smaller size channelized receiver can be built using SAW filters. Disadvantages of the SAW filters include time domain spurious signal response (e.g., triple-transit echoes), limited bandwidth, and high insertion loss.

Theoretically, a channelized receiver operates in such a way as to perform a Fourier transform on the input signal. For any given instant, the magnitudes of the detected slot outputs represent the frequency spectrum of the input signal. Unfortunately, the ability

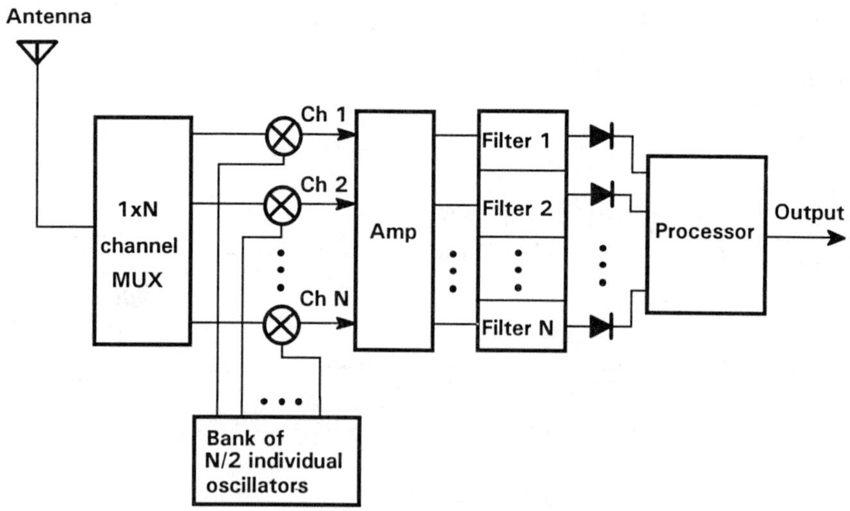


**Figure 16.5** Channelized receiver.

to extract the proper information from a channelized receiver is solely dependent on the method of processing used. For example, a narrow pulse signal will have a frequency response that results in detected outputs from a number of slots. If another, relatively small, signal occupies a slot near that of the narrow pulse, the processor must be able to determine that two signals are present and to which frequency each slot belongs. This problem is further aggravated by high signal densities (multiple time-coincident signals) and nonideal pulse characteristics.

### 16.2.6 Limitation of Receiver Techniques

Table 16.1 compares the performance and limitations of various receiver techniques. Up to now, all EW systems have been deployed based on combinations of the first three techniques, which include CVR, SHR, and IFM. As can be seen from the table, all of these receiver types are inadequate for modern EW systems due to their inability to detect simultaneous signals, a requirement necessary for the increasingly dense environment. Theoretically, both the compressive and channelized receivers are capable of meeting this critical requirement. In fact, a few developmental systems have been built that demonstrated the usefulness of these two techniques. The primary problem has been the complexity of the hardware and associated high cost. In addition, the development of cost-effective digital postprocessors for these two types of advanced receivers remains a technical challenge yet to be resolved.

## 16.3 PRINCIPLE OF ACOUSTO-OPTIC SPECTRUM ANALYSIS

The use of acousto-optics to perform signal processing is attractive because of its potential for wide instantaneous bandwidth and a large number of parallel channels. In the past two decades there has been significant progress in acousto-optic technology. The interested reader may refer to the review papers on acousto-optic devices and signal processing applications [2–4].

**Table 16.1**
Comparison of Receiver Techniques

| | *CVR* | *SHR* | *IFM* | *Microscan* | *Channelized* |
|---|---|---|---|---|---|
| Instantaneous bandwidth | Wide | Low | Wide | Moderate | Moderate |
| Frequency resolution | None | High | Moderate | Moderate | Moderate |
| Sensitivity | Low | High | Moderate | High | High |
| Probability of intercept | High | Low | High | High | High |
| Simultaneous signal capability | Poor | Poor | Poor | Good | Good |
| Dynamic range | Low | High | Moderate | Moderate | Moderate |
| Size | Very small | Small | Small | Medium | Large |
| Cost | Low | Medium | Low-medium | Medium-high | High |

Among the various acousto-optic processors, the RF spectrum analyzer is certainly the simplest and most important for practical applications. Functionally, it performs a finite Fourier transform of the input RF signal. Depending on the processing architecture, the acousto-optic technique can be implemented either as a channelized receiver or a compressive receiver. In an acousto-optic channelized receiver, RF signal channelization is accomplished by using the frequency dispersion of acousto-optic diffraction. Similar to the SAW implementation, the acousto-optic compressive receiver performs a Fourier transform via chirp Z transform.

In the following, the principle of acousto-optic receivers will be discussed. To begin with, a brief review of acousto-optic diffraction will be presented.

### 16.3.1 Acousto-Optic Diffraction

At the heart of all acousto-optic signal processing systems is an acousto-optic Bragg cell, which functions as the real time RF-to-optical input transducer. As shown in Figure 16.6(a), the Bragg cell consists of an optically transparent medium onto which a piezoelectric transducer is bonded. The transducer converts the input RF signal into a traveling acoustic wave inside the medium via the elasto-optical effect, and the acoustic wave produces a spatial modulation of the refractive index. This provides a phase grating, which may diffract a portion of the incoming optical beam into one or more directions. This phenomenon is known as *acousto-optic diffraction* [2].

Acousto-optic diffraction has certain salient characteristics that can be employed for RF spectrum analysis. Figure 16.6(b) shows the wave vector diagram for acousto-optic diffraction in an isotropic medium. For coherent building of the acousto-optic

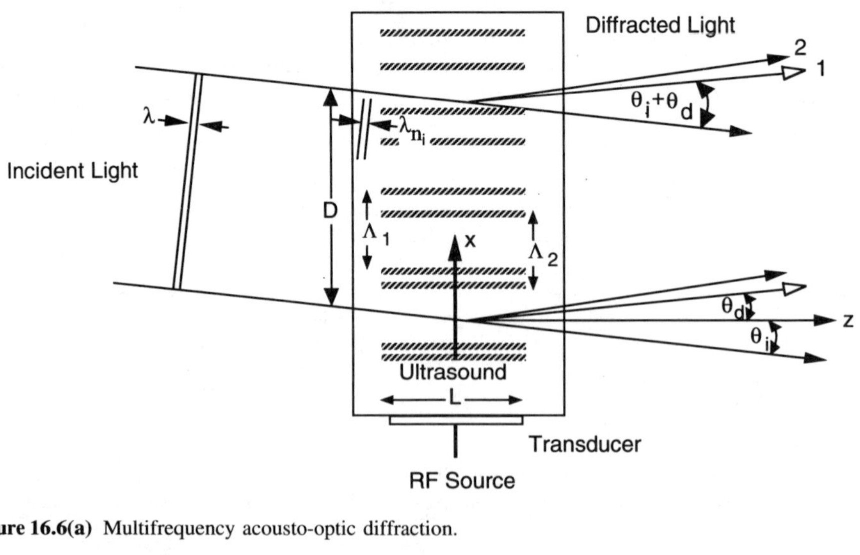


**Figure 16.6(a)** Multifrequency acousto-optic diffraction.

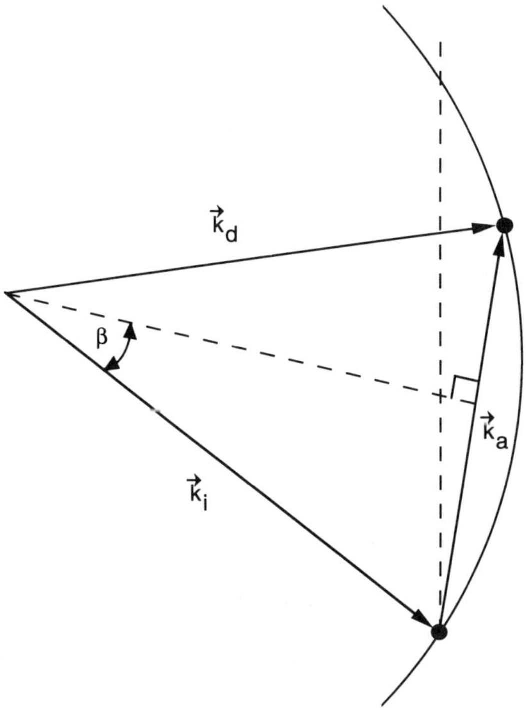


**Figure 16.6(b)** Wave vector diagram for isotropic acousto-optic diffraction.

diffraction, the incident, acoustic, and diffracted optical waves must satisfy the momentum-matching condition.

$$k_d = k_i + k_a \tag{16.1}$$

where $k_a$ is the acoustic wave vector, and $k_i$ and $k_d$ are the wave vectors for the incident and diffracted light, respectively.

Referring to Figure 16.6(b), the momentum-matching condition yields the deflection angle for the diffracted light beam.

$$\Delta\theta = 2\sin^{-1}\left(\frac{\lambda_o f}{2V}\right) \approx \frac{\lambda_o f}{V} \tag{16.2}$$

for $\Delta\theta \leq 0.1$ rad, where $\lambda_o$ is the optical wavelength, $V$ is the acoustic velocity, and $f$ is the acoustic frequency.

If a lens is placed behind the Bragg cell, it will transform the angular deviation into a displacement in the back focal plane (referred to as the *frequency plane*). The displacement of the first-order deflected spot from the undeflected spot in the frequency plane is thus proportional to the frequency of the input signal; that is,

$$X = F_0 \lambda_o f/V \tag{16.3}$$

where $F$ is the transform focal length.

Mathematically, the above operation yields a Fourier transform of the light distribution at the optical aperture of the Bragg cell

$$F(s) = \int w(x) \exp(-j2\pi sx)dx \tag{16.4}$$

where $w(x)$ is the weighting function, $s = f\tau$ is the normalized frequency ($\tau$ is the time aperture of the Bragg cell), and $F(s)$ is the desired Fourier transform. The photodetector array in the transform plane measured the magnitude square of $F(s)$ (i.e., the instantaneous power spectra of the signal read into the Bragg cell).

The frequency resolution and sidelobes inherent to the instantaneous power spectra are determined by the weighting function. In practice, the incident optical beam has a Gaussian distribution with a truncation ratio of 1.5. The corresponding power spectra distribution can be obtained from a numerical calculation of (16.4). A typical spectrum is shown in Figure 16.7. It shows a 3-dB width of $\delta f = 1.2/\tau$. This is the frequency resolution or minimum separations for resolving two equal signals.

At relatively low signal power levels, the amplitude of the diffracted light is linearly proportional to the input RF signal. Thus, when multiple RF signals are applied to the Bragg cell, the diffracted light will result in a spatial (angular) distribution that is a duplicate (except for a proportionally constant) of the spectrum of input RF signal.

### 16.3.2 Acousto-Optic Channelized Receiver

The frequency dispersion of acousto-optic diffraction described above can be used to perform RF channelization [5–9]. The simplest system architecture (shown in Fig. 16.8) is the power spectra analyzer (PSA), which measures the magnitude square of the Fourier transform. It consists of a laser, collimating/beam forming optics, acousto-optic Bragg cell, Fourier transform lens, and linear photodetector array for real-time electronic readout. The laser beam amplitude profile is tailored by the beam forming optics so as to match the interaction aperture of the Bragg cell. The Bragg cell diffracts a portion of the laser light into an angular distribution of intensities that are proportional to the input RF power spectra. The Fourier transform lens converts the angular diffracted beam into a linear intensity distribution coincident with the photodetector array. The output of the photodetector thus yields the real-time power spectra of the input RF signal.

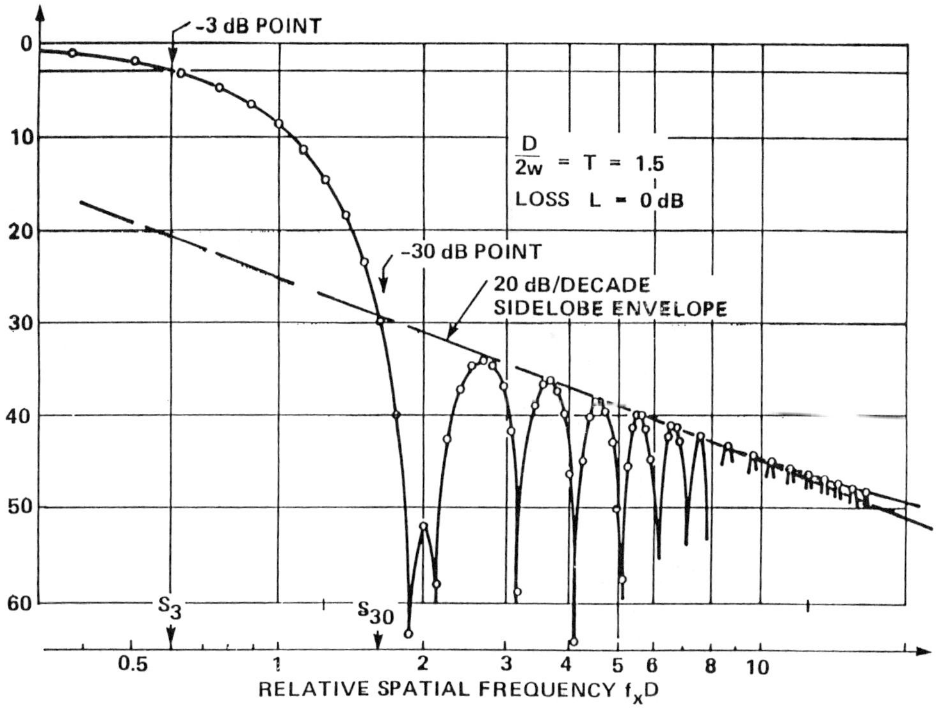


**Figure 16.7** Spectrum of truncated Gaussian beam.

In spite of its extreme simplicity, the PSA is equivalent to a large number of contiguous narrowband receivers realized in small size. Unity POI is ensured for analysis of a large number of simultaneous signals over a wide instantaneous bandwidth. The achievable bandwidth and frequency resolution are determined by the performance of acousto-optic Bragg cells.

At present, the major deficiency of an acousto-optic channelized receiver is its relatively low dynamic range. Dynamic range is a term generally used to indicate the signal range that a receiver can process properly. For EW receivers that perform spectral analysis of simultaneous signals, there are two dynamic ranges to ascertain. Figure 16.9 shows the single-tone instantaneous dynamic range and the two-tone spurious signal-free dynamic range. The single-tone instantaneous dynamic range is defined as the maximum amplitude separation between a strong signal at a 1-dB compression point and a weak signal at noise level. When two strong signals are simultaneously applied to a receiver, its nonlinear response will generate intermodulation products known as *spurious signals.* The maximum signal is then limited by the requirement that the spurious signals generated must be below or equal to the noise level.

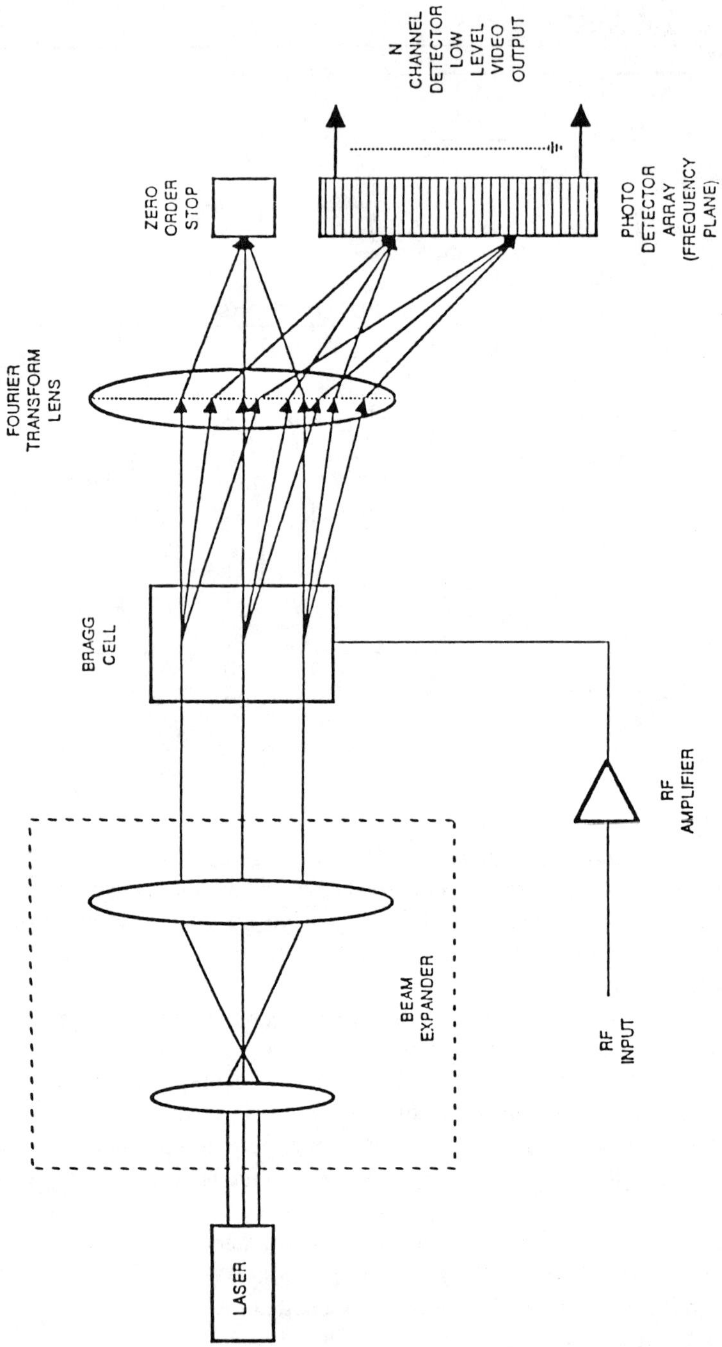


**Figure 16.8** Acousto-optic power spectra analyzer.

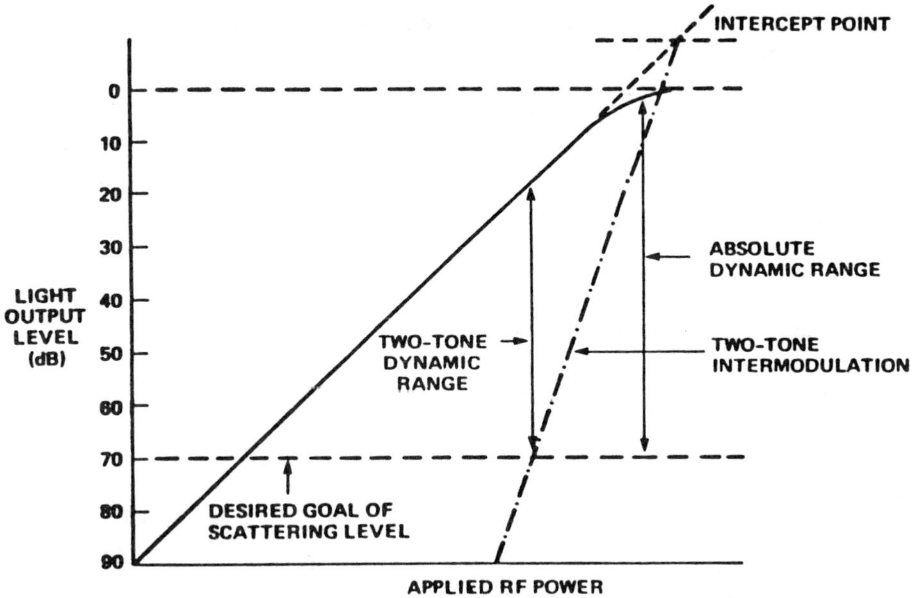


**Figure 16.9** Single- and two-tone dynamic range.

It is instructive to estimate the dynamic range of an acousto-optic PSA. Referring to Figure 16.9, the two-tone dynamic range is defined as the difference between maximum efficiency ($\eta_{max}$) and the efficiency at threshold ($\eta_{min}$). It can be expressed in decibels as

$$DR = \eta_{max} - \eta_{min} \tag{16.5}$$

From inspection of Figure 16.9,

$$DR = 2/3(\eta_{31} - \eta_{min}) \tag{16.6}$$

Where $\eta_{31}$ is the third-order intercept. For Bragg cells where the dominant third-order nonlinear response is caused by multiple acousto-optic diffraction, $\eta_{31} = 7.8$ dB. Equating the diffracted light at $\eta_{min}$ to the noise floor,

$$I_{th} = I_0 + T_o + \eta_{min} \tag{16.7}$$

where $I_0$ is the laser power and $T_o$ is the optical throughput. Substituting (16.7) into (16.6), we get

$$DR = 2/3(I_0 + T_o - I_{th}) + 5.2 \text{ dB} \tag{16.8}$$

As an example, if $I_0 = 20$ dBm, $T_o = -3$ dB, and $I_{th} = -50$ dBm, the two-tone dynamic range of the PSA estimated from (16.8) is about 50 dB.

The dynamic ranges shown in Figure 16.9 are defined with respect to the noise floor. For practical system operation, in order to keep the false alarm rate below a specified level, the detection threshold must be set above the noise floor. Typically, the SNR is chosen to be approximately 15 dB at the detector output. Another common practice is to specify the dynamic range with respect to tangential sensitivity (TSS), which is about 8 dB above noise level at the detector output.

### 16.3.3 Acousto-Optic Compressive Receiver

An alternative approach for RF spectrum analysis is to implement a compressive receiver using acousto-optics. Functionally, the acousto-optic compressive receiver performs the Fourier transform via the chirp Z transform. Using the identity $-2ft = (f - t)^2 - f^2 - t^2$, the Fourier transform can be written as

$$F(f) = e^{-j\pi f^2} \int f(t)[e^{-j\pi t^2} \cdot e^{j\pi(f-t)^2}]dt = e^{-j\pi f^2}\{[e^{-j\pi t^2} f(t)]*e^{j\pi t^2}\} \tag{16.9}$$

where * denotes convolution. The Fourier transform is accomplished by premultiplying the signal with a forward chirp, convolving with a reverse chirp, and then postmultiplying with another chirp. Since the algorithm involves premultiplication, correlation, and post-multiplication, it is referred to as the *MCM scheme*. For power spectrum analysis, the second multiplication can be neglected.

The desired convolution can be implemented using either a space-integrating [10] or a time-integrating [11] architecture. The space-integrating compressive receiver is illustrated in Figure 16.10(a). The signal to be analyzed, $x(t)$, is multiplied by the forward chirp and input into the first Bragg cell, BC1. The optical amplitude distribution is imaged onto the second Bragg cell, BC2, which is driven by the reverse chirp. The product is spatially integrated by lens L3 to produce the final output, $\tilde{x}(f)$, which is (except for a phase factor) the Fourier transform of $x(t)$. The single photodetector measures the desired power spectrum and displays it as a function of time. The frequency resolution of the space-integrating compressive receiver is set by the finite aperture of the Bragg cell, thereby making this type of architecture useful for a coarse frequency analysis.

An alternative method of implementing the compressive receiver is the time-integrating architecture shown in Figure 16.10(b). Like the space-integrating system, the product of the input and the forward chirp from the first Bragg cell is imaged onto the second Bragg cell (driven by the reverse chirp). However, the product light distribution leaving the second cell is now time-integrated by the photodetector array (instead of the spatial integration due to a lens). The time-integrating architecture has significant advantages. Since the integration time is now limited by the photodetector and is orders of magnitude

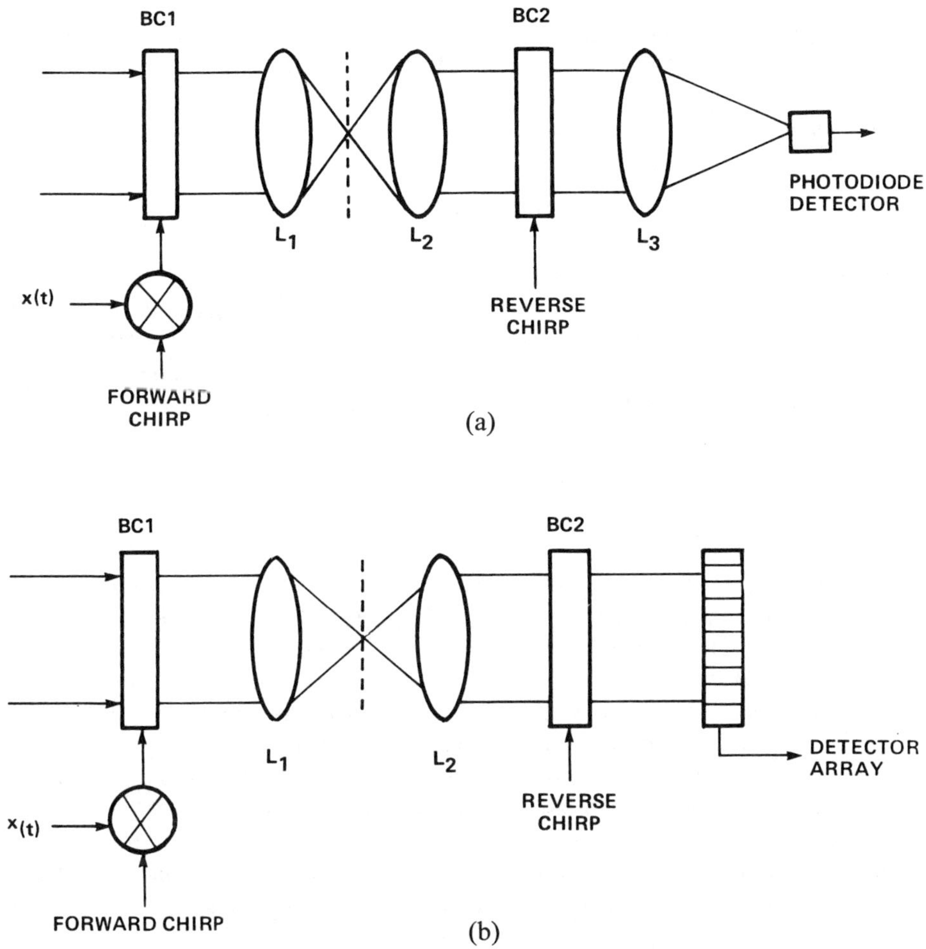


**Figure 16.10** (a) Space-integrating optical correlator; (b) time-integrating optical correlator.

larger than that obtainable from spatial integration, very long integration time is realizable, thereby achieving fine frequency resolution and large time-bandwidth products. In addition, the frequency resolution can be varied by changing the integration time and the chirp rate. Due to these features, the time-integrating compressive receivers are best suited to communication intelligent (Comint) applications that require extremely large signal processing gain.

Theoretically, the acousto-optic compressive receiver using the space- or time-integrating architecture appears to have great potential for advanced receiver applications. In practice, since the achievable dynamic range and bandwidth are limited by the performance of photodetectors, the acousto-optic implementation of the compressive receiver offers little performance advantage compared to that based on the more mature SAW technology. As a result, most of the previous efforts have been directed toward RF signal channelization. Therefore, in the remainder of the chapter, the discussion will be limited to acousto-optic channelized receivers.

## 16.4 KEY COMPONENT PERFORMANCE

The performance of the acousto-optic channelized receiver depends critically on the technologies of the key components, which include laser diode, Bragg cells, and photodetector arrays. The purpose of this section is to discuss the optimized selection of the components in meeting system requirements. Due to the rapid technological advances, the component performance parameters described in this section should be considered state-of-the-art values as of 1993.

### 16.4.1 Laser Source

Selection of the laser for an acousto-optic channelized receiver is based on output power, spatial distribution, spectral purity, noise characteristics, packaging, and lifetime. In early work on acousto-optic channelized receivers, HeNe gas lasers were used. The disadvantages of HeNe lasers are large size and low output power. Since 1979, semiconductor injection laser diodes have replaced the HeNe lasers for this application. Single-mode laser diodes with output power exceeding 100 mW are now commercially available. Besides high output power, these diodes also have near-Gaussian transverse mode profiles with negligible multimode structures. The only problem with these laser diodes appears to be the broad spontaneous emission level, which is about 30 dB below the dominant mode. This may produce undesired beam spread in the frequency plane. Special narrowband filters may be used to avoid degradation of simultaneous signal resolution.

### 16.4.2 Bragg Cells

In an acousto-optic signal processing system such as the channelized receiver, the Bragg cell plays the critical role of converting the input electrical signal into a spatial modulation of the optical beam. The input RF bandwidth and the frequency resolution of the acousto-optic channelized receiver are determined primarily by the Bragg cell. The minimum frequency difference of two RF signals that are resolvable is approximately equal to $1/\tau$, where $\tau$ is the acoustic transit time across the interaction aperture. Theoretically, the total number of signals that can be simultaneously resolved by the acousto-optic channelized receiver is approximately given by the time-bandwidth product of the Bragg cell [2]:

$$N = \tau \Delta f$$

where $\Delta f$ is the bandwidth of the Bragg cell.

To maximize the number of resolvable signals, it is desirable to increase either the time aperture or the frequency bandwidth. A wider optical aperture will provide fine frequency resolution. However, the realizable aperture is usually limited by the high-frequency acoustic attenuation in the Bragg cell. On the other hand, there exists a basic trade-off relation between bandwidth and efficiency. The bandwidth achievable for a Bragg cell is thus limited by the applied maximum power or power density.

As shown in Figure 16.6(a) a short interaction length is used in order to obtain a wide angular spread of acoustic radiation pattern, so that the momentum-matching condition can be met over wide bandwidth. A short length $L$ will reduce the diffraction efficiency of the Bragg cell. Several techniques have been developed to increase the interaction length while maintaining the bandwidth. These include the use of birefringent phase matching [12], phased-array beam steering [13], or a combination of the two [14]. Figure 16.11 shows the measured bandpass response of a shear wave $LiNbO_3$ cell using the birefringent phase-matching geometry. The cell had a 3-dB bandwidth of 2.1 GHz and a peak diffraction efficiency of 12%/RF watt when operated at 633 nm. Based on Rayleigh criteria, the two-signal frequency resolution was found to be about 3 MHz. The time-bandwidth product is about 630.

In choosing a Bragg cell for a specific application, the guideline is to satisfy the bandwidth and resolution requirement and to maximize the diffraction efficiency so that the drive power is within practical limits. Other secondary design considerations include optical scatters, spurious signals, size, and environmental effects. Table 16.2 shows the typical performance of several selected acousto-optic Bragg cells. These Bragg cells are classified into three types: (a) Type I—These $TeO_2$ shear wave cells are high-resolution (<100 KHz) devices usable for Comint receivers; (b) Type II—These are medium-bandwidth (>500 MHz), medium-resolution (< 1 MHz) devices that are best fitted to electronic intelligent (ELINT) receivers; (c) Type III—These wideband (>1,000 MHz), low-resolution (~10 MHz) cells have been designed to suit radar warning receiver (RWR) applications.

Besides efficiency, bandwidth, and resolution, another important performance parameter of a Bragg cell is its dynamic range. This is the capability of analyzing a weak signal in the presence of strong signals. For signals that are close in frequency, the instantaneous dynamic range is limited by the sidelobes of the acousto-optic diffraction, which can be effectively suppressed by the choice of the weighting window for the illuminated optical beam (apodization). For signals far apart in frequency, the dynamic range is limited either by optical scattering or by the inmodulation products (IMP) generated in the Bragg cell. The most dominant inband intermodulation products for two simultaneous signals at $f_1$ and $f_2$ are the third-order terms at $2f_1 - f_2$ and $2f_2 - f_1$.

The main contributions of IMPs are due to multiple acousto-optic diffractions occurring at high diffraction efficiencies [15]. For two equal intensity signals, the two-tone third-order IMP is equal to

$$I_3 = \frac{I_1^3}{36} \tag{16.10}$$

BANDSHAPE RESPONSE

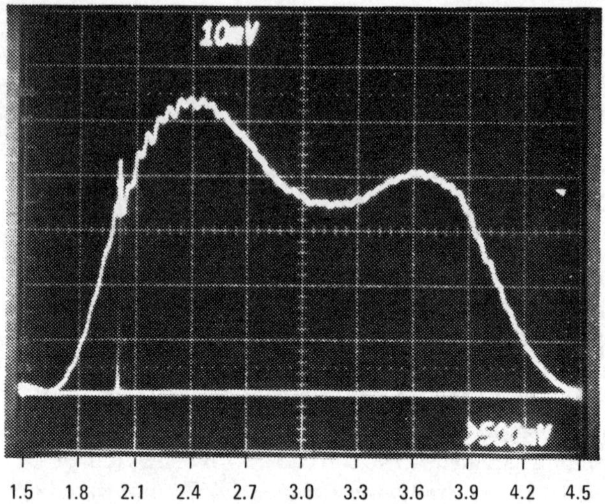


FREQUENCY (GHz)

3 dB BANDWIDTH = 2080 MHz (1940 TO 4020 MHz)

PEAK EFFICIENCY = 12 PERCENT/WATT

TWO SIGNAL FREQUENCY RESOLUTION

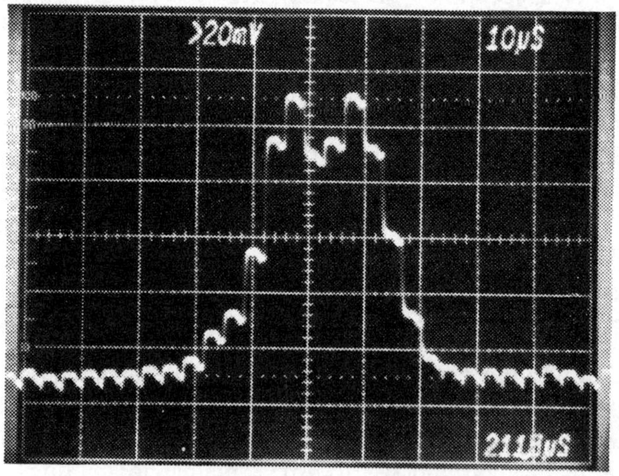


FREQUENCY RESOLUTION = 3 MHz

**Figure 16.11** Frequency response of wideband Bragg cell.

**Table 16.2**
Performance of Acousto-Optic Bragg Cells

| *Type* | *Material (Mode)* | *Center (MHz)* | *Bandwidth (MHz)* | *Aperture ($\mu s$)* | *TB Product* | *Efficiency (%/W)* |
|---|---|---|---|---|---|---|
| I | $TeO_2$(S) | 90 | 50 | 40 | 2,000 | 110 |
| I | $TeO_2$(S) | 160 | 100 | 10 | 1,000 | 95 |
| II | GaP(S) | 1,000 | 500 | 2.0 | 1,000 | 30 |
| II | GaP(S) | 2,000 | 1,000 | 1.0 | 1,000 | 12 |
| II | $LiNbO_3$(S) | 2,500 | 1,000 | 1.0 | 1,000 | 10 |
| III | GaP(L) | 2,500 | 1,000 | 0.25 | 250 | 44 |
| III | GaP(L) | 3,000 | 2,000 | 0.15 | 300 | 10 |
| III | GaP(S) | 3,000 | 2,000 | 0.25 | 500 | 8 |
| III | $LiNbO_3$(S) | 3,000 | 2,000 | 0.30 | 600 | 6 |

Note: $\lambda_o$ = 830 nm.

Where $I_1$ is the first-order diffracted light efficiency. Referring to Figure 16.9, the acousto-optic type IMP corresponds to a third-order intercept of 7.8 dB.

A second type of IMP is attributed to the acoustic nonlinearity of the Bragg cell medium [16]. This type of IMP becomes increasingly severe in wideband Bragg cells, where the acoustic power densities are high. Most of the recent Bragg cells have employed birefringent or phased-array designs to increase the interaction length. These techniques have practically eliminated the acoustic IMP problem.

### 16.4.3 Photodetector Array

Another key element in the acousto-optic channelized receiver is the photodetector array. It transforms the spatially modulated light pattern to analog electronic signals. Two methods have been developed to extract the video output in an acousto-optic channelized receiver. In the integrating detection method, self-scanned photodetector arrays are used. A number of linear photodetector arrays are commercially available with 64 to 2,048 detector elements. The associated circuit of the detector array also contains a shift register that converts the integrating charge on each detector into a sequential readout data stream. The sample time, or revisit time, is the time to cycle through all detector elements. It is also the period in which charge is integrating on the detector.

For large numbers of channels, the sample time is very long. For instance, with a system clock frequency of 10 MHz, it takes 0.1 ms to read out a 1,000-detector array. This does not meet the TOA resolution requirements for most EW applications. Furthermore, the dynamic range is severely degraded if the signal pulsewidth is shorter than the integration time. Because of the inability to handle short pulses, the integration detection method has been used only in receivers designed for ELINT applications.

The deficiencies of the integrating detection method can be overcome by using the peak detection method. In this case, the output voltages of individual detectors are parallel in readout.

Monolithic parallel readout detector arrays using silicon technology have been developed for acousto-optic channelized receiver applications. Each channel consists of a pin diode, a linear preamplifier, and a compressed amplifier section. Table 16.3 summarizes the performances of typically serial and parallel addressed photodetector arrays [17].

The most severe performance limitation for the photodetector arrays in acousto-optic applications has been the relatively low dynamic range. Therefore, considerable development efforts have been directed at increasing the dynamic range of photodetectors. The basic approach for realizing this is to obtain a compression of the output voltage. One-design utilizes the leakage effect of a short-channel MOSFET to realize a logarithmic conversion and thereby significantly increase the detector dynamic range [18]. However, the logarithmic photodetector amplifiers tend to have poor fall-time characteristics, resulting from an image lag effect at low light levels. To overcome this deficiency, an improved design of a charge-coupled device (CCD) detector used a piecewise linear gain compression to approximate the logarithmic output [19]. The design provides a large dynamic range while suppressing the temporal lag problem. Using this design, a 1,024-element detector was developed that demonstrated 56-dB dynamic range when operated at a 10-MHz rate.

## 16.5 POWER ACOUSTO-OPTIC RECEIVER

### 16.5.1 Integrating Acousto-Optic Channelized Receiver

The advantages of acousto-optical signal processing are best shown in an integrating-type acousto-optic channelized receiver. In this radiometric architecture, a serial readout photodetector array is used to measure the optical power spectra distribution in the focal plane. This type of acousto-optic receiver has been developed for detecting and analyzing signals over an instantaneous bandwidth of 500 MHz with a frequency resolution of 1 MHz. Functionally, it is analogous to a bank of 500 narrowband contiguous RF filters,

**Table 16.3**
Photodetector Arrays for Acousto-Optic Processors

| *Parameter* | *Serial Addressed* | *Parallel Addressed* |
|---|---|---|
| Elements | 6,000 | 27 |
| Pitch | 5 μW | 64 μm |
| Noise equivalent power | 0.1 nW | 5 nW |
| Dynamic range | 55 dB | 45 dB |
| Crosstalk | −48 dB | −35 dB |
| Clock rate | 20 MHz | — |
| Rise time | — | 1 ns |

Note: CVR = crystal video receiver; SHR = superheterodyne receiver; IFM = instantaneous frequency measurement (receiver).

followed by square-law detection and integration, and is equivalent to $N$ total power radiometers parallel with an integration time $T$ set by the revisit time. Advantages of the integrating acousto-optic channelized receiver include wide instantaneous bandwidth, a large number of channels, and high sensitivity. The simple serial readout scheme is particularly attractive to total power radiometers with processing gains.

The integrating acousto-optic receiver is grossly inadequate for detection of pulsed signals. It cannot provide TOA information. A more severe drawback is the low sensitivity for short pulses due to the small ratio of pulsewidth to integration time. The integrating acousto-optic channelized receiver is more appropriate for Comint receivers.

For ELINT receivers, the revisit time of the integrating acousto-optic receiver needs to be significantly reduced. This can be accomplished by dividing the detector array into a number of parallel segments. For instance, the use of a 480-element hybridized tapped photodetector array with 10 taps and an integration time of 5 $\mu$s has been demonstrated.

### 16.5.2 Parallel Output Acousto-Optic Channelized Receiver

To enhance the TOA resolution and to improve dynamic range, acousto-optic channelized receivers with fully parallel readout detector arrays have been developed. Figure 16.12 shows a block diagram of such a receiver. One way of implementing the system hardware is to use a fiber-optic array in the frequency plane that couples the diffracted light into discrete modules of detectors and amplifiers. This approach has the advantage of optimizing short-pulse sensitivity and of low crosstalk of individual shielded detector modules.

In airborne EW receiver applications, small size, weight, and low power consumption are important. In addition, the receiver must meet environmental requirements. The miniature instantaneous Fourier transform (MIFT) system being developed at Litton Applied Technology is an acousto-optic channelized receiver that addresses these objectives [8]. The optical arrangement of the MIFT is essentially the same as that shown in Figure 16.12. It includes a laser diode, GaP Bragg cell, and a 27-channel monolithic photodiode/preamp array. The laser diode is current-controlled to achieve constant output intensity. A thermoelectric-cooler-controlled circuit is also used to ensure operation of the MIFT over the specified temperature range. An additional feature is the incorporation of mechanical design for meeting the shock and vibration test requirements. No significant performance degradation of the MIFT was observed in the environment test. Figure 16.13 shows a photo of the 5-in$^3$ MIFT mounted on an environmental test bench.

## 16.6 COHERENT ACOUSTO-OPTIC RECEIVER

The dynamic range of a PSA is limited due to square-law direct detection and low dynamic range of the detector array. One method for increasing the dynamic range is to use an interferometric detection approach first proposed by King et al. [20]. In an interferometric spectra analyzer (ISA), the detector output current is proportional to the amplitude spectrum

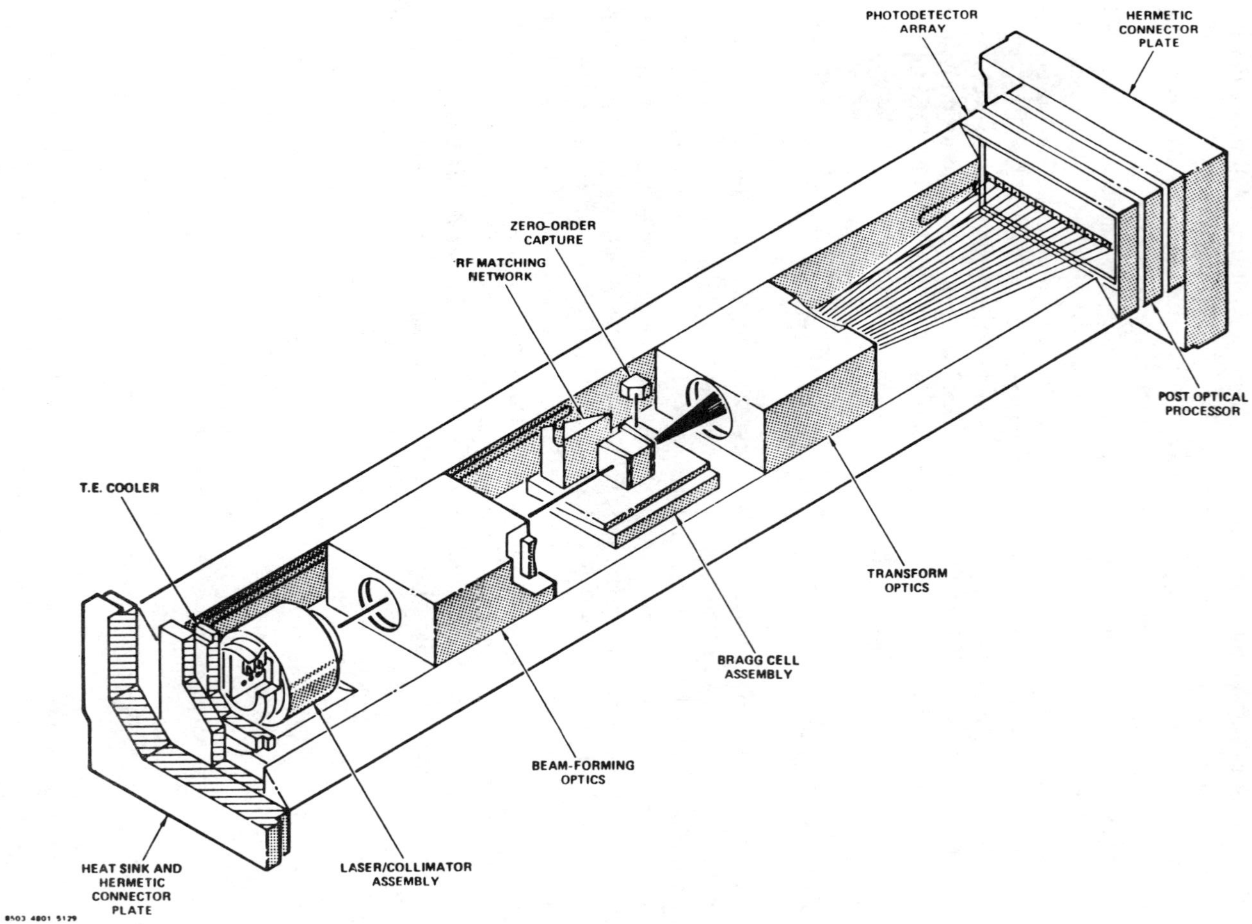


**Figure 16.12** Parallel output acousto-optic channelized receiver.

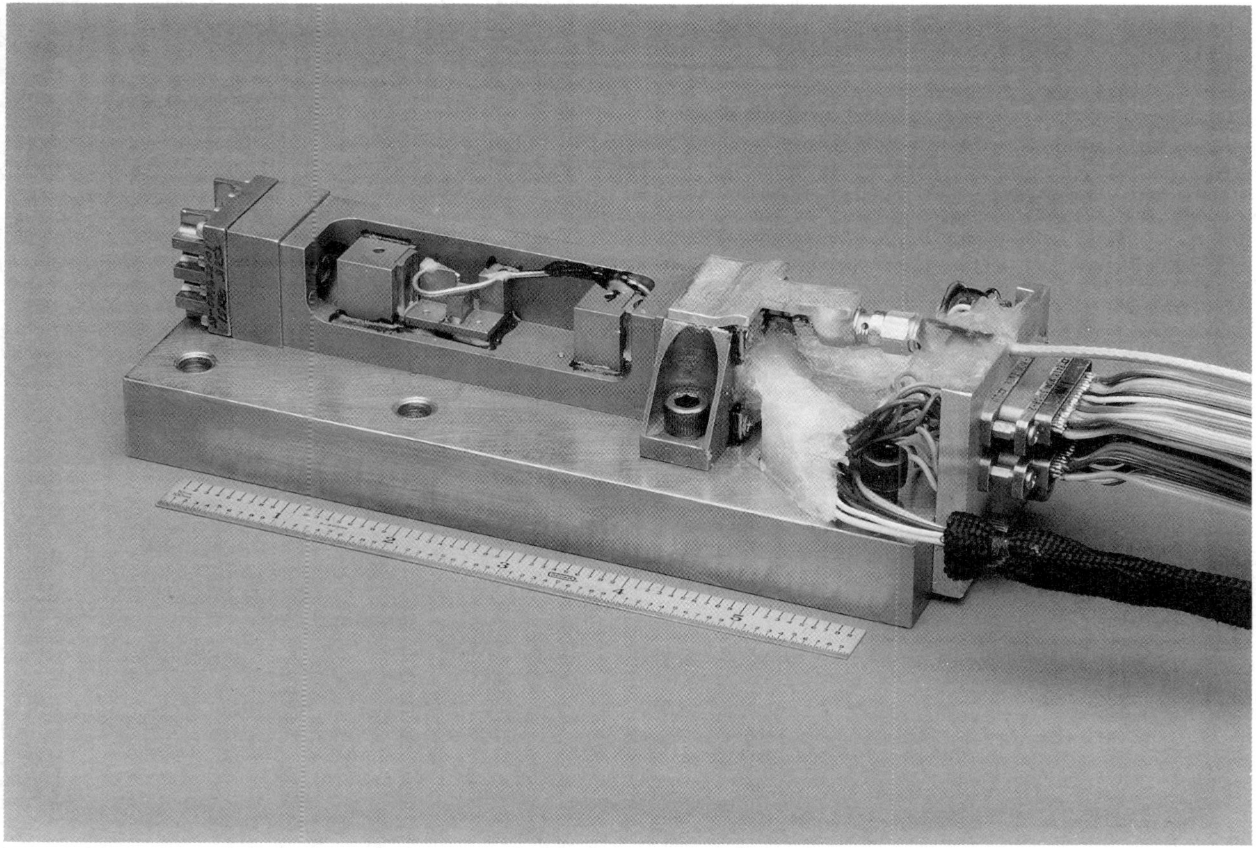

**Figure 16.13** MIFT mounted on an environmental test bench (courtesy of Litton Applied Technology).

of the RF signal. This is in contrast to a PSA, where the detector output current is proportional to the RF power. In principle, the dynamic range of an ISA is expected to be twice that of a PSA. In practice, due to the splitting of optical power in the signal arm and many channels in the reference arm, the overall improvement is less. It can be shown that the two-tone dynamic range of an ISA is given by [21]

$$\mathrm{DR} = \frac{4}{3}\left(I_0 + T_o - I_{\mathrm{th}} + 5\,\log\left(\frac{\eta_R}{8N}\right)\right) + 5.2\ \mathrm{dB} \tag{16.11}$$

where, as in (16.8), $I_0$ is the laser power, $T_o$ is the optical throughput of the signal arm, and $I_{\mathrm{th}}$ is the detector threshold. The last term in (16.11) represents the additional optical loss due to the reference arm; $\eta_R$ is the effective reference cell efficiency, and $N$ is the number of reference channels. We assume, as in the previous example of PSA, $I_0 = 20$ dBm, $T_o = -3$ dB, and $I_{\mathrm{th}} = -50$ dBm. In addition, let $\eta_R = 0.1$ and $N = 100$. The two-tone dynamic range estimated from (16.11) is about 68.5 dB, which is significantly less than twice that of a PSA.

Vander Lugt described a special ISA configuration that used two Bragg cells: a signal cell and a reference cell [22]. The reference cell is used to provide a spatially modulated reference beam. With proper arrangement, the interference term of the signal and reference beams will produce a heterodyning detector output with almost constant temporal IF (within one resolution over the entire spatial frequency plane). The constant IF implementation scheme has the advantage of significantly lowering the required bandwidth of the detector arrays.

Based on the fixed-offset IF scheme, a number of ISAs have been built and have demonstrated increased dynamic range [21,23,24]. Experimental results obtained from these ISAs have shown that the dynamic range is limited primarily by the spurious modulation associated with the wideband RF reference.

Figure 16.14 shows a schematic of an ISA using a Mach-Zehnder type of interferometer. The optical bench is referred to as the *acousto-optic interferometer,* since it uses two Bragg cells to modulate the optical path difference of the two equal path arms. The input signal is applied to one Bragg cell, the signal cell, through an RF amplifier. A wideband RF reference is fed to the second Bragg cell, the reference cell. The incident laser light is divided by the beam splitter and results in two diffracted laser beams out of the signal and reference cells. By appropriately adjusting the tilt between the two Bragg cells, the two diffracted beams are made to overlap at the Fourier plane and to mix collinearly at the desired IFs.

The signal cell is operated in the linear range at relatively low efficiency (about 5% or less) to avoid spurious signals due to IMPs and to achieve a high two-tone dynamic range. The reference is usually driven at high efficiency in order to improve receiver sensitivity. The IMPs generated from the reference cell will not severely degrade receiver performance.

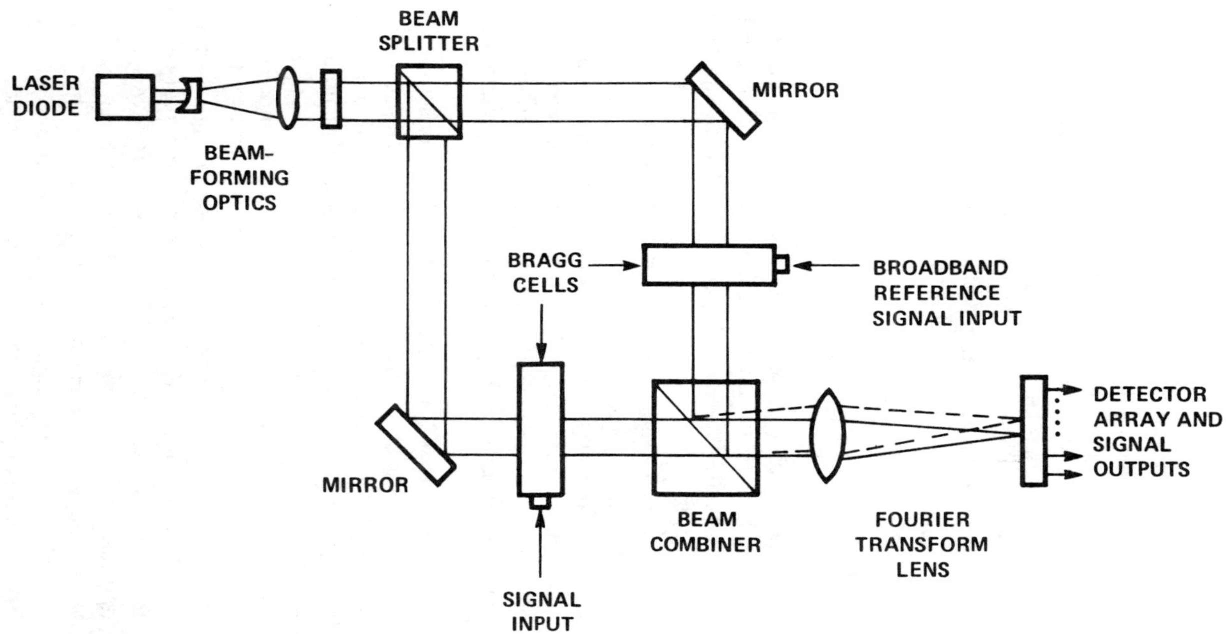


**Figure 16.14** Coherent acousto-optic receiver.

Theoretically, the ISA can significantly improve the dynamic range of an acousto-optic channelized receiver. Measured results from an experimental model [24] reveal that the improvement is realizable only for purely CW signals. For pulsed signals, the sidelobe rejection of the IF bandpass is severely degraded due to the inherent frequency gradient in the leading and trailing pulse edges. This introduces false signals in the sidelobe region and greatly increases the difficulty of processing the optical processor outputs. Other critical problems include optical stability, complexity of wideband reference and IF subsystem, and, most of all, the implementation of wideband detector arrays. Since these added complexities will certainly increase the size and power consumption, the acousto-optic channelized receiver using the ISA configuration appears to be inadequate for EW applications.

### 16.6.1 Acousto-Optic IFM

Another performance limitation of current acousto-optic channelized receivers is the achievable frequency resolution. To obtain a fine frequency resolution, a large number of parallel readout detector-amplifier arrays is required. The complexity of such a system makes implementation less practical.

In principle, signal channelization and frequency measurement are two separate requirements. In most scenarios, the frequency separation between time-overlapping signals is much larger than the frequency resolution needed for signal sorting. Moreover, since the complexity of a channelized receiver increases rapidly as the number of channels increases, it is desirable to use fewer channels as long as there is at most one signal within one channel at any time. The fine frequency resolution may be determined using other techniques, such as IFM.

Recently we have demonstrated a channelized IFM receiver using acousto-optic techniques [25]. A homodyne interferometric approach was developed wherein the reference is derived from the input signal by passing through a delay line, a frequency shifter and a limiting amplifier. A bank of IF phase discriminators was used to measure the signal phases, from which the fine frequency data were determined.

Figure 16.15 shows the schematic of an acousto-optic channelized IFM. An RF input which may include multiple time-overlapping signals is split by the power divider into branch A and branch B. The signal in branch A passes through a known time delay (as compared to branch B) and is fed to the signal channel of a dual cell acousto-optic interferometer. The reference is fed into an SSB modulator followed by a frequency-selective limiting amplifier. The SSB modulator shifts the RF signal frequency by a specified intermediate frequency ($f_m$) signal provided by an LO. The limiting amplifier produces the constant amplitude output required for the reference. By properly adjusting the tilt between the diffracted beams from the signal and the reference channels, the two beams will mix collinearly at the detectors (located at the Fourier transform plane) to produce the desired heterodyning component at the intermediate frequency $f_m$. A significant

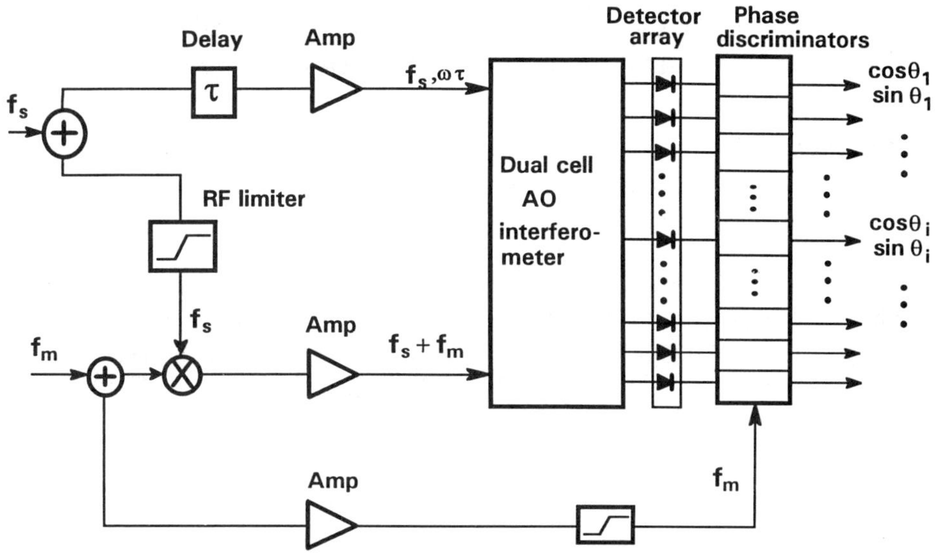


**Figure 16.15** Acousto-optic channelized IFM.

result of this arrangement is that the phase term of the channelized RF signal is preserved (in the IF signal). A conventional phase discriminator operating at the IF $f_m$ determines the inphase ($I$) and quadrature-phase ($Q$) components and the corresponding phase of the channelized RF signal.

A feasibility model of the acousto-optic channelized IFM was implemented using a compact inline configuration. The Bragg cells used in the interferometer were a pair of 500-MHz bandwidth, shear wave GaP devices centered at 1 GHz. The frequency resolution was about 25 MHz per channel. The frequency offset between the signal and reference arm (i.e., the IF) was set to be about 65 MHz. The signal and reference beams were focused onto the fiber array positioned at the Fourier transform plane and coupled into a separate detector-amplifier assembly. An SSB modulator operated at the IF of 65 MHz was used to provide the frequency shift for the reference beam. The heterodyned RF outputs were fed into an IF phase discriminator. From the measured results of $I$ and $Q$ output for each channel, the phase delays and corresponding signal frequencies can be calculated.

Experimentally, it was found that the phase measurement result was relatively stable and the phase error was estimated to be about 10 deg. This corresponds to a frequency accuracy of 1.0 MHz. Compared to the channel width of 20 MHz, the acousto-optic IFM provides a significant enhancement of frequency resolution.

For RWR tape systems, it appears reasonable to choose the channel width as wide as 100 MHz, while maintaining the frequency resolution to be on the order of 10 MHz.

Due to its relative simplicity, the acousto-optic channelized IFM appears to be a cost-effective approach for meeting the specified system requirements.

## 16.7 SUMMARY

An acousto-optic receiver can be considered as a channelized receiver using an acousto-optic implementation. Compared to other technologies, the most significant advantages of the acousto-optic approach are its small size and low-cost potential, particularly for systems with a large number of channels. The primary technical problem has been the optical-to-electronic interface which includes the photodetector arrays and the postdetection processors. If these bottlenecks are overcome, the acousto-optic receiver could realize its full potential for meeting the requirements of future EW receivers.

## REFERENCES


[1] Tsui, J.B.Y., *Microwave Receiver With Electronic Warfare Applications,* New York: John Wiley, 1986.

[2] Chang, I. C., ''Acousto-Optic Devices and Applications,'' *IEEE Trans. Sonics and Ultrasonics,* Vol. SU-23, No. 1, 1976, p. 2.

[3] Berg, N. J., and J. N. Lee, eds. *Acousto-Optic Signal Processing,* New York: Marcel Dekker, 1983.

[4] Lee, J. N., and A. Vander Lugt, ''Acousto-Optic Signal Processing and Computing,'' *Proc. IEEE,* Vol. 77, No. 10, 1989, p. 1528.

[5] Lambert, B., ''Wideband Instantaneous Spectrum Analyzers Employing Delay Line Modulators,'' *IRE National Convention Record,* Vol. 10, part 6, 1962, p. 69.

[6] Hecht, D. L., ''Spectrum Analysis Using Acousto-Optic Devices,'' *Opt. Eng.,* Vol. 16, No. 5, 1977, p. 461.

[7] Lennert, L. C., et al., ''A High Performance GHz Bandwidth Spectrum Analyzer,'' *Proc. of SPIE,* Vol. 352, 1982, p. 10.

[8] Chang, I. C., ''Acousto-Optics Channelized Receiver,'' *Microwave J.,* March 1986, p. 141.

[9] Anderson, G. W., et al., ''Advanced Channelization Technology for RF, Microwave and Millimeter Wave Applications,'' *Proc. of IEEE,* Vol. 79, No. 3, 1991, p. 355.

[10] Berg, N. J., et al., ''Real Time Fourier Transformation via Acousto-Optics,'' *Appl. Phys. Lett.,* Vol. 34, 1979, p. 15.

[11] Turpin, T. M., ''Time Integrating Optical Processors,'' *Proc. SPIE,* Vol. 154, 1978, p. 196.

[12] Chang, I. C., and S. Lee, ''Efficient Wideband Acousto-Optics,'' *IEEE Ultrasonics Symp. Proc.,* 1983, p. 427.

[13] Chang, I. C., ''Design of Wideband Acousto-Optic Bragg Cells,'' *Proc. SPIE,* Vol. 352, 1982, p. 352.

[14] Chang, I. C., ''Birefringent Phased Array Bragg Cells,'' *IEEE Ultrasonics Symp. Proc.,* 1985, p. 381.

[15] Hecht, D. L., ''Multifrequency Acousto-Optic Diffraction,'' *IEEE Trans. Sonics and Ultrasonics,* Vol. SU-24, 1977, p. 7.

[16] Chang, I. C., and R. T. Wererka, ''Multifrequency Acousto-Optic Diffraction,'' *IEEE Ultrasonics Symp. Proc.,* Oct. 1983, p. 445.

[17] Anderson, G. W., et al., ''Photodetector Arrays for Acousto-Optic Channelizers,'' *Proc. SPIE,* Vol. 936, 1988, p. 10.

[18] Nashkurak, W. D., et al., ''Wide Dynamic Range Tapered Linear Array Image Sesor,'' *Proc. SPIE,* Vol. 936, 1988, p. 14.

[19] Swarthout, J. D., et al., "High Dynamic Range CCD Detectors for Acousto-Optic Applications," *Proc. SPIE,* Vol. 1102, 1989, p. 61.
[20] King, M. C., et al., "Real-Time Electro-Optical Signal Processors With Coherent Detection,," *Applied Optics,* Vol. 6, 1967, p. 1367.
[21] Chang, I. C., et al., "High Dynamic Range Acousto-Optic Receiver," *Proc. SPIE,* Vol. 545, 1985, p. 95.
[22] Vander Lugt, A., "Interferometric Spectra Analyzer," *Applied Optics,* Vol. 20, 1981, p. 2770.
[23] Shah, M. L., et al., "Interferometric Bragg Cell Spectrum Analyzer," *IEEE Ultrasonics Symp. Proc.,* 1981, p. 743.
[24] Tsui, T. B., et al., "Interferometric Acousto-Optic Receiver," *Proc. SPIE,* Vol. 1102, 1989, p. 176.
[25] Chang, I. C., and R. Lu, "Acousto-Optic Phase Interferometer," *Proc. SPIE,* Vol. 936, 1988, p. 208.

# Chapter 17

# Fiber-Optic Delay Line Signal Processing Architectures

*Behzad Moslehi*
*Optivision*

## 17.1 OVERVIEW, BACKGROUND, AND MOTIVATION

The advent of the laser in 1960 [1] opened a new era in the field of optics by realizing a coherent light source. This realization in turn led to the development of low-loss and large-bandwidth optical fibers around 1970 [2,3]. In the following two decades, a variety of passive and active fiber- and integrated-optic components (such as directional couplers, tree and star couplers, polarizers, phase/amplitude modulators, and optical amplifiers) were developed and commercialized. A relatively recent practical development is the addition of optical amplifiers [4,5] to the menu of available fiber-optic components. Both erbium-doped fiber amplifiers (EDFA) and semiconductor fiber amplifiers (SOA) have been transferred from research laboratories to the commercial world. These optical amplifiers currently involve high cost, limiting the extent that they can be employed in the experimental prototypes to only a few at this time. But it is anticipated that the cost of these amplifiers will drop in the future as they find more applications and become mass producible. These developments and the progress in semiconductor technology which provided the necessary light sources and photodetectors established fiber optics as a promising new branch of optics.

The first significant application of fiber optics was in the telecommunication area (although there had been some medical applications before 1960 [6]). However, in subsequent years, several other important application areas have emerged, such as sensing, networking, and signal processing. The use of fiber-optic technology for signal processing was proposed in 1976 [7] by noting that fiber delay lines were attractive due to their low loss and low dispersion, the same properties that motivated the communications application.

Fiber-optic filters offer the unique capability of processing wideband and high-frequency (multigigahertz) baseband signals directly in the optical domain. The information to be processed intensity modulates a small high-speed semiconductor laser. The modulated optical signal is then launched into the fiber-optic filter, which performs a variety of linear functions on the signal (both in the time and frequency domains). The processed signal can then be sent into another optical network or detected by a small pin photodiode for further processing. The direct processing in the optical domain eliminates the need for inefficient and costly intermediate optoelectronic conversions and electronic filtering required by conventional approaches, and also extends the speed and bandwidth capabilities to levels beyond those of state-of-the-art electronics.

Most of the fiber-optic signal processing devices demonstrated so far are primarily based on feed-forward tapped delay line structures [8–12], which are equivalent to classical transversal filters [13]. This chapter, however, concentrates on new fiber-optic arrangements that are similar to the feed-forward (nonrecirculating) and feed-backward (recirculating) lattice structures of digital filtering theory [14–16]. A lattice structure is a structural form composed of cascaded interconnection of compatible sections having the same basic topology, as shown in Figure 17.1 [17,18]. These structures have two input ports and two output ports, and they assume either a feed-forward (Fig. 17.1(a)) or feed-backward (Fig. 17.1(b)) configuration. There is highly developed theory for the design and synthesis of filters based on such structures. These processors are capable of optically performing various broadband frequency-domain and time-domain operations, including frequency filtering [17,18] and matrix operations [17–19]. Integrated optic implementation of the filters discussed is also possible when the delay lengths involved are not too large (under several centimeters). This chapter is primarily concerned with lattice structures, their fiber-optic implementation, applications, and various theoretical formulations. We also describe new fiber-optic signal processing systems, which include the latest optical amplifier

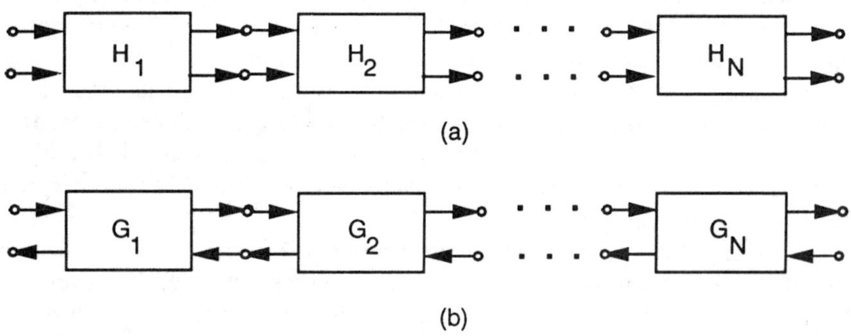


**Figure 17.1** Lattice structure topology composed of cascaded two-pair subsystems: (a) feed-forward configuration; (b) feed-backward configuration.

technologies which will enable the realization of processing systems with greater processing power than is possible using more conventional approaches to fiber-optic signal processing.

The major available hardware that can be used for the realization of basic fiber-optic lattice structures includes fiber, couplers, splitters/combiners, optical sources, and detectors with their associated electronics. More advanced lattice structure may employ fast adjustable couplers, switches, and optical amplifiers. Lattice structures can be implemented in either all-fiber-optic or all-integrated-optic forms, or their combination. While the all-fiber implementation is more practical at relatively lower speeds (<~5 GHz), the all-integrated-optic form becomes more attractive at higher speeds (>~10 GHz). Both fiber- and integrated-optic filters involve multiple optical paths of different delay lengths in their structure. The shortest delay (referred to as the *basic unit delay*) length is comparable to the wavelength of the modulating signal, which can be in the microwave range. As a result, higher modulation frequencies are associated with shorter basic unit delay lengths. It should also be noted that many useful filters need a large number of couplers (taps). Fabrication of many couplers requires long lengths of integrated-optic chips, which may not be practical. Although basic lattice structures can be fabricated in a monolithic integrated-optic form, more advanced versions demand a material base capable of integrating all the required passive and active components monolithically in a single substrate.

Fiber-optic filters can be used to perform both frequency-domain and time-domain signal processing operations in the microwave frequency range [20]. Examples of frequency-domain functions are low-pass filtering (LPF), bandpass filtering (BPF), and notch filtering (NF). Examples of time-domain operations are matrix-vector algebra, encoding/decoding, correlation/matched filtering, switching (time slot interchange), data rate transformation, analog-to-digital and digital-to-analog conversion, storage, and equalization.

Figure 17.2 shows a second-order (two loops) feed-backward fiber-optic lattice structure, which consists of three couplers and two loops. The couplers' coupling ratios can be adjusted to reconfigure the arrangement and to obtain different output responses. Figure 17.3(a,b) illustrates the frequency filtering application of the fiber filter of Figure 17.2 [17]. Figure 17.3(c) shows the time-domain application when the structure of Figure 17.2 is used as a systolic array matrix-vector multiplier [17–19]. These array types were originally developed for VLSI implementations [21]. The systolic concept is important in that it offers a simple way for balancing the I/O and computation requirements by pipelining operations. In an all-fiber implementation, all the operations are performed inside the fiber through realizing an all-fiber matrix-vector systolic multiplier. Architectures such as systolic arrays are very attractive in modern signal processing, since they are suitable structures to perform specific operations with high efficiencies (special-purpose architectures).

Because of the fact that most architectures used for implementing optical filters are intrinsically optically lossy, fiber-optic signal processing systems, which include the latest optical amplifier technologies, are also of interest. In addition to the optical loss compensation, optical amplifiers have the potential to provide extra flexibility for unique fiber-

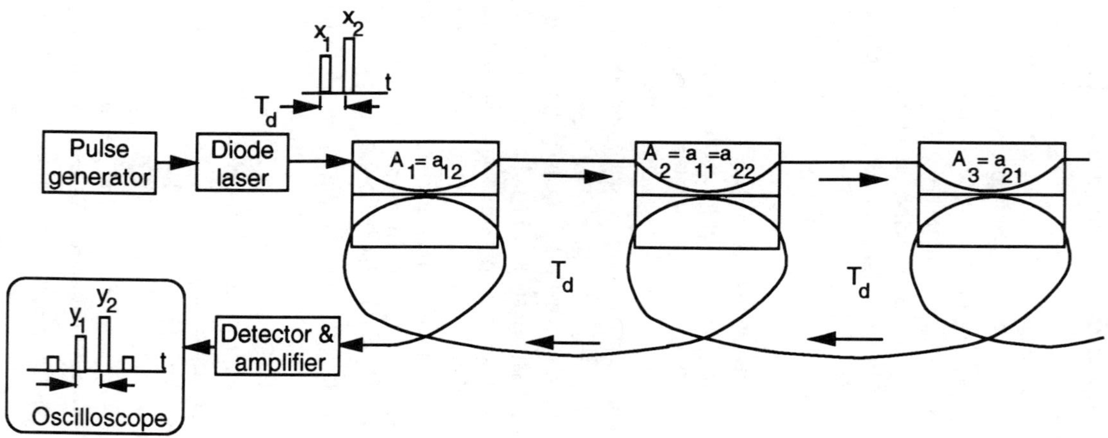


**Figure 17.2** Fiber-optic implementation of a second-order feed-backward lattice structure.

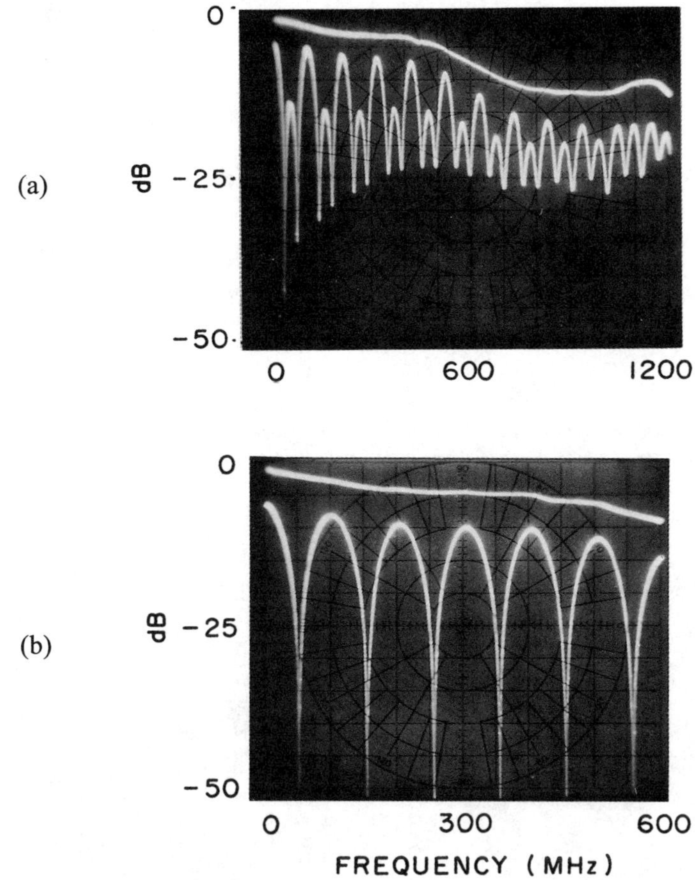


**Figure 17.3** (a) Frequency-domain (a) and time-domain (b) applications of lattice filter of Figure 17.2. Log-magnitude response (0 to 1.2 GHz) when the configuration is set as a second-order filter (coupling ratio of the middle coupler is between 0 and 1). The upper trace is the frequency response of the electronics; log-magnitude response (0 to 600 MHz) when the configuration is set as a first-order filter (coupling ratio of the middle coupler set to 1).

optic filter designs, not available in passive configurations [22–25]. The application of optical amplifiers and polymer-dispersed liquid crystal (PDLC) shutters to electrically reconfigurable fiber-optic delay line signal processors has been recently demonstrated (see Fig. 17.4) [26]. Here, two basic eight-tap finite-impulse response (FIR) fiber filter modules were fabricated. Each basic filter module was implemented with two 1 × 8 integrated-optic single-mode fiber couplers (one used as a splitter at the input and the other as a combiner at the output), liquid crystal shutter tunable weights, and fiber delays. The output signal from the FIR section is then optically amplified by an EDFA before

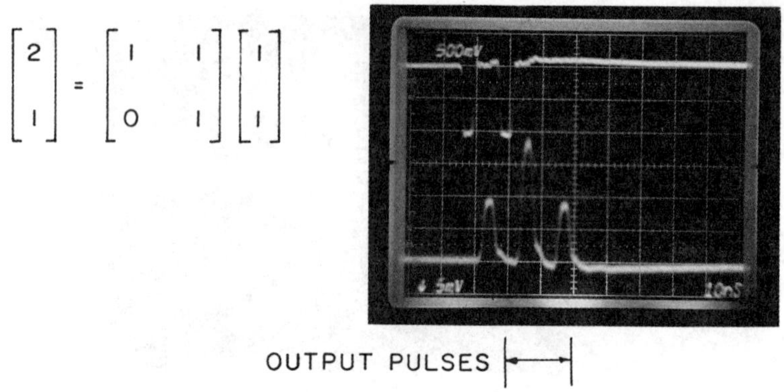


**Figure 17.3(c)** Pipelined systolic matrix-vector multiplier. The lower trace shows the output pulses representing the multiplication result of a $2 \times 2$ matrix by a $2 \times 1$ vector. The vector elements are encoded in the intensity of a laser diode (top trace). The values of the coupling ratios represent the matrix elements.

**Figure 17.4** Overall integrated packaging of the optically amplified, electrically reconfigurable fiber-optic filter subsystem, composed of input splitter, tunable weights (PDLC shutters), single-mode fiber delays, output combiners, and an optical amplifier (EDFA). The top cover was removed to show the internal hardware. The dimensions are $H \times W \times L = 10.47 \times 17.28 \times 19.66$ in$^3$.

detection or retransmission. These two basic modules can be interconnected in a cascaded or parallel configuration to implement 16-tap fiber FIR filters, or in a feedback configuration to realize infinite-impulse response (IIR) filters. Reconfigurability and tunability were obtained by varying the amplifier gain and/or the PDLC shutter weight values electrically. These optically amplified, electrically reconfigurable fiber signal processors have the potential to lead to the realization of complex programmable and adaptive optical systems.

The basic frequency-domain and time-domain operations mentioned above are fundamental to the realization of many signal processing systems, including those in radars. A specific successful application of fiber-optic delay lines is in PAA systems (Fig. 17.5), which consist of an array of radiating antenna elements. Here, optical fibers are used to realize wideband RF-frequency-independent true-time delays for adjusting the phases of RF signals in a beam forming network [27]. The radar signals corresponding to each antenna element intensity modulate an optical carrier. The modulated optical signals are fed through the switchable fiber-optic time-delay network. The delay network consists of a fiber delay line network and an optoelectronic switching array. The switching array scans through the possible fiber delays. Different delays induce different RF phase shifts, which in turn steers the radiation pattern of the PAA. Each state of the time-delay network corresponds to a particular orientational angle of the RF wavefront. Fiber optics can also be employed in communications/control analog and digital links. In addition to high bandwidth, other advantages of fiber optics in radar applications include small size, light weight, electrical passivity, and EMI/EMP. Optics has the potential to penetrate further into all levels of signal processing problems, as more optical processors are developed. It is anticipated that fiber-optic technology will significantly improve the performance and capabilities of next-generation advanced radars.

### 17.1.1 Signal Processing

Generally speaking, signal processing is concerned with the representation of signals by physical quantities, such as light intensity and phase or acoustic wave amplitude, and the

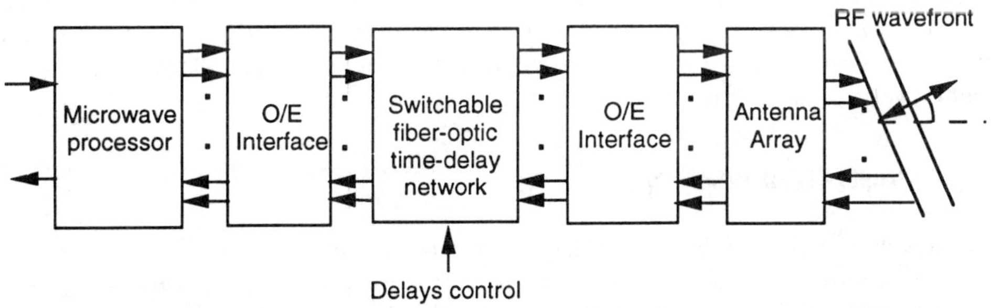


**Figure 17.5** Phased-array antenna in both receiving and transmitting modes. A switchable fiber-optic time-delay network is used to steer the RF radiation pattern.

processing of these quantities by physical structures [21]. Many processing functions involve repetitive computations on large amounts of sophisticated data signals. These requirements demand advanced processing techniques capable of handling such data. The function of the processing, for example, could be to transform a signal into a more desirable form, to perform an algebraic computation on numbers represented by the signals, or to estimate characteristic parameters of a signal [16,28]

Many diverse fields such as telecommunications, acoustics, speech communication, image transformation, radar, sonar, geophysics, and biomedical engineering involve tremendous amounts of signal processing [29]. There is a variety of different techniques (optical, acoustic, electronic) and methods (analog or digital) of realizing a signal processing device. These techniques and methods are briefly reviewed in the following.

### 17.1.2 Optical Processing Versus Other Techniques

Two major advantages of optics are its ultra-high-speed and wideband signal processing capabilities. Other advantages include parallelism and 3-D signal propagation capabilities, electrical passivity, and EMI/EMP immunity. Also, because of its high center frequency, an optical carrier can handle wideband modulating signals. The speed of optical signals is equal to the speed of light in the propagation medium. Further, optical devices and elements, such as lenses, mirrors, and prisms, can easily handle millions of resolvable points simultaneously. The difficulty is how to fully utilize the available parallelism and bandwidth. Moreover, optical signals can pass through each other in a nonplanar manner, without interference. These attractive features have stimulated an extensive amount of work to investigate the potential applications of optical systems in different areas. Among other advantages of optical processing systems are their compactness, modest energy requirements, and low cost. There are various ways of implementing optical processing systems, including conventional discrete optics and fiber optics.

As mentioned earlier, this work is concerned with the utilization of the (primarily single-mode) fiber-optic technology in signal processing. The technology offers the opportunity to perform high-speed broadband signal processing functions by providing the advantages that are briefly described in the following section. It should be noted that, in the guided-wave optics realization of the filters discussed in this chapter, the parallel nature of optics has not been utilized.

### 17.1.3 Properties of Fiber Optics

The low loss and large bandwidth-length product of optical fibers have provided an attractive delay line medium, which is capable of guiding broadband (well over 1 GHz) optical signals on flexible light paths. It is assumed that the reader is somewhat familiar with optical fiber. However, here we outline some of the properties of optical fibers that are relevant to this discussion.

Fibers may propagate a single mode or multiple modes, depending on the fiber core radius, core-cladding index difference, and the wavelength. Single-mode transmission results when the core radius of the fiber is on the order of the optical wavelength. Single-mode fibers outperform multimode fibers, since they have typically a larger information carrying capacity (bandwidth-length product exceeding 100 GHz·km [30]) and lower attenuation than the multimode types. Attenuations less than 0.5 dB/km at 1.27 $\mu$m [31,32] and less than 0.2 dB/km at 1.55 $\mu$m [33] have been reported. (The attenuation is essentially independent of the modulation frequency.) In addition to its minimum-loss property, the 1.55-$\mu$m wavelength window has a special significance with respect to its compatibility with EDFAs. In applications for which coherent propagation is essential, single-mode fiber is used, since the coherence of the light beam is preserved as it is guided along the fiber. In the following, we briefly review the major mechanisms involved in determining the loss and bandwidth of optical fibers.

### *Loss*

Loss [34] in singe-mode fibers is caused by scattering, absorption, splicing, and bending, all of which are functions of the optical wavelength. Intrinsic scattering, which results from microscopic inhomogeneities in the fiber material, is inversely proportional to the fourth power of the wavelength. The scattering loss is approximately 0.24 dB/km at 1.3 $\mu$m of wavelength and 0.12 dB/km at 1.55 $\mu$m. Light absorption is caused by hydroxyl ($OH^-$) ions trapped in the fiber material during processing. This absorption loss is dependent on the manufacturing process; however, it is relatively low except for the hydroxyl peaks (1.25 and 1.39 $\mu$m). Splicing loss results whenever fibers are interconnected. Fusion splicers can be used to interconnect fibers with a splicing loss of less than 0.1 dB. Connectors and chemical bonding agents can also be used to interconnect fibers. Finally, bending loss is produced when a fiber is bent to the extent that part (or all) of the fiber guided mode power is lost by radiation. For a given bending curvature, the bending loss increases substantially if the operating wavelength is larger than a critical wavelength. This wavelength is determined by the fiber parameters. A large core-cladding index difference and a small core radius increase the critical wavelength. Therefore, if the operating wavelength is between the single-mode cutoff wavelength (defined as the wavelength below which the fiber is no longer single-mode) and the critical bending wavelength, the bending loss will be relatively low. Of the four loss mechanisms discussed above, the intrinsic scattering loss defines the fundamental limit of attenuation. Loss values of commercially available fibers are approaching the theoretical limit.

### *Dispersion*

Among the factors that determine the pulse dispersion (or pulse broadening) and bandwidth of single-mode fibers are material dispersion, waveguide dispersion, the spectral width of the optical source, and polarization dispersion. Material dispersion results from wavelength dependence of the core refractive index, while waveguide dispersion arises from the relation of the fiber dimensions (core radius), refractive index profile, and the source wavelength. Polarization dispersion occurs because of birefringence in real optical fibers; the two polarization eigenmodes have different group velocities. This dispersion can be kept relatively low (less than 0.2 ps/km [35]) by taking special care in fabrication. However, at the wavelength where the net material and waveguide dispersion is minimum, polarization mode dispersion is sometimes important [35]. Single-mode fibers capable of preserving linear polarization are essential in some applications, including those involving integrated optics and interferometers. Polarization preservation can be achieved by designing fibers having a sufficiently large difference in the propagation constants of the two orthogonal polarization eigenmodes [36].

In general, material dispersion and waveguide dispersion are interrelated; the dispersive properties of the refractive index affect both of the dispersion mechanisms. However, unless a very exact value is desired, they can be calculated independently [37] and then added to the total dispersion. Another method is to use the concept of the group index, which includes the effects of both types of the dispersion. Using this concept, the group velocity $V_g$ can be related to the light velocity in vacuum ($c$) through a parameter known as the group index ($n_g$) as shown below:

$$V_g = c/n_g \tag{17.1}$$

where both the group velocity and the group index are functions of wavelength $\lambda$. The group index, which includes the combined effects of both the material and waveguide dispersions, is a weighted average of the group indexes of the different materials in the fiber [38]. The weights are determined by the amounts of optical power propagating through the fiber materials. At short wavelengths, almost all the optical power is inside the core, resulting in a group index that is essentially the same as that of the core. However, at large wavelengths, due the extension of the mode field beyond the core, the mode group index is affected by the outer claddings. Using (17.1), we can obtain the group delay ($\tau_g$) for an optical pulse propagating through a single-mode fiber of length $L$:

$$\tau_g = n_g L/c \tag{17.2}$$

which gives a value of approximately 5 ns for every meter (or 5 ps/mm) of typical fibers. This relatively low value of the group delay per unit length makes optical fibers very attractive by providing both short and precise time-delay functions. Equation (17.2) can also be used to calculate the time-bandwidth product ($\tau_g B$) of the fiber. The result is expressed [39] as

$$\tau_g B = n_g/[\Delta\lambda(\mathrm{d}n_g/\mathrm{d}\lambda)]_{\lambda=\lambda_0} \tag{17.3}$$

where $\lambda_0$ is the operating wavelength and $\Delta\lambda$ is the spectral width of the input optical signal. At $\lambda_0 = 0.85$ $\mu$m (where the material dispersion is the dominant dispersion mechanism), with typical values [39] of $n_g = 1.47$ and $dn_g/d\lambda = 2.5 \times 10^{-5}$, we get a (first-order) time-bandwidth product of

$$\tau_b B \approx 5.9 \times 10^4/\Delta\lambda(\text{nm}) \tag{17.4}$$

For silica-type fibers, material dispersion reverses its sign at a wavelength in the infrared region ($\lambda_0 \sim 1.3$ $\mu$m, called the *zero-dispersion wavelength*). At around 1.3 $\mu$m, where the first-order dispersion is zero, second-order dispersion effects become important. In this case, the time-bandwidth product is given by [39]

$$\tau_b B = 8n_g/[(\Delta\lambda)^2(d^2n_g/d\lambda^2)]_{\lambda=\lambda_0} \tag{17.5}$$

which for typical parameter values gives

$$\tau_b B = 3.9 \times 10^8/[\Delta\lambda(\text{nm})]^2 \tag{17.6}$$

For an input signal spectral width of $\Delta\lambda = 1$ nm, (17.4) and (17.6) give time-bandwidth products of $5.9 \times 10^4$ and $3.9 \times 10^8$ at 0.85 and 1.3 $\mu$m, respectively. At 1.55 $\mu$m, the value of the time-bandwidth product will be between the values at 0.85 and 1.3 $\mu$m.

### 17.1.4 Temperature Effects

By differentiating $\tau_g$, given in (17.2) with respect to temperature (T), and replacing $\partial L/\partial T$ with $\alpha L$ ($\alpha$ is the linear thermal expansion coefficient), one can also obtain the effects of temperature variation on the group delay of the fiber:

$$\partial\tau_g/\partial T = (L/c)[\partial n_g/\partial T + n_g\alpha] \tag{17.7}$$

Using typical values of $\alpha = 8 \times 10^{-7}/°\text{C}$ and $\partial n_g/\partial T = 1 \times 10^{-5}/°\text{C}$, we find that the second term on the right-hand side of (17.7) is small compared to the first term. Therefore, in silica fibers, refractive index variations with temperature have a larger effect on the time delay than does linear thermal length expansion [40]. Equation (17.7) may also be used to determine the maximum allowable temperature change for a given amount of fiber delay change. As an example, if for a fiber 1m long we allow a delay change of 1 ps, the maximum allowable temperature range will be approximately 30°C. This rather wide range is applicable in the case of incoherent (intensity-based) signal processing. For coherent signal processing (see next section for more details), however, the temperature changes must be kept well under 1°C [12]. This is due to the fact that the time period of the lightwaves in femtoseconds (fs) is about $3.3\lambda$ ($\mu$m), yielding 2.8-, 4.3-, and 5.2-fs periods for, respectively, 0.85, 1.3, and 1.55 $\mu$m. This implies that, for the coherent case, the delay changes must be kept under 1 fs.

### 17.1.5 Fiber Structures

The zero-dispersion wavelength of a fiber is a function of fiber composition and the refractive index profile. By tailoring the waveguide parameters (refractive index profile and core diameter), one can enhance the waveguide dispersion to cancel out the material dispersion at the minimum-loss wavelength ($\lambda_0 = 1.55$ $\mu$m). Using this canceling property, single-mode fibers with triangular index profile (to reduce the loss resulting from the core-cladding interface) have been manufactured having zero-dispersion wavelengths shifted (dispersion-shifted fibers) to 1.56 $\mu$m, with losses as low as 0.24 dB/km at $\lambda = 1.55$ $\mu$m [41]. Dispersion-shifted fibers have their minimum dispersion as well as their minimum loss simultaneously located at a single operating wavelength. As a result, the dispersion increases as the wavelength deviates from the optimum wavelength. In order to reduce the total dispersion over a wide wavelength range, including the low-attenuation region, fibers with more than one cladding (multilayered cladding) have been produced. Double-clad [42] and quadruple-clad [38] (or segmented core) fiber geometries have been used to flatten the dispersion spectra over an extended range. The latter geometry has the advantage that the light leaking from the inner cladding can be retrapped by the light guide formed by the three outer claddings, thereby reducing the bending loss. Using the quadruple-clad structures, a total dispersion of less than 2 ps/km-nm has been reported [39] in the wavelength range of 1.3 to 1.6 $\mu$m with 0.6 dB/km and 0.4 dB/km power attenuations at 1.3 and 1.6 $\mu$m, respectively.

The various fiber structures and designs described above are of importance in some applications. However, for most optical signal processing applications, there is no need for complex fiber structures. In fact, in the work presented here, ordinary step-index (conventional design) single-mode fibers are adequate.

### 17.1.6 Sources and Detectors

Single-mode fiber-optic systems usually include semiconductor laser diodes and detectors to generate and detect optical signals. For the so-called first *optical source window* (700 to 900 nm) the ternary alloy $Ga_{1-x}Al_xAs$ is the principal laser material used. The aluminum mole fraction $x$ determines the peak wavelength $\lambda_0$ ($\mu$m) = 1.24/$E_g$ (eV), where $E_g$ is the band gap energy (in electron volts) of the laser emission, and it is usually selected to yield a wavelength in the range 800 to 850 nm. At longer wavelengths (1.0 to 1.7 $\mu$m) the quaternary alloy $In_{1-x}Ga_xAs_yP_{1-y}$ is of prime importance. In this case, there are two degrees of freedom (rather than one in the ternary case), in that a range of wavelengths and lattice parameters [43] can be covered by varying the mole fractions $x$ and $y$. For the above alloys it is possible to match the crystal lattice parameters of the heterojunction interfaces [35] to increase the laser radiative efficiency and lifetime. Semiconductor laser diodes are of interest in fiber optics, since their output emission can be easily coupled into a fiber by pigtailing or "connectorizing", and they can also be directly modulated with high speeds by varying the electrical injection current. Modulation bandwidths over 8 GHz have been achieved [44]. Other advantages of semiconductor lasers are their small size and high efficiency.

Two types of photodetectors are commonly used to convert the optical power into a corresponding electrical current: pin photodetectors and APDs. These photodiodes meet the high-performance requirements of fiber optics, including size and area compatibility, sensitivity, reliability, and speed. For high-speed applications, small photodiodes as fast as 100 GHz have been demonstrated [45]. APDs are preferred for low-level optical signal detection. Silicon is the primary photodetector material used for wavelengths ranging from 800 to 900 nm. Above this range, in the wavelength range 1.0 to 1.65 $\mu$m, germanium and InGaAs materials are known to be the best candidates.

In most experimental studies reported to date, pin photodetectors and multimode GaAlAs laser diodes have been used. These optical sources can be amplitude modulated from 20 Hz to over 2 GHz. The output light intensities are detected by high-speed photodiodes whose outputs are connected to measurement systems such as an oscilloscope, network analyzer, or spectrum analyzer, and whose output voltage is proportional to input intensity.

### 17.1.7 Analog Versus Digital Signal Processing

The potential of optical techniques has been successfully demonstrated in signal processing applications such as convolution, correlation, and matrix operations. In virtually all of these applications, the data upon which computations are performed are obtained by continuously varying the magnitude of a physical quantity such as light intensity. This is because such optical processing techniques are analog. Analog optical processors provide powerful means for performing a variety of data processing operations (they have high speeds, large throughputs, and large dynamic range). However, in optical processors involving repetitive computations on large amounts of sophisticated data, each step may introduce some error. These errors (if not corrected) accumulate, resulting in an output deviating from its expected value. In a digital representation of data, signals may pass through a large number of digital computation elements without accumulating errors. Digital systems are also programmable. These inherent properties of digital processing have encouraged researchers to investigate the applications of optics to digital computations [46–48].

To perform computational operations such as logic functions, one needs nonlinear elements or logic gates. Lack of efficient nonlinear devices in optics at present is the main impediment to realizing a digital optical computer. However, the recent encouraging advances in optical bistability [49] may lead to the realization of efficient optical bistable devices capable of subpicosecond switching between stable states. The remaining problem is related to materials which tend to have low nonlinear coefficients and/or high losses.

Without relying on optical bistable devices, there are various ways to incorporate digital accuracies in optical processors. One way is to combine high-speed capabilities of optics with the accuracy of digital processing through the use of hybrid optical digital processing techniques [50,51]. These techniques have the disadvantage that the processor is not entirely optical and involves some electronics which may limit full utilization of the advantage promised by optics. Another approach is to implement digital operations, such as binary addition and multiplication, by analog processing. This technique, which is briefly described below, is attractive, since analog functions can be easily performed optically (this is not the case for binary operations). Using the binary representation ($N$ bits), we can write the sum of two data points $A$ and $B$ as

$$A + B = \sum_{k=0}^{N-1}(a_k + b_k)2^k \tag{17.8}$$

where $a_k$ and $b_k$, which respectively represent the $k$th-order bits of the numbers $A$ and $B$, are either zero or one. The above equation indicates that to obtain the sum of $A$ and $B$, we can algebraically add the corresponding bits of the binary representations [48] (subtraction is similarly performed). Similarly, analog optical processors can be employed to implement digital multiplication by analog convolution (DMAC) [46–48]. In this case, the product of two numbers $A$ and $B$ may be written as [48]

$$A \cdot B = \left[\sum_{n=0}^{N-1} a_n 2^n\right]\left[\sum_{m=0}^{N-1} b_m 2^m\right] = \sum_{k=0}^{2N-1}\left[\sum_{m=0}^{N-1} a_{k-m} b_m\right]2^k = \sum_{k=0}^{2N-1}[a_k \otimes b_k]2^k \tag{17.9}$$

As seen in (17.9), the term in brackets is the discrete convolution of $a_k$ and $b_k$. Therefore, we conclude that the product of two binary numbers can be obtained by calculating the convolution of the bits representing the two numbers. The output has a maximum of $2N - 1$ elements. The $k$th-order element is the coefficient of the term $2^k$ in (17.9), which is equal to the $k$th element of the sequence $[a_k \otimes b_k]$. Each element of this sequence can assume any value between zero and $N$, resulting in outputs that are not in full binary form (they are, rather, in a an intermediate mixed-binary form [46]). Since the output is not in a full binary form, there is a need for encoding schemes capable of converting the output to more common forms such as decimal or full binary. This conversion may be achieved through the use of a combination of operations such as A/D conversion, shifting, and adding [48]. In connection with optical processors, an all-optical decoding scheme would be very attractive. Also, the output dynamic range requirement increases as the input sequences become longer or as the number of computations increases. However, compared to the case when no coding is used, the dynamic range requirement is lower.

## 17.2 INCOHERENT VERSUS COHERENT REGIME

### 17.2.1 Coherence Effects

In general, both the amplitude and phase of an optical source vary in time in random fashion. Due to the different physical processes involved, the statistical nature of laser light emission is different from that of a thermal light source. This statistical nature, producing correspondingly statistical responses, gives rise to the concept of coherence, which can affect the behavior and performance of an optical system fed by such sources. Furthermore, in fiber-optic signal processing applications, it is of interest to investigate the various signal processing functions that can be performed in different optical coherence regimes (coherent as opposed to incoherent processing).

Coherence effects can be studied by classical or quantum mechanical methods. However, for most cases of interest, the classical approach predicts the outcome of the experiments more easily. The question of coherence arises when optical beams are combined or superimposed. Coherence is related to the ability of two light beams to generate interference fringe patterns upon superposition.

### 17.2.2 Coherence Functions

There are two types of coherence: spatial and temporal. Spatial coherence refers to the case when an optical beam is split into two different paths to arrive at two different points in space. Here, one may need to know how the lightwaves in the two paths are correlated. Similarly, temporal coherence refers to the case when an optical beam is split into two parts which, after being delayed with respect to each other, are recombined. Spatial and temporal coherence can be studied mathematically in terms of coherence functions known as *mutual coherence* (cross-correlation) and *self-coherence* (autocorrelation) functions [52,53]. Since in this chapter temporal coherence is of more significance, we limit our discussion to that case. The self-coherence function of a stationary (complex) optical disturbance $x$ is defined as

$$R_x(\tau) = \langle x(t + \tau)x^*(t)\rangle \tag{17.10}$$

where the angled brackets and the star sign represent the ensemble average and the complex conjugation sign, respectively, and $\tau$ is the time delay. Since the ensembles involved in optics are usually stationary and ergodic, the ensemble averages can be replaced by time averages. (Stationary means that all the ensemble averages are independent of the time origin, while ergodicity implies that ensemble averages can be replaced by their corresponding time averages.) The modulus of the self-coherence function, when normalized with respect to its value at the origin, has a value between zero and one. The value is 1 for a fully coherent light source, and 0 (except at $\tau = 0$) for fully incoherent light. For any value between the above two limits, the light is called partially coherent.

Using the self-coherence function, coherence time has been mathematically defined in a number of different ways. In one instance, the coherence time is defined [53] as the area under the squared modulus of the normalized self-coherence function, while in another case it is defined [52] as the rms width of the squared modulus of the normalized self-coherence function. Apart from any particular definition, there is a general relationship between spectral width $\Delta\nu$ of an optical source and the coherence time $\tau_c$ as follows. The difference in the number of maxima passing with the highest and lowest frequency components of the light beam over a time interval $\Delta t$ is equal to $\Delta\nu\Delta t$. For a time interval $\Delta t \leq \tau_c$, the above-mentioned difference is no greater than a certain number (e.g., unity). In this case, the intensity maxima of all the frequency components in the spectral width will have a virtually fixed phase relationship, implying that $\Delta\nu\tau_c \leq 1$. The above inequality is similar to the uncertainty relationships for photons. In fact, coherence can be related [54] to the accuracy with which one can localize the position of a photon. If the optical system does not include a relative time delay between two light paths, which is greater than the inverse of the source bandwidth, the uncertainty principle predicts that the photons cannot be localized in one path as opposed to the other. Conversely, if time delays are great enough, then it becomes possible to localize the photon in one or another of the paths, resulting in the disappearance of the temporal coherence (incoherent superposition).

### 17.2.3 Linearity

Considering the fact that the source light signal is a stochastic process, in the following we obtain the response of a fiber-optic system, with the baseband complex envelope impulse response $h(t)$, to an input light signal with its complex envelope represented by $\mathbf{E}$. We assume that all the signals are properly normalized and that all the stochastic processes involved are stationary and ergodic. Further, we avoid nonessential proportionality constants. Now, if the input signal is intensity modulated by the signal $m(t)$, the modulated complex envelope $\mathbf{E}_{\text{in}}$ entering the fiber-optic system can be written as

$$\mathbf{E}_{\text{in}}(t) = [1 + m(t)]^{1/2}\mathbf{E}(t) \tag{17.11}$$

(It is assumed that no phase modulation accompanies the amplitude modulation, an assumption which is not generally true.) Since the propagation of the light in the system is governed by Maxwell's equations, the input and output complex envelopes are linearly related through a convolution operation:

$$\mathbf{E}_{\text{out}}(t) = \mathbf{E}_{\text{in}}(t) \otimes h(t) = \int_{-\infty}^{\infty} h(\tau)\mathbf{E}_{\text{in}}(t-\tau)\mathrm{d}\tau \tag{17.12}$$

Assuming that the response of the photodetector at the output is the expected value of the output optical intensity $I_{\text{out}}(t)$, we get

$$<I_{out}(t)> = \int_{-\infty}^{\infty}\int_{-\infty}^{\infty} h(\tau_1)h^*(\tau_2)[1 + m(t - \tau_1)]^{1/2}[1 + m(t - \tau_2)]^{1/2}R_E(\tau_2 - \tau_1)d\tau_1 d\tau_2 \quad (17.13)$$

where $R_E$ is the self-coherence (autocorrelation) function of the input complex envelope as defined in (17.10). As can be seen, (17.13) is generally a nonlinear equation. However, in some special cases it becomes linear. In the following, we consider a few special cases. If the input light is fully coherent ($R_E(\tau) = 1$), we obtain

$$<I_{out}(t)> = |h(t) \otimes [1 + m(t)]^{1/2}|^2 \quad (17.14)$$

implying that, in the fully coherent case, the response of the system to a modulating signal and the modulating signal have a linear relationship on an amplitude basis. In contrast, for fully incoherent input light ($R_E(\tau) = \delta(\tau)$), we have

$$<I_{out}(t)> = \{|h(t)|^2 \otimes [1 + m(t)]\} \quad (17.15)$$

As can be seen in this case, the linearity operation is on signal intensity. This property also holds reasonably well when the coherence time of the input light is much less than the shortest relevant time delay in the system. The foregoing condition must be satisfied in all experimental studies involving incoherent signal processing. Lastly, in the partially coherent case under certain assumptions, (17.13) can be approximated by a linear equation. One such instance is when the duration of $h(\tau)$ is short with respect to the duration of $m(t)$. In this case, the term $[1 + m(t - \tau_1)]^{1/2}[1 + m(t - \tau_2)]^{1/2}$ in (17.13) can be approximated [55] by $1 + 1/2m(t - \tau_1) + 1/2m(t - \tau_2)$ to obtain the linear relationship

$$<I_{out}(t)> \approx h_0 + [h_1(t) \otimes m(t)] \quad (17.16)$$

where $h_0$ is a constant dc term and $h_1(t)$ is given by

$$h_1(t) = \mathrm{Re}\{h(t) \cdot [h^*(t) \otimes R_E(t)]\} \quad (17.17)$$

with Re representing the real part. Note that similar relationships also hold when the functions involved are discrete in time.

In the above discussion, it is implicitly assumed that the guided optical power densities are low enough to avoid the occurrence of nonlinear processes in the fiber. However, at high power densities, Raman and Brillouin scattering processes can be stimulated within fiber, resulting in loss mechanisms that are dependent on the optical power. Forward Raman scattering introduces a frequency down-shift in the transmitted radiation. On the other hand, backward Raman or Brillouin processes can attenuate the forward traveling wave significantly. Under reasonable assumptions [56], the critical optical powers beyond which the various stimulated processes become important in single-mode fibers, operated in the continuous wave mode, are approximately given by

$$P_{\text{crit}} \approx \kappa(\alpha A/\gamma_0) \tag{17.18}$$

where $\alpha$ is the attenuation constant ($\text{cm}^{-1}$), $A$ is the cross-sectional area ($\text{cm}^2$) of the fiber core, and $\gamma_0$ is the gain coefficient (cm/W) of the nonlinear process. The parameter $\kappa$ is a constant whose value is determined by the type of the nonlinear process. It is approximately equal to 16, 20, or 21 for the forward Raman, backward Raman, and backward Brillouin scattering processes, respectively. For a single-mode fiber with $A \approx 10^{-7}$ $\text{cm}^2$ and $\alpha = 1.15 \times 10^{-5}$ $\text{cm}^{-1}$ (5 dB/km), at $\lambda = 0.83$ $\mu$m and with gain constants of $3 \times 10^{-11}$ cm/W and $3 \times 10^{-9}$ cm/W for the Raman and Brillouin processes respectively, we obtain critical optical powers of approximately 0.6W for the forward Raman, 0.8W for the backward Raman, and 8 mW for the backward Brillouin processes (note that forward Brillouin scattering does not occur in the fiber geometry). As can be seen, the maximum power that can be launched is limited by the backward stimulated Brillouin scattering, which is usually the case due to the fact that the Brillouin gain constant is larger than that of the Raman process in glass fibers. Equation (17.18) holds approximately as long as the linewidth of the source is less than the linewidth of the spontaneous scattering of concern. In the Raman case, the linewidth is large enough to ensure the validity of the above equation for most laser sources. However, for Brillouin scattering it is usually necessary to consider the input power within any bandwidth equal to that of the spontaneous scattering [56]. In case of a very broadband source, the stimulated Raman process may set the upper power limit. Moreover, when relatively short pulses are fed into the fiber, the above critical powers are further increased. It should be noted that all fiber-optic linear filtering experiments must be operated with power levels in a range where the nonlinear processes mentioned can be neglected.

### 17.2.4 Incoherent Versus Coherent Processing

In an incoherent optical filter, signals to be processed are modulated as intensity variations on optical carriers whose coherence time is less than the shortest relevant time delay in the system. This restriction is to avoid environmentally sensitive coherent optical interference effects. As a result, an incoherent fiber-optic filter is linear in the intensities of the guided optical signals, quantities which are nonnegative in value. This positivity property of incoherent fiber-optic system affects the frequency filtering characteristics of such systems. A more detailed study of this subject is presented in later sections.

In principle, fiber-optic signal processing can be accomplished coherently (as opposed to incoherent methods). In this case, laser coherence is used to carry data which may be modulated in amplitude, frequency, or phase on a coherent optical carrier. The demodulation process can be performed by using optical heterodyne or homodyne detection systems [57]. These detection schemes are sensitive to phase, frequency, polarization states, or wavefront mismatch between the signal and detection reference waves. The optical heterodyne detection converts the carrier frequency in the optical region to an IF that is much lower. This IF signal is then used to demodulate the data. In the homodyne configuration, an alternative approach is an optical phased-locked-loop device, which works on the same concept as its electronic counterpart. While in optical fiber communications systems, coherent transmission is of interest because of the expected performance improvement toward long repeater spacing and large information carrying capacity [57], in signal processing, coherent methods have potential advantages such as bipolar and complex valued data representation by physical quantities. A disadvantage is that environmental effects such as temporal changes of temperature, mechanical vibrations, and stress might influence the system response by randomly varying the time delays and optical phases involved.

Except for experimental demonstrations of two-tap fiber transversal filters [12,58,59] and four-tap and eight-tap integrated-optic transversal filters by Nippon Telegraph and Telephone (NTT) [60], not much work has been reported in the area of coherent fiber-optic signal processing. In the former work, tapping is achieved by attaching external piezoelectric phase modulators to a single-mode fiber along its length. It is claimed that this filter, in addition to its capability of electronic control of taps, has increased dynamic range compared to incoherent fiber-optic transversal filters. In the NTT work to improve temperature stability, short delay lengths (4 cm) were fabricated in silica-based single-mode waveguides monolithically integrated with thermally tunable splitters and phase shifters on a silicon substrate. In order to gain further insight into its potential, coherent fiber-optic signal processing demands more attention.

## 17.3 FIBER FILTER ANALYSIS

This section contains some basic background required for better understanding of the forthcoming materials which are concerned with the processing of signals using fiber-optic technology. The application of the Z-transform to fiber-optic filters is introduced. The following subsections cover a brief discussion of Z-transform techniques. With the Z-transform, we deal with functions of $z$.

### 17.3.1 Z-Transform

In order to increase our ability to handle more complex systems problems, we use transform techniques for describing the signals and systems, particularly the Z-transform [28]. This usage is justifiable, since all the systems of interest in this chapter are, first, linear and time-invariant and, second, discrete in time. The latter is due to the fact that we can define a basic time delay T, such that any other relevant system delays are integer multiples of this time delay. That is, the impulse response (response to an input impulse) of the system is a series of impulses equally spaced in time, as shown in Figure 17.6. As a result, we can simplify the mathematical analysis of such systems by considering the values of the

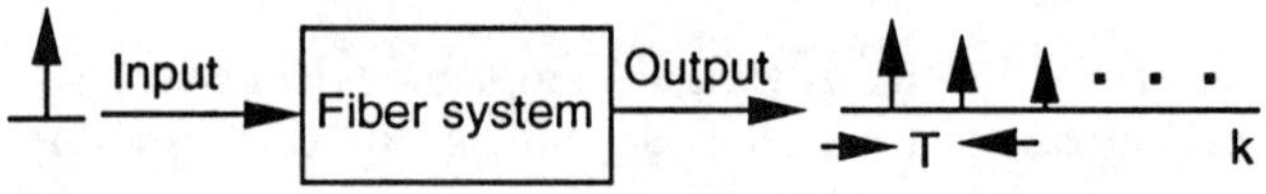


**Figure 17.6** Impulse response of a linear and time-invariant discrete time system with a basic time delay $T$; $k$ is the time index.

system signals only at discrete instants. Since the input and output are described in terms of discrete samples, these systems are called *sampled data systems* [28].

The mathematical analysis of discrete systems is essentially similar to that of continuous data systems, but it is simpler, in addition to having more intuitive physical interpretations.

The Z-transform $F(z)$ of a signal $f(k)$ is defined by the following expression:

$$F(z) = \sum_{k=-\infty}^{\infty} f(k)z^{-k} \tag{17.19}$$

where $k$ is the time index, which is an integer multiple of the basic time delay ($k = nT$), and $z$ is the transform variable, which represents a unit time advance (in contrast to $z^{-1}$, which represents a unit time delay).

We also describe the input-output relationship with a system transfer function $H(z)$, defined as the Z-transform of the impulse response $h(k)$, which can usually be expressed as a ratio of two polynomials in $z$ as shown below:

$$H(z) = Z\{h(k)\} = \frac{\prod_{i=1}^{m}(z - z_i)}{\prod_{i=1}^{n}(z - p_i)} \tag{17.20}$$

The roots of the numerator ($z_i$) and denominator ($p_i$) are referred to as zeros and poles, respectively. The locations of the poles and zeros in the complex $z$ plane are shown by small crosses and circles (see Fig. 17.7). Pole-zero locations are very important in the design and analysis of frequency-selective filters. The frequency response is obtained by evaluating the transfer function at $z = e^{j\omega T}$, which describes the unit circle (centered at the origin) in the $z$ plane, as illustrated in Figure 17.7. The frequency response can also be visualized graphically from a plot of pole-zero locations.

### 17.3.2 Stable and Minimum Phase Filters

The unit circle also plays an important role in problems regarding stability. To ensure stability, all the poles of the system transfer function must be inside the unit circle. Fiber lattice filters incorporating elements without gain are sure to be stable. Furthermore, if

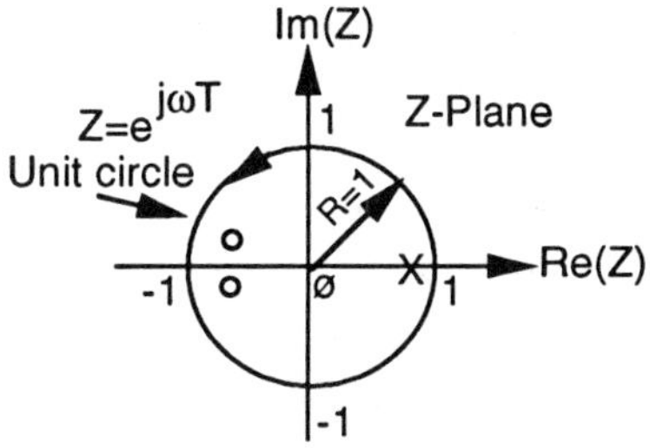


**Figure 17.7** Z-plane with the unit circle. The locations of the poles and zeros are shown by small crosses and circles, respectively.

all the zeros are also inside the unit circle, the system is called a *minimum phase* system, while if any zero lies outside the unit circle, the system is called a *nonminimum phase* system. It should be mentioned that, for a minimum phase filter, the phase response and the logarithm of the magnitude response are a Hilbert transform pair. This is in addition to the property of all real stable causal systems that the real and imaginary parts of the frequency response are Hilbert transformations of each other. Oppenheim and Schafer [28] give a more complete description of the Z-transform theory.

### 17.3.3 Block Diagram Representation and Mason's Rule

A block diagram representation can be used to describe a system composed of several typically elementary components interconnected in a prescribed fashion [61]. Using block diagrams, system transfer functions can be obtained graphically. This approach can sometimes be easier and provide more insight into the system than the alternative approach of manipulating algebraic equations in an abstract way. A block diagram can be easily translated into a signal-flow graph, which is a set of interconnected nodes representing the system signals. The nodes are connected by branches that show the signal flow in the system. The gain of a branch is defined as the transmittance between its two associated nodes. A path is defined as a continuous sequence of several unrepeated branches, with its direction shown by arrows. Two paths with a common component are said to touch. A forward path is a path from input to output with no node included more than once. A loop is a closed path returning to the original node with no branch repeated. The path gain and the loop gain are, respectively, the products of gains (transmittances) in a path or a loop. Using the preceding terminology, Mason's rule states that the system transfer function can be obtained from the following relationship

$$H(z) = \frac{Y(z)}{X(z)} = \frac{1}{\Delta}\sum_i G_i \Delta_i \tag{17.21a}$$

where $Y(z)$ is the Z-transform of the output signal, $X(z)$ the Z-transform of the input signal, $G_i$ the path gain of $i$th forward path, $\Delta = 1 -$ {sum of all individual loop gains minus sum of gain products of all possible nontouching two loops plus sum of gain products of all possible nontouching three loops minus . . . }, and $\Delta_i$ = {$\Delta$ evaluated for the part of signal-flow graph that does not touch the $i$th forward path}.

For the special case where all forward paths and loop paths touch, $\Delta = 1 -$ {sum of all individual loop gains} and $\Delta_i = 1$. In this case, Mason's rule is simplified to

$$H(z) = \frac{1}{\Delta}\sum_i G_i \tag{17.21b}$$

which simply states that the gain of a feedback system is equal to the sum of forward path gains divided by 1 minus the sum of loop gains.

Mason's rule is useful in calculating the system transfer function by hand without resorting to any computer. It can also be useful in getting insight into the effects of optical amplifier gains on the stability of fiber-optic filters. Applications of Z-transform, block diagram representation, and Mason's rule in fiber-optic filters are demonstrated in Section 17.3.5, after the next section which introduces the elementary signal processing functions required in realizing fiber-optic filters.

### 17.3.4 Elementary Signal Processing Functions

To perform the various signal processing operations introduced in Section 17.1, we need to employ, in general, time-delay, time-advance, tapping, summing, and branching elements. These basic elements are described in the following subsections.

#### *Time Delay*

The output of the delay element is the input delayed by a unit of time. That is, for a discrete time input signal $f(k)$, the output is $f(k-1)$, where $k$ represents the time index. A block diagram representation and the impulse response (response to an intensity impulse input) of this element are shown in Figure 17.8(a). Note that the $z^{-1}$ inside the block diagram is a Z-transform notation representing a unit time delay. Fiber lengths provide precise time delays that can be employed to perform the time-delay function. Assuming a refractive index of about 1.5 for the fiber, the propagation delay is ~5 ns/meter.

#### *Time Advance*

In this case, the signal is advanced one unit of time rather than delayed, as depicted in Figure 17.8(b). Note that this element has only mathematical meaning, and that systems requiring unit advance subsystems in their construction are not physically realizable.

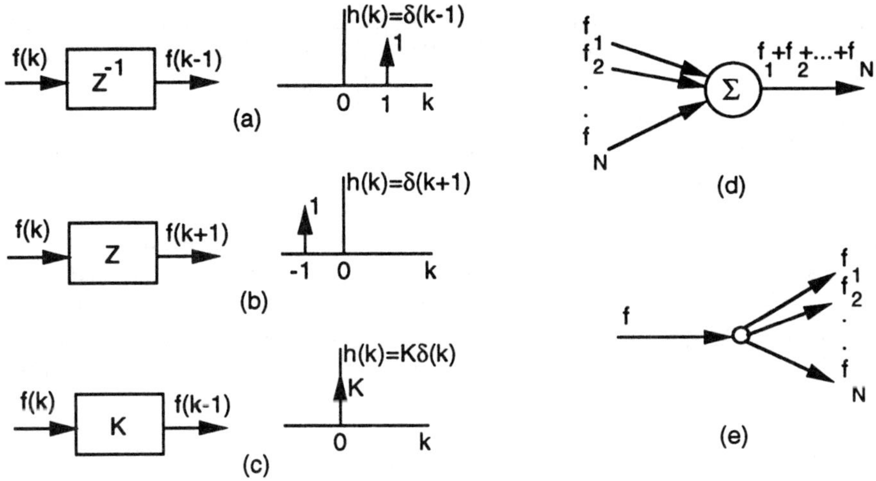


**Figure 17.8** Block diagram representations and the intensity impulse responses of: (a) time-delay function; (b) time-advance function; (c) tapping function ($K$ is the tapping factor); (d) summing element; (e) branching element.

*Tapping*

Various mechanisms, such as evanescent directional couplers/splitters or bends (''kinks''), can be used to accomplish tapping. Mathematically, the output of the tapping element is a constant fraction $K$ (tapping factor) of the input (see Fig. 17.8(c)). The impulse response in this case consists of an impulse of height $K$ at the origin.

*Summing*

This subsystem sums incoming signals (Fig. 17.8(d)). In fiber-optic systems, a directional coupler or a series of directional couplers can be used to do the summing operation.

*Branching*

This subsystem divides the incoming signal into several parts (Fig. 17.8(e)). A series of directional couplers can be used for this purpose.

### 17.3.5 Example: Fiber-Optic Recirculating Delay Line Filter

In this section, we demonstrate applications of the Z-transform, block diagram representation, and Mason's rule introduced in the previous sections in fiber-optic filters. Figure

17.9 demonstrates the component layout of the single-input single-output (SISO) unamplified fiber-optic recirculating delay line (UFORDL) and amplified fiber-optic recirculating delay line (AFORDL) [22,23]. As shown, one of the output leads of a $2 \times 2$ directional coupler is looped around and connected to one of the input leads. The inverse of the loop time delay $T$ is equal to the sampling frequency of the filter. The AFORDL contains an optical amplifier with power gain $G$ in the loop. The coupler's coupling ratio $K$ can assume any value between 0 and 1. Without loss of generality, it is assumed that the coupler is symmetric and the filter's operation is polarization-independent.

Figure 17.10 depicts a block diagram representation of the UFORDL and AFORDL of Figure 17.9 using the Z-transform approach. When $G = 1$, the block diagram will correspond to the UFORDL. This block diagram representation reveals the details of the structure and can be very useful in the design and analysis of the filter, its different

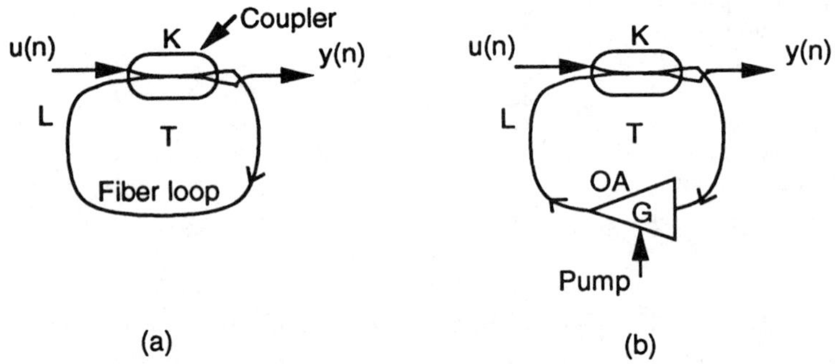


**Figure 17.9** Fiber-optic recirculating delay line layout: (a) passive UFORDL; (b) active AFORDL. $T$ is the loop time delay, $L$ the loop intensity transmittance, $K$ the coupler's coupling ratio, $u(n)$ the input intensity with the time index $n$, $y(n)$ the output intensity, and OA the optical amplifier with gain $G$.

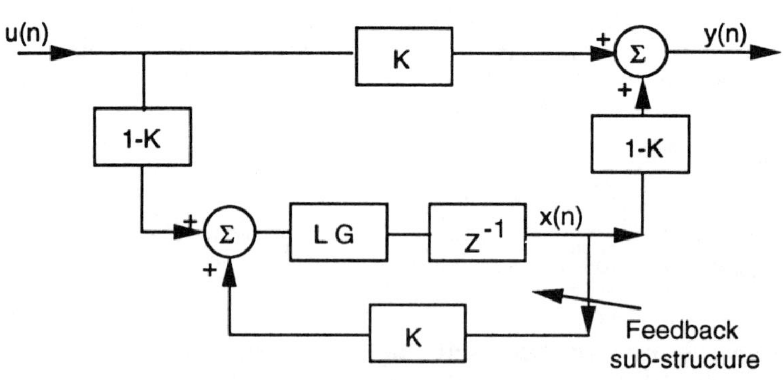


**Figure 17.10** Block diagram representation of the FORDL structures of Figure 17.9. $Z^{-1}$ represents a unit delay in the Z-transformation, and $x(n)$ is the internal state of the filter in a state space model.

variations, or its combination with other filters. Multiple loop recirculations have been taken into account in the model of Figure 17.10. The input is connected to the output via two feed-forward paths whose outputs are summed to produce the output. One of these paths goes through a simple constant scaling equal to the coupling ratio $K$, while the other path touches a feedback loop with the feedback value equal to $K$. As will be seen later, the location of the poles and the stability state are set by the feedback loop on the second feed-forward path, which includes the amplifier gain $G$ in the transfer function of the path.

A direct but somewhat more difficult approach is to apply the general case of Mason's rule. It is easier, however, to break down the block diagram of Figure 17.10 into two smaller and simpler block diagrams and apply the special case of Mason's rule. The two block diagrams, which are connected in parallel, are $H_1(Z)$, the top forward path, and $H_2(Z)$, consisting of the lower forward path and feedback loop. We then add the results to get the total transfer function $H(Z)$:

$$H_2(z) = \frac{1}{\Delta}\sum_i G_i = \frac{(1-K)^2 LGZ^{-1}}{1-KLGZ^{-1}} \tag{17.22a}$$

$$H(z) = H_1(z) + H_2(z) = K + \frac{(1-K)^2 LGZ^{-1}}{1-KLGZ^{-1}} = K\frac{Z - LG(2K-1)/K}{Z - LGK} \tag{17.22b}$$

The transfer function in (17.22b) corresponds to the difference equation between the input $u$ and the output $y$ described by $y(n) = Ku(n) + LG(1-2K)u(n-1) + LGKy(n-1)$, and has one zero, $Z_r$, and one pole, $P_l$, given by the following relations

$$Z_r = LG(2K-1)/K \tag{17.23a}$$

$$P_l = LGK \tag{17.23b}$$

Note that the pole value given by (17.23b) is equal to the loop net gain, which we will denote with $g$. As will be shown, extra delay line functions are realizable without requiring a unity loop gain.

In the following discussion, we explore the possibility of simultaneously placing the pole close to the unit circle ($P_l \rightarrow 1$) and the zero at $-1$ on the unit circle ($Z_r = -1$). A filter with a zero at $-1$ and a pole near $+1$ can be used to suppress one signal and pass another (RF or microwave interference rejection). We consider both the passive UFORDL and active AFORDL. For simplicity, and without loss of generality, we assume that $L = 1$ and the amplifier gain $G$ will then be the excess gain (in the general case, $LG$ is the excess gain). We will show that the active AFORDL structure is capable of realizing all-fiber filters not possible with the passive UFORDL. This result is significant because it shows that optical amplifiers can function beyond just the trivial optical loss compensation to provide extra flexibility not available in passive designs. An AFORDL design is presented which has a pole near +1 and a zero at −1 in the $Z$ plane, resulting in useful characteristics in the magnitude and phase responses. The results suggest that extension of the concept to higher order filters has the potential to lead to the realization of more complex optical processors.

*Case 1: Passive UFORDL.* In this case, $G = L = 1$, and $g = K$. Therefore, from (17.23a) and (17.23b), we find

$$Z_r = (2K - 1)/K \tag{17.24a}$$

$$P_l = K \tag{17.24b}$$

From (17.24a) and (17.24b), it is clear that it is impossible to satisfy the desired pole-zero placement. To place the pole near +1, the coupling ratio must be increased to approach unity. This, in turn, will yield a zero approaching unity, which cancels the pole. The resulting transfer function will also approach unity (constant) as expected. In other words,

$$P_l = g \to 1,\ K \to 1,\ Z_r \to 1,\ H(Z) \to 1 \tag{17.25}$$

The impulse response, pole-zero plot, and frequency response are shown in Figure 17.11.

To place the zero at −1, the coupling ratio must be set to the value 1/3. This, in turn, will yield a pole at 1/3. The resulting transfer function will not be constant. In other words,

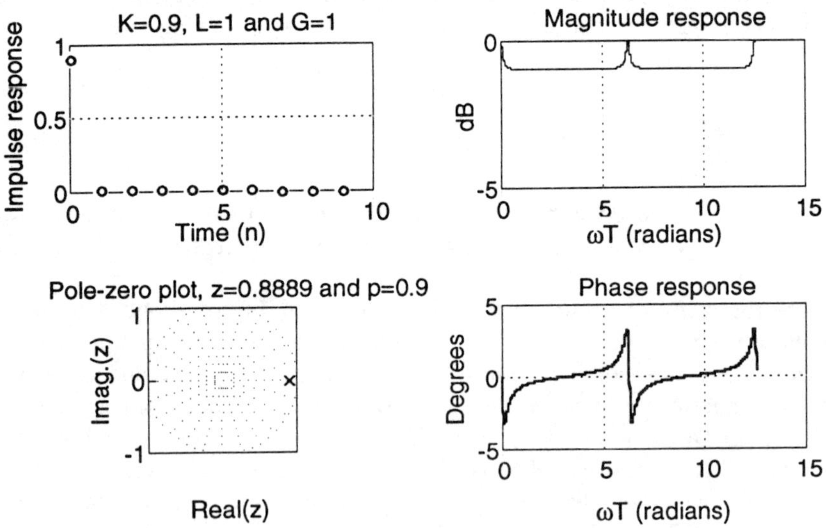


**Figure 17.11** Impulse response, pole-zero plot, magnitude response, and phase response for the passive UFORDL with the pole approaching +1. The loop net gain is $g = 0.9$.

$$Z_r \to -1,\ K \to 1/3,\ P_l = g \to 1/3,\ H(Z) \to (1/3)[Z + 1]/[Z - 1/3] \qquad (17.26)$$

The impulse response, pole-zero plot, and frequency response are shown in Figure 17.12.

*Case 2: Active AFORDL.* In this case, with $G > 1$ and $L = 1$, and from (17.23a) and (17.23b) we obtain

$$Z_r = G(2K - 1)/K \qquad (17.27a)$$

$$P_l = g = GK \qquad (17.27b)$$

From (17.27b), to place the pole close to +1, we get

$$P_l = g \to 1,\ G \to 1/K \qquad (17.28a)$$

$$Z_r \to -(1 - 2K)/K^2 \qquad (17.28b)$$

And from (17.27a), to place the zero at −1, we get

$$Z_r \to -1,\ G = K/(1 - 2K) \qquad (17.29a)$$

$$P_l = g \to K^2/(1 - 2K) \qquad (17.29b)$$

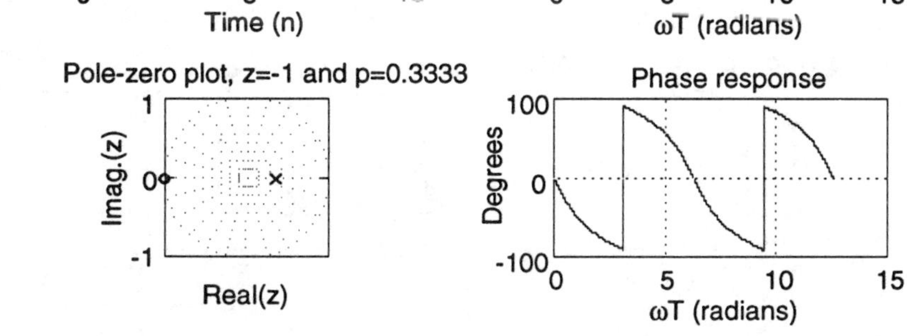


**Figure 17.12** Impulse response, pole-zero plot, magnitude response, and phase response for the passive UFORDL with the zero approaching −1. The loop net gain is $g = 0.3333$.

The relations for the gains in (17.28a) and (17.29a) have a single common solution, for which the values for the coupling ratio, amplifier gain, loop net gain, and the resulting transfer function are given by the following relations:

$$K \rightarrow \sqrt{2} - 1 \tag{17.30a}$$

$$G \rightarrow \sqrt{2} + 1 \tag{17.30b}$$

$$g = KG \rightarrow 1 \tag{17.30c}$$

$$H(Z) \rightarrow (\sqrt{2} - 1)[Z + 1]/[Z - 1] \tag{17.30d}$$

The impulse response, pole-zero plot, and frequency response are shown in Figure 17.13. The loop net gain in Figure 17.13 is $g = 0.988$. This value corresponds to a finesse of over 250, which is practical. Excellent long-term, pump power, and thermal stability can be achieved by the insertion of proper control circuitry and closed-loop dynamic gain adjustment. The use of exact extreme values shown in (17.30) would yield oscillatory characteristics. For stable behavior, the values of the coupling ratio and amplifier gain should be chosen to satisfy the relation $g < 1$. In the case of $g = 1$, the filter becomes marginally stable and is an RF oscillator which can be useful for certain applications to replace their electronic counterparts. In the case of $g > 1$, the filter becomes unstable (and unusable).

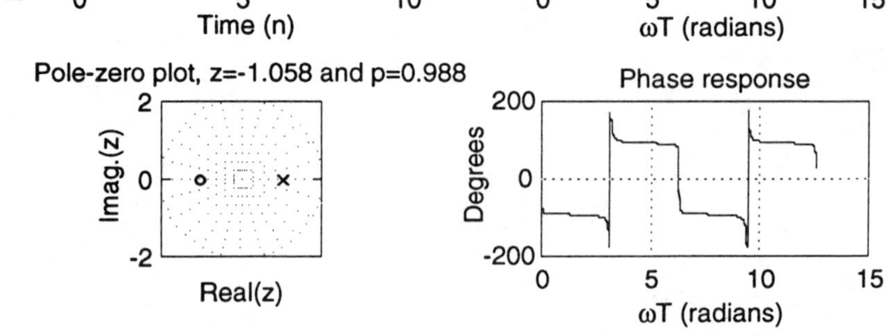


**Figure 17.13** Impulse response, pole-zero plot, magnitude response, and phase response for the active AFORDL, with the pole approaching +1 and the zero at −1.

## 17.4 OPTICAL AMPLIFIERS IN FIBER-OPTIC SIGNAL PROCESSING

Recent advances in the manufacturing of high-quality optical amplifiers and their subsequent commercialization by several manufacturers in Europe and the U.S. in response to the needs of communications systems have created a growing interest in the application of such amplifiers in optical signal processing and fiber-optic filters [22–26,62]. These amplifiers have been used in three different ways: (1) optical loss compensation [22–24], (2) design of unique filters [22,23], and (3) realization of tunable filters [24–26]. In this section, we briefly review the motivation for optical amplification, amplifier types, and characteristics and the effects of amplifier length and gain on filtering response.

### 17.4.1 Motivations for Optical Amplification

Most useful optical filters are characterized by high optical losses. Inherent sources of loss originate from fundamental limitations imposed by the architectures, waveguiding principles, and materials properties, particularly those arising from the combining/splitting of optical signals, mode-mismatched coupling joints, and material loss. In contrast, noninherent excess sources of loss, which are not fundamental in nature, include excess losses associated with the splitters/combiners, connectors, splices, and fiber-amplifier interfaces. Optical loss can result in degraded SNR. Shot-noise-limited operation is the preferred mode of operation. Signal amplification can compensate for the loss and improve the SNR if the noise sources external to the amplifier are originated after the amplifier. Any noise power within the amplifier bandwidth generated before the amplifier will experience the same gain as the signal power. Optical amplifiers provide a convenient way to amplify optical signals without the need for optoelectronic conversion, which is very limiting in both modulation bandwidth and multichannel amplification capabilities. Other issues such as efficiency and cost may also play some role in certain applications.

SOAs and EDFAs [4,5] are now commercially available. Traveling-wave SOAs are devices very similar to laser diodes, with the exception that their end-facet reflectivities have been suppressed, and thus there is no resonant laser cavity. EDFAs use a single-mode optical fiber with a core doped with erbium. By optically pumping this doped core with a pump laser, amplification can be obtained. SOAs have the advantages of small size and the possibility of being integrated with other SOAs, passive waveguide sections, and even drive electronics in a single monolithic device. A major difficulty is associated with the interfaces between fibers and SOAs. Because of difference in the mode sizes of optical fibers and SOAs, lenses are required to maximize the coupling efficiency. This fact, and the associated submicron alignment tolerances, dictates specialized packaging and results in coupling losses of over 3 dB per facet. EDFAs have the advantages of fiber compatibility (fiber coupling simply by fusion splicing), polarization insensitivity, low crosstalk, high saturation power, and low temperature sensitivity. The four major characteristic amplifier curves for an EDFA, gain spectrum, gain saturation, gain-pump current, and ASE spectra, are shown in Figure 17.14.

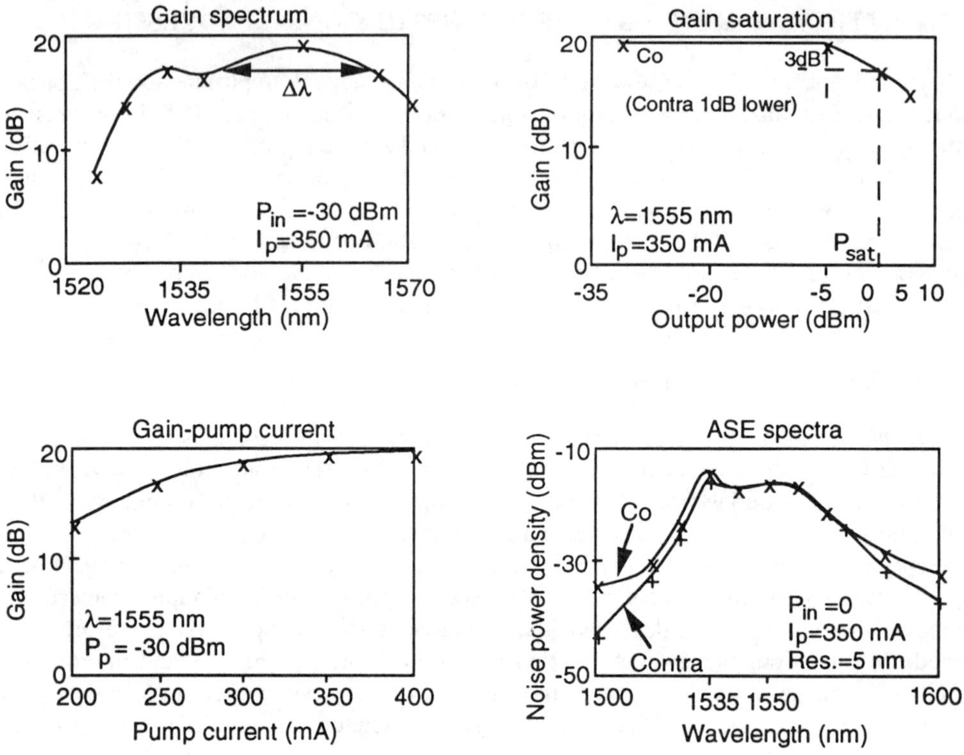


**Figure 17.14** Characteristic curves of an EDFA (values taken from the BT&D optical amplifier model EFA3000). Codirectional and contradirectional pumping are denoted by Co and Contra, respectively.

An ideal optical amplifier would faithfully reproduce its input optical signal in amplified form. However, real optical amplifiers are characterized by their associated nonlinearities, noise, and polarization sensitivity, which can affect the performance (including SNR, dynamic range, linearity, and distortion) of signal processing systems. We will briefly discuss these issues below for both SOAs and EDFAs.

### 17.4.2 Optical Amplifier Characteristics

The performance of optical signal processing systems incorporating optical amplifiers is affected by nonlinearity, noise, coupling losses, polarization sensitivity, and speed characteristics of amplifiers. These characteristics are briefly discussed in the following.

*Nonlinear Characteristics*

In both SOAs and EDFAs, the depletion of the excited state population results in decreasing gain as the incident optical intensity is increased. The output intensity, where the gain is substantially depleted, is characterized by the saturation intensity $I_{sat}$. This intensity is related to the saturation power, $P_{sat} = I_{sat} A_m$, where $A_m$ is the effective area of the optical mode in the amplifier. In double-heterostructure SOAs, $P_{sat}$ is typically 5 to 9 dBm. However, quantum-well SOAs have been demonstrated with $P_{sat}$ approaching 20 dBm, which is accomplished largely by increasing $A_m$. In EDFAs, $P_{sat}$ is pump-power-dependent, but is typically on the order of 20 dBm.

In addition to the saturation characteristics in the presence of unmodulated (dc) optical signals, there are also dynamic saturation effects. At output signal powers on the order of the saturation power or less, the gain can respond to a change in the signal power in a time of approximately the spontaneous lifetime $t_{sp}$. This is a major difference between the behavior of SOAs and EDFAs. Whereas $t_{sp}$ in SOAs is usually in the range of 1 to 10 ns, in EDFAs $t_{sp}$ is 10 to 15 ms. This difference of six to seven orders of magnitude in gain response time means that while SOAs may exhibit substantial nonlinearity in the presence of signals under about a gigahertz, EDFAs do not. Instead, they exhibit gain which is linear at the high frequency, but which can vary slowly with the average power over times on the order of 10 ms (100 Hz). Amplifier nonlinearity can generate harmonics (particularly the second) and intermodulation (particularly the primary) distortion components. These effects, which might impose a limit on the maximum amplifier output power and maximum modulation index, need to be considered in the design and realization of optical filters incorporating optical amplifiers. For maximum dynamic range, the amplifier is typically biased to the midpoint of its saturation power.

*Noise*

The stimulated emission responsible for the gain in optical amplifiers is accompanied by spontaneous emission. The spontaneous emission covers the entire gain bandwidth (20 to 50 nm) of the optical amplifier and is amplified to produce amplified spontaneous emission (ASE) power. ASE has two effects. First, it can contribute to gain saturation; this is particularly true of high-gain amplifiers or cascaded amplifiers. The second effect of spontaneous emission is its contribution to noise. Because of the square-law response of standard photodetectors, in addition to the thermal noise and shot noise (from signal and ASE) terms, spontaneous-spontaneous (sp-sp) beat noise and signal-spontaneous (s-sp) beat noise terms are also produced at the detector. The various noise components can be calculated from [63]

$$I^2_{\text{thermal}} = i^2_{\text{thermal}} B_e \tag{17.31a}$$

$$I^2_{s,\text{shot}} = 2eI_s B_e \tag{17.31b}$$

$$I^2_{\text{shot,ASE}} = 2eI_{\text{ASE}}B_e \tag{17.31c}$$

$$I^2_{\text{s-sp}} = 4I_{\text{ASE}}I_sB_e/mB_o \tag{17.31d}$$

$$I^2_{\text{sp-sp}} = 2I^2_{\text{ASE}}B_e/mB_o \tag{17.31d}$$

where the output signal and ASE currents are given by

$$I_s = RGP_s = (e\eta\lambda_s/hc)GP_s \tag{17.31e}$$

$$I_{\text{ASE}} = RP_{\text{ASE}} = Rh\nu mn_{\text{sp}}(G-1)B_o = e\eta mn_{\text{sp}}(G-1)B_o \tag{17.31f}$$

In the above relations, $\eta$ is the detector quantum efficiency, $\nu$ the optical frequency, $B_e$ the photodetection electrical bandwidth, $B_o$ the optical detection bandwidth, $G$ the amplifier gain, $m$ the number of amplifier modes, and $n_{\text{sp}}$ is the spontaneous emission factor (with a typical value in the range of 1 to 2). $n_{\text{sp}}$ increases rapidly for $\lambda < 1{,}520$ nm.

Another important amplifier parameter is the noise figure (NF), defined as the ratio of the SNR at the input for an ideal coherent source to the SNR at the output of the device. It can be shown that in the signal-spontaneous beat noise limit $\text{NF} = 2n_{\text{sp}}\chi c_{\text{in}}$, where $\chi$ is the excess noise factor and $c_{\text{in}}$ is the coupling efficiency. The excess noise factor is defined as $\chi = (1 - RG)/(1 - R)$, where $R$ is the fiber-end reflectivity, and $G$ is the gain. In an ideal fiber amplifier, $R = 0$, $c_{\text{in}} = 1$, $\chi = 1$, and $n_{\text{sp}} = 1$, which result in $\text{NF}_{\text{ideal}} = 2$ (= 3 dB). In practice, however, $n_{\text{sp}}$ is larger than unity, and input coupling losses are incurred in couplers, connectors, and splices. For example, for the BT&D EFA3000-1550 amplifier, NF = 7 dB at a 1,550-nm signal wavelength and maximum pump power (400-mA pump current).

An important factor to be considered in incoherent optical filters is the effect of optical source coherence length. The coherence length should be much less than the filter's unit delay in order to suppress coherence effects and environmental instabilities. Shorter coherence length is equivalent to wider field spectrum. This in turn requires a broader output optical filter at the amplifier output, which increases the noise power (trade-off between the coherence and SNR).

In general, the SNR can improve or degrade, depending on the location of amplifiers and noise sources. In direct detection systems, SNR can be improved by the proper insertion of amplifiers. In the case of a strong signal and/or the use of optical filtering, the signal-spontaneous beat noise will often be the dominant noise term. Since higher pumping energies result in a higher inverted population, the noise performance of amplifiers improves at higher pumping power.

*Coupling Losses*

Coupling from a signal fiber to an EDFA is relatively trivial; a standard fiber splicing technique (such as fusion) can be used to yield a very-low-loss and low-reflection connection. The size of the mode in an EDFA is usually somewhat smaller than that in a standard 1.55-$\mu$m single-mode fiber, but a coupling loss of less than 1 dB can be readily achieved. The mode in an SOA, however, is highly mismatched to the mode in a single-mode fiber, and very high loss results if a fiber is simply butt-coupled to the SOA. To achieve low-loss coupling (2 to 4 dB), some type of lens coupling must be used. Here, alignment tolerances are tight (<1 $\mu$m) and the lens and fiber interfaces can result in reflections back into the SOA, necessitating the use of AR coatings. These various practical factors lead to highly demanding packaging and assembly techniques.

The net effect of these considerations is that while coupling fibers to an EDFA is relatively easy, low-loss coupling to SOAs is involved and more inefficient. However, in more complicated systems, such as the high-order filters, multiple optical amplifiers may be required. In this case, SOAs might be more attractive than EDFAs, because integration of multiple SOAs with passive waveguides and even drive electronics is possible. Such integrated assemblies can reduce the number of fiber-to-SOA interfaces required.

*Polarization Sensitivity*

EDFAs exhibit gain which is independent of the polarization of the input light. Such behavior is ideal for use in a system which employs standard circular core fiber. SOAs generally employ an active region which is much thinner than it is wide. A consequence of such a geometry is that the TE mode, which is better confined to the active layer, experiences more gain per unit length than the TM mode. This leads to a gain which is dependent on the polarization of the light input to the amplifier. If circular core fiber is used at the SOA input, the light launched into the amplifier will be in an arbitrary elliptical state of polarization (SOP). This SOP is subject to variations due to changes in temperature and position of the input fiber, and the gain obtained from the SOA will vary. This is unacceptable in a system where we require the gain to be precise in order to implement a desired weight value. Many of the SOAs that have been demonstrated exhibit a gain differential of 5 dB or more between the TE and TM polarizations. More recently, specially designed devices have been demonstrated with negligible gain differential between polarizations.

*Switching Speed*

In applications where rapid change of the filter transfer characteristic is desired, the values of the weights must be varied rapidly. In cases where an optical amplifier serves the dual role of providing a variable weight and providing gain, the gain of the amplifier must be varied in order to effect a desired change in the filter transfer function. For optical amplifiers operating in a regime of limited saturation, the gain of the amplifier can be varied in a time on the order of the spontaneous lifetime $t_{sp}$ in the material. For SOAs, $t_{sp}$ is usually in the range of 1 to 10 ns. In sharp contrast, the spontaneous lifetime in EDFAs is approximately 10 to 15 ms. Thus, in applications where fast changes in the filter transfer characteristic are required, the use of EDFAs is limited to providing gain. The variation of the weight in such a system can be accomplished through the use of some other type of fast passive shutter. EDFAs and liquid crystal shutters have been used for low-speed tuning of filter weights [25,26]. A tunable filter can be used in a variety of applications, including encoding/decoding [25] and adaptive channel equalization [64].

### 17.4.3 Effect of EDFA Length on Filter Design

The EDFA length is a device parameter in the optical amplifier design and optimization. For a given pump power and wavelength, by optimizing the length of the EDFA, the amplifier gain characteristics can be adjusted for either maximum gain (on the long-wavelength peak at 1,555 nm) or better optical spectral flatness across the 1,530- to 1,565-nm band. Another effect of the length is on the frequency response of fiber filters. Typically, EDFAs have a length in the range of 10m to 40m. A 20m length will result in a $T$ = 100-ns time delay. If this is used as a unit delay, the filtering sampling rate will be limited to only 10 MHz (= $1/T$). But realizing the fact that, in FIR filters, it is the differential delays that determine the filtering speed and not the absolute delay values, we can insert optical amplifiers in all delay branches with a small differential delay to obtain high speeds (e.g., 20m, 21m, 22m, 23m, ... delays). This, however, will increase the latency by an amount equal to $T$. In the IIR filter case, the situation can be different. A fiber-optic recirculating delay line (FORDL) filter with an EDFA in the loop is limited to a relatively low sampling frequency set by the inverse of the loop length. This FORDL, however, can be attractive as a transient buffer memory with a long storage time [62].

### 17.4.4 Effect of Gain on Stability

Amplifiers can be placed at any position in the system. The criteria for the placement can be the trade-off between SNR considerations and saturation, as well as stability. From an SNR standpoint, it would be preferable to place all of the gain as early as possible in the signal path; an amplifier placed before the first splitter would yield optimal SNR. However, since the signal power is the highest there, such an amplifier placement is very likely to lead to substantial amplifier saturation. A trade-off between saturation and SNR is obtained by distributing the gain throughout the system. Following each loss, an amplifier is placed in the system which compensates for the loss. Though the use of distributed gain yields the best SNR performance, it is generally impractical due to the large number of amplifiers required. As monolithic devices with many SOAs and passive waveguide sections become available, distributed gain systems will be more practical. A different approach and objective for the amplifier placement can be to identify those locations where the realization of unique filters becomes possible.

In IIR filters, the output signal is fed back through the system. In such systems, amplifiers can play a role more significant than in FIR filters. In any linear time-invariant (LTI) filter, the locations of the poles and zeros of the transfer function determine the filtering characteristics. These locations are, in general, functions of variables and parameters such as optical losses, optical amplifier gains, and the couplers' (or splitters') coupling ratios. In most useful IIR filters, the pole should be placed inside and close to the unit circle in the $Z$ plane to allow large responses at or near the frequencies corresponding to the poles. This implies slowly decaying signal recirculations through the filter. The stability considerations require that the impulse response decay to zero with time. In the Z-transform domain, this translates into the requirement that all the poles of the transfer function must lie within the unit circle. If any pole is on the unit circle, the filter will oscillate with some natural frequency and is referred to as marginally stable. In all analyses presented here, stable (and in some cases marginally stable) filters are of interest.

## 17.5 FIBER-OPTIC FILTERING ARCHITECTURES

Most fiber-optic signal processing devices demonstrated so far are based on transversal filtering structures. However, this section concentrates on new fiber-optic arrangements which are similar to the feed-forward and feed-backward lattice structures of digital filtering theory. These structures currently form an important class of systems in the digital signal processing area due to their advantageous features, such as simplicity, modularity, and regularity. In addition, there is highly developed theory for the design and synthesis of filters based on such structures.

In a particular configuration, the fiber-optic lattice structure becomes similar to systolic arrays [65], which were originally developed for VLSI implementations for performing matrix operations. The systolic concept is important in that it offers a simple way for balancing the I/O and computation requirements by pipelining operations. Here, one or more input data elements are processed repeatedly by successive pipeline stages. An acousto-optic systolic array using incoherent light has already been proposed [66], which is capable of performing matrix-vector multiplication. This processor consists of a linear array of LEDs to realize the matrix elements, and a linear CCD array at the output to compute the elements of the output vector. In an all-fiber implementation, however, all the operations are performed inside the fiber through realizing an all-fiber matrix-vector systolic multiplier [19]. Architectures such as systolic arrays are very attractive in modern signal processing, since they are suitable structures to perform specific operations with high efficiencies (special-purpose architectures).

Most previous work has concentrated on feed-forward tapped delay line forms (transversal filters). Various mechanisms, such as radiation due to bending [9] and evanescent coupling by polishing the cladding down very close to the fiber core [10], have been used to realize the taps. In this chapter, attention is on different fiber-optic structures, namely, lattice (or ladder) forms, which can be used in optical signal processing [17,18]. This section is concerned with lattice structures, their fiber-optic implementation, and various theoretical formulations.

### 17.5.1 General Lattice Structures

Lattice (or ladder) structures have been the subject of study in the area of digital signal processing for more than a decade. Work on these structures is motivated by the advantageous characteristics of the analog versions (in the classical filter networks), which have been studied extensively. For example, inductor-capacitor ladder filters have the desirable feature of being relatively insensitive to slight variations in the circuit element values around the nominal value. In the following, we become familiar with the topology and advantages of lattice structures.

*Topology*

The topology of feed-forward (nonrecirculating) and feed-backward (recirculating) lattice structures is shown in Figure 17.1. In the feed-forward configuration, there are two signal lines which transmit the signals in the same directions (shown by the arrows in the figure). These two lines interact with each other through a number of cells distributed along them. In the feed-backward case, the two signal lines transmit the signals in the opposite directions.

*Advantages*

Lattice structures have found increasing applications in signal processing during the past decade [16]. Interest has arisen for the following reasons.

- They allow a systematic approach to filter synthesis.
- They have a very straightforward implementation.
- They are both regular and modular. That is, the implementation is accomplished by interconnecting compatible sections having the same basic topology, making the assembly easily expandable.
- By interconnecting only adjacent cells, global interconnections are eliminated, reducing signal delay times.
- They have convenient numerical properties that make them suitable for applications to digital signal processing.
- They provide a natural model for many important physical and mathematical phenomena, such as geophysical and speech signals [15], and are useful in estimation problems.

In fiber-optic form, we use single-mode fibers and directional couplers in conjunction with semiconductor laser sources and optical detectors to realize a fiber-optic lattice processor. Higher order lattices are simply linear extensions of the basic sections. The fiber-optic implementation of lattice structures and their applications in signal processing are discussed in later sections.

### 17.5.2 Fiber-Optic Implementation of Lattice Structures

Recirculating (feed-backward) and nonrecirculating (feed-forward) delay lines are two structures that are basic to fiber-optic signal processing. Using optical fibers and directional couplers, one can construct a two-coupler nonrecirculating delay line (Fig. 17.15(a)) and a two-coupler recirculating delay line (Fig. 17.15(c)). The corresponding impulse responses are shown in Figures 17.15(b,d). These basic delay lines are described below.

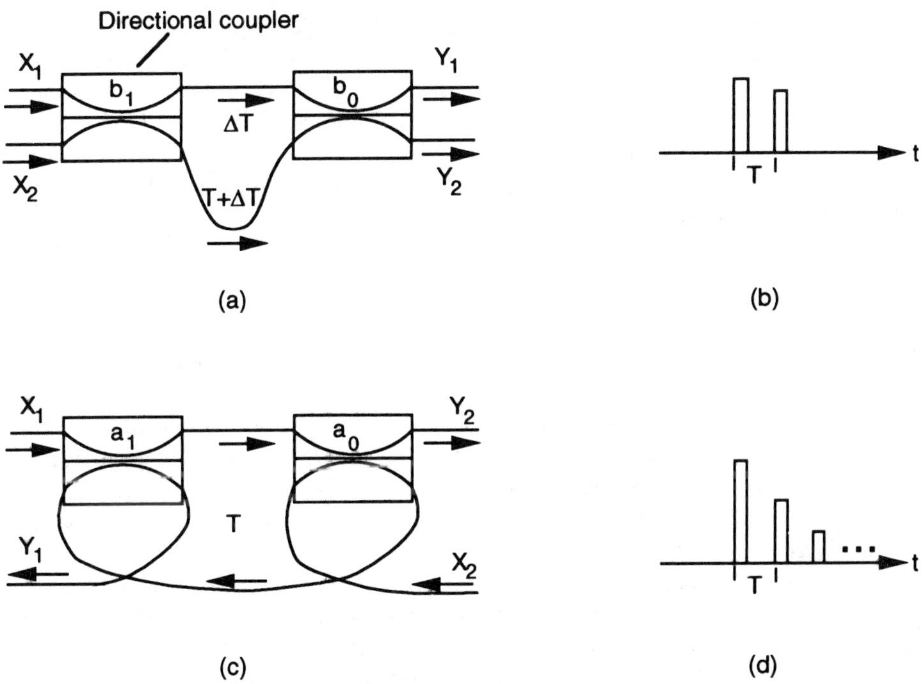


**Figure 17.15** Basic fiber-optic structures: (a) two-coupler nonrecirculating (feed-forward) delay line; (b) two-pulse impulse response of the two-coupler nonrecirculating delay line; (c) two-coupler recirculating (feed-backward) delay line; (d) multiple-pulse impulse response of the two-coupler recirculating delay line.

### *Two-Coupler Nonrecirculating Delay Line*

The two-coupler nonrecirculating delay line consists of two directional couplers and two feed-forward fiber lines (see Fig. 17.15(a)). In this case, the outputs of the first coupler are both fed forward to the second coupler, where they are recombined after a time delay $T$, which is the time-delay difference between the two feed-forward fiber lines (differential time delay). The impulse response of the system is composed of just two impulses spaced by the amount $T$ in time as depicted in Figure 17.15(b). The heights of the output impulses are dependent on the values of coupling coefficients and loss (or possible gain).

### *Two-Coupler Recirculating Delay Line*

In the two-coupler recirculating delay line, a single-mode fiber loop (with loop delay $T$) is closed upon itself by two directional couplers, with the second coupler feeding some portion of the light back to the first coupler through the fiber loop (see Fig. 17.15(c)). Thus, optical signals sent into the input port, say at $X_1$, circulate repeatedly around the loop, sending a portion of the recirculating intensity to the output ports $Y_1$ and $Y_2$. The impulse response of the system consists of a series of decaying impulses equally spaced in time by the loop time delay $T$, as shown in Figure 17.15(d). Here again, the heights of the output impulses depend on the values of coupling coefficients and loss or gain.

The basic nonrecirculating and recirculating delay lines described previously are used as elementary sections of feed-forward and feed-backward fiber lattice structures, respectively. These structures are discussed next.

### *Feed-Forward Configuration*

Figure 17.16(a) shows an equivalent diagram of the feed-forward lattice structure. Here, $b_i$ is the coupling coefficient of the coupler and $l_i$ is the intensity transmittance between adjacent couplers, with $z^{-1}$ representing a unit time delay between neighboring couplers. Figure 17.16(b) illustrates an $N$th-order feed-forward (nonrecirculating) fiber lattice structure, which is implemented by cascading $N$ two-coupler nonrecirculating delay lines (see Fig. 17.15(a)). We define the order of the lattice as the number of cascaded basic sections, which is equal to the number of couplers minus one.

Note that in this configuration, there are two feed-forward fiber lines with no feed-backward line. In the general case, different sections of the lattice can have differential time delays associated with them. A feed-forward fiber lattice can be visualized as a feed-forward tapped delay line with directional couplers operating as optical taps.

For an $N$th-order system ($N + 1$ couplers), the impulse response consists of $N + 1$ impulses equally spaced in time by the differential time delay $T$, as shown in Figure 17.16(c). The heights of the impulses are dependent on the values of the coupling coefficients $b_i$ and intensity transmittances $l_i$.

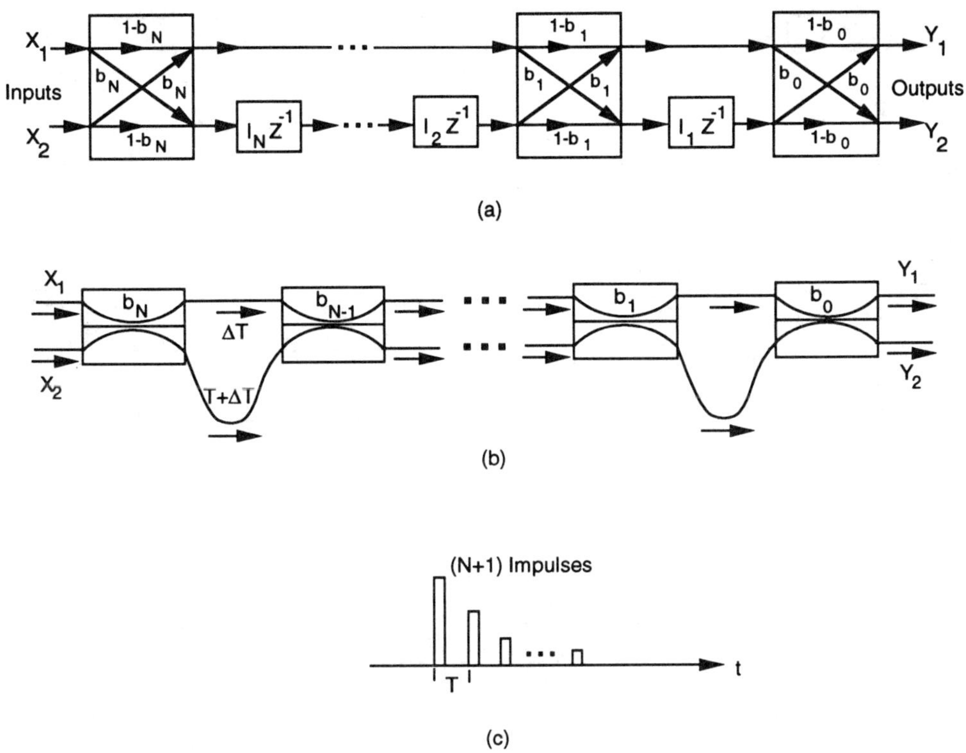


**Figure 17.16** *N*th-order feed-forward fiber-optic lattice structure with $N + 1$ couplers and $N$ delay elements: (a) block diagram representation; (b) fiber implementation; (c) intensity impulse response. $b_i$ is the intensity coupling coefficient of the *i*th coupler, and $l_i$ is the intensity transmittance of the *i*th section.

## *Feed-Backward Configuration*

Figure 17.17(a) illustrates an equivalent diagram of the feed-backward lattice structure. Here, $a_i$ is the coupling coefficient of the coupler and, again, $l_i$ is the intensity transmittance between adjacent couplers, with $z^{-1}$ representing a unit time delay between neighboring couplers.

The fiber-optic implementation of an *N*th-order feed-backward (recirculating) lattice structure is shown in Figure 17.17(b). *T* represents the loop delay of the system. Here we are interested in cases where *T* is the same for all sections, but this restriction would not be necessary in general.

This structure is obtained by interconnecting *N* two-coupler recirculating delay lines (see Fig. 17.15(c)) in tandem. As can be seen in Figure 17.17(b), this structure has one feed-forward and one feed-backward fiber line.

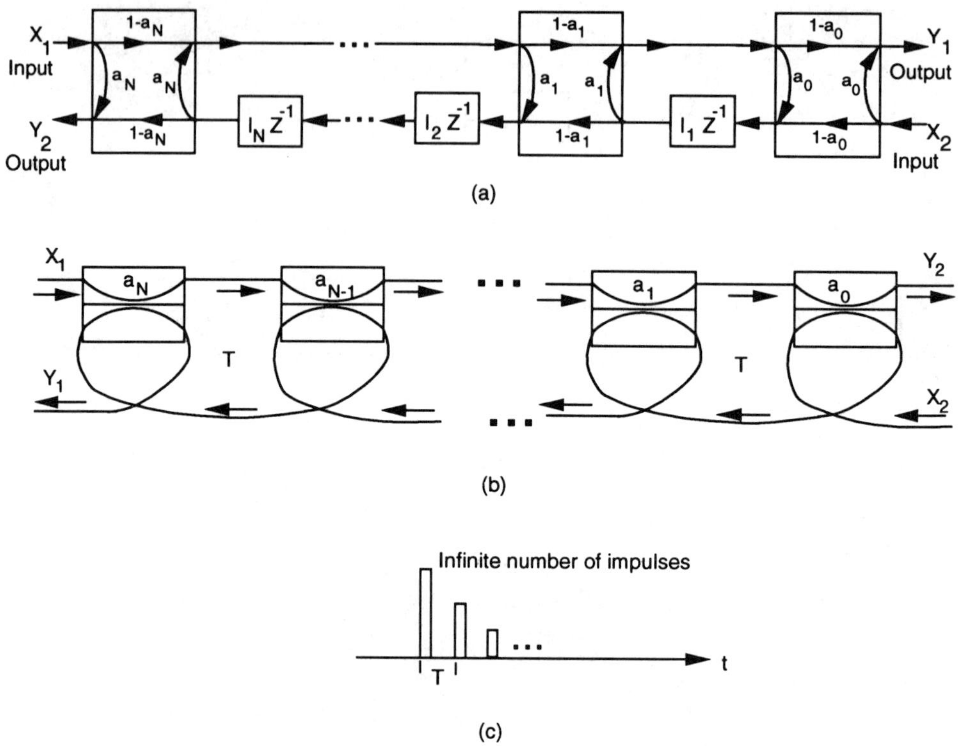


**Figure 17.17** *N*th-order feed-backward fiber-optic lattice structure with $N + 1$ couplers and $N$ delay elements: (a) block diagram representation; (b) fiber implementation; (c) intensity impulse response. $a_i$ is the intensity coupling coefficient of the *i*th coupler, and $l_i$ is the intensity transmittance of the *i*th section.

For an *N*th-order system ($N + 1$ couplers), the impulse response consists of a train of decaying impulses equally spaced in time by the loop delay $T$, as shown in Figure 17.17(c). Here again, the heights of the impulses are dependent on the values of the coupling coefficients $a_i$ and the intensity transmittances $l_i$.

### 17.5.3 Cascaded Fiber Lattice Structures

Feed-forward and feed-backward lattice forms can also be combined to construct more sophisticated fiber-optic structures. Figure 17.18 illustrates a scheme in which an *N*th-order feed-forward fiber lattice is cascaded with an *N*th-order feed-backward form (the order of the cascaded sections can, in general, be different). Here, the differential time delay is assumed to be equal to the loop delay. The impulse response in this case consists of a train of impulses equally spaced in time by the system unit delay $T$, with their heights again dependent on the values of the coupling coefficients and intensity transmittances.

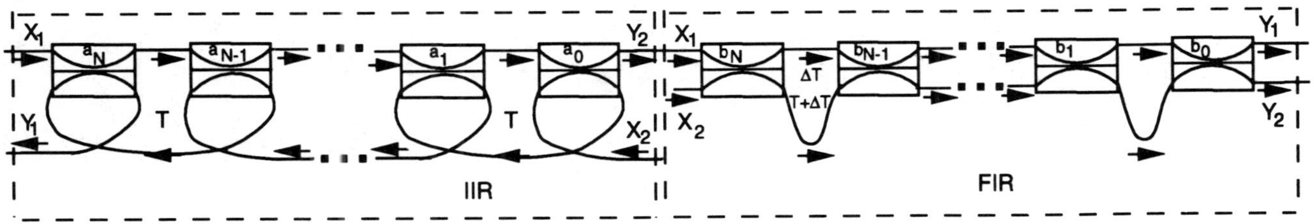


**Figure 17.18** Cascaded fiber-optic lattice structure composed of feed-forward and feed-backward lattice structures.

### 17.5.4 Amplified Lattice Filters

In both FIR and IIR lattice filters, optical amplifiers can be inserted between every two adjacent lattice sections (Fig. 17.19). The amplifiers can be used to realize large-size lattice filters. They can also be used to design unique filters not possible with their passive counterparts, as well as programmable adaptive filters.

### 17.5.5 Theoretical Formulations

This section is devoted mainly to establishing the further theoretical background needed to analyze fiber-optic systems (particularly lattice structures). In the following we show the applications of the Z-transform (see Sec. 17.3) using a few simple fiber-optic system examples. We then introduce the transfer matrix and chain matrix formulations, which we use to simplify the mathematical manipulations. Finally we present the modern control theory formulation. Its usage in the theory of positive fiber-optic systems is shown in Section 17.6.

*Z-Transform Formulation*

The system transfer function is defined as the Z-transform (or Laplace transform) of the impulse response. Using the definition of the Z-transform (17.19), we find the following expressions for the system transfer functions $H(z)$ of the two-coupler delay line introduced earlier (Fig. 17.15):

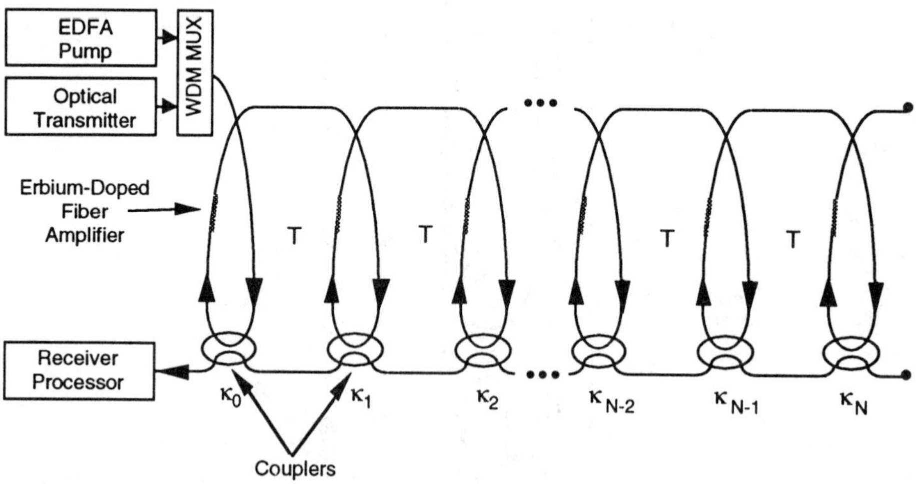


**Figure 17.19** Topology of the feed-backward recursive lattice filter integrated with EDFAs.

The expression for the two-coupler nonrecirculating delay line, with $X_1$ as the input and $Y_1$ as the output (Fig. 17.15(a)), is

$$H_{11}(z) = (1 - b_1)(1 - b_0)l_{11} + b_1 b_0 l_{12} z^{-1} \tag{17.32}$$

where $b_1$ and $b_0$ are the intensity coupling coefficients of the couplers, and $l_{11}$ and $l_{12}$ are the intensity transmittances in the two feed-forward fiber lines. This system has one pole at the origin and one zero at

$$z = -\frac{b_1 b_0 l_{12}}{(1 - b_1)(1 - b_0)l_{11}}$$

The expression for the two-coupler recirculating delay line, with $X_1$ as the input and $Y_1$ as the output (Fig. 17.15(b)), is

$$H_{11}(z) = \frac{a_1 + (1 - 2a_1)a_0 l_1 z^{-1}}{1 - a_1 a_0 l_1 z^{-1}} \tag{17.33}$$

where $a_1$ and $a_0$ are the intensity coupling coefficients of the couplers, and $l_1$ is the loop intensity transmittance of the system. As (17.33) shows, this system has one zero at $z = [(2a_1 - 1)/a_1]a_0 l_1$ and one pole at $a_1 a_0 l_1$.

The expression for the two-coupler recirculating delay line, with $X_1$ as the input and $Y_2$ as the output (Fig. 17.15(c)), is

$$H_{21}(z) = \frac{(1 - a_1)(1 - a_0)l_{11}}{1 - a_1 a_0 l_1 z^{-1}} \tag{17.34}$$

where $l_{11}$ is the intensity transmission factor of the forward fiber line. In this case, the system has one zero at the origin and one pole at $z = a_1 a_0 l_1$.

Zeros or poles at the origin do not affect the frequency response of the system (except for a linear phase factor). For this reason, we refer to the two-coupler recirculating delay line with $Y_2$ as the output as a first-order (one pole) all-pole system. The two-coupler nonrecirculating delay line and the two-coupler recirculating delay line with $Y_1$ as the output are referred to as first-order (one zero) all-zero, and first-order (one pole) pole-zero systems, respectively. The pole-zero diagrams of the systems discussed in this section are illustrated in Figure 17.20. These diagrams contain information about the frequency responses of the corresponding systems.

### *Matrix Formulations*

All the structures of interest in this chapter have, in general, two inputs and two outputs (two-pair systems). The transfer matrix and chain matrix formulations are two methods

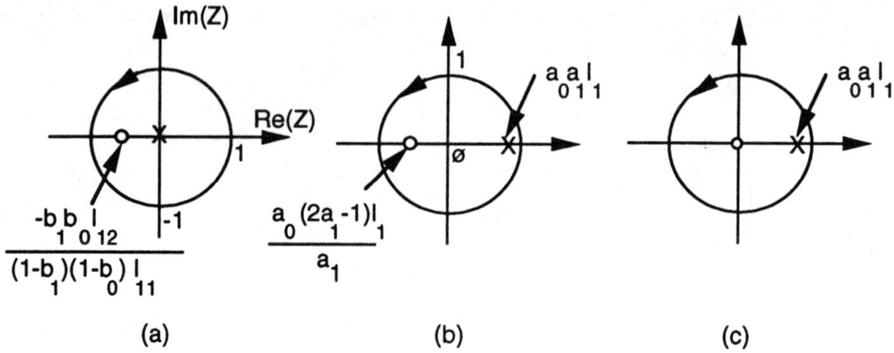


**Figure 17.20** Pole-zero diagrams of the (first-order) two-coupler delay lines showing locations of the poles and zeros in the $Z$ plane: (a) two-coupler nonrecirculating delay line; (b) two-coupler recirculating delay line, with $Y_1$ as the output; (c) two-coupler recirculation delay line, with $Y_2$ as the output.

of characterization which are suitable for cascaded two-pair systems. These methods are first explained very briefly, and then applied to the lattice structures to obtain the total transfer functions. A more detailed explanation of the chain matrix method can be found in [14].

*Transfer Matrix*

Consider a linear and time-invariant two-port system, with each port having two terminals. In a feed-forward two-pair system, as shown in Figure 17.21(a), $(X_1, X_2)$ and $(Y_1, Y_2)$ constitute the first and second ports. Note that the input variables are shown by $X$ and the output variables by $Y$. The inputs and outputs are related to each other by the transfer matrix

$$\begin{bmatrix} Y_1(z) \\ Y_2(z) \end{bmatrix} = \begin{bmatrix} H_{11}(z) & H_{12}(z) \\ H_{21}(z) & H_{22}(z) \end{bmatrix} \begin{bmatrix} X_1(z) \\ X_2(z) \end{bmatrix} \tag{17.35}$$

where $z$ is the Z-transform variable.

Each element of the transfer matrix is the transfer function from an input port to an output port. Note that $H_{mn}(z)$ is the transfer function from input $X_n$ to output $Y_m$.

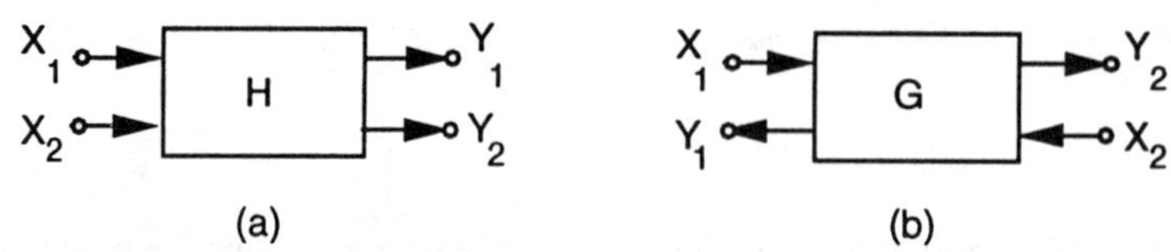


**Figure 17.21** A two-pair system: (a) feed-forward configuration; (b) feed-backward configuration.

### *Chain Matrix*

In a feed-backward two-pair system, as shown in Figure 17.21(b), the first and second ports consist of the variables ($X_1$, $Y_1$) and ($X_2$, $Y_2$), respectively. Here again, the input variables are shown by $X$ and the output variables by $Y$.

While the transfer matrix formulation is suitable for the analysis of cascaded feed-forward two-pair systems, there is another method of characterization which is more suitable for the analysis of cascaded feed-backward two-pair systems. This method makes use of the chain matrix which relates the pair ($X_1$, $Y_1$) to the pair ($X_2$, $Y_2$):

$$\begin{bmatrix} X_1(z) \\ Y_1(z) \end{bmatrix} = \begin{bmatrix} G_{11}(z) & G_{12}(z) \\ G_{21}(z) & G_{22}(z) \end{bmatrix} \begin{bmatrix} Y_2(z) \\ X_2(z) \end{bmatrix} \tag{17.36}$$

It should be noted that, unlike the transfer matrix, the elements of the chain matrix are not transfer functions from an input port to an output port. However, the equivalent transfer matrix of the system can be obtained through some conversion formulas (see next section).

Note that both the transfer matrix and the chain matrix are $2 \times 2$ matrices which are uniquely determined for specified input and output variables.

### *Properties*

It can be easily shown that elements of the transfer matrix of a feed-backward two-pair system are related to the elements of its chain matrix through the following expressions:

$$H_{11} = \frac{G_{21}}{G_{11}} \tag{17.37}$$

$$H_{12} = \frac{G_{11}G_{22} - G_{12}G_{21}}{G_{11}} \tag{17.38}$$

$$H_{21} = \frac{1}{G_{11}} \tag{17.39}$$

$$H_{22} = \frac{G_{12}}{G_{11}} \tag{17.40}$$

If $G_{11}G_{22} - G_{12}G_{21} = 1$, the system is called a *reciprocal* ($H_{12} = H_{21}$) two-pair.

A very important property of both transfer matrix and chain matrix formulations is that one can obtain the total transfer matrix or chain matrix of cascaded two-pair systems simply by multiplying the individual characteristics in the right order. Note that these multiplicative rules (see (17.43) and (17.44)) are order-dependent, since the multiplicands are matrices.

For cascaded feed-forward two-pair systems (see Fig. 17.1(a)), the total transfer matrix is the product of individual transfer matrices multiplied in the order opposite the one they are cascaded in, as shown below:

$$H_{\text{total}} = H_N, \ldots, H_3 H_2 H_1 \tag{17.41}$$

where $N$ is the number of cascaded systems.

Similarly, for cascaded feed-backward two-pair systems (see Fig. 17.1(b)), the chain matrix of the composite systems is the product of the individual chain matrices, but in this case multiplied in the same order as the systems are cascaded; that is,

$$G_{\text{total}} = G_1 G_2 G_3, \ldots, G_N \tag{17.42}$$

where here again $N$ is the number of cascaded systems.

These multiplicative properties of the transfer matrix and chain matrix formulations greatly simplify the mathematical analysis of cascaded two-pair systems.

### *Matrix Formulations of Fiber Lattice Elements*

Lattice structures are good examples of cascaded two-pair systems (see Fig. 17.1). As mentioned before, these structures are obtained by a cascade connection of individual two-coupler delay lines (basic lattice sections) of the types illustrated in Figure 17.15.

It can readily be shown that for a directional coupler with $a$ as the intensity coupling coefficient, the transfer matrix $H_c$ and the chain matrix $G_c$ are given by the following expressions:

$$H_c = \begin{bmatrix} 1 - a & a \\ a & 1 - a \end{bmatrix} \tag{17.43}$$

and

$$G_c = \frac{1}{1 - a} \begin{bmatrix} 1 & -a \\ a & 1 - 2a \end{bmatrix} \tag{17.44}$$

Similarly, for a simple two-line delay system with a unit time delay ($z^{-1}$) and intensity transmission factors $l_{11}$ and $l_{12}$, as depicted in Figure 17.22, we obtain

$$H_d = \begin{bmatrix} l_{11} & 0 \\ 0 & l_{12}z^{-1} \end{bmatrix} \tag{17.45}$$

and

$$G_d = \frac{1}{l_{11}}\begin{bmatrix} 1 & 0 \\ 0 & l_{11}l_{12}z^{-1} \end{bmatrix} \tag{17.46}$$

The two-coupler delay lines (see Fig. 17.15) are a cascaded connection of two couplers and a two-line delay line. Therefore, using (17.41) through (17.46), we can calculate the chain matrix and transfer matrix of, respectively, the two-coupler nonrecirculating and recirculating delay lines. The results are shown below.

For the two-coupler nonrecirculating delay line, we have

$$H_f = H_{c0}H_dH_{c1} = \begin{bmatrix} H_{11} & H_{12} \\ H_{21} & H_{22} \end{bmatrix} \tag{17.47a}$$

where the subscripts $c_0$, $d$, and $c_1$ refer, respectively, to the coupler with the coupling coefficient $a_0$, two-line delay system, and the coupler with the coupling coefficient $a_1$, and

$$\begin{aligned} H_{11} &= (1-b_1)(1-b_0)l_{11} + b_1b_0\,l_{12}z^{-1} \\ H_{12} &= b_1(1-b_0)l_{11} + (1-b_1)b_0\,l_{12}z^{-1} \\ H_{21} &= (1-b_1)b_0\,l_{11} + b_1(1-b_0)l_{12}z^{-1} \\ H_{22} &= b_1b_0\,l_{11} + (1-b_1)(1-b_0)l_{12}z^{-1} \end{aligned} \tag{17.47b}$$

Note that the $H_{11}$ element is the same as the one given in (17.32).

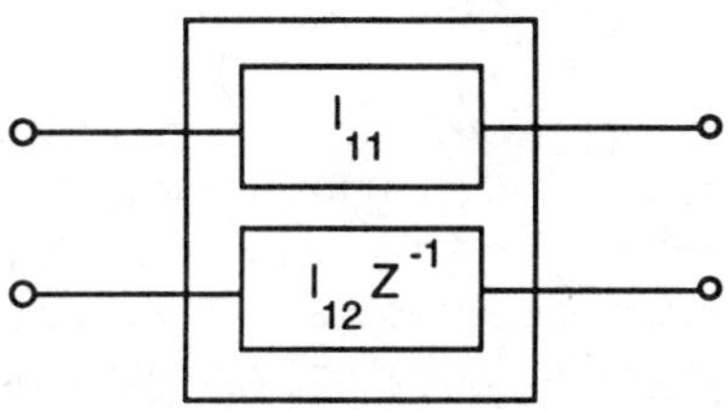


**Figure 17.22** A simple two-line delay system. $l_{11}$ and $l_{12}$ are intensity transmission factors of the first and second line, respectively.

For the two-coupler recirculating delay line, we obtain

$$G_b = G_{c1}G_dG_{c0} = \frac{1}{l_{11}(1 - a_1)(1 - a_0)}\begin{bmatrix} G'_{11} & G'_{21} \\ G'_{12} & G'_{22} \end{bmatrix} \tag{17.48a}$$

where

$$\begin{aligned} G'_{11} &= 1 - a_1a_0\, l_1z^{-1} \\ G'_{12} &= -a_0 - a_1(1 - 2a_0)l_1z^{-1} \\ G'_{21} &= a_1 + (1 - 2a_1)a_0\, l_1z^{-1} \\ G'_{22} &= -a_1a_0 + (1 - 2a_1)(1 - 2a_0)l_1z^{-1} \end{aligned} \tag{17.48b}$$

and $l_1 = l_{11}l_{12}$ is the loop intensity transmittance, $G_{c1}$ and $G_{c0}$ are coupler chain matrices (see (17.44)) with coupling coefficients equal to $a_1$ and $a_0$, respectively, and $G_d$ is the chain matrix of the two-line delay system given in (17.46). Moreover, in this case the elements of the corresponding transfer matrix can be found through (17.37) to (17.40), which are consistent with those of (17.33) and (17.34).

The transfer and chain matrices of the basic lattice sections, given in (17.47) and (17.48), can be used to calculate the total transfer and chain matrices of higher order lattice structures. Using this approach, together with Z-transform techniques, we can obtain general expressions for the transfer functions of the lattice structures.

*Modern Control Theory Formulations*

In modern control theory [67], a general (discrete time) linear $n$th-order system is described in a state space by the state update and the output equations as shown below:

$$\begin{aligned} \mathbf{x}(k + 1) &= A(k)\mathbf{x}(k) + B(k)\mathbf{u}(k) \\ \mathbf{y}(k) &= C(k)\mathbf{x}(k) + D(k)\mathbf{u}(k) \end{aligned} \tag{17.49a}$$

We represent the matrices with upper case notation and the vectors with bold lower case notation. In (17.49a), the vector $\mathbf{x}(k)$ is an $n$-dimensional state vector, and $\mathbf{A}$, $\mathbf{B}$, $\mathbf{C}$, and $\mathbf{D}$ are $(n \times n)$, $(n \times m)$, $(p \times n)$, and $(p \times m)$ matrices, respectively. In the case of shift-invariant (time-invariant) systems, the above matrices are independent of time. However, these matrices are, in general, functions of system parameters, such as coupling coefficients and system losses or gains.

It should also be noted that the state of a dynamic system is usually defined as the smallest set of variables (called *state* variables) such that the knowledge of these variables at the initial time $t = t_0$, together with the inputs for $t \geq t_0$, completely determines the behavior of the system for any time $t \geq t_0$. Usually, the state variables are taken to be the outputs of the delay elements.

In the mathematical model described by (17.49a), we have assumed the general case of a multiple-input ($m$ inputs) multiple-output ($p$ outputs) (MIMO) system. If the system is only SISO, the equations can be expressed as

$$\mathbf{x}(k+1) = \mathbf{A}(k)\mathbf{x}(k) + \mathbf{b}(k)\mathbf{u}(k)$$
$$\mathbf{y}(k) = \mathbf{c}^T(k)\mathbf{x}(k) + d(k)\mathbf{u}(k) \tag{17.49b}$$

where now **b** and **c** are vectors and $d$ is scalar. As an example, taking the outputs of the delay elements as the state variables, we get the following expression for the [$A$, $B$, $C$, $D$] matrices of the $N$th-order feed-backward lattice structure:

$$\mathbf{A} = \begin{bmatrix} a_1 a_0 l_1 & a_2(1-a_1)a_0 l_1 & \cdots & a_N \prod_{k=1}^{N-1}(1-a_k)a_0 l_1 \\ (1-a_1)l_2 & a_2 a_1 l_2 & \cdots & a_N \prod_{k=2}^{N-1}(1-a_k)a_1 l_2 \\ \vdots & \ddots & \ddots & \vdots \\ 0 & \cdots & \cdots & a_N(1-a_{N-1})a_{N-2}l_{N-1} \\ & & (1-a_{N-1})l_N & a_N a_{N-1} l_N \end{bmatrix} \tag{17.50a}$$

$$\mathbf{B} = \begin{bmatrix} \prod_{k=1}^{N}(1-a_k)a_0 l_1 & (1-a_0)l_1 \\ \prod_{k=2}^{N}(1-a_k)a_1 l_2 & 0 \\ \vdots & \vdots \\ (1-a_N)a_{N-1}l_N & 0 \end{bmatrix} \tag{17.50b}$$

$$\mathbf{C} = \begin{bmatrix} 0 & 0 & \cdots & 1-a_N \\ a_1(1-a_0) & a_2(1-a_1)(1-a_0) & \cdots & a_N \prod_{k=0}^{N-1}(1-a_k) \end{bmatrix} \tag{17.50c}$$

and

$$\mathbf{D} = \begin{bmatrix} a_N & 0 \\ \prod_{k=0}^{N}(1 - a_k) & 0 \end{bmatrix} \tag{17.50d}$$

where the input, output, and state vectors are shown below:

$$\mathbf{u} = \begin{bmatrix} X_1 \\ X_2 \end{bmatrix} \tag{17.50e}$$

$$\mathbf{y} = \begin{bmatrix} Y_1 \\ Y_2 \end{bmatrix} \tag{17.50f}$$

and

$$\mathbf{x} = \begin{bmatrix} x_1 \\ x_2 \\ \cdot \\ \cdot \\ \cdot \\ x_N \end{bmatrix} \tag{17.50g}$$

Note that all the indexes are numbered increasingly from the right to the left of the lattices' schematic diagrams.

As can be seen, all elements of the above matrices and vectors are nonnegative in value. Using this formulation, it can be shown that if the coupling coefficients of all the middle couplers (the first and last couplers excluded) are greater than 0.5, then all the system poles are real and nonnegative (see Sec. 17.6). The reason is that, in this case, all the minors of the $A$ matrix, which is an upper Hessenberg matrix, are nonnegative.

## 17.6 BIPOLAR, COMPLEX, AND POSITIVE OPTICAL SYSTEMS

Throughout this chapter, we have mainly considered systems in which the signals (optical intensities) and coupling coefficients are nonnegative quantities. These systems form a subclass, which is well matched to the requirements of several fields of application. These and many other linear systems of interest can be represented by a set of linear matrix-vector equations in which all the elements of the matrices and vectors involved are (real) nonnegative in value, as shown in Section 17.5. In this section, after a brief review of bipolar and complex systems, we shall concentrate on the properties of positive systems (i.e., systems in which the quantities involved are positively valued). We define a positive system using the state space model, described in Section 17.5, to represent the system.

After that we present the theoretical background, including some definitions and important theorems of positive systems theory. Using this theoretical background, we obtain some results regarding the frequency response and the locations of the poles and zeros (in the $Z$ plane) of the transfer function of a linear shift-invariant positive system. Finally, after discussing the results obtained and verifying their consistency with intuition, we conclude this section with a few remarks about the effects of the positivity feature on positive optical signal processing systems.

### 17.6.1 Bipolar and Complex Systems

In optical signal processing, coherent methods have potential advantages, such as bipolar and complex valued-data representation by physical quantities. A disadvantage is, however, that environmental effects such as temporal changes of temperature, mechanical vibrations, and stress might influence the system response by randomly varying the time delays and optical phases involved. These effects can be minimized by incorporating suitable feedback and stabilization circuits in the system. This approach, however, adds to the complexity and cost of the system. Further work is needed to fully investigate the practicality of coherent fiber- and integrated-optic signal processing systems in a real working environment.

In the case when the fiber-optic system is operated in an incoherent regime, all the parameters and signals involved assume only nonnegative values. As a result, bipolar or complex valued signals cannot be represented directly. However, there are various coding schemes [68–70] that can be used to include more general bipolar or complex data. In the bipolar case, one way is to add a constant dc bias that is larger than the absolute value of the most negative quantity. Due to the possibility of a dynamic range problem, dc biasing is not normally considered as a good choice [70]. For the general case of complex data, three schemes [68–70], known as the four-component, three-component, and two-component decompositions, are described below. The general idea here is to use a mapping function to transform the complex valued quantities involved into nonnegative physical quantities, and then, after processing, recover the complex output data using a suitable decoding algorithm. In the four-component decomposition, a complex quantity is decomposed into four nonnegative numbers ($x_R^+$, $x_R^-$, $x_I^+$, $x_I^-$) representing components angularly spaced by 90 deg in the complex plane; that is, for a given complex number $x$ we have

$$x = (x_R^+ - x_R^-) + j(x_I^+ - x_I^-) \tag{17.51}$$

where $x_R^+$, $x_R^-$, $x_I^+$, and $x_I^-$ are the positive real part, negative real part, positive imaginary part, and negative imaginary part, respectively. In order to make the decomposition unique, it is assumed that two of the components (one from the real part and one from the imaginary part) are always zero. Such representation is, however, redundant, since three nonnegative components would be enough to represent a complex valued quantity. Here the complex number is decomposed into three nonnegative numbers $(x_0, x_1, x_2)$ representing components angularly separated by 120 deg in the complex plane as

$$x = x_0 e^{j0} + x_1 e^{j(2\pi/3)} + x_2 e^{j(4\pi/3)} \tag{17.52}$$

Again, in order to make the decomposition unique, it is assumed that at least one of the components is always zero. Using this representation we need to perform nine nonnegative convolutions (or multiplications) to compute one complex valued convolution (or multiplication). The nine output data then have to go through some summation operations to obtain the three-component representation of the complex output. Another coding system is the two-component decomposition method. In this scheme, the real and imaginary parts of the complex data $x = x_R + jx_I$ are properly biased to ensure nonnegative values. All the coding algorithms described above usually require preprocessing and postprocessing circuits, which may involve the subtraction operation. A particular choice of the coding system is based on factors such as system complexity and SNR. In general, while the two-component scheme is the simplest, the three-component decomposition yields the maximum SNR [69].

Another potential way to realize bipolar systems has been reported [71] based on the fact that a signal can be inverted by applying a $\pi$ phase shift to it. If this phase shift is independent of the RF frequency, the exact equivalent of a negative weight is obtained. In an incoherent system, a phase shift is readily obtained by using a length of fiber. Unfortunately, this length of fiber actually acts as an ideal time delay $\tau'$ and thus results in a phase shift of $\varphi = \omega\tau'$, which is a linear function of frequency; a phase shift of exactly $\pi$ is obtained only at $\omega = \pi/\tau'$. The possibility of "synthesizing" narrowband negative weights thus exists. Where the desired transfer function requires a negative weight, the time delay may be increased (or decreased) by $\tau'$ and the weight value reevaluated. That is, a branch with a negative weight $w_k$ and a total delay $\tau_\kappa$ is substituted by a branch with a positive weight $v_k$ and a total delay of $\tau_\kappa + \tau'$.

### 17.6.2 Positive Systems

In general, the impulse response $h(k)$ of a discrete linear time-invariant system can assume both positive and negative values. However, for a positive system, the heights of the impulse response at all possible times are limited to only positive values, and an important theory exists for the study of such systems.

The theory of positive systems is closely associated with the well-developed theory of positive matrices. This enables us to use the results of the theory of positive matrices in studying positive systems. Real matrices with nonnegative elements have important application in the theory of probability for Markov chains [72] (stochastic matrices) and in the theory of oscillatory properties of elastic vibrations in mechanics [73] (oscillation matrices). After looking at the theory of positive systems, we will explore its applicability to optical systems that incorporate positively valued parameters. A very interesting and advantageous feature of positive systems theory is that it provides us with some fairly strong and important conclusions about the behavior of a positive system. In addition, it is usually consistent with intuition.

Based on the state space representation of a linear system, a linear positive system is defined as a linear system in which the state variables (and the outputs) are always positive (or at least nonnegative) in value. Such systems arise frequently in practice since the state variables of many real systems describe quantities which are always nonnegative. For example, in many economic systems, the variables which describe quantities of goods remain nonnegative. The same is true for the subset of linear optical systems for which the quantities involved are positively valued, such as incoherent optical processors, which have applications in signal processing and filtering operations. In the following section, we present some facts from positive systems theory.

*Theoretical Background*

If $A = [a_{ij}]$ and $B = [b_i]$, we adopt the following definitions for our later discussions.

DEFINITION 17.1. $A > B$ (or $B < A$) if and only if $a_{ij} > b_{ij}$ for all $i, j$.

DEFINITION 17.2. $A \geq B$ (or $B \leq A$) if and only if $a_{ij} \geq b_{ij}$ for all $i, j$.

DEFINITION 17.3. $A \geqq B$ (or $B \leqq A$) if and only if $a_{ij} \geq b_{ij}$ for all $i, j$ with $a_{ij} \neq b_{ij}$ for at least one $(i, j)$.

DEFINITION 17.4. For a complex-valued matrix $C$, $C^+$ denotes the matrix modulus $C$, which is obtained from $C$ when all the elements are replaced by their moduli. For the special case when $B = 0$, we call $A$ strictly positive, positive (or strictly nonnegative), and nonnegative when $A > 0$, $A \geq 0$, and $A \geqq 0$, respectively.

DEFINITION 17.5. Two matrix inequalities:

(a) the Cauchy-Schwarz inequality:

$$[A \cdot B]^+ \leq A^+ \cdot B^+ \tag{17.53}$$

where the plus sign (the matrix modulus)and the inequality signs are as defined earlier. The equality sign in (17.53) holds for the $i,j$th elements if $a_{ik}$ and $b_{kj}$ are real and have like signs for all possible values of $k$. For the special case when $A$ and $B$ are both nonnegative matrices $[A \cdot B]^+ = A \cdot B$.

(b) The triangle inequality:

$$[A + B]^+ \leqq A^+ + B^+ \tag{17.54}$$

Here the equality sign holds for the $i,j$th elements when $a_{ij}$ and $b_{ij}$ are real and have like signs.

Now we proceed to describe three of the major properties of linear positive systems [67]. The first important property is described by the Frobenius [74] and Perron [75] theorems. This property is related to the fact that for any linear positive system there exists a dominant eigenvalue and eigenvector. In the following, the Frobenius-Perron theorem and some of the related theorems are stated without proof. For more complete information on this topic, see [67,76].

THEOREM 17.1. *The Frobenius-Perron Theorem:* if $A > 0$ (strictly positive), then there exists $\lambda_0 > 0$ and $\boldsymbol{x}_0 > 0$ such that

1. $A\boldsymbol{x}_0 = \lambda_0\boldsymbol{x}_0$.
2. If $\lambda \neq \lambda_0$ is any other eigenvalue of $A$, then $|\lambda| < \lambda_0$.
3. $\lambda_0$ is an eigenvalue of geometric and algebraic multiplicity 1 (i.e., a simple eigenvalue with only one independent corresponding eigenvector).

THEOREM 17.2. If $A \geq 0$ (positive or strictly nonnegative) and $A^q > 0$ for some positive integer $q$, then all the conclusions of Theorem 17.1 apply to $A$.

THEOREM 17.3. If $A \geqq 0$ (nonnegative), then there exists $\lambda_0 > 0$ and $x_0 \geq 0$ such that:

1. $A\boldsymbol{x}_0 = \lambda_0\boldsymbol{x}_0$;
2. If $\lambda \neq \lambda_0$ is any other eigenvalue of $A$, then $|\lambda| \leq \lambda_0$.

The second property of positive systems is concerned with the connection between stability and positivity. It can be shown [67] that for positive systems there is a direct correspondence between the existence of a positive equilibrium point and stability. The third major property involves the effects produced by changes in the values of the system parameters (comparative statics). For positive systems, positive changes (such as increasing an element of the system matrix $A$) produce corresponding positive changes (increases) in the components of the equilibrium points (the steady-state state vector).

Therefore, even if the precise values of the parameters of a linear positive system are not known in advance, we can draw some significant qualitative conclusions about the behavior of the system. It should also be noted here that the structure of continuous-time positive systems is slightly different from that of discrete-time positive systems, because in the former case the system matrix relates the state to the derivative of the state, and the derivative need not be positive, but in both cases the results are virtually the same in character [67].

Furthermore, in addition to the theorems stated previously, we will make use of Lemma 17.1 and Theorem 17.4, which are described below.

LEMMA 17.1 *Series Expansion of Inverse:* if $A$ is a matrix with all eigenvalues strictly inside the unit circle ($|\lambda(A)| < 1$), then

$$[I - A]^{-1} = I + A + A^2 = A^3 + \ldots \tag{17.55}$$

THEOREM 17.4. If $A \geq 0$ (nonnegative) with associated Frobenius-Perron eigenvalue $\lambda_0$, then the matrix $[\lambda I - A]^{-1}$ exists and is positive if and only if $\lambda > \lambda_0$.

*Results*

In this section we utilize the theory of positive systems to draw some important conclusions about the transfer function of a linear time-invariant positive system. These results are presented below.

First, the modulus of the transfer function (or frequency response) of a linear time-invariant positive system takes its maximum value at the origin (zero frequency); that is,

$$[H(e^{jwT})]^{+} \leq H(1) \tag{17.56}$$

Note that $H(1)$ is real and positive, which also means that the phase of the transfer function is zero at $\omega = 0$. These features are clearly seen in the frequency responses in Figure 17.3.

Secondly, for stable linear time-invariant positive systems, there cannot be an odd number of zeros on the right-hand sign of the pole with the largest magnitude (Frobenius-Perron pole). For example, to realize a first-order all-pass filter, we need one zero (an odd number) on the right-hand side of the pole. According to what we have said so far, this pole-zero arrangement cannot be obtained from a totally positive system. However, by incorporating an electronic subtractive detection system, one can realize the desired arrangement. In this form, the whole system is no longer a positive system, as is evident from its bipolar impulse response.

Finally, we show that for feed-backward fiber lattice filters, if the values of all the coupling coefficients between the two boundary couplers (excluding the first and last couplers) are greater than 0.5 ($a_i > 0.5$, $i = 1, 2, \ldots, N - 1$), then all the poles of the corresponding transfer function are real and positively valued. This statement implies that in order to realize complex-valued poles, it is necessary to have at least one of the middle coupling coefficients be less than 0.5. These conclusions, together with their effects on the frequency response, are shown in Figures 17.23 and 17.24 for the third-order and fifth-order fiber lattice filters.

*Conclusion*

The theory of positive systems has the attractive feature of being consistent with intuition. For example, for the conclusion given in (17.56), we can have the following explanations.

- At frequencies for which $\omega T = 2n\pi$ ($n = 0, 1, 2, \ldots$) the signals add constructively in the system, whereas at other frequencies this is not the case. Therefore, the magnitude of the transfer function should take its maximum value at frequencies which are integer multiples of $1/T$.
- Since the impulse response of a positive system is real and positively valued at all times, it can be considered analogous to a power spectrum (which is positive for all frequencies); therefore, the Fourier transform of the impulse response (or the

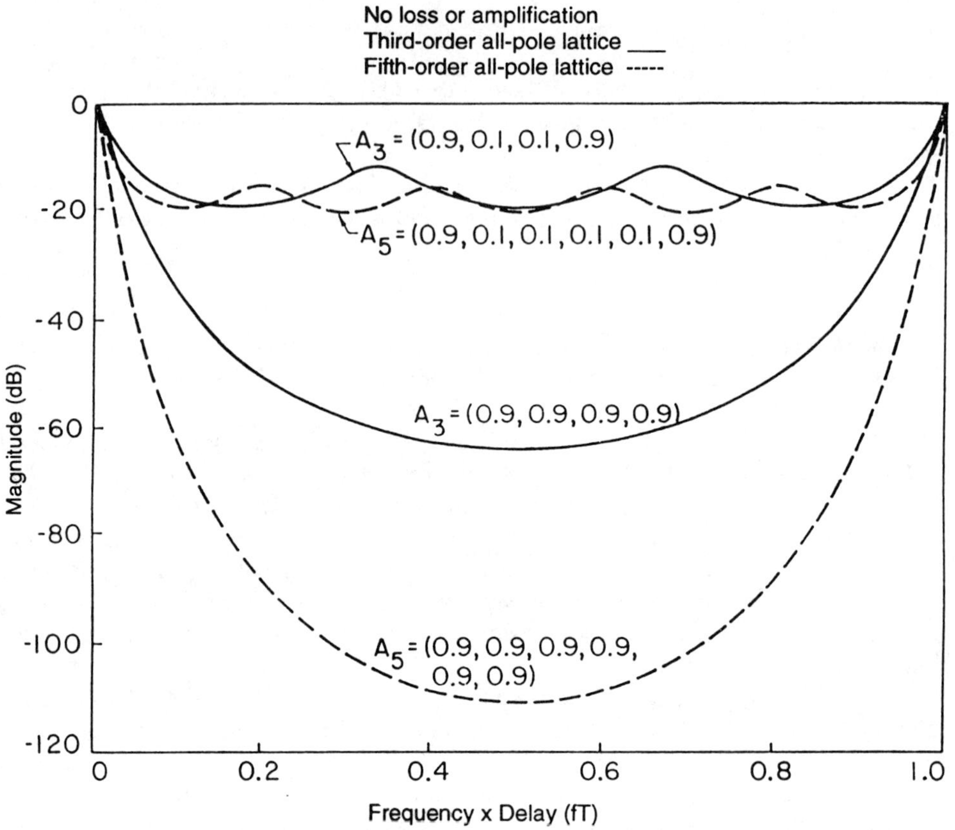


**Figure 17.23** Log-magnitude frequency responses of third-order and fifth-order recursive fiber-optic lattice filters for two different sets (shown above the figure) of the couple coefficient vector.

system transfer fuhction) can be regarded as an autocorrelation function, and it is well known that the value of an autocorrelation function at the origin is maximum and positive [77].

In conclusion, we have briefly described the theory of positive systems and used it to obtain some important information about the transfer function of such systems. These results are useful in predicting the general shape and some features of the frequency response. For instance, we can conclude that we cannot realize high-pass filters with positive systems, since these filters should be able to reject the low-frequency components of the input signals. Further flexibility is possible by combining positive sections with sections allowing negativity. One such example is in the design of an all-pass filter [17]. We emphasize the point that just knowing the fact that the system is positive enables us to make some fairly strong statements about the behavior of the system regardless of the

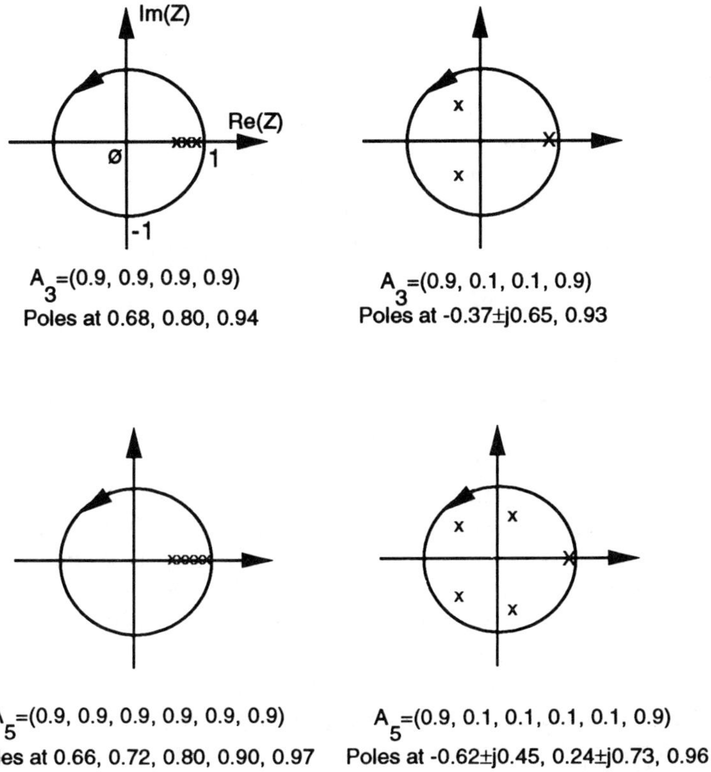


**Figure 17.24** Pole diagrams in the Z-plane corresponding to the filters shown in Figure 17.23.

precise values of the system parameters. More detailed studies are necessary to reveal other potential applications of positive systems theory in incoherent optical systems, such as iterative optical processors and optical baseband filters.

In radar systems which employ switchable fiber-optic delay networks to function only as RF phase shifters, positivity effect does not have any implications. If, however, the fiber delay line network is to perform more sophisticated signal processing functions, then the positivity property can be restrictive, as discussed earlier. In applications requiring bipolar or complex operations, one needs either an incoherent processor with a suitable coding scheme or a properly stabilized coherent processor.

## 17.7 CONCLUSIONS

The subject described in this chapter involves signal processing using single-mode optical fibers and waveguides, fiber- and integrated-optic components such as directional couplers, optical amplifiers, and switching arrays. Integrated-optic switching arrays can be fabricated in lithium niobate and semiconductor materials. Semiconductors are more attractive in that they have the potential to integrate many optical and electronic components on a single small substrate. Both experimental and theoretical studies to investigate the potential uses, performance, and properties of such signal processors were described.

The low loss (fractional decibels/kilometer) and large bandwidth length product (on the order of 100 GHz·km for single-mode fibers) of optical fibers, together with advances in the manufacturing of fiber-optic components, provide an attractive technology for processing broadband signals [7,78]. To exploit the attributes of single-mode fiber technology, a number of fiber-optic components have been developed which are compatible with this technology. Stable fiber components, such as directional couplers [79,80], phase modulators [81,82], polarization controllers [83], polarizers [84,85], frequency shifters [86,87], and amplifiers [4,5,88–90], have been demonstrated, with work continuing on improving them. These components manipulate the guided lightwaves inside the fiber, a property which makes their operation very efficient.

A careful study of dominant noise sources in fiber-optic processors is important in estimating the performance and dynamic range of such processors. The quantum nature of the radiation field of an optical source (photon noise) and fluctuations in the emission of photoelectrons in a photodetector (shot noise) constitute two important sources of noise in optical systems, leading to fluctuations in the photo currents. In general, the photoelectron statistics measured directly by a photodetector involve a convolution of laser photon and detector statistics [91]. Moreover, intensity and phase fluctuations of the laser emission can affect the performance of optical processors. It has been demonstrated [92–96] that in relation to many fiber processors, laser phase noise is the dominant source of noise. Note that the electronic components also contribute to the total output noise.

The following sections include a brief discussion of the impact and applications of this work, as well as the outlook for fiber- and integrated-optic signal processing.

### *Impact*

Research has demonstrated the feasibility of the fiber-optic technology in performing high-speed and broadband signal processing functions. This objective has been achieved by constructing prototype devices and systems, as well as developing a systematic approach for the analytical study of such structures [17,18].

More specifically, new fiber-optic arrangements, namely, feed-forward and feed-backward lattice structures, have been introduced. These structures are simple, regular, and easily expandable. Directional couplers constitute the building blocks of such structures. The couplers serve as the processing elements capable of optically performing basic operations such as addition and multiplication. These couplers communicate with each other through optical fiber or integrated-optic waveguide interconnections. All the operations are synchronized by the proper choice of fiber or waveguide lengths and data clock rates. Fiber- and integrated-optic lattice structures can be used to perform a variety of time- and frequency-domain functions. We have used digital filtering theory to design and analyze optical filters. In addition, we have extended the applications of the theory of positive systems to the study of the subset of optical systems for which the quantities involved are positively valued. Only knowledge of the fact that all the system quantities are positively valued enables us to make some strong statements about the properties of such systems. Also, the effects of laser phase noise on the performance of optical processors was mentioned. It has been shown [92–96] that the dynamic range of optical processors incorporating recirculating loops is limited by laser phase fluctuations.

*Applications*

Fiber- and integrated-optic signal processing devices can be constructed to perform various functions, such as convolution, correlation, matrix operations [19], and frequency filtering [9,17,18,97]. Pulse train generation, data rate transformation [98], code generation, pulse compression, matched filtering [40,99–101], optical differential amplifiers [102], and bit-serial digital computing [103] are also achievable.

The use of lattice structures in performing two important signal processing functions using temporally incoherent light, namely, matrix vector multiplication and frequency filtering, have been demonstrated. Although the speed of most experimental devices has been limited by the electronic components involved in the generation and detection of the optical signals, these processors are potentially capable of handling data frequencies into the multigigahertz region. Further applications may be found in diverse areas, such as special-purpose computing, matrix processing, and radar signal processing, as well as in optical communications (telephone network), local area networks, and interconnects.

*Outlook*

This chapter may result in stimulating further improvements in the design and fabrication of fiber-optic components, leading to faster and more efficient systems.

Fast switchable couplers and online fiber amplifiers would increase the capability and versatility of the fiber-optic processors presented in this work in addition to opening ways for the construction and demonstration of new processors. For instance, the inclusion of fast switchable couplers in the fiber-optic multiplier, described in [19], would make it possible to handle non-Toeplitz matrices. Iterative processing techniques could be applied to the fiber-optic matrix-vector multiplier described in this work to perform more complex matrix algebra, such as solving a linear matrix equation or computing the eigenvalues of a matrix. Also, by incorporating optical amplifiers, one could efficiently realize sharper and higher Q-factor optical filters. The techniques and experimental prototypes to be developed will depend heavily on the latest advances in optical amplifiers (in particular, EDFAs). These optical amplifiers currently involve high cost (~$20,000 per unit), limiting the extent that they can be employed in the experimental prototypes. But it is anticipated that the cost of these amplifiers will drop significantly in the future as they find wide applications and become mass producible. Moreover, the superior characteristics of fiber-optic technology will be utilized to greater advantage as faster optical sources and detectors become available.

Continued research is necessary so that new processing structures as well as data flow schemes can be developed which exploit the unique features of optics. One open area for research is to investigate the potential of fiber- and integrated-optic signal processing using coherence properties of light (coherent signal processing). In this case, data are transmitted by the phase, frequency, and the amplitude of a coherent optical carrier (rather than the light intensity). Coherent signal processing, in addition to providing better performance, might be able to achieve operations that are rather difficult or impossible to achieve with incoherent light. Further work is needed to explore the potentials of optical amplifiers in optical signal processing. Also, the implementation of optical devices which can perform nonlinear operations, such as logic functions, is of great importance in optical computing applications. Finally, it is hoped that the existing optical processing devices can be incorporated into larger systems capable of processing sophisticated data streams. Bandpass and notch filters, encoders/decoders, multiplexers/ demultiplexers, matrix algebraic processors, correlators, RF oscillators, and buffer memory are among some of the possible functional components that can be realized.

## ACKNOWLEDGMENTS

The author is grateful to Professor H. J. Shaw of the Edward L. Ginzton Laboratory, Stanford University, for his valuable comments and many helpful discussions. I also thank Professor C. C. Cutler, also of Edward L. Ginzton Laboratory, for his initial encouragement in writing this chapter and his constructive comments, and I thank Professor J. W. Goodman of the Electrical Engineering Department, Stanford University, for his valuable comments and support. The contributions of Professors Shaw, Goodman, and Cutler to parts of the work performed under their direction are also acknowledged.

## REFERENCES

[1] Maiman, T. H., "Stimulated Optical Radiation in Ruby," *Nature,* Vol. 187, 1960, pp. 493–494.

[2] Kao, K. C., and G. A. Hockman, "Dielectric Fiber Surface Waveguides for Optical Frequencies," *Proc. IEEE,* Vol. 133, 1966, pp. 1151–1158.

[3] Kapron, F. P., D. B. Keck, and R. D. Maurer, "Radiation Losses in Glass Optical Waveguides," *Appl. Phys. Lett.,* Vol. 17, 1970, pp. 423–425.

[4] Poole, S. B., D. N. Payne, and M. E. Fermann, "Fabrication of Low-Loss Optical Fibers Containing Rare-Earth Ions," *Electron. Lett.,* Vol. 21, 1985, pp. 737–738.

[5] Kobayashi, S., and T. Kimura, "Semiconductor Optical Amplifiers," *IEEE Spectrum,* June 1984, pp. 26–33.

[6] Kapany, N. S., *Fiber Optics, Principles and Applications,* New York: Academic Press, 1967.

[7] Wilner, K., and A. P. Van Den Heuvel, "Fiber Optic Delay Lines for Microwave Signal Processing," *Proc. IEEE,* Vol. 64, 1976, pp. 805–807.

[8] Chang, C. T., J. A. Cassaboom, and H. F. Taylor, "Fiber Optic Delay Line Devices for R.F. Signal Processing," *Electron. Lett.,* Vol. 13, 1977, pp. 678–680.

[9] Jackson, K. P., and H. J. Shaw, "Fiber-Optic Delay-Line Signal Processors," Ch. 7.1 in *Optical Signal Processing,* J. L. Horner, ed., 1987, pp. 431–476.

[10] Newton, S. A., K. P. Jackson, and H. J. Shaw, "Optical fiber V-Groove Transversal Filter," *Appl. Phys. Lett.,* Vol. 43, 1983, pp. 149–151.

[11] Newton, S. A., "Optical Fiber Delay Line Signal Processing," Ph.D. dissertation, Applied Physics Department, Stanford University, Stanford, CA, G.L. Report No. 3674, Jan. 1984.

[12] Jackson, K. P., "Fiber-Optic Delay Line Signal Processing: Coherent and Incoherent Systems," Ph.D. dissertation, Electrical Engineering Department, Stanford University, Stanford, CA, G.L. Report No. 3948, Oct. 1985.

[13] Tretter, S. A., *Introduction to Discrete Time Signal Processing,* New York: John Wiley, 1976.

[14] Mitra, S. K., and R. J. Sherwood, "Digital Ladder Networks," *IEEE Trans., Audio Electro-Acoustic,* Vol. AU-21, 1973, pp. 30–36.

[15] Markel, J. D., and A. H. Gray, *Linear Prediction of Speech,* New York: Springer-Verlag, 1976.

[16] Turner, J. M., "Use of the Digital Lattice Structure in Estimation and Filtering," in *Signal Processing: Theories and Applications,* M. Kunt and F. de Coulon, eds., North-Holland, EURASIP, 1980, pp. 33–41.

[17] Moslehi, B., J. W. Goodman, M. Tur, and H. J. Shaw, "Fiber-Optic Lattice Signal Processing," *Proc. IEEE,* Vol. 72, 1984, pp. 909–930.

[18] Moslehi, B., "Fiber-Optic Lattice Signal Processing," Ph.D. dissertation, Edward L. Ginzton Laboratory of the W. W. Hansen Laboratories of Physics, Electrical Engineering Department, Stanford University, Stanford, CA, G.L. Report No. 3755, Aug. 1984.

[19] Tur, M., J. W. Goodman, B. Moslehi, J. E. Bowers, and H. J. Shaw, "Fiber Optic Signal Processor With Applications to Matrix Vector Multiplication and Lattice Filtering," *Opt. Lett.,* Vol. 7, 1982, pp. 463–465.

[20] Taylor, H. F., "Optoelectronic Techniques for Microwave Signal Processing," *Proc. SPIE,* Vol. 1102, 1989, pp. 121–133.

[21] Mead, C. A., and L. A. Conway, eds., Ch. 9 in *Introduction to VLSI Systems,* Addison-Wesley, 1980.

[22] Moslehi, B., "Fiber-Optic Filters Employing Optical Amplifiers To Provide Design Flexibility," *Electron. Lett.,* Vol. 28, No. 3, 1992, pp. 226–227.

[23] Moslehi, B., and J. W. Goodman, "Novel Amplified Fiber-Optic Recirculating Delay Line Processor," *J. Lightwave Tech.,* Vol. JLT-10, No. 8, 1992, pp. 1142–1147.

[24] Vazquez, M. C., B. Vizoso, M. Lopez-Amo, and M. A. Muriel, "Single and Double Amplified Recirculating Delay Lines as Fiber-Optic Filters," *Electron. Lett.* Vol. 28, No. 11, 1992, pp. 1017–1019.

[25] Capmany, J., and J. Cascon, "Optical Programmable Transversal Filters Using Fiber Amplifiers," *Electron. Lett.* Vol. 28, No. 13, 1992, pp. 1245–1246.

[26] Moslehi, B., K. Chau, and J. W. Goodman, "Optical Amplifiers and Liquid Crystal Shutters Applied to Electrically Reconfigurable Fiber-Optic Signal Processors," *Optical Engineering,* Vol. 32, No. 5, 1993, pp. 974–981.

[27] Tangonan, G. L., "Optoelectronic for Radar Applications," *Proc. SPIE,* Vol. 1102, 1989, pp. 34–38.

[28] Oppenheim, A. V., and R.W. Schafer, *Digital Signal Processing,* Prentice-Hall, 1975.

[29] Oppenheim, A. V., ed., *Applications of Digital Signal Processing,* Prentice-Hall, 1978.

[30] Ogawa, K., "Considerations for Single-Mode Fiber Systems," *Bell Syst. Tech. J.,* Vol. 61, 1982, pp. 1919–1931.

[31] Lazay, P. D., and A. D. Pearson, "Developments in Single-Mode Fiber Design, Materials and Performance at Bell Laboratories," IEEE J. Quantum Electron., Vol. QE-18, 1982, pp. 504–510.

[32] Kawachi, M., A. Kawana, and T. Miyashita, "Low Loss Single-Mode Fiber at the Material Dispersion Free Wavelength of 1.27 $\mu$m," *Electron. Lett.,* Vol. 13, 1977, pp. 442–443.

[33] Miya, T., T. Terunuma, T. Hosaka, and T. Miyashita, ''Ultimate Low Loss Single-Mode Fiber at 1.55 $\mu$m,'' *Electron. Lett.*, Vol. 15, 1979, pp. 106–108.

[34] Beck, D. B., ''Single-Mode Fibers Outperform Multimode Cables,'' *IEEE Spectrum,* March 1983, pp. 30–37.

[35] Mochizuki, K., Y. Namihira, and H. Wakabayashi, ''Polarization Mode Dispersion Measurements in Long Single-Mode Fibers,'' *Electron. Lett.*, Vol. 17, 1981, pp. 153-154.

[36] Ramaswamy, V., W. G. French, and R. D. Standley, ''Polarization Characteristics of Non-circular Core Single-Mode Fibers,'' *Applied Optics,* Vol. 17, 1978, pp. 3014–3017.

[37] Keiser, G., *Optical Fiber Communications,* New York: McGraw-Hill, 1983.

[38] Cohen, L. G., and W. L. Mammel, ''Low Loss Quadruple Clad Single-Mode Lightguides With Dispersion Below 2 ps/km nm Over the 1.28 $\mu$m–1.65 $\mu$m wavelength range,'' *Electron. Lett.,* Vol. 18, 1982, pp. 1023–1024.

[39] Chang, C. T., D. E. Altman, D. R. Wehner, and D. J. Albares, ''Noncoherent Radar Moving Target Indicator Using Fiber-Optic Delay Lines,'' *IEEE Trans., Circuits Syst.,* Vol. CAS-26, 1979, pp. 1132–1135.

[40] Rausch, E. O., R. B. Efurd, and M. A. Corbin, ''A Fiber-Optic Pulse Compression Device for High Resolution Radars,'' *IEEE Conf. Radar,* London, 1982.

[41] Ainslie, B. J., K. J. Beales, D. M. Cooper, C. R. Day, and J. D. Rush, ''Monomode Fiber With Ultra Low Loss and Minimum Dispersion at 1.55 $\mu$m,'' *Electron. Lett.,* Vol. 18, 1982, pp. 842–844.

[42] Cohen, L. G., W. L. Mammel, and S. Lumish, ''Tailoring the Shapes of Dispersion Spectra to Control Bandwidths in Single Mode Fibers,'' *Opt. Lett.,* Vol. 7, 1982, pp. 183–185.

[43] Olsen, G. H., ''In GaAsP Laser Diodes,'' *Opt. Eng.,* Vol. 20, 1981, pp. 440–445.

[44] Lau, K. Y., N. Bar-Chaim, I. Ury, C. Harder, and A. Yariv, ''Direct Modulation of Short Cavity GaAs Lasers up to X-Band Frequencies,'' *Applied Physics,* Vol. 43, 1983, pp. 1–3.

[45] Wang, S. Y., and D. M. Bloom, ''100 GHz Bandwidth Planar GaAs Schottky Photodiode,'' *Electron. Lett.,* Vol. 19, 1983, pp. 554–555.

[46] Caulfield, H. J., J. A. Neff, and W. T. Rhodes, '' Optical Computing: The Coming Revolution in Optical Processing,'' *Laser Focus,* Vol. 19, 1983, pp. 100–110.

[47] Whitehouse, H. J., and J. Speiser, *Aspects of Signal Processing With Emphasis on Underwater Acoustics,* Part II, G. Tacconi, ed., Boston: Reidel, 1977.

[48] Psaltis, D., D. Casasent, D. Neft, and M. Carlotto, ''Accurate Numerical Computation by Optical Convolution,'' *Proc. SPIE,* Vol. 232, 1980, pp. 160–167.

[49] Smith, P. W., and W. J. Tomlinson, ''Bistable Optical Devices Promise Subpicosecond Switching,'' *IEEE Spectrum,* June 1981, pp. 26–33.

[50] Goodman, J. W., F. I. Leonberger, S. Kung, and R. A. Athale, ''Optical Interconnection for VLSI,'' *IEEE Proc.,* Vol. 7, 1984, pp. 850.

[51] Stephens, W. E., and W. H. Steier, ''Hybrid Optical Digital Signal Processing Applied to an Optimal Nonlinear Phase Estimator,'' *Applied Optics,* Vol. 22, 1983, pp. 787–795.

[52] Born, M., and E. Wolf, Ch. 10 in *Principles of Optics,* New York: Pergamon Press, 1975.

[53] Goodman, J. W., Chapter 5 in *Statistical Optics,* New York: John Wiley, 1984.

[54] Maitland, A., and M. H. Dunn, Ch. 10 in *Laser Physics,* Amsterdam: North-Holland, 1969.

[55] Van Etten, W., ''The Ergodicity of Laser Light in Connection With Optical Fiber Transmission,'' *Optics and Quantum Electron.,* Vol. 13, 1981, pp. 519–521.

[56] Smith, R. G., ''Optical Power Handling Capacity of Low Loss Optical Fibers as Determined by Stimulated Raman and Brillouin Scattering,'' *Applied Optics,* Vol. 11, 1972, pp. 2489–2494.

[57] Yamamoto, Y., and T. Kimura, ''Coherent Optical Fiber Transmission Systems,'' *IEEE J. Quantum Electron.,* Vol. QE-17, 1981, pp. 919–935.

[58] Davies, D.E.N., and G. W. James, ''Fiber Optic Tapped Delay Line Filter Employing Coherent Optical Processing,'' *Electron. Lett.,* Vol. 20, 1984, pp. 95–97.

[59] Jackson, K. P., G. Xiao, and H. J. Shaw, ''Coherent Optical Fiber Delay Line Processor,'' *Electron. Lett.,* Vol. 22, No. 25, 1986, pp. 1335–1337.

[60] Sasayama, K., M. Okuno, and K. Habara, ''Coherent Optical Transversal Filter Using Silica-Based Waveguides for High-Speed Signal Processing,'' *J. Lightwave Tech.,* Vol. JLT-9, No. 10, 1991, pp. 1225–1230.

[61] Franklin, G. F., J. D. Powel, and A. Emami-Naeini, Ch. 2 in *Feedback Control of Dynamic Systems,* Menlo Park, CA: Addison-Wesley, 1986.

[62] Kringlebotn, J. T., P. R. Morkel, C. N. Pantell, D. N. Payne, and R. I. Laming, ''Amplified Fiber Delay Line With 27000 Recirculations,'' *Electron. Lett.*, Vol. 28, No. 2, 1992, pp. 201–202.

[63] Giles, C. R., E. Desurvire, and J. L. Zyskind, ''Erbium-Doped Fiber Amplifiers for High-Speed Fiber-Optic Communication System,'' *SPIE Proc.,* Vol. 1171, 1989, pp. 318–327.

[64] Ghosh, A., J. Barner, and P. Paparao, ''Design of Least-Mean-Square Based Adaptive Optical Equalizers,'' *Optics Communications,* Vol. 91, 1992, pp. 280–292.

[65] Kung, H. T., and C. E. Leiserson, ''Highly Concurrent Systems,'' Section 8.3 in *Introduction to VLSI Systems,* C. A. Mead and L. A. Conway, eds., MA: Addison-Wesley, 1980; H. T. Kung, ''Why Systolic Architectures?'' *Computer,* Vol. 15, 1982, pp. 37–46.

[66] Caulfield, H. J., W. T. Rhodes, M. J. Foster, and S. Horvitz, ''Optical Implementation of Systolic Array Processing,'' *Optical Communication,* Vol. 40, 1981, pp. 89–90.

[67] Luenberger, D. G., *Introduction to Dynamic Systems: Theory, Models and Applications,* John Wiley, 1979.

[68] Goodman, J. W., and L. M. Woody, ''Methods for Performing Complex Valued Linear Operations on Complex Valued Data Using Incoherent Light,'' *Applied Optics,* Vol. 16, 1977, pp. 2611–2612.

[69] Goodman, J. W., A. R. Dias, L. M. Woody, and J. Erickson, ''Application of Optical Communication Technology to Optical Information Processing,'' *Proc. SPIE,* Vol. 190, 1979, pp. 485–496.

[70] Glaser, I., ''Representing Bipolar and Complex Imagery on Noncoherent Optics Image Processing Systems: Comparison of Approaches,'' *Opt. Eng.,* Vol. 20, 1981, pp. 568–573.

[71] Gookin, D. M., and M. H. Berry, ''Finite Impulse Response Filter With Large Dynamic Range and High Sampling Rate,'' *Applied Optics,* Vol. 29, No. 8, 1990, pp. 1061–1062.

[72] Romanovskii, V. I., *Discrete Markov Chains*, Moscow: Gostekhizdat, 1948.

[73] Gantmacher, F. R., and M. G. Krein, *Oscillation Matrices and Kernels and Small Vibrations of Dynamical Systems,* 2d ed., Moscow: Gostekhizdat, 1950.

[74] Frobenius, G., ''Uber matrizen aus positiven elementen,'' *Sitzungsberichte, Konigl. Preussichen Akad. Wiss,* Vol. 8, 1908, pp. 471–476.

[75] Perron, O., ''Zur theorie der matrizen,'' *Math. Ann.,* Vol. 64, 1907.

[76] Gantmacher, F. R., Ch. 13 in *The Theory of Matrices,* Vol. 2, New York: Chelsea, 1960.

[77] Bracewell, R. N., Ch. 3 in *The Fourier Transform and its Applications,* 2d ed., New York: McGraw-Hill, 1978.

[78] Taylor, H. F., ''Fiber and Integrated Optical Devices for Signal Processing,'' *Proc. SPIE,* Vol. 176, 1979, pp. 17–27.

[79] See, for example, R. A. Bergh, G. Kotler, and H. J. Shaw, ''Single-Mode Fiber-Optic Directional Coupler,'' *Electron. Lett.,* Vol. 16, 1980, pp. 260–261; or S. K. Sheem and T. G. Giallorenzi, ''Single-Mode Fiber Optical Power Divider: Encapsulating Etching Technique,'' *Opt. Lett.,* Vol. 4, 1979, pp. 29–31.

[80] Digonnet, M.J.F., and H. J. Shaw, ''Analysis of a Tunable Single Mode Optical Fiber Coupler,'' *IEEE J. Quantum Electron.,* Vol. QE-18, 1982, pp. 746–754.

[81] Davies, D.E.N., and S. A. Kingsley, ''Methods of Phase Modulating Signals Optical Fibers: Applications to Optical Telemetry Systems,'' *Electron. Lett.,* Vol. 10, 1974, pp. 21–22.

[82] Trowbridge, F. R., and R. L. Phillips, ''Metallic Glass Fiber-Optic Phase Modulators,'' *Opt. Lett.,* Vol. 6, 1981, pp. 636–638.

[83] Lefevre, H. C., ''Single Mode Fiber Fractional Wave Devices and Polarization Controllers,'' *Electron. Lett.,* Vol. 16, 1980, pp. 778–780.

[84] Bergh, R. A., H. C. Lefevre, and H. J. Shaw, ''Single-Mode Fiber-Optic Polarizer,'' *Opt. Lett.,* Vol. 5, 1980, pp. 479–481.

[85] Eickhoff, W., ''In-Line Fiber-Optic Polarizers,'' *Electron. Lett.,* Vol. 16, 1980, pp. 762–763.
[86] Nosu, K., S. C. Rashleigh, H. F. Taylor, and J. F. Weller, ''Acousto-Optic Frequency Shifter for Single-Mode Fibers,'' *Electron. Lett.,* Vol. 19, 1983, pp. 816–818.
[87] Risk, W. P., R. C. Youngquist, G. S. Kino, and H. J. Shaw, ''Acousto-Optic Frequency Shifting in Birefringent Fiber,'' *Opt. Lett.,* Vol. 9, 1984, pp. 309–311.
[88] Injeyan, H., O. M. Stafsudd, and N. G. Alexopoulos, ''Light Amplification by Evanescent Wave Coupling in Multimode Fiber,'' *Applied Optics,* Vol. 21, 1982, pp. 1928–1932.
[89] Periasamy, N., and F. P. Schafer, ''Laser Amplification in an Optical Fiber by Evanescent Field Coupling,'' *Applied Physics,* Vol. 24, 1981, pp. 201–203.
[90] Sorin, W. V., K. P. Jackson, and H. J. Shaw, ''Evanescent Amplification in a Single-Mode Optical Fiber,'' *Electron. Lett.,* Vol. 20, 1983, pp. 820–822.
[91] Sargent, M., III, M. O. Scully, and W. E. Lamb, Jr., Ch. 20, *Laser Physics,* MA: Addison-Wesley, 1974.
[92] Tur, M., B. Moslehi, J. E. Bowers, S. A. Newton, K. P. Jackson, J. W. Goodman, C. C. Cutler, and H. J. Shaw, ''Spectral Structure of Phase Induced Intensity Noise in Recirculating Delay Lines,'' *Proc. SPIE,* Vol. 412, April 1983.
[93] Tur, M., and B. Moslehi, ''Laser Phase Noise Effects in Fiber-Optic Signal Processors With Recirculating Loops,'' *Opt. Lett.,* Vol. 8, 1983, pp. 229–231.
[94] Tur, M., B. Moslehi, and J. W. Goodman, ''Theory of Laser Phase Noise in Recirculating Fiber-Optic Delay Lines,'' *J. Lightwave Tech.,* Vol. JLT-3, No. 1, 1985, pp. 20–31.
[95] Moslehi, B., ''Analysis of Optical Phase Noise in Fiber-Optic Systems Employing a Laser Source With Arbitrary Coherence Time,'' *J. Lightwave Tech.,* Vol. JLT-4, No. 9, 1986, pp. 1334–1351.
[96] Moslehi, B., ''Noise Power Spectra of Optical Two-Beam Interferometric Induced by the Laser Phase Noise,'' *J. of Lightwave Tech.,* Vol. JLT-4, No. 11, 1986, pp. 1704–1710.
[97] Bowers, J. E., S. A. Newton, W. V. Sorin, and H. J. Shaw, ''Filter Response of Single Mode Fiber Recirculating Delay Lines,'' *Electron. Lett.,* Vol. 18, 1982, pp. 110–111.
[98] Newton, S. A., R. S. Howland, K. P. Jackson, and H. J. Shaw, ''High Speed Pulse Train Generation Using Single Mode Fiber Recirculating Delay Lines,'' *Electron. Lett.,* Vol. 19, 1983, pp. 756–758.
[99] Ohlhaber, R. L., and K. Wilner, ''Fiber-Optic Delay Lines for Pulse Coding,'' *Electro-Optical System Design,* Feb. 1977, pp. 33–35.
[100] Marom, E., ''Optical Delay Line Matched Filters,'' *IEEE Trans. Circuits and Systems,* Vol. CAS-25, 1978, pp. 360–364.
[101] Jackson, K. P., S. A. Newton, and H. J. Shaw, ''A 1 Gigabit/sec Code Generator and Matched Filter Using an Optical Fiber Tapped Delay Line,'' *Appl. Phys. Lett.,* Vol. 42, 1983, pp. 556–558.
[102] Berry, M. H., and D. M. Gookin, ''Integrated Optical Differential Amplifier With Large Dynamic Range for Gigahertz Bandwidth Electrical Signals,'' *IEEE Photonics Tech. Lett.,* Vol. 3, No. 3, 1991, pp. 276–277.
[103] Benner, A. F., K. F. Jordan, and V. P. Heuring ''Digital Optical Computing With Optically Switched Directional Couplers,'' *Opt. Eng.,* Vol. 30, No. 12, 1991, pp. 1936–1941.

# Chapter 18
# Summary, Future Opportunities, and Directions

***Michael L. VanBlaricum***
*Toyon Research Corporation*

Future advanced military and commercial systems will require antenna and phased-array radar and communication systems with strict performance requirements. These include bandwidths of an octave or more, dramatic reduction in size and weight, low observability, increased efficiency, and greater isolation from electromagnetic interference and crosstalk between modules and subarray feeds. These requirements are driving the search for innovative antenna elements, antenna feeds, and control interfaces, as well as novel signal distribution, control, and processing systems.

This book has introduced photonics, or optoelectronics, from a radar point of view. Photonic links and systems can be used for array functions as simple as the transmission of digital signals to control the phase and amplitude weights for each element of an array. A more challenging application of photonics is the actual distribution of RF signals to and from antenna elements via optical-fiber systems. Finally, once the RF signal is in the optical domain, one can use optically based space-time signal processing to perform beam forming and signal processing functions. This will ultimately allow some of the massive processing required from large digital beam forming phased arrays to be handled before the signal is ever digitized and put into the computer.

For well over a decade now, research groups around the world have been investigating the use of direct and external modulation devices connected to optical-fiber systems for distribution of analog RF signals in antenna and electromagnetic field sensor systems. Applications have included shipboard electromagnetic emission monitoring, nuclear and lightning-generated electromagnetic pulse measurements, anechoic chamber calibration, near-field probing, and antenna remoting, as well as novel and sophisticated antenna and array designs. As discussed in this book, optical-fiber and photonic devices offer many advantages to the antenna and phased-array designer:

- Low transmission loss over long distances compared to conventional coaxial and waveguide feeds;
- Light weight;
- Small size;
- Broad bandwidth;
- Immunity from electromagnetic interference and crosstalk;
- Low observability due to the dielectric nature of the fibers;
- Suitable for implementing beam steering and signal processing in the optical domain.

Fiber-optic links were initially used for long-haul digital telecommunications applications. The requirement for long cables focused the research on high-speed, low-loss cable development. Antenna and array applications rarely have cable runs of more than a few hundred meters, with high-frequency array interconnects being as short as a few centimeters. Therefore, phased-array applications of photonics are driven more by the performance of the optical devices: lasers, modulators, detectors, switches, and amplifiers. The development of temperature-stable devices with broad bandwidth and low noise has been of critical importance to antenna and microwave applications. Fortunately, the search for high-speed interconnection using optical devices for the optical digital telecommunications markets and to overcome the interconnect bottleneck in high-speed digital computing architectures has helped push the high-frequency characteristics of lasers, modulators, and detectors. It was only a few years ago that optical photodetectors were limited to a few gigahertz. Now it is possible to obtain good detectors which work above 30 GHz.

Recent developments in laser technology allow both direct and externally modulated links to have very low noise figure and high dynamic range. In fact, it is now possible to build analog fiber-optic receive links with intrinsic RF power gain and very low noise figure without having to use RF amplifiers [1]. Having RF links with intrinsic gain allows for the possible reduction of the antenna element size and potential elimination of preamplifiers at the antenna interface.

Optoelectronic devices, which have historically and are still predominantly built in inorganic crystalline materials such as $LiNbO_3$, are now being built in III-V semiconductors (GaAs, InP). The use of these materials will ultimately lead to optoelectronic integration with MMIC devices [2]. Therefore, one can see the merging of photonic and microwave circuitry on one MMIC substrate. The MMIC devices can then be controlled via optical fiber and, likewise, signals can be retrieved via optical fibers. Hence, the control and RF connectivity to an MMIC-based T/R module can be made via optical fiber. This will certainly increase the speed and bandwidth as well as reduce the weight and size of interconnections to phased-array elements.

Similarly, breakthroughs in electro-optical polymer technology will allow optical interconnects to move into another direction. Nonlinear optical (NLO) polymers have many advantages over inorganic crystals for use in electro-optical devices [3,4]. The main advantage is the low dielectric constant, which will reduce the required RF drive signal power and increase the bandwidth. In addition, polymer fabrication techniques are inexpensive and compatible with well-established device fabrication methods. Polymers can be spin-coated over a surface. One could imagine a 2-D array of metallic antenna elements deposited on a dielectric substrate and then overcoated with a layer of electro-optical polymer. The optical modulators could be placed in the polymer to line up with the terminals of the antenna elements. Optical waveguides would be deposited in the polymer to carry the signals. Hence, a thin-film, surface-conformal optical interconnected array could be made very cheaply. This technology bodes well for the phased-array "smart skins" the military has been talking about for some time.

Other new developments are increasing the capabilities and utility of photonic RF links. The linear dynamic range of externally modulated photonic systems is limited by the noise floor at the low end and the linearity at the high end. For Mach-Zehnder interferometric modulators, the nonlinearity is known and can be compensated for in a number of ways, including postdetection digital correction and feed-forward and feed-backward techniques [5].

The introduction of erbium-doped fiber amplifiers has started a revolution in optical communications, opening new possibilities in the domains of high bit rate, long-distance transmission, analog and digital video distribution, and optical switching [6]. While most currently envisioned applications of fiber amplifiers are for telecommunications and cable TV applications, it is just a matter of time until the antenna and radar communities take advantage of the opportunity. It has already been suggested that the use of fiber amplifiers can reduce the total number of source lasers needed for high gain, high dynamic range, and low noise figures for externally modulated antenna links. In the limit, a single laser could be used to drive all the optical links, with fiber amplifiers being used in each individual line to increase the optical power level and to overcome optical insertion losses.

As reiterated above, using optical fiber for RF signal distribution in active phased-array antennas offers a number of advantages. The natural extension of optical signal distribution is to carry out phase shifting operations required for antenna beam forming in the optical domain as well. The book has discussed methods for performing beam forming for phased-array antennas. Beam forming, based on differential delay networks, can be used to produce simple, frequency-independent beam steering, and exploits the low-loss and dispersion advantages of optical transmission. Coherent beam forming techniques, while more complex, allow greater flexibility and appear to have a promising future. Using photonic beam forming techniques gives the ability to form multiple, simultaneous, independent beams with nearly arbitrary frequency relationships. In addition, photonic techniques allow the incorporation of signal processing operations, with beam forming allowing the possibilities of optical pattern recognition for image processing and target identification, remote sensing, and electronic countermeasures.

The ultimate use of optics in radar is laser radar, LIDAR or LADAR. Of course, in this case, the transmittted and received signal is a laser pulse instead of an RF pulse. Besides the military applications of laser radar, it is currently being used by the commercial sector as well [7]. Cloud and aerosol LIDARS are being used throughout the world. LIDARS operating from spacecraft platforms have been discussed for many years. A laser altimeter instrument is included on the Mars Observer. One of the most important emerging uses of LIDAR is for remote sensing applications, such as detection of air pollution and toxic chemical and gas detection. Growth of LIDAR use in these areas will be driven by the newly emerging tunable laser sources.

Photonic technologies for use in phased-array systems has benefited greatly from research and product development for the fields of telecommunications, cable television, and optical computing. So it is important that we continue to track the optoelectronics development in these fields. At the same time, however, it is necessary that we continue to carry our own load. If anything has held back the application of photonics to antenna and phased-array systems (besides general component availability and cost), it is the lack of cross-training of antenna and microwave engineers in the photonics technologies. This book goes a long way toward meeting that need [8].

As the telecommunications and cable television technologies turn more toward the use of digital signal transmission, it is possible that the requirements of the photonic technologies developed for those applications may diverge from those needed in phased-array systems. The development of techniques to propagate optical soliton waves in long-haul telecommunications is a case in point.

Soliton propagation depends on the nonlinear properties of optical fibers. This same nonlinearity is deleterious to amplitude-modulated signals commonly used in array links. On the other hand, the development of soliton techniques has simultaneously boosted the development of the erbium-doped fiber amplifiers, which, as stated above, may prove very useful to phased-array applications.

It is clearly true that photonic systems will revolutionize the way in which antenna systems will be designed and built in the future. It is not only important that antenna and radar engineers understand the opportunities afforded them by implementation of optoelectronics into their systems, it is also important that they make use of these enhanced capabilities to create new antenna and radar designs. Antenna and matching networks are already the bandwidth-limiting devices in most photonic phased-array system designs being considered.

## REFERENCES

[1] Cox, C. H., III, "Analog Fiber-Optics Links With Intrinsic Gain," *Microwave J.,* Vol. 35, No. 9, Sept. 1992, pp. 90–99.

[2] Herczfeld, P., "Monolithic Microwave-Photonic Integrated Circuits: A Possible Follow-up to MIMIC," *Microwave J.,* Vol. 35, No. 1, Jan. 1992, pp. 64–78.

[3] Horak, L. A.. ed., Polymers for Lightwave and Integrated Optics: Technology and Application, New York: Marcel Dekker, 1992.

[4] Chen, M., L. Yu, L. R. Dalton, Y. Shi, and W. H. Steier, "New Polymers With Large and Stable Second Order Nonlinear Optical Effects," *Macromolecules,* Vol. 24, 1991, pp. 5421–5428.

[5] Farwell, M. L., and W.S.C. Chang, "An Electro-Optical Intensity Modulator With Improved Linearity," *Proc. 2nd Annual DARPA/Rome Laboratory Symp. on Photonic Systems for Antenna Applications,* Monterey, CA, 10–12 Dec. 1991, pp. 77–79.

[6] Zaglankis, C. D., and A. J. Seeds, ''The Use of Optical Amplifiers for Signal Distribution in Optically Controlled Phased Array Antennas,'' *J. Communications,* Vol. 43, Nov. 1992, pp. 6–13.
[7] Killinger, D. K., and R. T. Menzies, ''Recent Advances in LIDAR/Laser at Atmospheric Sensing,'' *LEOS Newsletter,* Vol. 7, No. 2, April 1993.
[8] Haus, H. A., ''Molding Light Into Solitons,'' *IEEE Spectrum,* March 1993, pp. 48–53.

# *List of Acronyms*

| | |
|---|---|
| AO | acousto-optic |
| A/D | analog-to-digital |
| ABMDA | Advanced Ballistic Missile Defense Agency |
| AFORDL | amplified fiber-optic recirculating delay line |
| APC | angled polished connector |
| APD | avalanche photodiode |
| AR | antireflection |
| ARPA | Advanced Research Projects Agency |
| ASE | amplified spontaneous emission |
| AWACS | Airborne Early Warning and Aircraft Control System |
| BIFODEL | binary fiber-optic delay line |
| BPF | bandpass filtering |
| BSO | bismuth silicon oxide |
| BW | bandwidth |
| CDM | central data mixing |
| CDR | compression-limited dynamic range |
| CFAR | constant false alarm rate |
| Comint | communication intelligent |
| CTP | charge-transfer plate |
| CVR | crystal video receiver |
| CW | continuous wave |
| DDL | dispersive delay line |
| DEMUX | demultiplexer |
| DFB | distributed feedback |
| DiBi | binary delay lines |
| DKDP | deuterated potassium dihydrogen phosphate |
| DMAC | digital multiplication by analog convolution |
| DMD | deformable mirror devices |

| | |
|---|---|
| DR | dynamic range |
| DRAM | dynamic random-access memory |
| e-LCLM | electron-beam-addressed liquid crystal light modulator |
| EDFA | erbium-doped fiber amplifier |
| ELINT | electronic intelligent |
| EMI | electromagnetic interference |
| EO | electro-optic |
| EOM | electro-optical modulators |
| EW | electronic warfare |
| FAA | Federal Aviation Association |
| FFT | fast Fourier transform |
| FIR | finite-impulse response |
| FLC | ferroelectric liquid crystal |
| FO | fiber-optic |
| FORDL | fiber-optic recirculating delay line |
| FP | Fabry-Perot |
| GRIN | gradient index |
| HPT | heterojunction phototransistors |
| IC | integrated circuit |
| IF | intermediate frequency |
| IFM | instantaneous frequency measurement |
| IIR | infinite-impulse response |
| IL | insertion loss |
| IM | intermodulation |
| IMP | inmodulation products |
| I/O | input/output |
| IO | integrated optical |
| ISA | interferometric spectra analyzer |
| ISAR | inverse synthetic aperture radar |
| JSTARS | Joint Stars |
| LCLV | liquid crystal light valve |
| LED | light-emitting diode |
| lidar | light detection and ranging |
| LMS | least mean square |
| LO | local oscillator |
| LP | linearly polarized |
| LPE | liquid-phase epitaxy |
| LPF | low-pass filtering |
| LPI | low probability of intercept |
| LTI | linear time-invariant |
| MBE | molecular beam epitaxy |
| MIC | microwave integrated circuit |

| | |
|---|---|
| MIFT | miniature instantaneous Fourier transform |
| MIMO | multiple-input multiple-output |
| MLM | membrane mirror SLMs |
| MMIC | monolithic microwave integrated circuit |
| MO | master oscillator |
| MOSLM | magneto-optic SLM |
| MQW | multiple QW |
| MSLM | microchannel SLM |
| MSM | metal-semiconductor-metal |
| MTD | moving target detection |
| MTI | moving target indicator |
| MUX | multiplexer |
| NA | numerical aperuture |
| NF | noise figure |
| NF | notch filtering |
| NLO | nonlinear optical |
| OCPAR | optically controlled phased-array radar |
| OEIC | optoelectronic integrated circuit |
| OTH | over the horizon |
| PAA | phased-array antennas |
| PADEL | parallel delay line |
| PD | photodiodes |
| PD | pulsed Doppler |
| PDLC | polymer-dispersed liquid crystal |
| PIC | photonic integrated circuits |
| PICOC | photorefractive incoherent-to-coherent optical converter |
| POI | probability of intercept |
| PRF | pulse repetition frequency |
| PRI | pulse repetition interval |
| PROM | Pockels readout optical modulator |
| PSA | power spectra analyzer |
| QW | quantum well |
| RC | resistance-capacitance |
| RDM | remote data mixing |
| RF | radio frequency |
| RIN | relative intensity noise |
| rms | root mean square |
| RRE | radar range equation |
| RS | reference set |
| RWR | radar warning receiver |
| SAR | synthetic aperture radar |
| SAW | surface acoustic wave |

| | |
|---|---|
| SDC | Strategic Defense Command |
| SEED | self-electro-optic device |
| SFD | signal flow diagram |
| SFDR | spurious-free dynamic range |
| SHR | superheterodyne receiver |
| SIR | shuttle imaging radars |
| SISO | single-input single-output |
| SLM | spatial light modulator |
| SMD | segmented mirror device |
| SMF | single-mode fiber |
| SNR | signal-to-noise ratio |
| SOA | semiconductor optical-fiber amplifiers |
| SOP | state of polarization |
| SPDT | single-pole double-throw |
| SPIE | Society of Photo-optical Instrumentation Engineers |
| SRAM | static random-access memory |
| SRODEL | square-root fiber-optic delay line |
| SSB | single sideband |
| STC | sensitivity time control |
| T/R | transmit/receive |
| TDU | time-delay unit |
| TE | transverse electric |
| TEM | transverse electromagnetic |
| TGG | terbium gallium garnet |
| TLD | tunable laser diodes |
| TM | transverse magnetic |
| TMD | theater missile defense |
| TOA | time of arrival |
| TRL | through-reflect line |
| TSS | tangential sensitivity |
| TTD | true time delay |
| UFORDL | unamplified fiber-optic recirculating delay line |
| VCSEL | vertical cavity surface-emitting laser diodes |
| VLSI | very-large-scale integration |
| VSTEP | vertical-to-surface transmission electrophotonic |
| VSWR | voltage standing wave ratio |
| WDM | wavelength division multiplexing |
| YAG | yttrium aluminum garnet |
| YIG | yttrium iron garnet |

# *About the Authors*

*Edward Ackermann* received his B.S. degree from Lafayette College and his M.S. degree from Drexel University, both in electrical engineering. He is presently pursuing his Ph.D. degree from Drexel. He is currently employed as a microwave engineer by the Martin Marietta Electronics Laboratory in Syracuse, New York, where he is developing high-fidelity photonic systems for microwave and millimeter-wave radar and communications applications.

*David Barton* received his A.B. in physics from Harvard University. He is currently vice-president for engineering at ANRO Engineering. He also lectures at George Washington University and is series editor for the Artech House Radar Series. During his career, he has been associated with the Raytheon Corp., the RCA Corp., and the U.S. Army Signal Corps Engineering Laboratories. His fields of research include radar systems, propagation of radar waves, radar tracking and measurement, and radar guidance of missiles. He holds numerous awards and honors including the IEEE Microwave Theory and Technique Distinguished Lecturer and the IEEE Centennial Medial. He is also a fellow of the IEEE.

*I. C. Chang* received his Ph.D. in electrical engineering from Stanford University. He is the former director of electro-optic technology at Litton Applied Technology, where he led the effort on the development of the miniaturized AO channelized receiver. He is currently the president of Aurora Associates, an R&D company specializing in acousto-optic devices and signal systems. He is a fellow of the Optical Society of America.

*D. Kenneth Davies* received his B.Sc. and Ph.D. degrees in physics from the University of Wales (UK). He is currently with the Westinghouse R&D Center, where he is engaged in research in the areas of atomic physics, spectroscopy, vacuum discharges, gaseous electronics, optical computing, and optical signal processing. His current research interests include precision measurements of electron and ion transport in gases, long-delay acousto-optic Bragg cells, and high-speed fiber optic signal processing. He is a fellow of the Institute for Physics and a member of APS.

*Barry Fell* received his B.S. from Case Institute of Technology and his M.S. from the State University of New York at Buffalo, both in physics. He is currently vice president of remote sensing and advanced signal processing at GORCA Systems Inc. His work is in the development and demonstration of a wide range of advanced radar and sonar system concepts, including distributed aperture radar and sonar, airborne early warning systems, and space-based surveillance systems. He leads a team of experts that are developing leading-edge technology advances in SAR/ISAR motion compensation, adaptive phased array radar self-calibration, and adaptive space-time processing. He also serves on the faculty of Drexel University, teaching radar at the graduate level.

*Anastasios P. "Akis" Goutzoulis* received his B.S. degree in physics from the University of Ioannina, Greece, his M.S. degree in optoelectronics from the University of Essex, England, and his Ph.D. in electrical engineering from Carnegie-Mellon University. He is currently with the Westinghouse Science and Technology Center, where he is involved in research in the areas of optics, acousto-optic devices, acousto-optic and electro-optic signal processing, optical computing, and optical communications. His current research interests are the areas of wideband fiber optic signal processing for phased-array antennas, microwave- and millimeter-wave analog fiber optic links, and ultrawideband Bragg cells. He is a member of the OSA, SPIE, and IEEE.

*Paul C. Hamilton* received his B.S. in engineering from the U.S. Naval Academy and his Ed.M. and Ed.D. degrees in mathematics and science from Harvard University. He is currently vice-president for radar studies at ANRO Engineering. During his career, he has been associated with the Raytheon Corp. and the Hughes Aircraft Corp. His fields of research include radar systems, air defense systems, airborne radar and avionics technology, and missile seekers. He is also a senior member of the IEEE.

*William Jemison* received his B.S. in electrical engineering from Lafayette College, his M.Sc. from Pennsylvania State University, and his Ph.D. in electrical engineering from Drexel University. He is currently employed by Martin Marietta Government Electronic Systems in Moorestown, NJ, where he is involved in the development of solid-state phased array antenna systems. He was previously employed by the Naval Air Warfare Center Aircraft Division, where he was involved with the development of phased-array radar for airborne early-warning applications and photonic techniques for phased-array beamsteering. He is a member if the IEEE and vice chairman of the Philadelphia Microwave Theory and Technique and Antennas and Propagation Society.

*Kam Y. Lau* received his B.S., M.S., and Ph.D. in electrical engineering from the California Institute of Technology. After serving seven years as chief scientist for the Ortel Corporation, he joined the faculty of Columbia University. He is presently professor of electrical engineering at the University of California at Berkeley.

*Emmett Leith* received his M.S. in physics and Ph.D. in electrical engineering from Wayne State University. He has been on the faculty of the University of Michigan for over thirty years and is currently the Schlumberger Professor of Electrical Engineering. His work is primarily in the areas of optical processing, holography, interferometry, and synthetic apeture radar. He holds numerous honors and distinctions, including the National

Medal of Science. He is a fellow of the IEEE, OSA, and SPIE, and a member of the National Academy of Engineering.

*Robert J. Mailloux* received his B.S. in electrical engineering at Northeastern University and his M.S. and Ph.D. from Harvard University. He is currently a senior scientist at the U.S. Air Force Rome Laboratory. He has served as the chief of the antennas and components division at the Rome Air Development Center and as a physicist at the Air Force Cambridge Research Laboratory. He is an expert on antenna systems and a fellow of the IEEE.

*Behzad Moslehi* received his B.S. in electrical engineering from Arya-Meher (Sharif) University of Technology, Iran, an M.S. in applied physics and an M.S. and Ph.D. in electrical engineering from Stanford University. He has been associated with Quantex, Xerox, Litton, and Physical Optics Corp. and been a visiting research scholar at Stanford's Ginzton Laboratory. He is currently a senior member of the technical staff at Optivision where he performs research in the area of fiber-optic signal processing using novel methods. He is a member of OSA and SPIE and a senior member of the IEEE.

*Douglas Norton* received his B.E. from Clarkson University and his M.S. from Syracuse University, both in electrical engineering. He is presently employed by the Rome Laboratory Photonic Center, where he is involved with research in optical signal processing for microwave and communications applications. He is also the founder and president of Otter Solutions Inc., a company providing specialized advanced software products to the optical- and radar-engineering industry.

*Joe Paslaski* received his B.S., M.S., and Ph.D. in applied physics, all from the California Institute of Technology. He is presently with Ortel Corporation, where he is the principal investigator for research projects in high-speed photodiodes, optoelectronic mixers, mode-locked laser devices, among many others. He has extensive experience in high-speed optoelectronic devices, including techniques for picosecond photoconductive sampling, as well as in fiber amplifiers for CATV systems.

*Art Paolella* received his B.S. from Monmouth College, his M.S. from Fairleigh Dickinson University, and his Ph.D. from Drexel University. He presently leads the Microwave Photonics Team within the Electronic and Power Sources Directorate at the U.S. Army Research Laboratory at Fort Monmouth, where he is responsible for research in the area of optical microwave interactions, including research to develop two- and three-terminal optical detectors, analog microwave fiber optic links, fiber optic delay lines, and photonic integrated circuits. He is a senior member of the IEEE.

*Dennis R. Pape* received his B.A. in physics from Cornell University and his Ph.D. in physics from Duke University. He was a member of the technical staff in the Optical Information Processing Branch at Texas Instruments' Central Research Laboratory, where he designed and developed an integrated silicon-based optically addressed membrane spatial light modulator. He was group leader of the Acousto-Optic Device Development Group at Harris Corporation where he was responsible for the development and application of advanced acousto-optic devices for military optical signal processing systems. He is currently the president of Photonic Systems Inc. His research interests include wideband–

high resolution acousto-optic spectrum analysis for instantaneous frequency measurement for signal analysis, spectroscopic, and test- and measurement-equipment applications. He is a member of the Optical Society of America and the American Physical Society.

*William Penn* received his B.S. from Washington University and his Ph.D. from Syracuse University, both in electrical engineering. Most of his career has been spent at the General Electric Corp. as manager of the signal processing and optical technology units and as a consulting engineering in laser applications and optics. His research areas include electronic, radar, and sonar signal-processing systems, laser and optical applications, and optical holography. He is presently on the faculty of Syracuse University and a consultant at Martin Marietta.

*Richard Soref* received his B.S. and M.S. from the University of Wisconsin and his Ph.D. from Stanford University, all in electrical engineering. He has spent two years at the Massachusetts Institute of Technology's Lincoln Labs, eighteen years at Sperry Research Center, and ten years at Rome Labs, where he is presently a research scientist in the Photonic Device Technology Division. His research interests include III-V and Group-VI quantum-well photonics. He is a member of APS, OSA, MRS, SPIE, and a senior member of IEEE.

*Denzil Stilwell* received his A.B. in physics from West Virginia University and his M.S. in physics from American University. He is currently employed by the Naval Research Laboratory and works in the areas of remote sensing, signal processing, antennas and radar. His specialty is the use of optical alternatives to the more common RF and microwave approaches in system implementations. These include photographic and video means of ocean-wave spectral determination, high-speed analog optical methods of processing digital data, and the use of analog bulk optics for beam forming and adaptive beam synthesis in array antennas.

*Edward Toughlian* received his B.S. from Arizona State University and his M.S. and Ph.D. from Syracuse University, all in electrical engineering. He is currently an officer in the U.S. Air Force, assigned to the Rome Laboratory Photonics Center, where he is the director of the Photonic Microwave Systems Division. A specialist in Air Force radar and communications systems, his research interests are in the use of photonic methods for system implementation of RF, microwave and millimeter-wave radar and telecommunications systems. He is a member of the IEEE and the Air Force Electronics and Communications Association.

*Cardinal Warde* received his B.S. from Stevens Institute of Technology and his M.Phil. and Ph.D. from Yale University. He is presently professor of electrical engineering and computer science at the Massachusetts Institute of Technology. His research activities are in the areas of spatial light modulators, electro-optic and photorefractive optical information processing and computing.

*Michael VanBlaricum* received his B.S., M.S., and Ph.D. in electrical engineering from the University of Illinois. He is vice president and director of advanced technology for Toyon Research Corp. and has over twenty years of research experience in the following areas: antenna analysis, electromagnetic scattering, interaction, and coupling; signature prediction and control; and applications of photonics to antenna design. Prior to joining Toyon, he managed the Engineering Research Group at General Research Corp., where he worked on the development of photonic electromagnetic field sensor. He is a member of the IEEE and presently the IEEE distinguished lecturer on photonic systems for antenna applications. He is also the creator of the ARPA Symposium on Photonic Systems for Antenna Application.

*Henry Zmuda* received his B.E. from Stevens Institute of Technology and his M.S. and Ph.D. from Cornell University, all in electrical engineering. He is currently an associate professor of electrical engineering at Stevens Institute of Technology and teaches extensively in the areas of system theory, circuit theory, and microwave electronics. His research interests are in the areas of optical signal processing for microwave systems, broadband impedance-matching methods for optical telecommunications, and mathematical modelling of RF/microwave systems. He is a member of the IEEE.

# Index


Absorption loss, 467
Acoustical crosstalk, 217
Acoustic near field, 215
Acoustic slowness surface, 203, 204
Acousto-optic (AO) Bragg cells, 201–2, 213–14
  acousto-optic effect in, 202–5
  in active aperture array, 307–11
  Bragg interaction in, 205–12
  commercially available devices, 219–21
  commonly used materials for, 206, 213
  as deflectors, 213–15, 219
  device types, 214–18, 219–21
  dynamic range of, 225–26
  electrical-to-optical intensity transfer function of, 224–25
  electrical-to-optical spatial transfer function of, 223–24
  impedance-matching network for, 213
  as modulators, 215–17, 220–21
  for moving window, 415, 417–18
  as multichannel devices, 217–18, 220, 221
  performance of selected, 445, 446
  in receivers, 444–47
  signal processing characteristics of, 218–26
  tandem, 309–11
Acousto-optic diffraction, 436–38, 439
Acousto-optic (AO) effect, 202–5
Acousto-optic IFM receiver, 454–56
Acousto-optic interferometer, 452–54
Acousto-optic materials, commonly used, 206
Acousto-optic receiver
  acousto-optic diffraction in, 436–38, 439
  Bragg cells in, 444–47
  channelized, 438–42
  coherent, 449–56
  compressive, 442–44
  IFM, 454–56
  key component performance for, 444–48
  laser source for, 444
  parallel-output, 449, 450, 451
  photodetector array in, 447–48
  power, 448–49, 450, 451
  principles of, 435–44
Active aperture array, 297
  beam forming for, 229–304
  conventional vs. photonic hardware for, 298–99, 311
  cost of, 297–98
  differential phase systems in, 311–16
  new or enhanced capabilities of, 318–20
  photonics for, 304–11
  time-delay systems in, 316–18
Adaptive systems, using optical correlators, 421–26
Airborne radar, 7
  detection of targets in clutter for, 45–47
Amplified fiber-optic recirculating delay line (AFORDL), 482, 483–84
  active, 485–86
Amplified lattice filters, 500
Amplified spontaneous emission (ASE), 489
Amplifiers
  characteristics of, 488–92
  coupling losses with, 491
  effect of gain on stability with, 492–93
  erbium-doped fiber, 487–88
  length of, 492
  motivation for, 487–88
  noise with, 489–90
  nonlinearity in, 489
  polarization sensitivity of, 491
  semiconductor fiber, 487
  in signal processing, 461–64, 487–93
  switching speed of, 491–92

Amplitude errors, 284–88
Amplitude-only modulator, 167–68
Analog beam forming feeds, 289–96
Analog beam forming systems, 288–89
Anamorphic lens, 389–90
Angle, 8
Anisotropic Bragg interaction, 211–12
Antenna, 9–11
    focused, 384–85
Antenna aperture, effective, 58–60
Antenna gain, 21, 58
Antenna principal beamwidths, 9
Antenna radiation efficiency, 9
Antenna scan rate, 60
Antenna sidelobe level, 21
Antireflection (AR) coating, 81
APDs. *See* Avalanche photodiodes
Aperture efficiency, 60
Applications, 53, 54
Array(s)
    active aperture. *See* Active aperture array
    directivity of, 283–84, 285, 301
    effect of uncorrelated error in phase and amplitude on, 284–88
    feeds and controls for, 288–96
    one-dimensional (linear), 281–82
    patterns of, 278–83
    properties of, 300–3
    sidelobe levels of, 284, 285–88
    tapered, 280
    two-dimensional, 282–83
ASE (amplified spontaneous emission), 489
Availability, 57
Avalanche photodiodes (APDs), 124, 131
    in signal processing, 471
Azimuth resolution, 58

Band gap engineering, 258
Band gap grading, 136
Bandwidth
    of array, 300–1
    radiofrequency, 53
Basic unit delay, 461
Beam broadening factor, 280
Beam forming, 299–304
Beam forming feeds, analog, 289–96
Beam forming systems, analog vs. digital, 288–89
Beam optics, Gaussian, 77
Beam splitters, 81, 307, 308
Beam squint, 282
Beam steering processor, 52–53
Beamwidth
    of antenna, 9
    of array, 300
Bending loss, 467
Bias, optical, 405
Binary fiber-optic delay line (BIFODEL), 360–65
Binary optics, 83–84
Bipolar signal processing systems, 509–51
Birefringence, 81, 236–38
Bistable optical devices, optoelectronic, 185–86
Blind speeds, 41
Block diagram, of radar system, 9–12, 52–53
Bragg angle, 417
Bragg cells. *See* Acousto-optic receiver, Bragg cells in
Bragg interaction
    anisotropic, 211–12
    isotropic, 205–8
    phased array, 208–11
Branching, 481
Brewster's angle, 234
Broadband array feeds, 294–96
Butt coupling, 244–45

Canceler
    delay-line, 55
    two-pulse, 41–42, 43
Cancellation, multiple pulse, 41–42, 43
Canonical cascade, of lenses, 306
Carrier frequency, 231
Carrier lifetime reduction, 157
Cascaded fiber lattice structures, 498, 499
Cascaded optical correlators
    least mean squares algorithm adaptive system using, 422–26
Cauchy-Schwarz inequality, 511
Central data mixing (CDM) architecture, 327, 328
CFAR (constant false alarm rate), 24
Chaff, 40
Chain matrix, 503–4
Channelized receiver, 434–35
    acousto-optic, 438–42
Channel waveguide, 239, 240
Charge-transfer plate (CTP), 182–84
Charge trapping, at heterojunction, 136
Chebyshev illuminations, 280, 281
Chirp pulse compression, 390–91, 392
Chirp waveform, self-focusing correlator for, 411
Closed-loop correlator-convolver system, 422–25
Clutter, 16–19
    detection of targets for airborne or space-based radar in, 45–47
    detection of targets for ground-based radar in, 36–45
Coherence effects, 473

Coherence functions, 473–74
Coherent acousto-optic receiver, 449–56
Coherent integration, 32, 33–35, 55
Coherent processing, 476–77
Collection efficiency
dc, 143–46
of photodiodes, 137–42
Command circuits, optical, 266, 267
Complex signal processing systems, 509–10
Compressive receiver, 433–34, 435
acousto-optic, 442–44
Conjugate filter, 31
Constant false alarm rate (CFAR), 24
Continuous wave (CW) radars, 55
Contrast ratio (CR), of SLM, 171
Converter, photorefractive incoherent-to-coherent optical, 177–79
Copolarization terms, 11
Corporate feeds, 288–91
Correlation
adaptive systems using, 421–26
cascaded, 422–26
direct imaging, 405–7
fixed- vs. moving-window configurations for, 414–18
liquid crystal light valves in, 419–20
optical imaging and optical diffraction in, 404–5
photorefractive devices in, 420–21
purpose and mathematical description of, 403–4
self-focusing, 411
spatial, 405–12
time-domain, 412, 413
time-integrating devices for spatial output functions, 418–21
transform, 407–9
Correlator-convolver system, closed-loop, 422–25
Coupling
fiber-to-circuit, 263, 265
optical, 243–47, 340–42
Coupling losses, with amplifiers, 491
CPU-level data mixing, 327
CR (contrast ratio), of SLM, 171
Critical angle, 234
Cross-polarization terms, 12
Crystal video receiver (CVR), 430, 435
CTP (charge-transfer plate), 182–84
CW (continuous wave) radars, 55

Data mixing, 327–30
Data processor, 12
dc collection efficiency, 145–46
De-embedding, 339
Deflectors, acousto-optic Bragg cell, 213–15, 219
Deformable mirror devices (DMDs), 186–87
Delay line, fiber-optic
amplified, 482, 483–84, 485–86
binary, 360–65
laser diode switch parallel, 358–59, 360
optical switch parallel, 356–58
Raman-Nath, 415–17
square-root, 365–66
two-coupler nonrecirculating, 495, 496
two-coupler recirculating, 495, 496
unamplified, 482, 484–85
Delay-line canceler, 55
Depletion waveguide, 242–43
Design
example of preliminary, 57–64
factors influencing, 56–57
Detectability factor, 62, 63
Detection
in clutter for airborne or space-based radar, 45–47
in clutter for gound-based radar, 36–45
direct, 337, 343–49
integrating, 447
optical/RF techniques for, 336–38
peak, 447
of signals in noise, 24–28
valid, 24
Detectors
down-converting, 147–57
pin. *See* Pin detectors
in signal processing, 470–71
types of, 124,131
Deuterated potassium dihydrogen phosphate (DKDP) light valve, 174
Development time, 57
DFB (distributed feedback) lasers, noise in, 119–22
Differential phase systems, for active aperture array, 311–16
Differential quantum efficiency, 331
Diffraction, acousto-optic, 436–38, 439
Diffraction gratings, 246–47
Diffraction theory, 71–74
Diffractive optical elements, 83–84
Diffusion tails, 134–36
Digital beam forming systems, 288–89
Digital signal processor, 53
Diode-pumped YAG lasers, noise in, 118–19
Direct detection, 337, 343–49
Direct imaging correlator, 405–7
Directivity, of array, 283–84, 285, 301
Direct modulation fiber-optic link, 331, 332, 333–35, 343–47

Dispersion, 467–69
Dispersion-shifted fibers, 470
Distributed feedback (DFB) lasers, noise in, 119–22
DKDP (deuterated potassium dihydrogen phosphate) light valve, 174
DMDs (deformable mirror devices), 186–87
Doping, remote, 258
Doppler, unambiguous, 33, 41
Doppler effect, 36–39
Doppler filtering, 42–45, 46
Doppler radars, 55
Doppler resolution, 8
Double-clad fibers, 470
Down-converting detector, 131, 147–57
Duplexer, 52
Duty factor, 60
Dynamic range
- of acousto-optic channelized receiver, 438–42
- of array, 303–4
- of Bragg cell, 225–26, 445
- of fiber-optic link, 326
- of photodetector arrays, 448

E-beam-addressed liquid crystal charge-transfer plate light modulator (e-LCLM), 181–83
EDFAs (erbium-doped fiber amplifiers), 487, 488
Effective antenna aperture, 58–60
Effective radiated power (ERP), 14
Electrically addressed silicon VLSI liquid crystal SLMs, 183–84
Electrical-to-optical intensity transfer function, 224–25
Electrical-to-optical spatial transfer function, 223–24
Electronically controlled optical waveguide, 242–43
Electro-optic SLMs, 172–74
Element pattern, 278
Embedding, 339
Erbium-doped fiber amplifiers (EDFAs), 487, 488
ERP (effective radiated power), 14
Evanescent-wave region, 71
External modulation fiber-optic link, 331–33, 334, 335–36, 347–49
Extraordinary wave, in anisotropic Bragg interaction, 211–12

Fabry-Perot (FP) interferometer, 245
Fabry-Perot (FP) lasers, noise in, 118, 119–22
False alarm, 24
Far-field radiation pattern, 278
Far-field region, 71
Feed(s), array, 288–96
Feed-backward lattice structures, 494, 495, 497–98
Feed-forward lattice structures, 494, 495, 496, 497
Ferroelectric liquid crystal (FLC) SLMs, 181
Fiber-optic delay line. *See* Delay line, fiber-optic
Fiber-optic links. *See* Links, fiber-optic
Fiber-optic waveguide, 233–38
Fiber-to-circuit coupling, 263, 265
Filter(s)
- architectures for, 493–508
- conjugate, 31
- Doppler, 42–45, 46
- example of, 481–86
- finite-impulse response, 463–65
- infinite-impulse response, 465, 493
- lattice structures for, 494–508
- linear time-invariant, 493
- matched, 30–32, 79–80
- stable and minimum phase, 478–79
- transversal, 493
- Vander Lugt, 408, 409, 410
- Z-transform techniques in, 477–86

Finite-impulse response (FIR) filters, 463–65
First blind speed, 41
Fixed-window configuration, for correlator, 414, 415
FLC (ferroelectric liquid crystal) SLMs, 181
Flip-chip bonding, 263
Focused antenna, 384–85
Four-component decomposition, 509–10
Fourier-beam-forming transmitter system, 165
Fourier hologram, 409
Fourier transform, two-dimensional, 67–71
FP (Fabry-Perot) interferometer, 245
FP (Fabry-Perot) lasers, noise in, 118, 119–22
Framing speed
- of photorefractive device, 176
- of SLM, 170

Fraunhofer diffraction, 74
Fraunhofer region, 71
Free-space radar detection range, 61
Frequency bands, 5, 53
Frequency plane, 437
Frequency response, of waveguide photodiode, 146–47
Fresnel diffraction, 71–73
Fresnel kernel, 73
Fresnel region, 71, 215
Frobenius-Perron theorem, 512
Functions, of radar system, 5, 6, 7–8, 53, 54, 56
Future directions, 523–26

GaAs MQW self-electro-optic device, 186
Gain
- antenna, 21, 58
- of array, 301

Gain compression, with lasers, 92–94

Gain-lever effect, 110–13
Gaussian beam optics, 77
Glass waveguides, 247
Gradient-index (GRIN) lenses, 340, 341
Grafting, 263, 264
Grating lobes, 282, 291–93, 302
Gray-scale dynamic range, of SLM, 171
Ground-based radar, 7
  detection of targets in clutter for, 36–45
Ground clutter, 40
Group delay, 468
Group index, 468
Guided lightwaves, 256

Half-wave voltage, 173
H-control, 374
Heterodyne process, 84–86
Heteroepitaxy, 258, 262–63
Heterojunction, charge trapping at, 136
Heterostructures, 258
High-frequency intensity noise, 100–1
Holographic complex spatial filter, 399–400
Holographic devices, volume, 175–76
Holographic image, in synthetic aperture radar, 386–91, 392, 398–400
Holography, 77–79
Hybrid integration, 255, 261–62
Hybrid radars, 23–24

IFM (instantaneous frequency measurement) receiver, 432, 435
  acousto-optic, 454–56
IIR (infinite-impulse response) filters, 465, 493
IL (insertion loss), for true-time-delay system, 355
IM (intermodulation), 98–100
  translation of low-frequency noise to high frequency by, 105–7
IMP (inmodulation products), 445–47
Impedance matching, 107–10
Impedance-matching network, for acousto-optic Bragg cells, 213
Incoherent processing, 476
Index profiles, in optical fibers, 235, 237
Infinite-impulse response (IIR) filters, 465, 493
Inmodulation products (IMP), 445–47
Insertion gain, of fiber-optic link, 324
Insertion loss (IL), for true time-delay system, 355
Instantaneous frequency measurement (IFM) receiver, 432, 435
  acousto-optic, 454–56
Integrated circuits, optoelectronic *See* Optoelectronic integrated circuits (OEICs)
Integrated optical waveguide, 238–43
Integrating detection method, 447
Integration
  hybrid, 255, 261–62
  monolithic, 255, 261, 262, 263
  of radar returns, 32–35
Integrator, for adaptive system using optical correlators, 425–26
Intensity modulator, 169
Intensity noise, 100–7
Intensity transfer function, electrical-to-optical, 224–25
Interferometer
  acousto-optic, 452–54
  Fabry-Perot, 245
  Mach-Zehnder, 304, 305
Interferometric spectra analyzer (ISA), 449–54
Intermodulation (IM), 98–100
  translation of low-frequency noise to high frequency by, 105–7
Inverse Fourier transform, 68–69
Inverse velocity surface, 203, 204
Invisible portion, of spectrum, 70
Ion bombardment, 157
ISA (interferometric spectra analyzer), 449–54
Isolators, 82–83
Isotropic Bragg interaction, 205–8

Jamming power, 19, 20
Johnson noise, 231

Laser(s)
  for acousto-optic receiver, 444
  bandwidth of, 87–88
  intensity noise of, 100–7
  intermodulation of, 98–100
  microwave phase noise of, 113–16
  mode-coupling enhancement of, 95–97
  modulation efficiency and signal throughput of, 107–13
  quantum well and strained-layer quantum well, 88–94
  SNR of fiber links in, 118–22
  theoretical modulation response of, 88, 89
Laser diode (LD) switch parallel fiber-optic delay line, 358–59, 360
Lattice structures, 460, 461
  advantages of, 494–95
  amplified, 500
  cascaded, 498, 499
  feed-backward, 494, 495, 497–98
  feed-forward, 494, 495, 496, 497
  fiber-optic implementation of, 495–98

Lattice structures (cont.)
general, 494–95
second-order, 461, 462, 463
theoretical formulations of, 500–8
topology of, 494, 495
two-coupler nonrecirculating, 495, 496
two-coupler recirculating, 495, 496
LCLV (liquid crystal light valve), 179–85
correlators using, 419–20, 425
LD (laser diode) switch parallel fiber-optic delay line, 358–59, 360
Least mean squares algorithm adaptive system, using cascaded optical correlators, 422–26
Lens(es)
anamorphic, 389–90
canonical cascade of, 306
GRIN, 340, 341
transfer function of, 74–77
LIDAR (light detection and ranging) systems, 5, 525–26
Life cycle cost, 57
Light valves. *See* Spatial light modulators (SLMs)
Linear array, 281–82
Linearity, 474–76
Linear time-invariant (LTI) filter, 493
Links, fiber-optic, 324
advantages of, 323–24
block diagram of, 325
design criteria for, 324–26
detection techniques in, 336–38
direct modulation/direct detection, 343–47
external modulation/direct detection, 347–49
mixing architectures for, 326–30
modulation techniques in, 330–36
performance models for, 342–49
performance optimization of, 338–42
performance parameters of, 324–26
SNR of, 118–22
Liquid crystal light valve (LCLV), 179–85
correlators using, 419–20, 425
Lithium niobate waveguides, 247–48
Loss, 467, 487
Low-frequency mode-partition noise, 101–5
Low-frequency noise translation factor, 105–7
Low-probability-of-intercept (LPI) radar, 53
LTI (linear time-invariant) filter, 493
Lumped elements, in acousto-optic Bragg cells, 213

Mach-Zehnder interferometer, 304, 305
Magneto-optic SLMs (MOSLMs), 191–92
Marcum target, 30
Mason's rule, 479–80
Matched filtering, 30–32, 79–80
Material dispersion, 468
Matrix formulation, 501–2
of fiber lattice elements, 504–6
Maximum unambiguous range, 33
Maxwell's equations, 234–35
MCM scheme, 442
Membrane mirror light modulators (MLMs), 187–91
Memories, photorefraction-based 3-D optical, 177
Microchannel SLM (MSLM), 170, 174
Microscan receiver, 433–34, 435
acousto-optic, 442–44
Microwave phase noise, 113–16
Microwave radar, frequency bands for, 5
Millimeter-wave OEICs, 270–75
Millimeter-wave radar, 53
Miniature instantaneous Fourier transform (MIFT) system, 449, 451
Minimum phase system, 479
Mirror rotation, for time delay, 317–18
Missed target detection, 24
Mixing architectures, for fiber-optic links, 326–30
MLMs (membrane mirror light modulators), 187–91
Mode-coupling enhancement, 95–97
Mode-hopping noise, 103, 118
Mode locking, 96
Mode-partition noise
in FP lasers, 120, 121–22
low-frequency, 101–5
Modulation efficiency, 107–13
Modulation techniques, RF/optical, 330–36
Modulation transfer function (MTF), of SLM, 170–71
Modulators
acousto-optic Bragg cell, 215–17, 220–21
spatial light. *See* Spatial light modulators (SLMs)
Monolithic integration, 255, 261, 262, 263
MOSLMs (magneto-optic SLMs), 191–92
Moving target indicator (MTI) systems, 40–42, 55
Moving-window configuration, for correlator, 414–18
MSLM (microchannel SLM), 170, 174
MTF (modulation transfer function), of SLM, 170–71
Multichannel Bragg cell, 217–18, 220, 221
Multimode fiber, 235–36
Multiple-order wave plate, 82
Multiple pulse cancellation, 41–42, 43
Mutual coherence, 473

Narrowband modulation, by mode-coupling enhancement, 95–97
Nematic liquid crystal SLMs, 180–81

Noise
in amplifiers, 489–90
detection of signals in, 24–28
laser, 100–7
microwave phase, 113–16
with optical waveguides, 231–32
out of receiver, 12
shot, 231–32
thermal (Johnson), 231
Noise figure (NF)
of amplifier, 490
of fiber-optic link, 324
Noise power, of fiber-optic link, 324–26
Noise spikes, 101, 102
Noncoherent integration, 32, 35, 36, 37, 55
Nonlinearity, in amplifiers, 489
Nonminimum phase system, 479
Normalized frequency, 235
*n*th blind speed, 41
Numerical aperture, 234

OCPAR (optically controlled phased-array radar) systems, 163
typical signal-processing architectures for, 164–65
OEICs. *See* Optoelectronic integrated circuits (OEICs)
Off-axis holography, 399
One-dimensional array, 281–82
Operability, 57
Optical bias, 405
Optical clock distribution, 256
Optical command circuits, 266, 267
Optical coupling techniques, 243–47, 340–42
Optical heterodyne process, 84–86
Optically controlled phased-array radar (OCPAR) systems, 163
typical signal-processing architectures for, 164–65
Optical converter, photorefractive incoherent-to-coherent, 177–79
Optical memories, photorefraction-based 3-D, 177
Optical-microwave circuits, 264–66
Optical phase-shift circuits, 266–67, 268
Optical reflection, microwave phase fluctuation due to, 115–16
Optical source window, 470
Optical strip line, 239, 240–42
Optical switch parallel fiber-optic delay line, 356–58
Optical system, passive components of, 80–84
Optical time-delay circuits, 267–70, 271, 272
Optical waveguides. *See* Waveguide(s), optical
Optics, and radar, 4–5
Optoelectronic bistable optical devices, 185–86
Optoelectronic integrated circuits (OEICs)
applications of, 256
advantages of, 255–56
categories of, 261
fiber-to-circuit coupling of, 263, 265
grafting and related techniques in, 262–63, 264
guided and unguided lightwaves in, 256
hybrid, 255, 261–62
materials in, 257
millimeter-wave, 270–75
monolithic, 255, 261, 262, 263
optical-microwave circuits, 264–66
phased-array antenna, 266–75
photonic, 258–61
semiconductor techniques in, 258
SLMs, 184–85
wafer-scale, 270
Opto-optic devices, 257
Ordinary wave, in anisotropic Bragg interaction, 211
Overlapped subarraying system, 294, 295

PAAs. *See* Phased-array antennas (PAAs)
Parallel delay line (PADEL)
laser diode switch, 358–59, 360
optical switch, 356–58
Parallel output acousto-optic channelized receiver, 449, 450, 451
Parameter measurement, 8
Paraxial approximation, 72
Passive components, of optical system, 80–84
PD (pulsed Doppler) radars, 55–56
Peak detection method, 447
Performance, 56
Performance models, for fiber-optic links, 342–49
Performance optimization, of fiber-optic link, 338–42
Phase-amplitude distribution system, 164, 165
Phase-conjugating mirrors, self-pumped photorefractive, 176–77
Phased-array antennas (PAAs)
directivity of, 283–84, 285
effects of ucorrelated error in phase and amplitude on, 284–88
feeds and controls for, 288–96
OEICs, 266–75
one-dimensional (linear), 281–82
patterns of, 278–83
sidelobe levels of, 284, 285–88
signal processing in, 465
tapered, 280
true time-delay architecture for. *See* True time-delay (TTD) architecture
two-dimensional, 282–83

Phased array Bragg interaction, 208–11
Phase errors, 284–88
Phase noise, microwave, 113–16
Phase-only modulator, 168, 169
Phase quantization, grating lobes due to, 291–93
Phase-shift circuits, optical, 266–67, 268
Photoconductive mixing, down-converting photodetector by, 147–57
Photoconductors, 123, 124
Photodetector(s), in signal processing, 470–71
Photodetector array, in acousto-optic receiver, 447–48
Photodiodes
  avalanche, 124, 131, 471
  collection efficiency of, 137–42
  dc collection efficiency of, 145–46
  frequency response of, 146–47
  speed considerations for, 131–37
  structure of, 142–45
Photonic integrated circuits (PICs), 258–61
Photorefraction-based 3-D optical memories, 177
Photorefractive incoherent-to-coherent optical converter (PICOC), 177–79
Photorefractive integrator, 425–26
Photorefractive materials and devices, 175–79
  correlator architectures using, 420–21
Physical constraints, 57
PIC(s) (photonic integrated circuits), 258–61
PICOC (photorefractive incoherent-to-coherent optical converter), 177–79
Pigtailed isolators, 83
Pin detectors, 124, 131
  collection efficiency of, 137–42
  dc collection efficiency of, 145–46
  frequency response of, 146–47
  in signal processing, 471
  speed considerations with, 131–37
  structure of, 142–45
Planck distribution, of noise, 25, 26
Platform, 53
Pockels readout optical modulator (PROM), 174, 179
Polarization dispersion, 468
Polarization-maintaining fiber, 236–38
Polarization-only modulator, 168–69
Polarization sensitivity, of amplifiers, 491
Poles, 478
Positive systems, 510–15
Postdetection integration, 32, 35, 36, 37
Power acousto-optic receiver, 448–49, 450, 451
Power-aperture product, 23, 51
Power gain, of antenna, 11
Power spectra analyzer (PSA), 438–42
Predetection integration, 32, 33–35
PRF (pulse repetition frequency), 32, 33, 53
PRI (pulse repetition interval), 32–33
Prism couplers, 246
PRIZ, 172, 174
PROM (Pockels readout optical modulator), 174, 179
PSA (power spectra analyzer), 438–42
Pulse broadening, 467–69
Pulsed Doppler (PD) radars, 55–56
Pulse dispersion, 467–69
Pulse integration, 32–35
Pulse repetition frequency (PRF), 32, 33, 53
Pulse repetition interval (PRI), 32–33
Pulsing, in synthetic aperture radar, 387

Quadruple-clad fibers, 470
Quantum confinement, 258
Quantum well gain-lever effect, 110–13
Quantum well (QW) lasers
  gain compression with, 92–94
  high-speed, 88–92
  strained-layer, 88–91

Radar, optics and, 4–5
Radar action, 52
Radar range equation (RRE), 13–24
Radar system(s)
  anatomy of, 51–53
  block diagram of, 9–12, 52–53
  factors influencing design of, 56–57
  functions of, 5, 6, 7–8, 53, 54
  preliminary design process for, 57–64
  subsystem functions of, 9–12, 52–53
  types of, 5–7, 53–56
Radar target models, 28–30
Radiated power, effective, 14
Radiation efficiency, of antenna, 9
Radiofrequency (RF) bandwidth, 53
Rain, 40
Raman-Nath optical delay line, 415–17
Raman-Nath regime, 202
Range, 8
  free-space radar detection, 61
  maximum unambiguous, 33, 60
Range interval, 385
Range rate, 8
Range resolution, 60
Rayleigh distribution, of noise, 26–27
Ray optics, 233–34
RDM (remote data mixing) architecture, 327–30
Readout wavelength range, of SLM, 172

Real space, 282
Receiver(s), 12
    acousto-optic. *See* Acousto-optic receiver
    channelized, 434–35
    compressive, 433–34
    conventional, 429–35
    crystal video, 430
    instantaneous frequency measurement, 432
    superheterodyne, 431
Reciprocal two-pair, 503
Rectangular dielectric waveguides, 239, 240
Reference function, 403
Reflectarray, 289, 290
Reflection, optical, microwave phase fluctuation due to, 115–16
Relative intensity noise (RIN), 100
Relaxation oscillation frequency, 87
    of strained-layer quantum well laser, 89–91
    and well dimension, 90
Remote data mixing (RDM) architecture, 327–30
Remote doping, 258
Resolution, 8
Retardation plates, 81–82
RF (radiofrequency) bandwidth, 53
RF fiber-optic links. *See* Links, fiber-optic
Rib waveguide, 239, 242
Ricean distribution, of noise, 27–28
Ridge guide, 239, 242
RIN (relative intensity noise), 100
RRE (radar range equation), 13–24

Sampled data systems, 478
SAR. *See* Synthetic aperture radar (SAR)
Saturation characteristics, 489
Scan rate, of antenna, 60
Scattering loss, 467
Schottky barrier detectors. *See* Pin detectors
Sea clutter, 40
Search coverage, 58
Search radar, 24
Segmented mirror device (SMD), 317–18
Self-coherence, 473
Self-collimating modes, 218
Self-electro-optic device (SEED), 186
Self-focusing correlator, for chirp waveform, 411
Self-pumped photorefractive phase-conjugating mirrors, 176–77
Semiconductor optical-fiber amplifiers (SOAs), 487
Semiconductor techniques, in optoelectronic integrated circuits, 258
Semiconductor waveguides, 248–50
Series feed, 289, 290
SFD (signal flow diagram) technique, 346
Shipboard radar, 7
Shot noise, 231–32
SHR (superheterodyne receiver), 431, 435
Sidelobe levels, 284, 285
    antenna, 21
    average, 285–87
    peak, 287–88
Signal detection
    in clutter for airborne or space-based radar, 45–47
    in clutter for ground-based radar, 36–45
    in noise, 24–28
Signal flow diagram (SFD) technique, 346
Signal function, 403
Signal processing, 12, 53, 465–66
    amplifiers in, 461–64
    analog vs. digital, 471–72
    applications of, 517
    bipolar and complex systems for, 509–10
    block diagram and Mason's rule for, 479–80
    branching in, 481
    dispersion in, 467–69
    elementary functions in, 480–81
    example of, 481–86
    fiber-optic vs. integrated-optic, 461
    fiber structure for, 470
    filtering architectures for, 493–508
    frequency-domain vs. time-domain, 461
    impact of, 516–17
    incoherent vs. coherent, 473–77
    lattice structure in, 460, 461, 462, 463, 494–508
    loss in, 467, 487
    optical amplifiers in, 487–93
    optical vs. other techniques of, 466
    outlook for, 517–18
    overview, 459–72
    positive systems for, 510–15
    properties of fiber optics in, 466–69
    single-mode vs. multiple-mode fibers in, 467
    sources and detectors in, 470–71
    summing in, 481
    tapping in, 481
    temperature effects in, 469–70
    time advance in, 480
    time delay in, 480
    Z-transform in, 477–86
Signal throughput, 107–13
Signal-to-noise ratio (SNR)
    with amplifiers, 490
    enhancement of, 30–35
    of fiber links consisting of Fabry-Perot and distributed feedback lasers, 118–22

Signal-to-noise ratio (cont.)
probability of detection vs., 28, 29
Silicon waveguides, 248–49
Single-mode fiber, 236
Single side-band (SSB) ratio, 115
Slab waveguide, 238–40
SLMs. *See* Spatial light modulators (SLMs)
Slowness surface, 203, 204
SMD (segmented mirror device), 317–18
SNR. *See* Signal-to-noise ratio (SNR)
SOAs (semiconductor optical-fiber amplifiers), 487
Space-bandwidth product, of SLM, 171
Spaceborne radar, 7
detection of targets in clutter for, 45–47
Space feeds, 289–91
Spatial coherence, 473
Spatial (space-integrating) correlators, 405
direct imaging, 405–7
radar applications of, 409–12
time-integrating devices for, 418–21
transform, 407–9
Spatial filter, holographic complex, 399–400
Spatial frequency domain variables, 68
Spatial light modulators (SLMs), 163
in active aperture array, 312, 313
amplitude-only, 167–68
cascadable, 172
characterization of, 169–72
classification of, 167–69
deformable mirror devices, 186–87
electro-optic, 172–74
future directions for, 192
intensity, 169
liquid crystal, 179–85, 419–20, 425
magneto-optic, 191–92
membrane mirror, 187–91
microchannel, 170, 174
optoelectronic bistable, 185–86
phase-only, 168, 169
photorefractive, 175–79
polarization-only, 168–69
structures, principles, and limitations of, 166–69
Spatial nonuniformities, of SLM, 171–72
Spatial optical processing. *See* Active aperture array
Spatial resolution, of SLM, 170–71
Spatial transfer function, electrical-to-optical, 223–24
Splicing loss, 467
Spurious signals, 439
Square-root fiber-optic delay line (SRODEL), 365–66
SSB (single side-band) ratio, 115
Staggered PRF waveforms, 42
State variables, 506
Steady radar target, 30
Stimulated Brillouin scattering, 232
Storage time, of SLM, 170
Strained layers, 258
Strip line, optical, 239, 240–42
Strip-loaded waveguide, 239, 240–42
Stroboscopic effect, 418
Subarraying system, overlapped, 294, 295
Subarray level, time delay at, 293–94
Subsystem functions, 9–12, 52–53
Summing, 481
Superheterodyne receiver (SHR), 431, 435
Surveillance equation, 19–23
Surveillance radar, 24
Swath, 385
Swerling models, 28–30
Switching energy per pixel, 170
Switching speed, with amplifiers, 491–92
Synthetic aperture radar (SAR)
development of, 381–82
digital processing for, 396–98
holographic image in, 386–91, 392, 398–400
legacy of, 398–400
principle of, 382–85
radar signal in, 385–89
tilted-plane processor in, 391–96, 397, 398
Systolic array, 461, 493

Tangential phase matching, 212
Tapered array, 280
Tapered couplers, 245–46
Tapping, 481
Target, Marcum or steady, 30
Target cross section, 11
Target detection, 7–8
in clutter for airborne or space-based radar, 45–47
in clutter for ground-based radar, 36–45
missed, 24
in noise, 24–28
Target models, 28–30, 60–61
Temperature effects, in signal processing, 469–70
Temporal coherence, 473
Test and evaluation, 57
Thermal noise, 231
Three-component decomposition, 509, 510
Three-dimensional (3-D) clutter, 16, 18
Three-dimensional optical memories, photorefraction-based, 177
Thyristor-type devices, 186

Tilted-plane processor, 391–96, 397, 398
Time advance, 480, 481
Time-bandwidth product, 9, 468–69
Time delay
  in signal processing, 480, 481
  at subarray level, 293–94
Time-delay circuits, optical, 267–70, 271, 272
Time-delay quantization, grating lobes due to, 291–93
Time-delay systems
  for active aperture array, 316–18
  true. *See* True time-delay (TTD) architectures
Time-domain (time-integrating) correlators, 412, 413
  for spatial output functions, 418–21
Time shift, 267–68
Timing and control subsystem, 9
Tracking equation, 23
Tracking radar, 24
Transfer function, of thin lens, 74–77
Transfer matrix, 502, 503–4
Transform correlation, 407–9
Translational invariance, 71
Transmit/receive (T/R)-level data mixing, 327
Transmit/receive (T/R) switch, 9
Transmitter, 9
Transversal filter, 407
Triangle inequality, 511
T/R-level data mixing, 327
T/R (transmit/receive) switch, 9
True time-delay (TTD) architectures, 165, 166
  binary fiber-optic delay line in, 360–65
  characterization parameters for, 355–56
  concept of, 352–55
  delay- and element-compressive, 367–79
  delay-compressive, 359–66
  hardware complexity of, 354–55
  laser diode switch parallel delay line in, 358–59, 360
  need for, 351–52
  noncompressive, 356–59
  optical switch parallel delay line in, 356–58
  overall insertion loss for, 355
  square-root fiber-optic delay line in, 365–66
  wavelength-multiplexed, 367–79
Two-component decomposition, 509, 510
Two-coupler nonrecirculating delay line, 495, 496
Two-coupler recirculating delay line, 495, 496
Two-dimensional (2-D) array, 282–83
Two-dimensional clutter, 16, 17–18
Two-dimensional Fourier transform, 67–71
Two-pulse canceler, 41–42, 43

Unambiguous Doppler, 33, 41
Unamplified fiber-optic recirculating delay line (UFORDL), 482
  passive, 484–85
Unguided lighwaves, 256
Unit circle, 478–79

Valid detection, 24
Vander Lugt filter, 408, 409, 410
V-control, 374
Velocity surface, inverse, 203, 204
Vertical-to-surface transmission electro-photonic (VSTEP) device, 186
Vignetting, 77
Visible portion, of spectrum, 70
Volume holographic devices, 175–76

Wafer-scale optoelectronic integration, 270
Waveform, characteristics of, 9
Waveform generator, 9
Waveguide(s), optical, 229–30
  advantages of, 230–32
  butt coupling for, 244–45
  channel, 239, 240
  coupling techniques for, 243–47
  depletion, 242–43
  diffraction gratings for, 246–47
  electronically controlled, 242–43
  fiber-optic, 233–38
  glass, 247
  integrated, 238–43
  lithium niobate, 247–48
  materials for, 247–50
  noise with, 231–32
  nonlinear effects with, 232
  prism couplers for, 246
  radar system applications of, 229–32
  rectangular dielectric, 239, 240
  rib, 239, 242
  semiconductor, 248–50
  silicon, 248–49
  slab, 238–40
  strip-loaded, 239, 240–42
  tapered couplers for, 245–46
  types of, 233–43
Waveguide dispersion, 468
Wavelength-multiplexed TTD architectures
  fiber-optic, 368–72
  hybrid electronic fiber-optic, 372–74
  for 1-D PAAs, 367–74
  for 2-D PAAs, 374–79
Wave plates, 81–82
Wideband array feeds, 294–96
Write light exposure, 169

Zero(s), 478
Zero-dispersion wavelength, 469
Zero-order wave plate, 82
Zone plates, 84
Z-transform, 477–86, 500–1, 502

## *The Artech House Optoelectronics Library*

Brian Culshaw, Alan Rogers, and Henry Taylor, *Series Editors*

*Acousto-Optic Signal Processing: Fundamentals and Applications*, Pankaj Das

*Amorphous and Microcrystalline Semiconductor Devices, Optoelectronic Devices*, Jerzy Kanicki, editor

*Electro-Optical Systems Performance Modeling*, Gary Waldman and John Wootton

*The Fiber-Optic Gyroscope*, Hervé Lefèvre

*Field Theory of Acousto-Optic Signal Processing Devices*, Craig Scott

*High-Power Optically Activated Solid-State Switches*, Arye Rosen and Fred Zutavern, editors

*Highly Coherent Semiconductor Lasers*, Motoichi Ohtsu

*Germanate Glasses: Structure, Spectroscopy, and Properties*, Alfred Margaryan and Michael A. Piliavin

*Introduction to Electro-Optical Imaging and Tracking Systems*, Khalil Seyrafi and S. A. Hovanessian

*Introduction to Glass Integrated Optics*, S. Iraj Najafi

*Optical Control of Microwave Devices*, Rainee N. Simons

*Optical Document Security*, Rudolf L. van Renesse

*Optical Fiber Amplifiers: Design and System Applications*, Anders Bjarklev

*Optical Fiber Sensors, Volume I: Principles and Components*, John Dakin andBrian Culshaw, editors

*Optical Fiber Sensors, Volume II: Systems and Applicatons*, John Dakin andBrian Culshaw, editors

*Optical Interconnection: Foundations and Applications*, Christopher Tocci and H. John Caulfield

*Optical Network Theory*, Yitzhak Weissman

*Optical Transmission for the Subscriber Loop*, Norio Kashima

*Photonic Aspects of Modern Radar,* edited by Henry Zmuda and Edward N. Toughlian

*Principles of Modern Optical Systems, Volumes I and II,* I. Andonovic and D. Uttamchandani, editors

*Reliability and Degradation of LEDs and Semiconductor Lasers*, Mitsuo Fukuda

*Semiconductors for Solar Cells*, Hans Joachim Möller

*Single-Mode Optical Fiber Measurements: Characterization and Sensing*, Giovanni Cancellieri

*For further information on these and other Artech House titles, contact:*


Artech House
685 Canton Street
Norwood, MA 01602
617-769-9750
Fax: 617-762-9230
Telex: 951-659
email: artech@world.std.com

Artech House
Portland House, Stag Place
London SW1E 5XA England
+44 (0) 71-973-8077
Fax: +44 (0) 71-630-0166
Telex: 951-659